Handbook of Neurosociology

Handbooks of Sociology and Social Research

Series Editor:
Howard Kaplan, *Texas A & M University, College Station, Texas*

HANDBOOK OF SOCIOLOGY OF AGING
Edited by Richard A. Settersten, Jr. and Jacqueline L. Angel

HANDBOOK OF COMMUNITY MOVEMENTS AND LOCAL ORGANIZATIONS
Edited by Ram. A Cnaan and Carl Milofsky

HANDBOOK OF DISASTER RESEARCH
Theory, Science, and Prevention
Edited by Zili Sloboda and William J. Bukoski

HANDBOOK OF THE LIFE COURSE
Edited by Jeylan T. Mortimer and Michael J. Shanahan

HANDBOOK OF POPULATION
Edited by Dudley L. Poston and Michael Micklin

HANDBOOK OF RELIGION AND SOCIAL INSTITUTIONS
Edited by Helen Rose Ebaugh

HANDBOOK OF SOCIAL PSYCHOLOGY
Edited by John Delamater

HANDBOOK OF SOCIOLOGICAL THEORY
Edited by Jonathan H. Turner

HANDBOOK OF THE SOCIOLOGY OF EDUCATION
Edited by Jan E. Stets and Jonathan H. Turner

HANDBOOK OF THE SOCIOLOGY OF GENDER
Edited by Janet Saltzman Chafetz

HANDBOOK OF THE SOCIOLOGY OF MENTAL HEALTH
Edited by Carol S. Aneshensel and Jo C. Phelan

HANDBOOK OF THE SOCIOLOGY OF THE MILITARY
Edited by Giuseppe Caforio

For further volumes:
www.springer.com/series/6055

David D. Franks • Jonathan H. Turner
Editors

Handbook of Neurosociology

Springer

Editors
David D. Franks
Department of Sociology
Virginia Commonwealth University
Richmond, VA, USA

Jonathan H. Turner
Department of Sociology
University of California, Riverside
CA, USA

ISSN 1389-6903
ISBN 978-94-007-4472-1 ISBN 978-94-007-4473-8 (eBook)
DOI 10.1007/978-94-007-4473-8
Springer Dordrecht Heidelberg New York London

Library of Congress Control Number: 2012942914

Printed on acid-free paper

Springer is part of Springer Science+Business Media (www.springer.com)

Preface

In the late 1990s, I was asked by the Editor of the *Annual Review of Sociology*, Karen Cook, to write a short essay for the millennial issue of that journal in which I would reflect on "what I don't know about my field but wish I did" (see Massey 2000). In the resulting article I wrote that "I have come to the reluctant conclusion that sociologists have gone too far in privileging the social over the biological," and went on to conclude that "we need to educate ourselves in the exciting work now being done on brain functioning, cognition, the regulation of emotion, and the biological bases of behavior." The hook line at the end of the essay was "I really wish I knew more about human beings as biological rather than social organisms and have begun reading to catch up."

Over the next several years I did a lot of reading on the subjects of behavioral ecology, evolutionary anthropology, genetics, neuroscience, and the emotional brain, all of which only reinforced my belief that sociologists needed to achieve a firmer understanding of human beings as biological and not just social beings. My goal then became to convince other sociologists of this need. I endeavored to accomplish this goal in my Presidential Address to the American Sociological Association, which was entitled "A Brief History of Human Society: The Origin and Role of Emotions in Social Life" (Massey 2002), which I later expanded into a book-length treatise on the evolution of urbanism (Massey 2005). The main point of my address was "to illustrate the creation and workings of the emotional brain and show how it operates independently of and strongly influences the rational brain." I argued that if sociology were to advance, "research and theory must grapple with both rational and emotional intelligence and focus particularly on the interplay between them."

I have been told by candid informants that in the wake of my ASA address, some colleagues were overheard muttering that I had "gone off the deep end" as they left the auditorium; and in the ensuing years, sociologists have charged me with being "reductionist," "eugenic," and even "racist." I have also discovered, however, that other colleagues in the field have come to the same conclusion as me and are pursuing parallel intellectual agendas. Many of these people came together in 2004 to form the ASA's Section on Evolution, Biology, and Society, with the stated goal of improving "dialogue between sociology and the biological sciences." The section is now well-established and offers a regular forum to consider the interface between the social and the biological.

The current volume is the fruition of efforts by many of those responsible for creating the new section. The *Handbook of Neurosociology* brings together work by leading social scientists who have been thinking about and studying the neurological bases of human sociality. The chapters offer a roll call of topics in which neurosociology has gained traction in understanding the dynamic interplay between the brain and the environment, covering such subjects as identity, rationality, interaction, sociality, prejudice, stereotyping, status, emotions, health, attachment, conformity, and the mind.

In my view, coming advancements in neuroscience to do to the social sciences what DNA did for the biological sciences: break down traditional boundaries between disciplines and promote work across sub-fields. It will do this by firmly establishing the neural bases for cognition, emotion, and behavior. Whereas models of sociality, rationality, and behavior in the past were grounded in convenient but unsubstantiated assumptions about the nature of human beings, in the future our theories increasingly will be based on actual knowledge not only of how human cognition works to shape behavior but also of how the environmental works to shape cognition. It is essential that sociologists be a part of the larger intellectual conversation now going on among neuroscientists because the critical environment forces shaping the expression of human proclivities is social and not physical.

The social environment is especially important in understanding human outcomes given the complexity of their genome and the importance of learning in shaping behavior. Human beings do not interact directly with the physical environment, but through the intervening filters of culture and society; and since cultural practices and social categories are transmitted through interaction with others, understanding the effects of the social settings that we inhabit on gene expression and human development becomes critical in comprehending the human condition. It is within specific social contexts that learning occurs and human proclivities play out, and to explain human outcomes one therefore must consider the series of social environments that people inhabit at different stages of the life cycle and at different historical stages of societal development.

The imperative for sociological and biological scientists to work together has assumed new urgency with the rise of epigenetics, the study of how the environment influences gene expression (Allis et al. 2007). For many years scientists had a rather static view of genetic inheritance in which specific genes were passed on by parents and duly inherited and expressed by progeny, irrespective of environmental conditions. Genes were thought invariably to be revealed biologically and the principal debate was over which was more important – genes or the environment – in accounting for observed traits and behaviors in the phenotypes of living organisms (Ceci and Williams 2000).

In recent years, however, this static view of gene expression has given way to a more dynamic model in which the environment itself determines whether and how specific genes are expressed (Ridley 2004). As a result, scientific debates now tend not to be over which is more important – genes or the environment – but about how genes and the environment interact to bring about the expression of certain inherited traits. The focus of current work in both the biological and behavioral sciences has thus shifted to gene–environment interactions (Rutter 2006). It is now understood that the environment – and for humans this means the social environment – not only shapes behavior through learning and conditioning, but also by determining which certain genes get turned on or off, and hence, expressed or not (Costa and Eaton 2006).

By exploring the dynamic interrelation between human brains, behavior, and the social environment, the *Handbook of Neurosociology* provides a welcome and very timely addition to the biosocial research literature. Its chapters offer a compelling introduction to basic precepts and pressing questions in the nascent field of neurosociology and lays the groundwork for future thinking and research. I am honored and pleased to commend it to all sociologists, for as I said in my essay for the *Annual Review*, "the beginning point in coming to terms with our situation must be the realization that we are indeed biological organisms.... We need to give up our historical resistance to the idea that social behavior has biological roots and accept the fact that we, as human beings, have inherited certain predispositions to thought and behavior that influence and constrain the social structures that we unconsciously evolve and rationally select." The publication of the *Handbook of Neurosociology* provides concrete evidence that the field of sociology is definitely moving in this direction.

Douglas S. Massey

References

Allis, C. D., Jenuwein, T., Reinberg, D., & Caparros, M.-L. (2007). *Epigenetics*. Cold Spring Harbor: Cold Spring Harbor Laboratory Press.
Ceci, S. J., & Williams, W. M. (Eds.). (2000). *The nature-nurture debate: The essential readings*. New York: Blackwell.
Costa, L. G., & Eaton, D. L. (Eds.). (2006). *Gene-environment interactions: Fundamentals of ecogenetics*. New York: Wiley-Liss.
Massey, D. S. (2000). What I don't know about my field but wish I did. *Annual Review of Sociology 26*: 699–701.
Massey, D. S. (2002). A brief history of human society: The origin and role of emotions in social life. *American Sociological Review 67*: 1–29.
Massey, D. S. (2005). *Strangers in a strange land. Humans in an urbanizing world*. New York: Norton.
Ridley, M. (2004). *The agile gene: How nature turns on nurture*. New York: Harper.
Rutter, M. (2006). *Genes and behavior: Nature-nurture interplay explained*. New York: Harper Collins.

Contents

Contributors

Lisa Aziz-Zadeh The Brain and Creativity Institute, University of Southern California, Los Angeles, CA, USA

Gary G. Berntson Chicago Center for Cognitive and Social Neuroscience, The University of Chicago, Chicago, IL, USA

Department of Psychology, The Ohio State University, Columbus, OH, USA

John T. Cacioppo Center for Cognitive and Social Neuroscience, The University of Chicago, Chicago, IL, USA

Justin M. Carré Department of Psychology, Wayne State University, Detroit, MI, USA

Jeff Davis Department of Sociology, California State University, Long Beach, CA, USA

Anne F. Eisenberg Department of Sociology, New York State University, Geneseo, NY, USA

David D. Franks Department of Sociology, Virginia Commonwealth University, Richmond, VA, USA

Stefan M. Goetz Department of Psychology, Wayne State University, Detroit, MI, USA

Stanford W. Gregory Jr. Ph.D. Department of Sociology, Kent State University, Kent, OH, USA

Michael Hammond Department of Sociology, University of Toronto, Toronto, ON, Canada

Louise C. Hawkley Chicago Center for Cognitive and Social Neuroscience, The University of Chicago, Chicago, IL, USA

Rosemary L. Hopcroft Department of Sociology, University of North Carolina, Charlotte, NC, USA

Will Kalkhoff, Ph.D. Department of Sociology, Kent State University, Kent, OH, USA

George Lakoff Department of Linguistics, University of California, Berkeley, CA, USA

Sook-Lei Liew The Brain and Creativity Institute, University of Southern California, Los Angeles, CA, USA

Maiken Luhmann Chicago Center for Cognitive and Social Neuroscience, The University of Chicago, Chicago, IL, USA

Alexandra Maryanski Department of Sociology, University of California, Riverside, CA, USA

Douglas S. Massey Department of Sociology, Princeton University, Princeton, NJ, USA

Allan Mazur Department of Sociology, Maxwell School of Citizenship and Public Affairs, Syracuse University, Syracuse, NY, USA

Pranjal H. Mehta Department of Psychology, University of Oregon, Eugene, OR, USA

Istvan Molnar-Szakacs Semel Institute for Neuroscience and Human Behavior, University of California, Los Angeles, CA, USA

Tennebaum Center for the Biology of Creativity, University of California, Los Angeles, CA, USA

Todd D. Nelson Department of Psychology, California State University, Stanislaus, CA, USA

Richard Niemeyer Department of Sociology, University of California, Riverside, CA, USA

Greg J. Norman Chicago Center for Cognitive and Social Neuroscience, The University of Chicago, Chicago, IL, USA

John R. Shook EdM Program Science and the Public, University at Buffalo, New York, NY, USA

Center for Neurotechnology Studies, Potomac Institute for Policy Studies, Arlington, VA, USA

Tibor Solymosi Department of Philosophy, University of Southern Illinois, Carbondale, IL, USA

Thomas S. Smith Department of Anthropology, University of Rochester, Rochester, NY, USA

Warren D. TenHouten Department of Sociology, University of California, Los Angeles, CA, USA

Jonathan H. Turner Department of Sociology, University of California, Riverside, CA, USA

Lucina Q. Uddin Psychiatry & Behavioral Science-Child and Adolescent Psychiatry, School of Medicine, Stanford University, Stanford, CA, USA

Chapter 1
Introduction: Summaries and Comments

David D. Franks and Jonathan H. Turner

David Franks: A Short History

In 2008, Howard Kaplan, sociology series editor for the Springer Press, requested that Professor Franks undertake this volume. At that time, the latter's response was that a handbook in neurosociology would be premature. Sociology was behind other disciplines in embracing neuroscience because anything biological was associated with the radical reductionism of E. O. Wilson's (1975) position that would have rendered sociology unnecessary. Things are obviously different now, but it has been an uphill battle, however rewarding for these editors. While early usage of the term neurosociology should be credited to Warren TenHouten (Bogen et al. 1972), the trek started in earnest with an issue of Franks' annual series with the JAI Press (1999) titled *Mind, Brain and Society* coedited with Thomas S. Smith. At that time, a reviewer wrote that all sociologists ought to read this volume, but that few would. According to our royalty checks, he could not have been more correct. A year later, Jonathan H. Turner came out with his *Origins of Human Emotions* (2000) that was heavily embellished with biological insights into the development of the Hominin brain processes underlying human emotion.

A great deal of the encouragement that existed then came from outside of sociology in the writings of neuroscientists like Gazzaniga who previously had written *The Social Brain* (1985) and others like Damasio in *Descartes' Error* (1994: 124–126). Both took an explicitly non-reductionist and pro-sociological approach in their work. Judging by the full house attracted by a didactic seminar on the social nature of the brain held at the annual meetings of the American Sociological Association in 2010, things were changing quickly even in sociology, and late in 2010 David Franks accepted Professor Kaplan's call for editing this handbook. It took only a week and a half to get enough sociological authors to start the next journey. Soon after that I was fortunate enough to recruit Jonathan Turner as my coeditor. Nonetheless, we have a few gaps in this volume that will be covered briefly here. One is of a methodological nature and the other is theoretical.

D.D. Franks (✉)
Department of Sociology, Virginia Commonwealth University,
Richmond, VA 23284, USA
e-mail: daviddfranks@comcast.net

J.H. Turner
Department of Sociology, University of California,
Riverside, CA 92521, USA
e-mail: jonathan.turner@ucr.edu

D.D. Franks and J.H. Turner (eds.), *Handbook of Neurosociology*, Handbooks of Sociology and Social Research, DOI 10.1007/978-94-007-4473-8_1,

Brain Activity Measures and Limitations

The most sophisticated scanner is the fMRI. While the MRI (magnetic resonance imaging) takes static pictures of the *structure* of the brain, the fMRI literally gives us "movie pictures" of its functioning or *activity*. The images of both however are indirect: that is, they do not depict neuronal activity *per se*. What they measure is the *oxygen* level in different *areas* of the brain. Neuronal activity absorbs oxygen (energy from the body) and this is as far as fMRIs go in what they actually measure. Thus, what they actually tap into is distanced from neuronal activities in at least two important ways. First, they are removed from direct neuronal activity and second, they do not depict single, actual neurons, but statistical averages of brain *areas*. Part and parcel of doing this is a technique referred to as BOLD (Blood Oxygen Level Dependence). This measures the *ratio* of oxygenated to deoxygenated hemoglobin in the blood stream between nonactive brain circuits and active ones. BOLD therefore is the critical measure in all fMRIs but it has limitations. It is far from clear which kinds of neuronal activities can cause BOLD responses. Possibilities include excitatory neurons, mixed neuronal populations, astroglia and axonal tracts, or fibers of passage. Churchland (2011: 124), a leading philosopher of neuroscience, joins others in warning that the bright colors of the BOLD resolutions are highly deceiving. They suggest well-defined processing areas which support the modular theory of the brain. This assumes that the brain is organized like a Swiss army knife where each tool is autonomous. She points out in contrast that the brain's neural activity is probably distributed in "loosely defined networks." Everywhere in the prefrontal lobes, for example, she says there are looping pathways creating the convergence and divergence of information. This issue between the brain's modularity and a broader systems approach is not resolved as yet in neuroscience, but influences our interpretations. She warns that the differences in brain activities are really tiny, but coloring them red and subtracting everything else out, which is often done, results in an exaggerated impression that the colored area is all that is active when this may not be true. Not only that, but those unnoticed areas may be important to the modular-appearing activity shown in red. She also reminds us that any one brain circuits can be active in a certain behavior but that circuits can do numerous other things as well. Brain areas light up under many different states. We never know if the brain activity is unique to that behavior or state.

Another commonly recognized caution in interpreting fMRIs is that BOLD signals do not tap into actual single neurons but *averages* of deoxinations in tiny spaces called voxels. Because of the minute volume of these spaces, they cannot pick up the activity of axons that are very long in length. Churchland says this is like hearing a large noise in the rumpus room but not knowing what each child is up to. As we stated when first describing BOLD, we have little clue about the micro level of neurons and their networks (Churchland 2011: 125).

Her last warning goes beyond the limitations of BOLD and has to do with the vast difference between the language of humans and the language of the brain, that is, do we have an accurate vocabulary for what particular brain processes do? One can reasonably hypothesize that a neuron "can respond to a vertical stimuli" while this is less true of brain activities like "self-control," "delayed satisfaction," or "strength of will" (Churchland 2011: 126).

Next, a common dissatisfaction in brain research of any kind is that we are stuck with correlations instead of explanatory causes. This can be minimized by transcranial magnetic stimulation (TMS)—a highly intrusive procedure: a magnetic field is created under a copper coil placed in the head of the subject. This causes an electric pulse called a TMS pulse. A rapid series of these pulses disrupts the functioning of the brain area under the copper coil indicating whether the area is a necessary and sufficient cause of the activity being studied. Needless to say this has quite limited use.

Another consideration has to do with the relationship between fact and theory. If a fact is seen as an empirical statement about phenomena in terms of a conceptual scheme, its scientific worth is dependent on one's theory. According to Brothers (2001: 67), there is no unifying theory in neuroscience as there is in vastly more mature fields like the atomic theory of matter, the germ theory of disease, or the natural selection that drives evolution. Again, this does not negate the enormous strides neuroscience has made in the last three decades, but it behooves us from the very onset to be cautious. After all, according to Edleman (2004: 15–16), the cerebral cortex makes up approximately two thirds of the human brain. "If you counted its connections or synapses" (single neurons are just pieces of meat) and started right now, at a rate of one per second, you would finish counting them "32 million years from now". Caution, patience, and determination are obviously in order here.

Jonathan H. Turner: Coming on Board as an Editor

By the time that I became coeditor of *Neurosociology*, a good deal of the work in soliciting chapters had already been done. I added a few authors, but David Franks had done most of the heavy lifting, made even more necessary when I became seriously ill for several months. What strikes me most about the chapters assembled in this book is their diversity. Neurosociology is now just a label—but a good one—for thinking about the human brain and its relationship to human interaction and social organization. Curiously, in their efforts to escape anything biological, sociologists often make the case that the large human brain allowed for language and culture, and thus, the social universe is constructed by human agency rather than by bioprogrammers and mechanisms in the brain. But, if all of these social constructions that sociologists study are the consequence of a larger neocortex enveloping older subcortical areas of the brain, it would seem not only logical but also necessary to understand how this brain evolved and how it works. Human capacities for thought, language, and production of arbitrary symbols that build up culture do not obviate the study of the brain; rather, they demand that we understand its evolution and operation.

Still, as David Franks mentioned earlier, sociologists fear reductionism; and I can only paraphrase George Homans' comment about such fears: if sociology had a set of clear explanations for the operation of interaction and social organization, they would not fear that some of these might be deducible from the laws of behavior or biology. There are emergent properties of the social universe that are clearly emergent and not reducible, but any science recognizes that these emergent properties are built from more elementary properties and processes and that understanding of the latter can increase the robustness in explanation of the former. My own work on the brain followed this path. When I first began studying emotions some 15 years ago, I soon recognized that I needed to know something about the brain since emotions are generated in the subcortical areas of the brain, as these interact with the prefrontal cortex. Learning something about the brain and then comparing the size and, more importantly, the organization of the human with the brain of great apes led me, in turn, to try to explain the selection processes during the course of hominin evolution that could explain these differences. Not only are humans wired to produce culture and language, but we are wired to be highly emotional. And the evidence for this conclusion is in the rather jury-rigged manner in which subcortical areas of the human brain are organized and connected to the neocortical parts of the brain. Selection was clearly working rapidly to make humans more emotional; and this led to the question of why such should be the case. Perhaps some saw this as "reductionism" and were threatened by such research, but I saw it as giving sociological explanations of the social universe more power. We are strengthened by interdisciplinary work, not weakened. Only sociologists' collective insecurity makes at least some to believe that reductionism threatens sociology.

I go even further, however. A sociology that is willing to study the biological basis of human interaction and organization is not only stronger as a discipline but it also can inform other disciplines. Almost all of the selection pressures working on the hominin brain were sociological in nature; they were the direct result of selection pressures to increase the power and duration of social bonds of humans so that they could construct groups and eventually larger scale sociocultural formations. The so-called "modules" trumpeted by evolutionary psychology are a consequence of selection pressures that are social. The result is that not only can sociology learn something from neurology, we can provide needed information on why and how the human brain came to be wired the way that it is. I hesitate to quote August Comte, but to some extent, a neurosociology is one strategy for fulfilling Comte's prophecy that sociology as "the queen science" (o.k., a bit delusional, I admit) emerged from biology. It would then eventually begin "to inform" biology. We should not, therefore, be fearful of what seems like reductionism, but rather, we should see it as an opportunity to colonize other disciplines.

What Does Neurosociology Have to Offer?

The human brain became wired to increase social bonds among humans, first through enhancing interpersonal behaviors and then through using these enhanced interpersonal capacities to form more stable social groupings that, over the course of human history, were used to build up the macrostructures and cultures that we see today. Sociologists, in contrast to Homans' derisive comments, know a great deal about the dynamics of micro-level interaction as well as meso-and macro-level social organization. This knowledge of the outcomes of the great rewiring of the hominin and human brain compared to the brain of the great apes can inform researchers and theorists. If we see the subject matter of microsociology—or interaction dynamics—as a partial outcome of alterations of the hominin and human brain compared to the brains of the great apes, then we have a set of guidelines for studying the brain: find out how the brain affects the complex set of microdynamic forces driving interaction. Sociology can thus set up a research agenda for using the imaging methodologies discussed by David Franks to discover how the brain affects interaction, and vice versa. These last two words are critical because we know the brain to be plastic, and thus, individuals' experiences in interaction in group contexts have large effects on the brain, particularly in its formative stages of development. Thus, here is another research agenda: see how the sets of interpersonal processes that make human interaction, and group formation possible affect the development of the brain.

Many of the chapters in this book implicitly address some of these issues, but I would argue that they do not go far enough. We need to go beyond specific mechanisms driving interpersonal behavior—say, exchange, identity formation, emotional arousal, etc.—and explore *the full range of interpersonal mechanism,* which only sociology has done. Psychology has a limited view of these mechanisms, as does economics and even behaviorally oriented political scientists; their analysis is not so wrong as it is incomplete. And thus, a neurosociology offers real hope to understand more fully the relationships between brain systems—or "modules," if you insist—and the interpersonal behaviors that made human survival possible by allowing for more permanent group formations among weak-tie apes and hominins and that now make macrostructural, sociocultural formations possible. Both sociology and neuroscience will benefit from a mature neurosociology; and it is for this reason that I joined David Franks in editing this volume. It is a good starting point for further advancement, especially as the older generations of sociologists with all their reductionist fears give way to the younger generation of sociologists, more of whom seem willing to engage biology in general and neurology in particular.

References

Bogen, J. E., Dezure, R., Tenhouten, W. D., & Marsh, J. F., Jr. (1972). The other side of the brain IV. The A/P ratio. *Bulletin of the Los Angeles Neurological Societies, 37*, 49–61.
Brothers, L. (2001). *Mistaken identity: The mind-brain problem reconsidered.* Albany: State University of New York Press.
Churchland, P. S. (2011). *Braintrust: What neuroscience tells us about morality.* Princeton: Princeton University Press.
Damasio, A. R. (1994). *Descartes' error: Emotions, reason and the human brain.* New York: Avon Books.
Edelman, G. M. (2004). *Wider than the sky: The phenomenal gift of consciousness.* New Haven: Yale University Press.
Franks, D. D., & Smith, T. S. (1999). In: D. D. Franks (Series Ed.), *Mind, brain, and society: Toward a neurosociology of emotion* in *social perspectives on emotion* (Vol. 5). Stamford: JAI Press.
Gazzaniga, M. S. (1985). *The social brain: Discovering the networks of the mind.* New York: Basic Books.
Turner, J. H. (2000). *On the origins of human emotions: A sociological inquiry into the evolution of human affect.* Stanford: Stanford University Press.
Wilson, E. O. (1975). *Sociobiology: The new synthesis*. Cambridge, MA: Harvard University Press.

Part I
Large Issues

Chapter 2
Neural Social Science

George Lakoff

It is obvious: Reason is central to the social sciences. It is so obvious that it is not discussed. And it is not discussed because it is assumed that all social scientists, being human, are endowed with the capacity for reason. We can take reason for granted and go on.

Or can we? The past three decades in the brain and cognitive sciences have vastly changed our understanding of the nature of reason. What has emerged is the empirical study of "real reason"—how people really think, whether they are people studied by social scientists or social scientists themselves.

The social sciences, of course, study the material causes of social and political effects: poverty, hunger, illness, homelessness, lack of education, joblessness, disparity of wealth, and so on. But how people think also has social effects: How do people understand morality, markets, the proper role of government, the nature of institutions, and so on?

How social scientists understand reason will affect their theories, both their theories of material causes and cognitive causes of social effects. It is therefore vital that social scientists get reason right. The Brain and Cognitive Sciences have shown that Real Reason—the way we really reason—is a matter of neural circuitry and has effects that are far from obvious. The way the brain shapes real reason therefore makes all social science into neural social science.

Reason Is Neural

Because we think with our brains, all reason is neural in character. That's not a surprise. What is a surprise is the effect this simple truth has on how the social sciences are studied.

Neural systems work to structure ideas physically in such a way as to produce rich, largely unconscious thought, with elements like conceptual image-schemas, frames, metaphors and narratives, and categories defined by many kinds of prototypes, not necessary and sufficient conditions. What this means is that critical social thought must go beyond logic and Enlightenment Reason to look at Real Reason, as revealed by the Brain and Cognitive Sciences. Real critical thinking requires an understanding of Real Reason.

G. Lakoff (✉)
Department of Linguistics, University of California,
Berkeley, CA 94720, USA
e-mail: lakoff@berkeley.edu

D.D. Franks and J.H. Turner (eds.), *Handbook of Neurosociology*, Handbooks of Sociology and Social Research, DOI 10.1007/978-94-007-4473-8_2, © Springer Science+Business Media B.V. 2013

Back to the Future

There is a long history in Sociology of studying how the ideas of individuals shape social life, from Max Weber's *Ideal Types* to Erving Goffman's *Frames,* which are structural configurations of mind that shape everyday understanding.

Weber recognized that the Protestant ethic—a system of ideas—had everything to do with the social and material causes of capitalism and the shape that capitalism took in the Northern Europe of his day. Goffman recognized that social institutions—from asylums to casinos—have their causal roots in "frames" that structure the mind and determine how institutions are cognitively structured in many ways: in the roles that people in institutions play (e.g., nurse, croupier), and in their understanding of what does and does not happen in social institutions within the given frames.

What's in a frame? Look, Goffman told us, for where the conventional frame "breaks." In the Surgery Frame, surgeons operate on patients not patients on surgeons. The powerful oppress the powerless, not the reverse. Casinos, not customers, set house rules. Judges use gavels, not ping pong paddles. It is through structured frames that social life operates, exerting very real forms of power through systems of frames. As Goffman was fond of repeating, "Social life is no joke!"

We now know from neuroscience that ideas are physical, that they are neural circuits. Fixed ideas are fixed brain circuits, with synapses strong enough to make them permanent. The causal effects of ideas are neural effects.

But neurons are meaningless in themselves. How do hundreds of billions of neural connections forming trillions of circuits become meaningful—and meaningful in ways that have social effects?

How Brain Circuits Become Meaningful

Social scientists are usually trained on Weber's favorite metaphors:

- Time is Money (from Benjamin Franklin)
- Devotion to a Useful Craft is Devotion to God
- In Calvinism and related forms of Protestantism: Prosperity is a Sign of Righteousness and God's approval.

Metaphors, we now know, are conceptual and hence neural in nature. Just as frames are neural structures allowing us to understand the structure of the material world and social life—the roles we play, the norms of actions, the expectations we form—so, metaphors are neural circuits that map frames to frames, preserving social values, emotions, inferences, and hence normal expectations.

How does this work? Via the differentiation between associative circuitry and body circuitry—and the way they are connected. Body circuitry includes motor neurons, perceptual neurons, emotional neurons, temperature neurons, pain neurons, etc. Associative circuitry consists of complex "cascades" DeHaene (2009) made up of simple neural circuitry that "bottoms out" at the body circuitry, the circuitry that extends throughout the body. The cascades of associative circuitry link together a myriad of forms of inherently meaningful embodied experiences, like experiences of moving, seeing, grasping, etc.

As a result, phenomena that appear to be objective and material—matters of the external world in itself—are not. They could not be, because all out understanding comes via embodied neural circuitry for frames, metaphors, and narratives. We take common framing to be "objective." Because we do not notice our unconscious neural understanding, we take the material world-as-neurally-understood to be objectively material. Even the external, material sociology of how many racial hirings, sex education courses, college admissions, etc. are frame-dependent and often metaphor-dependent. These are studied because the social sciences are essentially moral in nature: they are the study of the right things to do. But morality itself is anything but objective and material. The ideas characterizing morality arise from framing and metaphor, that is, frame-circuits and metaphor-circuits, circuitry that determines what we consider the right thing to do. In short, embodied frame-circuits and metaphor-circuits determine the very goals of the social sciences. Social science can benefit from an understanding of how this works.

Reason and Social Science

By *Neural Social Science*, I mean the approach to social science research that is based on, and integrates, methods and results from the cognitive and brain sciences—including cognitive linguistics, embodied cognition, experimental social psychology, neural computation, social neuroscience, and neuroeconomics. The line between the social sciences and the cognitive and brain sciences is disappearing—fast. The most fundamental reshaping concerns the concept of reason itself.

Reason Itself: Enlightenment Fallacies

Frankly, I find it scandalous.

The social sciences are supposed to be committed to a rational understanding of social life. Rationality and science are seen as going hand-in-hand. If you believe in reason, you should believe in science—hence social *science!* Materialist social science takes on the trappings of physical science: objectivity of method, facts and figures, classical categories, logic, statistics—as well as the values of science: making the world a better place by eliminating superstition and fallacious, harmful myths. These are the Enlightenment values, and in the Enlightenment context of the seventeenth and eighteenth centuries, they led to marvelous advances.

We are taught in the social sciences that Enlightenment Reason is the hallmark of critical thinking and at the core of liberal democratic thought. If you accept the Enlightenment view of rationality, then the application of rationality and science *will* make the world a better place. And we certainly need to make the world a better place. Unfortunately, right now, America and much of the world are beset with a disastrous form of false "rationality." We do need to be rational to make the world a better place, but we need to replace Enlightenment rationality with real rationality—with the way that rationality really works.

Rationality is crucial to improving the world, but the cognitive and brain sciences show us that the *Enlightenment theory* of rationality is so flawed that it is helping to create the problems that threaten to destroy us! The brain and cognitive sciences do not just improve marginally on reason—the scientific results change our understanding of reason in a way that is absolutely crucial to saving what is most valuable in our world. Neural social science is absolutely necessary, not just nice.

Where are the fallacies of Enlightenment Reason most in use? In the social science departments of our universities (political science, sociology, classical economics, law, and public policy) and in our public policy institutions, both public and private, both governmental and corporate.

I am not kidding! Our best and most socially committed thinkers are being trained in a flawed system of thought.

The Enlightenment Fallacies

The First Fallacy: Reason Is Conscious

Consciousness is linear while conceptual thought uses brain circuitry that is massively parallel. For this reason, most conceptual thought could not be conscious—and it isn't. Andrea Rock, in *The Mind at Night* (New York: Basic Books 2005) quotes neuroscientist Michael Gazzaniga as estimating that reason is *98% unconscious*! That estimate seems about right to most brain and cognitive scientists. Consider a random chunk of consciously-focused-on text, for example. Imagine writing down everything *not* in that chunk of text that is needed to understand it—every concept and every piece of background information, including every rule of grammar and phonology. The ratio of 50-to-1 nontext-to-text is a reasonable approximation.

Consciousness is only the tiny tip of the iceberg of reason.

The Second Fallacy: One Can Reason Directly About the World

Because we think with our physical brains, which are connected to our bodies, we can only comprehend what our bodies and brains pick out, structure, conceptualize, and categorize. You can only reason about the understandings of the world that the embodied neural circuitry of your brain permits. The relationship between reason and the world is always mediated by the brain and body.

The Third Fallacy: Thought Is Disembodied

All thought is physical, a matter of the activation of neural circuitry that is grounded in the body. What makes thought meaningful is the body, and how we function in the world with our bodies. The content of concepts is determined by the way we interact in the world with our bodies. Conceptual thought always has a bodily component.

The Fourth Fallacy: Words Are Defined Directly in Terms of Features of the External World

All words in all natural languages are defined in terms of embodied conceptual frames, not the external world. There is no direct way in which words can fit the world independent of the framing provided by body and brain.

The Fifth Fallacy: Reason Is Unemotional

The opposite is true. If you have a stroke or brain injury that makes it impossible to feel emotions, then you don't know what to want because *like* and *not like* mean nothing, and you can't tell whether others will like or not like what you do. For this reason, you couldn't set rational goals. You can't act rationally without emotion. Rationality requires emotion.

The Sixth Fallacy: Reason Is Literal and Logical

Real reason makes use of frames, image-schemas, mental images, conceptual metaphors, prototype categorizations, mental spaces, blends, emotions, and narratives. These are embodied conceptual structures that have "logics" of their own, which, for the most part, do not fit traditional mathematical logics. This does not make reason "subjective," since the real world—both physical and social—places constraints on your experience. The very structure of reason is interactive. It requires both you and the world outside of you.

The Seventh Fallacy: Categories Are Defined by Necessary and Sufficient Conditions

Categories are structured by prototypes of many kinds: social stereotypes, which have major social effects; ideal, typical, and nightmare cases, which define social standards, norms, and social disasters; and salient exemplars—well-known cases that raise probability judgments significantly and change social behavior and social policy.

The Eighth Fallacy: Reason Exists Primarily to Serve Self-interest

That is partly true. But we know from mirror-neuron research that empathy is physical, that the capacity to put oneself in someone else's shoes is built into our bodies and brains. That capacity is at the center of social life. Social and interpersonal relationships are mainly served by our capacity for real reason.

The Ninth Fallacy: Conceptual Systems Are Monolithic

It is commonplace for human beings to have different inconsistent value systems in the same brain. For example, consider the Saturday night and Sunday morning value systems. Saturday night (party) and Sunday morning (church) moral systems are very different, yet most people shift readily between them with barely a notice.

The reason is that each value system is realized in neural circuitry, and two contradictory neural circuits inhibit each other, so that the activation of one inactivates the other. When one is turned on, the other is automatically turned off.

A great many Americans have conservative values on some issues and progressive values on others—shifting back and forth in different contexts without notice, unless a contradiction appears that is both very conscious and very troubling. That is called "cognitive dissonance." It does occur, but it is rare, and it rarely leads to significant change in itself. Pointing out logical contradictions to those with political positions rarely changes those positions.

The Tenth Fallacy: Words Have Fixed Meanings, and Concepts Have Fixed Logics

We now know that most important concepts can be essentially contested. Concepts may have limited agreed-on central cases that immediately come to mind, but are relatively unimportant. The important cases of contested concepts arise where there are major value differences across people (or even within the same brain). Then, what seems like a single concept named by a single word can vary widely in its meaning, depending on value systems. The effect of a single word, expressing a single simple idea in uncontroversial cases, can be deadly in the controversial cases when opposite meanings of the word are held by populations with different value systems.

The most detailed study of a single important contested concept is my book *Whose Freedom?*, which surveyed the vast differences of meaning in the word "freedom" under conservative and progressive value systems. It matters what freedom means, and the meaning of the concept has triggered a life-and-death struggle, not just in America, but in many parts of the world.

The Eleventh Fallacy: The Truth Will Set You Free; If Enough People Know the Truth About Social Issues, They Will Change Their Attitudes, to Society's Benefit

Actually, worldviews in the form of frames and metaphors are physically realized in the brain so strongly that, when the facts don't fit the frames, the frames stay and the facts are ignored or disputed or just plain not seen.

Those are among most basic properties of the Enlightenment theory of Reason. There are more, and we will discuss them before long.

The point is straightforward: the Enlightenment theory of Reason is inadequate for the social sciences. The social sciences need to incorporate what the cognitive and brain sciences have shown us about the nature of Reason. Reason itself, as it has been traditionally taught, is—or should be—a major issue in the social sciences. These results will require a rethinking of certain tools defined by old reason: the rational actor model, cost-benefit analysis, polling, and surveys based on old views of language and reason, and so on.

The inadequacies must be made conscious and replaced with an adequate theory of reason and rationality. That is part of the job of an adequate Neural Social Science.

Some Brain Basics

Color

There is no color in the world—no green in grass, no red in blood, no blue in the sky.

Color is determined by

1. Wavelength reflectances of objects; but wavelengths are not colors
2. Nearby lighting conditions
3. Color cones in the retina
4. Neural circuitry in the brain connecting to retinal color cones

(1) Wavelength reflectances of objects and (2) surrounding lighting conditions are in the external world. (3) The color cones and (4) the neural circuitry are in you—in your body! Without your body, there is no color—no experience of color no color concepts, and no words for color concepts. Colors and color concepts are embodied; they subsist in the relation between you and the external world, not in the external world alone.

Perception and Action

Perception and action are not different from the brain's perspective. Perception and action are determined by mirror-neuron systems. Complex actions (like taking a drink) are coordinated—choreographed—by the premotor cortex, which connects to the motor cortex, which in turn carries out the combinations of simple motor actions (grasping, lifting, opening the elbow, etc.) needed to carry out the "same" complex action like taking a drink: about 30% of these drink-taking premotor neurons also fire when you see someone else taking a drink. The other 70% perform interesting complex correlations between perception and action. The mirror-neuron system allows us to connect with others via empathy, by putting ourselves in the shoes of others. "Super-mirror neurons" in the forebrain do part of the job of distinguishing our actions from those of others (Iacoboni 2008).

That's Why There Are Basic-Level Concepts

Basic-level concepts like *chair* and *car* come with motor programs (like driving a car), mental images (of what a chair looks like), and gestalt perception (the ability to perceive a chair or a car as a whole). The existence of basic-level concepts is a consequence of mirror-neuron circuitry, which is the same for perception and action.

That's Why Verb Roots Are the Same for First- and Third-Person Experiences

Action is a first-person experience (I drink). Perception is a third-person experience (I see that he drinks). Yet in language after language, the expression of those experiences via verb roots is the same—because the neural basis of the experiences is the same. Sometimes affixes differ—*drinks* versus *drink*. Sometimes the vowel shifts (*run* versus *ran*) with the consonantal root preserved—as with triliteral roots in Semitic languages. And sometimes, there is a historical explanation for root differences (*be* versus *are*).

Imagining and Doing Use the Same Brain Circuitry

The brain circuitry used when you *actually see* something in the world overlaps considerably with the brain circuitry used when you *imagine* seeing the same thing. The same is true of actually moving your body and imagining moving your body, as when you kick your foot and imagine kicking your foot. The same is true of remembering and doing, dreaming and doing, and speaking and doing. There is a unique portion of brain circuitry for performing specific actions/imaging them/remembering tem/dreaming about them/and speaking about them.

That is why, there is an overlap between gestalt perception and mental imagery, both of which are carried out by brain circuitry.

Neural Computation and Simulation

In 2005, Vittorio Gallese, of the University of Parma Neuroscience Group, and I published "The Brain's Concepts," http://www.google.com/url?sa=t&source=web&cd=1&sqi=2&ved=0CBoQFjAA&url=http%3A%2F%2Finst.eecs.berkeley.edu%2F~cs182%2Fsp07%2Freadings%2FGallese_Lakoff.doc&rct=j&q=the%20brain's%20concepts&ei=UkTsTZ7QKYy0sAO2suj4DQ&usg=AFQjC

NE1RZagC3j_yniSL-kJg8a1M-JqSQ&sig2=QDwFIvg4y99tqdNGlimNeQ, in which we reviewed the primary data on mirror neurons by Parma researchers. The data had been gathered via neuron-by-neuron probes of macaque monkeys' brains. The macaques were trained to perform discrete tasks—grasping and releasing, pushing buttons, peeling bananas, eating peanuts, etc. Each task made use of between several dozen and several hundred neurons—called a "cluster" or "node." Each neuron in the node has between 1,000 and 10,000 connections to other neurons along existing pathways. From the perspective of neural computation, each node can be seen as acting like a single, big, complex neural element, with many neural inputs and outputs. Though each neuron, at any instant, fires or does not fire, the node has many neurons firing or not, and hence appears to fire to some degree, as if the *probability* of the firing of individual neurons was the *degree* of firing of a node. Mathematically, in the theory of neural computation, the calculus of Bayesian probability is used to calculate what happens along a neural circuit made up of nodes. According to the Bayesian calculus, changes in activation in a neural circuit leads, via the Bayesian rules of computation, to other activations and inhibitions along the circuits. This permits Bayesian networks to model the "best fit" of certain changes to other changes, and so to model neural learning. Modeling is a theoretical enterprise, and so it is not known at present how well such modeling fits actual neural learning.

The Centrality of Metaphor in Social Life

The old Enlightenment Reason saw all meaning as literal, as abstract logical reason fitting the external world directly. Social policy studies have largely been based on this now-discredited view. The fact is that a huge area of reason—especially everyday thought about social concerns is—is metaphorical. Social policy is often made on the basis of metaphor, which by itself would not discredit the policy if the metaphor is apt, that is, if its entailments fit the social situation.

Always bear in mind that metaphor is a mode of thought. Linguistic expressions that are metaphorical are surface manifestations of metaphorical reasoning that shapes much of our social life. This is the basis of experimental work by Landau, Mark J.; Meier, Brian P.; Keefer, Lucas A.., "A metaphor-enriched social cognition." *Psychological Bulletin*, Vol 136(6), Nov 2010, 1045–1067. Landau and his colleagues have been producing experimental results that support the centrality of metaphor in social life, for example, "Evidence That Self-Relevant Motives and Metaphoric Framing Interact to Influence Political and Social Attitudes," Psychological Science 1, November 2009: 1421–1427.

Neural Metaphor

We now have a neural theory of how metaphorical thought arises and functions—and it is being confirmed experimentally, often by sociologists and social psychologists. Take the metaphor system of Moral Accounting, in which fairness and justice is defined in terms of the metaphor of Well-Being as Wealth. Receiving a favor is like receiving metaphorical money. You say *I owe you one, How can I repay you?, I'm in your debt.* Returning the favor is Restitution; it is balancing the moral books.

Consider harm—a decrease of well-being. Justice can either be a matter of Restitution, making up for the harm, or Retribution, harm balanced with harm. Revenge is based on the metaphor of Moral Arithmetic: Just as creating a debit is equivalent to removing a credit, so taking away something of value is a form of harm. These forms of moral accounting are central to our social life.

Just as eating pure food leads to satisfaction and eating rotten food leads to disgust, so, via the metaphor of Morality as Purity, immoral behavior is seen as disgusting. These metaphors characterize our emotional reactions to moral and immoral social behavior. And language follows: We speak of an immoral act as a disgusting, or rotten, thing to do. We speak of moral behavior as moral purity.

We metaphorically think of achieving a purpose as reaching a destination—a goal, and purposeful action as motion toward such a destination. Difficulties are conceptualized as things that get in the way of reaching a destination—encountering a roadblock, getting bogged down, being held back, being weighed down. In many cultures, including ours, people are expected to have life goals, with life seen as a journey toward such goals. In our culture, we even have special documents recording our progress on that journey—CVs, "curriculum vitae" (the "course of life"). To get certain jobs, one is expected to have an impressive CV, showing that one has made better than normal progress toward certain life goals. And married couples are expected to have compatible life goals.

These are just a handful of ways in which embodied conceptual metaphors define central aspects of our social lives.

The Narayanan-Johnson-Grady Neural Theory of Metaphor

One of the deepest results in theoretical neural cognition is the Neural Theory of Metaphor. It was arrived at via three interrelated dissertations at UC Berkeley in the mid-1990s by Srini Narayanan, Christopher Johnson, and Joseph Grady (Lakoff and Johnson 1999). Here is the basic idea.

Our conceptual systems are structured by hundreds of "primary" metaphors, conceptual mappings from one conceptual domain to another that are learned mostly in childhood just by functioning in the everyday world. We are usually not conscious of these metaphors, though we learn hundreds of them.

How Are Neural Circuits Learned?

By "recruitment."

At birth, our brains are structured to run our bodies, with existing neural pathways. We have 100 billion neurons and trillions of neural connections between them (between 1,000 and 10,000 connections per neuron). At birth, most of these connections are not yet organized into neural circuits that can perform particular functions. Functional circuits are formed when the right kinds of "strengthening" occur at synapses. Synaptic strengthening occurs when the neurons fire during experiences. As Donald Hebb noticed, "Neurons that fire together wire together." In Hebbian learning, two-way connections are formed slowly, as strengthening gradually occurs over time. Strengthening occurs as neurons fire and get used. Synapses that are unused die off. Between birth and the age of roughly 5, about half of the neural connections we are born with die off—the unused half. That is why, early childhood education is so important. If you do not hear music till the age of 5, you will not become a musician. A huge number of the ideas we use in later life are learned by recruitment due to early experience. Much of that experience is common—sometimes around the world, where More Is Up for everyone;—and sometimes in a society, where wealth may or may not be taken metaphorically as a sign of God's approval.

The Feldman Functional Circuitry Hypothesis

Jerome Feldman founded the International Computer Science Institute at UC Berkeley in 1986. I joined with him in setting up the Neural Theory of Language Project at the Institute. I brought with me fellow researchers and basic results of Cognitive Linguistics: the details of such embodied conceptual structures as image-schemas, frames, and conceptual metaphors, with language as the pairing of linguistic form (sounds, signs, gestures, writing, images) with such embodied structures. Feldman brought detailed computational theories of functional circuitry that gave promise of both describing how the cognitive structures got that way and providing a theoretical scientific account of how it all worked. Between us, we recruited a remarkable team of cross-disciplinary researchers: Charles Fillmore (and his whole FrameNet group), Eve Sweetser, Terry Regier, David Bailey, Lokendra Shastri, Srini Narayanan, Dan Jurafsky, Adele Goldberg, Benjamin Bergen, Vittorio Gallese, Lisa Aziz-Zadeh, Nancy Chang, Christopher Johnson, Joseph Grady, Carter Wendelken, Ellen Dodge, Steve Sinha, Joe Makin, Leon Barrett, Mett Gedigan, Behrang Mohit, John Bryant, Jenny Lederer, and others.

Over more than two decades, computational models of functional circuitry have drawn upon well-known computational techniques of PDP connectionism, localist connectionism, Petri nets, Bayes Nets, accounts of neural binding, models of neural modulation, models of basal ganglia, and so on. What has emerged is a sophisticated account of the theoretical–computational neuroscience of thought and language, based on the idea of functional circuitry unified with cognitive linguistics.

These ideas are surveyed in Feldman's MIT Press book *From Molecules to Metaphors.* The general point is that the brain functions according to simple computational structures defined over functional brain circuitry—and that is how all of thought and language works!

The details give rise to a theory of Neural Social Science—how meaningful social ideas arise in various cultures, providing a theory of significant cross-social similarities and differences.

Primary Metaphors

Primary metaphors are learned via synaptic strengthening of synapses in functional circuits. Consider the conceptual metaphor *More Is Up*.

Whenever a child sees a liquid poured into a container or things put in a pile, his or her brain "notices." Two areas of the brain are activated, one registering an increase in quantity and the other registering an increase in verticality.

Each time these areas are activated together, the neural synapses in both areas are strengthened. Because the neuronal groups are connected, neuron by neuron, each to thousands of other neurons, activation spreads along existing pathways, getting stronger each time the pairing of quantity and verticality occurs.

The spreading of increasingly stronger activation keeps on until a common pathway is found and the activations meet. The synapses along this pathway then get doubly strengthened from activation in both directions, until a permanent circuit is formed. That circuit is the physical realization of the *More Is Up* metaphor.

Narayanan on Spike-Time-Dependent Plasticity

The Hebbian learning account is a central part of the story, but not enough. Hebbian learning is bidirectional. But conceptual metaphors are unidirectional. We understand Affection as Warmth, not Warmth as Affection. Moreover, the subject matter of a metaphorical idea (e.g., Morality) can provide some activation of the range of metaphorical understandings of that concept: Morality can be Purity, Uprightness, Light, Obedience, Nurturance, Balancing Moral Books, and so on.

Moreover, primary metaphors—the irreducible ones—are all embodied: they link two brain areas with bodily connections. How can we make sense of this? Why *must* it be true? And why are the primary metaphors most likely to be the ones found all over the world, in culture after culture? Why should children know them early in life? And how can they be learned, in many cases, *even before language*!

Here is the intuitive idea.

Because the brain computes Verticality constantly, more than it computes Quantity, the synapses in the direction spreading from Verticality will be stronger than those spreading from Quantity. Since the neural circuit is formed by spreading in both directions, there will exist, at each point on the pathway, cases where the axon of neuron A forms synapses on neuron B, and conversely, where the axons of neuron B forms synapses on neuron A.

This situation gives rise to a phenomenon called "spike-time-dependent plasticity." Neurons "fire" in a series of "spikes." The neuron with the strongest input will spike first, and as a result, there will be synaptic strengthening in its direction and weakening in the opposite direction. This produces directionality in the metaphor. The stronger activation will define the source domain, the weaker will define the target domain. That is why metaphorical mappings are asymmetric. This correctly predicts the directionality of primary metaphors. For example, in More Is Up, Verticality is the source domain because the brain is always computing Verticality, even when we are asleep, but not always computing the concept of Quantity. In Affection is Warmth, Temperature is the source domain because the brain is always computing Temperature, but not always computing Affection. Thus, there is a neural explanation for the system of primary metaphors—the scaffolding on which our social concepts are built.

Neuromodulators and "Rewards"

Neuromodulators are chemicals, like dopamine, norepinephrine, acetylcholine, etc. that can powerfully change synaptic strengths, both positively and negatively, in a very short time. These constitute what is called the "reward system" of the brain. They play subtle roles in setting goals, changing attention, and producing emotional satisfaction and dissatisfaction. They are therefore central to decision-making. Decisions in the brain are made on the basis of vast numbers of neural circuits of the sort we have been discussing, for frames, metaphors, and so on.

Integrating Multiple Neural Systems

Contemporary research on real reason takes all of this, and much more, into account. Everything we understand uses frames, metaphors, and narratives that are characterized by neural circuitry, which in turn gets its meaning via embodiment. The circuits, as we currently understand them, have "gates"—points where they can be inhibited (turned off) or activated (turned on). The brain also has "binding circuitry" which, when turned on in context, can identify a concept in one circuit as being the "same" as that in another circuit. For example, a restaurant frame-circuit is composed of other frame-circuits: business, eating, and hosting. The customer in the business frame-circuit is neurally bound to the eater in the eating frame-circuit and to the guest in the hosting frame-circuit. When the binding circuitry is turned off, the frames for business, eating, and hosting can operate independently. Gating and binding allow for enormous compositionality of frame-circuits. And it allows for imagination and fictional entities, like flying pigs—where wing frame-circuits are neurally bound to the sides of the body of a pig in a pig frame-circuit.

The embodied brain keeps us connected to the world and each other, via all the embodied neural systems that we have. Mirror-neuron systems connect us to each other. Canonical neural systems connect our most normal actions to our perceptions of the world. Emotional systems characterize our goals and avoidances. Speech, hearing, reading, and gesture systems give meaning to communication. Temperature systems regulate the body. And association cortex links all these up in billions of complex ways.

Embodiment Evidence in Social Psychology

Over the past couple of decades, experimental social psychologists have provided massive evidence not only for the existence of fixed metaphorical brain circuits, but also for their effects on producing and understanding social behavior.

Here are a few examples:

A recent study in *Biological Psychology* showed that that when subjects leaned forward, the body posture activated desire. The metaphor activated is Achieving a Purpose (Desire) Is Reaching a Destination. Leaning forward activates motion to a destination, which is the source domain of the metaphor and which in turn activates the target domain of desire and purpose.

Eddie Harmon-Jones, Philip A. Gable, Tom F. Price. "Leaning embodies desire: Evidence that leaning forward increases relative left frontal cortical activation to appetitive stimuli." *Biological Psychology* 87 (2011) 311–313.

At *Yale*, researchers found that subjects holding warm coffee in advance were more likely to evaluate an imaginary individual as warm and friendly than those holding cold coffee. This is predicted by the conceptual metaphor that Affection is Warmth, as in *She gave me a warm greeting.*

Williams, L. E., & Bargh, J. A. "Experiencing physical warmth influences interpersonal warmth." *Science,* 322, 2008, 606–607.

At Toronto, subjects were asked to remember a time when they were either socially accepted or socially snubbed. Those with warm memories of acceptance judged the room to be 5° warmer on the average than those who remembered being coldly snubbed. Another effect of Affection Is Warmth.

Zhong, C. B., & Leonardelli, G. J. "Cold and lonely: Does social exclusion feel literally cold?" *Psychological Science*, 19, 2008, 838–842.

Subjects asked to think about a moral transgression like adultery or cheating on a test were more likely to request an antiseptic cloth after the experiment than those who had thought about good deeds. The well-known conceptual metaphor *Morality is Purity* predicts this behavior.

Zhong, C. B., & Liljenquist, K. (2006). Washing away your sins: Threatened morality and physical cleansing. Science, 313, 1451–1452.

Students told that that a particular book was important judged it to be physically heavier than a book that they were told was unimportant. The conceptual metaphor is *Important is Heavy.*

In a parallel study with heavy versus light clipboards, those with the heavy clipboards were more likely like to judge currency to be more valuable and their opinions and their leaders more important.

Nils B. Jostmann, Daniel Lakens, and Thomas W. Schubert. "Weight as an Embodiment of Importance," *Psychological Science,* September 1, 2009: 1169–1174.

Why does this happen? Because such conceptual metaphors are permanent physical circuits in the brain. In each case, the metaphor circuit is activated—turned on—by the context of the experiment, which in turn activates circuitry making the given behavior more likely. In short, we really do live by metaphor.

Imagine thinking about problems in your long-term love life. It would be normal for the Love As A Journey Metaphor to be activated, leading to thinking in terms of love-relationship goals in terms of long bumpy roads, going in different directions, things standing in the way, and so on.

Love as a Journey is a special case of a Journey metaphor, with long-term goals as destinations to be reached, say, by working out God's Plan for you. It is also one of many Love metaphors: Love can be conceptualized as a source of Light and Heat ("Juliet is the Sun"), as sacrifice (see Abraham and Isaac with respect to love of God, as well as military sacrifice for love of country), and so on.

Metaphors are not mere linguistic expressions. They are complex modes of thought, with systems of primary metaphors arising naturally just by living in the everyday world, based on systems of frames that also arise naturally just by living in the world.

Real Social and Political Life

The science behind Neural Social Science is sufficiently well understood to be taken very, very seriously. And that science allows us insights into our social and political life that were not available before (Lakoff 2009).

The individual issues are all too real: assaults on unions, public employees, women's rights, immigrants, the environment, health care, voting rights, food safety, pensions, prenatal care, science, public broadcasting, and on and on.

Budget deficits are a ruse, as we've seen in Wisconsin, where the Governor turned a surplus into a deficit by providing corporate tax breaks, and then used the deficit as a ploy to break the unions, not just in Wisconsin, but seeking to be the first domino in a nationwide conservative movement.

Deficits can be addressed by raising revenue, plugging tax loopholes, putting people to work, and developing the economy long-term in all the ways the President has discussed. But deficits are not what really matters to conservatives.

Conservatives really want to change the basis of American life, to make America run according to the conservative moral worldview in all areas of life.

In the 2008 campaign, candidate Obama accurately described the basis of American democracy: Empathy—citizens caring for each other, both social and personal responsibility—acting on that care, and an ethic of excellence. From these, our freedoms and our way of life follow, as does the role of government: to protect and empower everyone equally. Protection includes safety, health, the environment, pensions, and empowerment starts with education and infrastructure. No one can be free without these, and without a commitment to care and act on that care by one's fellow citizens.

The conservative worldview rejects all of that.

Conservatives believe in *individual* responsibility alone, not social responsibility. They don't think government should help its citizens. That is, they don't think citizens *should* help each other. The part of government they want to cut is not the military (we have 174 bases around the world), not government subsidies to corporations, not the aspect of government that fits their worldview. They want to cut the part that helps people. Why? Because that violates individual responsibility.

But where does that view of individual responsibility alone come from?

The way to understand the conservative moral system is to consider a strict father family. The father is The Decider, the ultimate moral authority in the family. His authority must not be challenged. His job is to protect the family, to support the family (by winning competitions in the marketplace), and to teach his kids right from wrong by disciplining them physically when they do wrong. The use of force is necessary and required. Only then will children develop the internal discipline to become moral beings. And only with such discipline will they be able to prosper. And what of people who are not prosperous? They don't have discipline, and without discipline they cannot be moral, so they deserve their poverty. The good people are hence the prosperous people. Helping others takes away their discipline, and hence makes them both unable to prosper on their own and function morally.

The market itself is seen in this way. The slogan, "Let the market decide" assumes the market itself is The Decider. The market is seen as both natural (since it is assumed that people naturally seek their self-interest) and moral (if everyone seeks their own profit, the profit of all will be maximized by the invisible hand). As the ultimate moral authority in economics, there should be no power higher than the market that might go against market values. Thus, the government can spend money to protect the market and promote market values, but should not rule over it either through (1) regulation, (2) taxation, (3) unions and worker rights, (4) environmental protection or food safety laws, and (5) tort cases. Moreover, government should not do public service. The market has service industries for that. Thus, it would be wrong for the government to provide health care, education, public broadcasting, public parks, and so on. The very idea of these things is at odds with the strict-father conservative moral system. No one should be paying for anyone else. It is individual responsibility in all arenas. Taxation from this perspective is thus seen as taking money away from those who have earned it and giving it to people who don't deserve it. Taxation cannot be seen as providing the necessities of life, a civilized society, and as necessary for business to prosper.

In the conservative ideal of family life, the strict father rules. Fathers and husbands should have control over reproduction; hence, parental and spousal notification laws and opposition to abortion. In conservative religion, God is seen as the strict father, the Lord, who rewards and punishes according to individual responsibility in following his Biblical word.

Above all, the authority of conservatism itself must be maintained. The country should be ruled by conservative values, and progressive values are seen as immoral and evil. Science should *not* have authority over the market, and so the science of global warming and evolution must be denied. Facts that are inconsistent with the authority of conservatism must be ignored or denied or explained away. To protect and extend conservative values themselves, the devil's own means can be used against conservatism's immoral enemies, whether lies, intimidation, torture, or even death, say, for women's doctors.

Freedom is defined as being your own strict father—with individual not social responsibility, and without any government authority telling you what you can and cannot do. To defend that freedom as an individual, you will of course need a gun.

This is the America that conservatives really want. Budget deficits are convenient ruses for destroying American democracy and replacing it with conservative rule in all areas of life.

What is saddest of all is to see Democrats helping them. Democrats often help conservatives when they use Enlightenment reason—when they take concepts and words as neutral, and when they ignore the centrality of framing, metaphor, emotion, and narrative. This leads Democrats to use conservative language and the frames and values that come with it.

Democrats help radical conservatives by accepting the deficit frame and arguing about what to cut. Even arguing against specific "cuts" is working within the conservative frame. What is the alternative? Pointing out what conservatives really want. Point out that there is plenty of money in America, and in Wisconsin. It is at the top. The disparity in financial assets is un-American—the top 1% has more financial assets than the bottom 95%. Middle class wages have been flat for 30 years, while the wealth has floated to the top. This fits the conservative way of life, but not the American way of life.

Democrats help conservatives by not shouting out loud over and over that it was conservative values that caused the global economic collapse: lack of regulation and a greed-is-good ethic.

Democrats also help conservatives by what a friend has called Democratic Communication Disorder. Republican conservatives have constructed a vast and effective communication system, with think tanks, framing experts, training institutes, a system of trained speakers, vast holdings of media, and booking agents. Eighty percent of the talking heads on TV are conservatives. Talk matters because language heard over and over changes brains. Democrats have not built the communication system they need, and many are relatively clueless about how to frame their deepest values and complex truths.

And Democrats help conservatives when they function as policy wonks—talking policy without communicating the moral values behind the policies. They help conservatives when they neglect to remind us that pensions are deferred payments for work done. "Benefits" are pay for work, not a handout. Pensions and benefits are arranged by contract. If there is not enough money for them, it is because the contracted funds have been taken by conservative officials and given to wealthy people and corporations instead of to the people who have earned them.

Democrats help conservatives when they use conservative words like "entitlements" instead of "earnings," when they speak of government as providing "services" instead of "necessities," and when they fail to point out how government contributes massively to corporate profits, which is why progressive taxation is justified.

The Conservative Advantage

Radical Republicans have their act down cold.

- They have a general conservative moral frame with language that fits all issues and that is repeated constantly.
- They have a communication system that operates all day every day in every city.
- The common conservative moral frame, when heard day after day no matter what the issue, activates that frame in listeners' heads, making the frame stronger. The constant repetition strengthens the conservative moral system in the brains of biconceptual (or so-called "independent") voters and weakens the progressive system in their brains. Even impoverished conservatives adopt and stick to that frame, because it comes to characterize their identity, and promises that everyone can be their own strict parent in their own domain, no matter how poor. It promises the conservative version of freedom.

The result is that in any contested districts, radical conservatives have an advantage because they can put pressure—linguistic pressure—on their office holders and candidates.

What Can Progressives and Democrats Do?

First, be rational: adopt real reason. Just stating the facts and arguing logically about policy is not enough. Progressives and Democrats do have a general moral worldview:

> Democracy is about empathy (citizens care about each other) and social as well as personal responsibility (to act on that care). Government has the moral mission of protecting and empowering all citizens equally.

These ideas can be applied to any issue. General language can be created for these ideas. They need to be stated positively over and over. And Progressives need to set up a communications network—with training institutes, and lists of volunteer speakers ready to go on every subject.

There is a reason why social scientists tend to be Democrats and Progressives—because the democratic vision of the good society that defines social science is exactly the moral worldview of Progressives and Democrats just discussed.

As important as it is to be positive, it is also important to call radical conservatives what they are—radical—and what their moral philosophy is—greed, for power in politics, for money in business, and for domination in social and religious life.

And remember to point out that government is not just carried out by the state. Corporations govern us just as much—for their profit, not for our benefit. Corporations determine the health care we can get, the news we can hear, the form of energy we are dependent on, the nutritive value of the food available to us, and so much more. Making government "smaller" does not eliminate government. It just moves government to the private sector, enabling corporations to govern us for their profit, not for our benefit.

Systems Thinking

Progressives and Democrats also need to add a crucial idea: systems thinking. In politics, the environment, and economics, systems prevail. Causation is not limited to a single category. Causation is systemic, not direct.

A well-functioning system is homeostatic, self-correcting.

Systems have feedback, both positive and negative. Feedback can be controllable, but can become uncontrollable and catastrophic, as in global warming and global financial crisis.

Systemic causes are not linear: small causes can have huge effects.

Systemic causes are not local: causes can have effects over very long distances.

Systemic causation is not limited to single categories. The problem of the deficit is not just about the deficit, but about all aspects of the economy.

Conservatism tends to use direct, not systemic, causation, and so it is vital for the public to hear about systemic causation all day, every day.

The Point

Neural Social Science is not just another approach to social science. Instead, it both gets to the heart of the values that define what the social sciences are about and it allows one to defend those values.

Remember that neural social science comes out of real science—the brain and cognitive sciences. It is a way that science can actually do a huge amount to get us out of the current conservative threat and to heal the world. But it has to be taken seriously by social scientists, despite all current academic institutional arrangements.

CODA

Solving a Social Science Puzzle

In 2009, when President Barack Obama chose the policy provisions for his health care plan, polls showed that most provisions (e.g., no preconditions, choice of plans) were supported by 60–80% of Americans. Yet, when the whole plan was polled, fewer than 50% supported it. Why? Why the disparity between the parts and the whole, when the whole literally equals the sum of the parts?

The answer is straightforward from the perspective of real reason. When President Obama came out with the provisions of his health care plan in early 2009, the conservatives decided to attack it not on policy grounds but on moral grounds. They chose two areas of morality: Freedom ("government takeover") and Life ("death panels"). And they repeated over and over that "Obamacare" (naming matters) was a government takeover that was a threat to individual freedom, with death panels that were a threat to life itself.

Note that the policy provisions were about the everyday details of dealing with one's HMO. They were in the Practical Health Care Details frame. The conservative attack was in the Morality frame, activating freedom and life. The conservatives understood that all politics is moral, that political leaders all say they are doing what is right, not what is wrong.

The policy details and the moral attack were in different frames, located in different parts of the brain. From the perspective of real reason, the whole health care act was, for those with a conservative worldview, not equal to the sum of its policy parts. Conservatives and independents (actually biconceptuals, who are progressive in some respects and conservative in others) had their conservative moral worldview activated by the conservative moral attack. This separated the moral whole from the practical parts.

For progressives, their morality and the practical details fit together; for conservatives and biconceptuals (aka "independents"), they were different subject matters.

Such an explanation is natural when you think in terms of the brain and frame-circuitry. It is not possible when you think in terms of the logic of Enlightenment reason, where the whole is necessarily (logically) the sum of the policy parts.

References

DeHaene, S. (2009). *Reading in the brain.* New York: Penguin Viking.
Feldman, J. (2008). *From molecule to metaphor.* Cambridge, MA: MIT Press.
Iacoboni, M. (2008). *Mirroring people* (p. 202). New York: Farrar, Straus, and Giroux.
Lakoff, G. (2009). Chapter 4: The Brain's Role in Political Ideologies. In *The Political Mind.* New York: Penguin.
Lakoff, G., & Johnson, M. (1999). *Philosophy in the flesh.* New York: Basic Books.

Chapter 3
Why We Need Neurosociology as Well as Social Neuroscience: Or—Why Role-Taking and Theory of Mind Are Different Concepts

David D. Franks

In this chapter, I compare George Herbert Mead's concept of role-taking in neurosociology with psychology's social neuroscience notion of theory of mind (ToM). (See Chap. 12.) My goal is to illustrate how the unit of analysis in the two fields is different and yet show how each has great potential for an innovative sharing of their substance and methodologies. I will also suggest some specific ways that the differences can complement each other. Next I use a hypothesis about how Mead's concept of role-taking is connected with social structure and power to show how typical instruments of social neuroscience could be used in replicating and verifying this relationship. I argue that in order to come close to explaining a satisfactory portion of the variance in such a study, we need both perspectives. In this case, neurosociology is useful in formulating hypotheses to be tested by social neuroscience using Read Montague's (Montague et al. 2002) fMRI method of "Hyperscanning" which allows persons in separate scanners to interact with each other simultaneously.

History of the Terms Neurosociology and Social Neuroscience

John Cacioppo and Gary Berntson used the term "social neuroscience" in 1992 to describe psychology's inquiry into the social nature of our brains (Cacioppo and Bernston 1992). Twenty years earlier, neurosurgeon Joseph Bogen and Warren TenHouten, a sociologist, coined the term "neurosociology" (Bogen et al. 1972: 49). Bogen had worked with Gazzaniga on the first split-brain research. Here they referred to "a confluence of neurologic and sociologic observations."

The term social neuroscience was comparatively standardized and accepted in the field of psychology by the beginning of this century while at that time neurosociology was accepted only with ambivalence by the field of sociology. There had been a vigorous battle over E. O. Wilson's reductionist version of sociobiology, so sociology had significantly more issues to sort out in the area of social and biological connections than did psychology.

This raises the question of why neurosociology is needed at all when a more developed and vigorously growing field called social neuroscience already exists. One way to answer this is to look at how the two fields deal with similar but different concepts and yet keep within their distinct academic boundaries.

D.D. Franks (✉)
Department of Sociology, Virginia Commonwealth University, Richmond, VA 23284, USA
e-mail: daviddfranks@comcast.net

D.D. Franks and J.H. Turner (eds.), *Handbook of Neurosociology*, Handbooks of Sociology and Social Research, DOI 10.1007/978-94-007-4473-8_3, © Springer Science+Business Media B.V. 2013

For example, although there are similarities in the sociological concept of "role-taking" and the concept of "theory of mind," a closer look will reveal differences that define the boundaries between the two fields and suggest why their differences should be maintained. The social neuroscience theory of mind, or "ToM," refers to the everyday person's informal "theories" of the minds of others as to what they are thinking, feeling and motivated to do.[1] ToM has nothing to do with the observing scientist's general theory of mentation though that may seem to be implied to the uninitiated.

Distinguishing Between the Two Fields Using Role-Taking and ToM

Role-taking, as developed by G. H. Mead, contains much more than "assumptions made about the mind of another" even if this "much more" is not always appreciated by sociologists. Exploring the full implications of role-taking can shed light on how the fields differ and why both fields can work together in mutually beneficial ways. Assuming a pre-established communicative context of shared symbols, role-taking refers to the capacity for one's verbal gestures to "call out" in one's self the same response that one's gesture *calls out in the other*. Put differently, the self assumes the perspective of the other and in effect apprehends what the other hears. Experiencing the self from the other's standpoint is called "taking the role of the other." This social and reciprocal shift creates intersubjectivity as well as fitting the actor into the social context and generating subsequent communication (See Mead 1934: 254–55). A major difference in role-taking and ToM is that in role-taking, actors experience not only what they *think* the other is thinking, but experience their own *selves* from their construction of the other's *different cognitive* perspective and emotional predispositions. The actor actually incorporates the imagined response of the other into his own emerging behavior. Frequently sociologists fail to emphasize the fact that actors have to *take on different social perspectives from their own* and then use these to assess how they will appear to others. This stresses interactional processes in a way that ToM does not necessarily do.

This process has important theoretical implications. Foremost is that it offers one of the few voluntaristic theories of the *self*-control of social conduct.[2] This alone would distinguish it from social neuroscience which is more often embedded in a different theory of social control, namely, a more individualistic learning theory. Granted, Mead's framework is compatible with learning theory since it applies to any mammal including human beings, but his intent is to go beyond that to what is distinctively human. Role-taking assumes two persons in interaction mutually influencing each other. We become social in a thoroughly sociological sense when *the other person's anticipated response is incorporated into one's own developing lines of action.* Another difference for sociologists using Mead's theory of the act is that we are *pulled* along in a teleological fashion by our anticipations of our own consummation of the act. Learning theory assumes the actor's motivation comes from external stimuli that happen *to* the person while Mead sees motivation as emanating from the actor him or herself—thus its voluntaristic quality.

---

[1] Valerie Stone (2006:106) is more specific: ToMs require "metarepresentations." Some inferred mental states involve metarepresentational abilities and some do not. According to her, for example, inferring another's emotional state does not have to involve representing a person's mental state as a ToM. Insofar as mirror neurons simulate emotions unconsciously, this may well be correct.

[2] Read Montague, a social neuroscientist, is no stranger to voluntaristic frameworks as indicated by his work on social trust. In this context, he uses more of a rational choice psychology than learning theory. Talcott Parson's theory of social action is also voluntaristic but Parsons is working on a structural level rather than a Meadian social psychological one.

Finally, role-taking meets the critical defining nature of a distinctively sociological concept because two actors who are role-taking can produce emergent behavioral outcomes that are not reducible to the persons taken singularly (Franks 2010).

It bears repeating that this does not make neurosociology and social neuroscience incompatible in any way because role-taking, which captures what is distinctively human, is built on top of learning theory.

Some Ways Role-Taking and Power Can be Explored Experimentally

The articulation of role-taking with the structural concept of power helped to move it outside of the exclusive confines of Mead's foundational symbolic interaction. For example, in 1972 Thomas et al. examined the ability of family members to predict accurately the responses of others in the family. They asked members to predict how each other family member would respond to a series of vignettes. An inverse relationship was found between the social status held by family members and each family member's accuracy in predicting the responses of other members. The most accurate role-takers were little girls, then little boys, with the least accurate role-takers being fathers and after them, mothers (See Franks 1989).

This research emphasizes an important distinction between *interpersonal* resources, which stem from one's social position and those stemming from one's *personal* capacities. Resources accruing from the self-system are available to all persons depending on their individual competencies in interpersonal relations. *Power* relates to the potential for control given by the impersonal structural position, in contrast to *influence* which is based on one's interpersonal skills and personality. The research by Thomas et al. suggests other investigations for neurosociology and social neuroscience, which separate personal dimensions from social structural ones in order to study their relationships. For example, would departmental chairpersons in a university be more accurate role-takers of deans than vice versa? Would deans be more accurate role-takers of vice presidents than of their departmental chairpersons? Exploring these questions would provide us with some clear demarcations of the social structural and the personal.

To date, however, these theoretically important findings have not been replicated although some unpublished attempts to test the hypothesis have been made. This leads us to ask how social neuroscience can help in this regard.

What Social Neuroscience Can Offer Sociological Research on Role-Taking and Power

On the social neuroscience side, the amount of research and replication is far greater than in neurosociology, further demarcating the differences in the two fields. The important places in the brain where a neurosociologist might look for correlates of role-taking start with the parts that are newest from an evolutionary point of view thus, one might expect uniquely human behavior to be related to uniquely human areas of the brain. That would lead us to the neocortex and particularly to the prefrontal lobes. There we would find dense, two-way paths to the thalamus, amygdala, and other subcortical structures.

In like fashion, social neuroscientists trace the significant evidence for relatively dedicated brain systems that are significantly active when subjects have to think about what others are thinking. These findings remain constant in numerous different research contexts and in ways that would be the envy of empirically oriented social psychologists from sociology. As Mitchell et al. (2006: 65) phrase it:

> Data from neuroimaging and patient studies…provide an efficient means for addressing the question of whether social cognition relies on its own set of mental processes or piggybacks on other more general processes of memory, inference, planning, and so forth.

They present a wealth of evidence that "social cognition" relies on a discrete set of brain regions and that this imaging data carries over specifically to "mind reading". Predominantly, this includes the medial prefrontal cortex, but as is usual with the brain, many other regions also contribute critically such as the temporoparietal junction, amygdala, superior temporal junction, and the orbitofrontal cortex. These results do not answer the important questions about how these areas work, or even if they should be conceived of as "areas," but from an evolutionary viewpoint, they do strongly suggest that mind-reading, however one looks at it, is a necessary brain process for human survival and that these processes are vital parts of current social behavior.[3] Despite their equivocal status, the results do allow some exploratory guidelines of what brain phenomena might be related to role-taking.

One methodological strategy in the context of communication involves isolating mind-reading tasks from non-mind-reading tasks and then showing how distinguishing between the two makes a difference in brain activity which seems relatively "dedicated" to this social process (See Mitchell et al. 2006: 66). For example, participants who were scanned with fMRIs were given three semantic categories to consider: names of a number of known people (Frank, John, Mary), names of fruits followed by names of articles of clothing. Each participant was presented with an adjective such as curious, bored or sad for people, pitted for fruit, and woolen for clothes. Then they were asked if the adjective could be appropriately applied to each item. The researchers found that judgments about the mental states of the named people were associated with modulations in a different brain area than were found in the responses about the inanimate objects. The areas used in contemplating persons were the MPFC, ventrolateral PFC and the right temporoparietal juncture, superior temporal sulcus, and the fusiform gyrus. As the authors (2006: 66) stated:

> (Subjects) making semantic differences about (mental) characteristics of other people appeared to engage a qualitative set of cognitive processes than did similar decisions about inanimate objects. (Parens. added).

Other studies, some using stories and cartoons that could only be understood by attributing mental states to their characters, found consistent correlations with greater activity in the medial PFC during tasks requiring the subjects to consider the mental states of others.

Empirically Testing the Role-Taking and Power Hypothesis

Given the differences between the units of analysis of sociology and psychology, it comes as little surprise that the variations in brain activity in intrapersonal rather than *inter*personal cogitations have been given most of the attention in social neuroscience. This is also to be expected when the

---

[3] In the first chapter of Cacioppo et al. (*Social Neuroscience* 2006), Berntson is more cautious than the quotes above and than other authors in the book. He warns that it is too early to tell whether social processes reflect special brain processes. "We do not know enough about either social psychological process or brain mechanisms to answer this question at the present time. Brain localization can inform neuropsychological theories, but meaningful neurological theories will not be about places nor will…they be couched in the language of space. Rather they will have to incorporate fundamental underlying processes that subserve social processes." Charles Kaplan in a personal communication suggests replacing "areas" with "circuitry."

major data collection techniques are scanners that isolate people from one another. However, as noted before, Montague (Montague et al. 2002) has developed imaging devices wherein several persons can communicate with each other across the space of the two "hyperscanners." Unfortunately, the technology does not allow for normal verbal interaction. It must be limited to movements of the fingers on keyboards and the subject's head cannot move which takes on significance when we consider the important role that body posture plays in human communication.[4]

Even with these limitations, hyperscanning might well be used for the purpose of testing hypotheses derived from Mead's concept of the role-taking process and especially in the context of role-taking and power. Social neuroscience has given us the consistent finding that more brain energy as measured by blood flow is needed when a person is involved in constructing theories of mind than when ToM is not a part of the person's activity. It would be expected that a nurse communicating with a well-known doctor would be more pressured to role-take than would a participant answering the question of whether subjective adjectives applied to people are more appropriate than adjectives applied to inanimate objects. In such settings, nothing is at stake personally. Identifying the brain processes differentially involved and comparing them with the previous social neuroscience findings might well benefit both fields. On the other hand, scanned indicants of brain activity in measures of doctors who are role-taking with nurses may be expected not to involve the same intensity of brain processing. In no sense would this imply exhaustive evidence but it would be a start.

Conclusions

The hypotheses about role-taking and power demonstrate differences in the ostensibly similar concepts of role-taking and ToM. At the same time, they also illustrate how the fields can be useful to each other. A joint study of the conditions where situational factors cause more brain activity as opposed to conditions where personality factors predominate would be a case in point. Another old and important question concerns the part that projection enters into mind reading. Those who study mirror neurons contend that understanding others' behavior is fostered because we actually *do* in our motor cortexes the behaviors we observe in others. There seems to be a great deal of evidence that such is the case (Iacoboni 2008).[5] In contrast, questions as to whether the medial PFC contributes differently when one takes the perspective of a role different from one's own also deserve attention. There is much to explore here and discovering some of the secrets of the human brain would move neurosociology to the next level.

References

Bogan, J. E., Dezure, R., TenHouten, W. D., & Marsh, J. F. (1972). The other side of the brain IV: The A/P ratio. *Bulletin of the Los Angeles Neurological Societies, 37*, 49–61.

Cacioppo, J. T., & Bernston, C. C. (1992). Social psychological contributions to the decade of the brain: Doctrine of multilevel analysis. *American Psychologist, 47*, 1019–1028.

Cacioppo, J. T., Visser, P. S., & Pickett, C. L. (2006). *Social Neuroscoence: People thinking about other people*. Cambridge, MA: The MIT Press.

[4] Charles Kaplan tells me in a personal communication that he is experimenting with virtual reality solutions to this problem and that Read Montague is aware of this.

[5] But see Philip Gerrands in John Cacioppo and Berntson, *Handbook of Social Neuroscience* (2009). He is not so sure of this evidence but he does not go in depth into his reasons in this particular source.

Franks, D. D. (1989). Role-taking, social power and imperceptiveness: The analysis of rape. In Denzin Norman (Ed.), *Studies in symbolic interaction* (pp. 229–259). Stamford: JAI Press.

Franks, D. D. (2010). *Neurosociology: The nexus between neuroscience and social psychology*. New York: Springer.

Iacoboni, M. (2008). *Mirroring people: The new science of how we connect with others*. New York: Farrar, Straus and Giroux.

Mead, G. H. (1934). In C. Morris (Ed.). *Mind, self and society*. Chicago: Chicago University Press.

Mitchell, J. P., Mason, M. F., Macrea, C. M., & Banaji, M. R. (2006). Thinking about others: The neural substrates of social cognition. In J. Cacioppo, P. S. Visser, & C. L. Pickett (Eds.), *Social neuroscience: People thinking about people* (pp. 63–82). Cambridge, MA: The MIT Press.

Montague, P. R., Burns, G. S., Cohen, J. D., et al. (2002). Hyperscanning: Simultaneous *f*MRI during linked social interactions. *NeuroImage, 16*(4), 1159–1164.

Montague, P. R., King-Casas, B., & Cohen, J. D. (2006). Imaging valuation models in human choice. *Annual Review of Neuroscience, 29*, 417–448.

Stone, V. (2006). Theory of mind and the evolution of social intelligence. In J. T. Cacioppo, P. S. Visser, & C. L. Pickett (Eds.), *Social neuroscience: People thinking about people* (pp. 103–130). Cambridge, MA: MIT Press.

Thomas, D. L., Franks, D. D., & Calonico, J. M. (1972). Role-taking and power in social psychology. *American Sociological Review, 37*, 605–614.

Chapter 4
Social Cognition and the Problem of Other Minds

John R. Shook

Where in the World Are Minds?

A classic problem of philosophy is the "problem of other minds," a special case of the general problem of knowing the "external" world. The problem of other minds is this: by what justification, if any, can we know that other minds exist? Bertrand Russell well-formulated the empiricist version of this problem, saying that "there seems no reason to believe that we are ever acquainted with other people's minds, seeing that these are not directly perceived" (Russell 1905, 480). This empiricist formulation, bequesting an eighteenth-century philosophical problem upon the twentieth century, is basically Cartesian – minds have such discrete and separate existences that experience cannot unite them or even bring them into contact with each other. Even more mysterious than the external world, of which experience must only be a thin inner representation, those other external minds cannot be reached by experience at all.

This Cartesian legacy established the terms for the modern "problem of other minds." Philosophers and psychologists within that tradition deal with the question of how other minds are to be known by first taking all minds to be strictly separate and individuated things. From the perspective of a mind, therefore, any other minds are entirely theoretical matters, as any knowledge of them can only be inferred from what is perceived. Other minds are even more remote and undetectable than atoms, since atoms are not merely theorized but they can be instrumentally detected; no inspection of anything in nature will expose a mind to view. Still, we can theorize about what other minds could be like. Cartesianism takes one's own mind to be the best known thing, so it follows that a person's knowledge about other minds is dependent on one's conception of one's self as a mind. Essentially, we have to theoretically project another mind of the same sort that we take ourselves to already have. On this line of reasoning, we can only project what we already can conceive, after all, and our best conception about whatever minds do exist has to be completely reliant on the only mind we know best, namely, our own.

The Cartesian theory of mind therefore leads to a radically individualistic presumption that self-knowledge of one's mind is both logically and developmentally prior to knowledge of other

J.R. Shook (✉)
EdM Program Science and the Public, University at Buffalo,
New York, NY, USA

Center for Neurotechnology Studies, Potomac Institute
for Policy Studies, Arlington, VA, USA
e-mail: jshook@pragmatism.org

D.D. Franks and J.H. Turner (eds.), *Handbook of Neurosociology*, Handbooks of Sociology and Social Research, DOI 10.1007/978-94-007-4473-8_4, © Springer Science+Business Media B.V. 2013

minds. Where's the problem, then? The "problem of other minds" erupts within Cartesianism when the fallibility of theoretical inferences from mere perceptions is taken seriously. It is not necessary to go to great skeptical lengths to simply show that knowledge of other minds must be a quite fallible matter, since other minds are so thoroughly hidden from natural view. How could any knowledge that other minds even exist receive adequate empirical justification? Perhaps we know little or nothing about other minds. Early Cartesianism linked itself to metaphysical dualism, holding that the self-certainty of one's own mind guaranteed a mental reality in addition to any natural reality. However, Cartesianism's individualistic agenda could be carried on by a materialist worldview. The school of "Cartesian materialism," as philosopher Daniel Dennett (1991) labeled it, continued to maintain that some parts of the brain are responsible for what we know best: our inner veridical experience of our own mind.

Philosophers and psychologists operating within this Cartesian legacy have had too easy a time developing their positions without dealing with the actual development and practical workings of human minds. How could self-knowledge of one's mind really be developmentally prior to knowledge of other minds? Not even folk psychology accords well with this legacy. Think about how folk psychology works. For example, folk psychology takes minds as forming intentions to do things – now, does a child first learn that she forms intentions, and only then begins to project intentions into other people's minds? Minds also have beliefs – but does a child first learn that he has beliefs, and only thereafter ascribes beliefs to others' minds? Minds use language to express propositional beliefs – does a child first acquire some language, and only thereafter judges that other people are using language to express their beliefs? Minds use concepts to categorize things in the world – does a child first acquire some concepts, and only thereafter attribute those conceptions to other minds? Minds take on social roles (being a parent, a child, a neighbor, etc.) – does a child first know what it is like to take on some social role, and only thereafter understand that someone else is minding their social role?

When Cartesian mental theorists attempt to explain how people behave socially, coordinating their conduct for mutually agreed aims, they inflate these developmental puzzles into paradoxes. Having assumed that minds must have quite discrete and separate ontological existences, and that they can only know each other by inferences from what each can observe of others' conduct, this "methodological individualism" of modern Cartesianism sets up huge obstacles for explaining how minds can be coordinated. For most of the twentieth century, psychologists and philosophers have struggled with this narrowly empiricist and Cartesian framework as they tried to explain knowledge of others' minds, and indeed the possibility of knowledge in general.

The Psychology of Individual Minds

Twentieth-century psychology inherited this Cartesianism in various forms even as it proposed to break away from armchair introspection. By regarding the subjective and personal as logically and ontologically more basic and real than anything else in the human world, the individual is given explanatory priority. For example, Skinner's methodological individualism regarded all social entities and events to be entirely explainable in terms of aggregate actions of intelligent agents (Skinner 1938). Skinner never regarded the cognition of others' minds as a serious possibility, instead taking any cognition of others as amounting to just the anticipations of habitual behaviors (Skinner 1953, 1983). This Cartesian individualism encouraged innumerable investigators to experimentally isolate subjects to try to discern their genuine capacities as lone thinkers and agents. For generations, behaviorists (and many in other psychology and psychiatry paradigms as well, including Freudians) preferred to study children individually apart from any dynamic social context. Piaget regarded the infant as a solipsist, who is "unaware of himself as subject and is familiar only with his own actions."

(Piaget 1954, 352). Why should the inherent contradiction within Piaget's claim here, between the infant's ignorance of himself and the infant's knowledge of "his own" actions, be so hard to detect? The well-intentioned view that mind is acquired socially goes immediately awry if an excessively Cartesian notion of infant experience is presumed.

Indeed, it is the case that infants lack the cognitive capacities to understand themselves as subjects, but all the same, infants rapidly become familiar with the actions of others, within weeks of birth. Only within the past two or three decades have many researchers really questioned what it could mean for an immature mind to gradually learn that it has a subjective existence and individualized capacities. Evidently, an infant's social interactions with mature minds are responsible for the growth of the infant mind in the first place; both the infant mind and its understanding of other minds grow together. Social cognition must be essential to mental growth – a fact mostly unappreciated by both philosophy and psychology during most of the twentieth century. Not even Piaget, nor Vygotsky (1978), supplied plausible sketches of the development of social cognition. The founder of American psychology, William James, was among the first to describe the several social selves we all use throughout life. His pragmatism unified cognition with action – "My thinking is first and last and always for the sake of my doing" (James 1890, 333) – yet James did not elaborate a developmental social psychology. The most philosophical of social psychologists writing in the English language, John Dewey (1922, 1938) and George Herbert Mead (1934), did offer some empirical-based theories of childhood learning and social cognition. However, they were largely ignored by academic psychology during most of the twentieth century, and the debate between two kinds of individualism – reductive behaviorism and non-reductive Cartesianism – preoccupied Anglo-American academic philosophy instead.

Behaviorists generally continued to assume that behaviors acquired and displayed by the organism in isolation are the core structures permitting interactive behaviors. This individualism, taking the situation of "self confronting world" as primary, was also central for main opponents of behaviorism. If the self's experiences after birth are insufficient for accounting for matured social skills, then there must be native cognitive structures doing the real work all along. Noam Chomsky proposed that infants are born with something akin to an adult theory of mind consisting of basic structures of social cognition, especially a universal grammar for semantic interpretation permitting fast language facility (Chomsky 1968). Jerry Fodor agrees, finding that the possession of an innate computational theory of mind is needed to explain the development of social skills during young childhood (Fodor 1987, 132). As befits its dualism, Cartesian individualism can also aid resistance to empirical science. Dale Jacquette represents an anti-naturalistic continental tradition by pronouncing that intentionality, as the essence of personal mentality, cannot even be conceptually analyzed or explained in terms of anything else (Jacquette 2011).

Respect for shared mentality and social cognition has only been partially restored in the philosophical field of social ontology. John Searle's attempts at social ontology and the origin of social facts and institutions still harbor lingering Cartesianism. Searle's Cartesian individualism has never been disguised: "Collective mental phenomena of the sort we get in organized societies are themselves dependent on and derived from the mental phenomena of individuals" (Searle 2010, 4). For society, the most important kind of mental phenomena are internal symbolic representations. Since an individual cannot directly display his symbolic representations, he uses language to express them, and that behavioral conduct is observed by others, so that they can theoretically infer what symbolic representations are intended. On Searle's account, our capacity to see the match between someone else's representations and our own permits what he called "collective intentionality": two or people intending the same thing. However, Searle's individualism remains strict: "all human intentionality exists only in individual human brains" (2010, 44). Since group plans depend on shared commitments, and one's commitments are in the form of symbolic representations requiring linguistic expression, then one can only tell whether another person has the same commitment through that person's statements. However, not just any sort of statements will do. I can form a theory of your mental commitments, according to Searle, only after you utter "declarations" that state your intentions. Searle rightly denies that "We-intentions" are reducible without meaningful remainder down to sets of "I-intentions." However, his Cartesian Theory of Mind assumptions obstruct any explanation of how "We-intentions" could form.

Cartesian individualism obstructs practical collective intentionality. I could learn that "You intend X" and match that up with "I intend X" but so far all that adds up to is "Both you and I intend X," not "We intend to do X together." Matters are not improved even if I learn that "You intend that we do X" and I add that "I intend that we do X." Logically, they add up to "We both intend that we do X together" but from a developmental perspective, what could really happen here? How do children begin to participate in collective intentionality? Searle's Cartesianism assumes far too much. What if one of us is quite immature, and has little notion of how we could do X? If I have little idea how we could do X, can I really intend that we do X even if I want to do X and I understand that you intend that we do X? A small child can intend to "help Daddy bake a cake" in some vague way, but the child will not actually be able to participate in the baking a cake even after understanding that Daddy intends for the child to help.

Developmentally, there is a vast practical gap between "We intend X" and "We are doing X." Searle assumes that "We intend X" must always be logically and epistemically prior to "We are doing X" but that can't always be the case in the real world, and it certainly cannot be developmentally correct. Logically, people can form "collective intentions" but they are not automatically practical intentions leading to group conduct. A small child will not know what her parent practically intends to do when making a cake, not does the child have much of a conception of what he can do to help. During young childhood, none of us began engaging in any sort of group activity by first accurately conceiving what we are supposed to do, verifying that all of us intend to do this together by hearing each other's utterances, and then undertaking our specific conduct to complete the group activity.

Social Psychology and Social Cognition

Serious studies of developmental social behavior did begin to propose alternative explanations in the 1970s and 1980s. Social cognition has come a long way from its long period of neglect during most of the twentieth century. However, the theoretical stance that developmental psychology must be thoroughly social and that most of adult cognition remains social is still thwarted by narrowly Cartesian definitions of social cognition. Social psychology and social neuroscience, along with psychology in general, often design inquiries using an excessively Cartesian framework. Perhaps this is due to social cognition's heavy reliance on the cognitive psychology of the 1970s when social cognition was struggling for recognition. Comparative social cognition, especially pioneering primate studies that revealed how adult primates and human children can have similar developmental stages for comprehending others' mental states, also perpetuated Cartesian assumptions.

In social psychology, the "theory of mind" theory appears to innocently propose that immature brains acquire understanding of others' beliefs, intentions, and plans through observation of others. According to folk psychology, minds possess inner mental states, among them such things as beliefs, intentions, and plans, which cannot be directly observed. Not surprisingly, the "folk" involved are almost always modern Western peoples whose tacit or explicit Cartesian views on personality and mind are well-entrenched. The developmental psychology of brains acquiring folk psychology can remain thoroughly infected with that same individualistic Cartesianism, since it will be assumed that the proper aim of immature brains is to develop into Cartesian minds and understand others as Cartesian minds as well. Definitions of things like the "problem of other minds," "theory of mind," and "metacognition," and even definitions of social cognition, can remain thoroughly infected with this Cartesianism.

Consider how definitions of social cognition make individuals central to the point of exclusivity. A typical definition runs like this: "human social cognition encompasses all cognitive processes relevant to the perception and understanding of conspecifics" (Jacob and Jeannerod 2005, 21). Is this kind of individualism, in which a lone individual is responsible for understanding just one other lone individual, still warranted? Should social cognition embrace only comprehension of other nearby individuals, and not anything else in the local human environment too, such as kitchens, libraries, race tracks, town halls, or parks? Is it possible to understand others apart from that environing context, and is it possible to understand that environing context without comprehending others? A narrow Cartesian approach won't disagree that knowledge of the environment is required too, but it does imply that such knowledge does not come from social cognition, but other modes of cognition instead. A narrowly individualistic definition of social cognition abstracts people away from the lived and built human environment. Compare that narrow definition above with a broader definition:

> Social cognition, more generally, constitutes the process by which people think about and make sense of other people, themselves, and social situations. (Fiske 2010, 128)

This broader definition of social cognition locates it within its proper home of social situations containing people modifying their environments.

Broad definitions of social cognition, and expansive depictions of social cognition's core role in explaining human psychology, behavior, and culture, have become more common (for just a sampling, consult Gazzaniga 1985; Bandura 1986; Wertsch 1991; Bogdan 1994, 2000; Schulkin 2000; Frith and Wolpert 2003; Schmitt 2003; Easton and Emery 2005; Cacioppo et al. 2006; Vogeley and Roepstorff 2009; Rochat 2009; Franks 2010). Investigations of "the social brain" have been underway in a variety of interrelated fields, from developmental and abnormal psychology to cognitive science and social neuroscience. There are robust empirical resources for locating such things as intentionality, agency, and intelligence firmly within the context of social cognition. It is becoming more widely recognized that both Cartesianism in general and individualistic reductionism – Leslie Brothers calls it "neuroism" (2001) – are theoretically inadequate as both philosophies and scientific research programs. Bold surveys of recent research have reached the stage of judging that social cognition cannot be an emergent phenomena from aggregate individual cognitions, but rather that any individualizable cognitions, including much of self-consciousness and agency, are emergent from a more fundamental field of embodied and dynamic social interaction (see, e.g., Rakison 2005; Robbins 2008; Smith 2008; Adolfs 2009; De Jaegher and Froese 2009; De Jaegher et al. 2010; Kitayama and Park 2010; Theiner et al. 2010; Krueger 2011; Schulkin 2011). Smith and Conrey (2009) neatly summarize their approval of this research trend:

> (a) Cognition is for the adaptive regulation of action, and mental representations are action-oriented. (b) Cognition is embodied, both constrained and facilitated by our sensorimotor abilities as well as our brains. (c) Cognition and action are situated in the sense of being contingent on specific aspects of the agent's social environment. (d) Cognition is distributed across brains and the environment and across social agents (e.g., when information is discussed and evaluated in groups).

Once taken to be the exclusive possessions of individualized minds taken singly, our complex cognitive processes are proving to be capacities only of brains taken collectively.

Consider how social cognition has always started from our perception of the environment around us. Perception at first glance can seem the most localizable and personal of matters. However, people don't just observe people who are in turn observing them. People observe people doing things, and usually we are observing people doing things together. We don't simply track and comprehend peoples' behavior – we closely observe and understand what they are trying to do with their surroundings. Strictly speaking, we aren't really interested in understanding people, but rather in understanding what people think they are accomplishing. Considering how such diverse things as operating a research laboratory, working in an office, shopping in a mall, vacationing in Paris, singing in a choir, constructing a building, filing tax returns, and passing legislation are all social situations, social cognition is practically everywhere in the human world. None of those social situations will proceed well, if at all, without elaborate understandings of others and what they are doing to their environments.

Comprehension of anything that involves human know-how or any social institution will involve social cognition. Trying to navigate a grocery store from entrance to exit with purchases is thoroughly a series of social cognition tasks, even while deciding on which cereal one's children like or figuring out how to justify buying an expensive cut of meat. Remembering what others have done, recognizing what other people are doing, communicating with others, and anticipating what others may do in the future are cognitive tasks at the heart of most anything cognitive that we do every minute of every day. This portrayal of pervasive social cognition is consistent with a broader understanding of all cognition as originally acquired and ultimately serving our conduct through life's many activities.

What Do Minds Learn to Mind?

It is a hard social fact that humans have to learn how to engage in group activities before they have any adequate conception of what each of them is supposed to do. We are not born with internal representations of how social activities are to be done. This fact of human development is not that different from a related social fact that also must be correct: We do not first understand our own internal minds and then experimentally project them on others. This can't be true of infants, and it can't become true thereafter through childhood. In fact, our capacity to understand the minds of others is developmentally linked with the development of our capacity to engage in group activities. Indeed, all we have to go on, from birth, is our engagement with the behaviors of others. However, those behaviors, and not any internal representations or intentions, are the genuine beginning place for mental capacities. Not only must humans learn how to engage in group activities before they have any adequate conception of what each of them is supposed to do, humans must engage in group activities before forming any ideas about any minds whatsoever, their own or others.

Furthermore, it is only through successfully engaging in many sorts of group activities before the age of four that young humans acquire minds. Only a more mature mind can help form another mind, and the more mature mind can only do this by engaging a younger mind in coordinated habitual activities – this much social behaviorism must be right. But Cartesian individualism is not needed. The younger mind need not form adequate conceptions of what the mature mind intends first, before successfully participating. Quite the opposite: it is only by attending to what the mature mind is doing and achieving that the less mature mind could acquire some partial conception of the mature mind's intentions and beliefs. In effect, "We are doing X together" must developmentally come prior to "You are intending that we do X," which in turn is developmentally prior to "I intend that we do X" – and only after that does the child go on to form additional notions such as "I believe X should be done" or "I believe that X is done this way." The child's internal mind only possesses intentions and beliefs after having successfully engaged in many group activities and formed notions of others' minds. Understanding group activities, one's social roles, and others' minds has developmental leadership guiding the acquisition of one's own mind.

Social cognition is not an intellectual capacity that develops at some particular age or specific stage. Many different kinds of cognitive tasks are involved with intelligently dealing with other people and social situations. These various cognitive tasks have their own time frames for normal emergence, as the kinds of age-specific social activities needed for acquiring the correlated cognitive abilities are successfully managed by the child during growth. As the complexity of the social activities grow, so grows the brain's structures. This is a very long process involving many areas of the brain: "the neural structures implicated in social cognition develop over an extended period from infancy until young adulthood" (Payne and Bachevalier 2009, 52). The primary setting for this protracted period of social development is the family. Children acquire facility with familial social conduct, activities, and roles, which in turn prepare them in turn for learning about adult social interactions and roles (Adolfs 2003; Flinn and Ward 2005).

The developmental stages needed for the emergence of social cognition begin in early infancy. The processing of visual perception is inaugurated in the first few months after birth. Acquiring familiarity with 3-Dimensional objects and their basic geometrical properties, learning notions of object permanence, and coming to expect that objects will follow certain motion trajectories (under conditions of inertial movement and the effect of gravity, e.g.) occur during the first year. Infants are also gaining facility with face recognition in general and remembering individual faces and individual identities during this period. They start to imitate the facial expressions of others and keep track of their imitative responsiveness, they acquire preferences for familiar faces and prosocial interactions with others, and they can track the identities of individuals by their characteristic behaviors (see surveys of infancy cognition in Ellis and Bjorklund 2005; Csibra and Gergely 2009; Hamlin and Wynn 2011; Meltzoff 2011, and see also a discussion of the role of "mirror neurons" in Sinigaglia and Rizzolatti 2011). By the end of the first year, infants cognize others as following familiar behavior patterns and expect older family members to be participating in social interactions in only certain habitual ways. By 8 months, infants recognize normal routines based on what they have been observing in prior months, and by 12 months they anticipate the appropriate goals and means of accomplishing them if they themselves have been successfully doing those actions too. Infants understand how daily social interactions are supposed to proceed in proportion to their opportunity to be regularly observing and participating in those interactions (such as recognizing the "right way" for people to be feeding after they have been successfully feeding themselves; see Gredebäck and Melinder 2010).

Infants acquire robust capacities for cognizing everyday social interactions in which they are directly and successfully involved, and their direct engagement in the full-blooded 3-D world of dynamic objects and supportive agents is the developing brain's cognitive priority. This all makes solid evolutionary sense; the brain's infantile cognitive processes are fixated on managing tasks most relevant to its survival through infancy. The infant's field of perception is the environing 3-D world, and not any sort of inner Cartesian theater. The infant brain is not cognizing phantasmal projections that have reached some ghostly internal field of consciousness. For the infant, perceptual consciousness and the enveloping field of social action are identical. After 12–15 months, anything that significantly departs from the perceptual and causal norms of the dynamic realm of 3-D objects and agents requires additional cognitive work and developmental time. That is why, for example, infant cognition in perception of 2-D representations through artifacts such as picture books and television screens requires another further stage of cognitive development, which is not identical with 3-D cognition. The "video deficit effect" is one notable instance of this cognitive divergence, as infants in their second year learn less from 2-D observation than 3-D observation. Until extra cognitive capacity for interpreting and translating 2-D representations and demonstrations into familiar 3-D modes is acquired, the total cognitive load is too high for smooth transfer of learning, but during the third year the deficit is largely erased (Strouse and Troseth 2008; Barr 2010). A similar effect is discovered among adults of a culture which has never used 2-D representations, as they are unable to fully understand information conveyed in photographs or films, for example.

Along with these many cognitive developments, infants are simultaneously attending to the gesturings and the vocal sounds and intonations of others, building up the cognitive abilities needed later on for successful verbal communication about goings-on in the environment. Infants follow others' gazes to nearby objects, they quickly learn how to look at things others are looking at, and later on, they can attend to objects handled and then gestured at by others. Simpler modes of communication such as gesturing and using vocal clues function as signaling by the infant to obtain what it wants from others. During the second year, infants are signaling and communicating in robust and flexible ways with a proto-vocabulary and they understand many simple communicative intentions of others (Aureli et al. 2009; Grosse et al. 2010).

Early language acquisition occurs almost entirely within the dynamics of mutual engagement in daily activities; language is not acquired simply by the infant's hearing of spoken words in isolation from any other ongoing interaction or activity. The meaning of language from the start is entirely embedded within significant activities. Words are first connected with ongoing events, not any things in isolation, and intentional reference exists not for the sake of picking out static objects, but for making desired things happen. Words mean intended results, not anything particular involved in that process – for a 2-year-old, for example, "cup" mostly means the speedy delivery of the seen cup's contents to the child's own mouth. That is how words acquire their primeval intentionality, by being parasitic on the goal-oriented intentions of actions and gaining powers to deliver successful results. Words are not essentially about static correspondences between just two things, vocal sounds and individuated objects, but rather about triadic dynamic engagements in which at minimum two people use words together to habitually initiate and guide a desired plan of action (Zlatev et al. 2008).

Through focused attention on others' conduct near them, infants become intense imitators of others' actions and manipulations of things. Infants will play with most anything in their reach, but they show preference for imitating actions on familiar objects that they observe others' doing first. The opportunity to repeatedly observe the same actions on objects by others and to practice those actions themselves results in the greatest transfer of manipulative ability (Yang et al. 2010). Furthermore, observing the practical success of object manipulation to some intended result increases the attention given to imitating and practicing that mode of manipulation by the infant throughout the second year and thereafter. Infants pay attention to reliable models of practical success and center their imitation there (Zmyj et al. 2010).

Although infants detect goal-oriented conduct by others and practice successful activities in dynamic learning, only imitation, and not creative emulation or novel construction, characterizes their behavior during early childhood. Through age 5 or so, children attend to older peoples' conduct largely in order to duplicate both their means and their ends, rather than to invent novel means to the same ends. The earlier emphasis on careful imitation and later development of experimental emulation indicates that children are remembering both the appropriate means to be used in the short term, and also the preferred goals in the long term (Simpson and Riggs 2011). Attending to one's own imitation of others, and attending to whether others are imitating one's self, is a constant cognitive preoccupation during early childhood. Mutual imitation is an essential aspect of the development of social cognition; we naturally appreciate and gravitate toward responsive behaviors intentionally coordinated with our own, and we acquire better coordination and communication skills through such processes (Meltzoff 2011; Tasker and Schmidt 2008).

The infant's capacities for sustained joint attention to, and successful participation in, interesting and engaging activities with others are capacities that develop further into what has been called "executive control." The heightened ability to focus on and figure out coordinated activities that culminate in successful practices, whether in free play or assigned tasks, demonstrates how many cognitive processes are functioning well together. Cognitive functions like "working memory," "impulse inhibition," "cognitive shifting," and "information integration," as specifiable yet complex aspects of developing social cognition, indicate how such things as attention, recognition, recall, prediction, coordination, language, and joint planning are all tightly integrated together as they develop through childhood (Wiebe et al. 2011). It is unnecessary to postulate a newly developed center of superior cognition, where this "executive" command sits as task master, since the emergent coordinated functioning of many lower-level processes is sufficient to explain the improved performance during childhood.

Avoiding neural Cartesianism is also wise in light of research on adult brains, which cannot find cortical evidence of a supreme command center constantly initiating or directing action. Social action, and not personal action, appears to have priority. Although the medial prefrontal cortex is proving to be a significant region serving functions for personal executive control, that region is thoroughly interlinked with cognition about social situations done by processes widely distributed across the brain (Graybiel 2008; Bogdan 2009; Krueger et al. 2009; Nummenmaa and Calder 2009). Executive control over tasks and management of social situations are two ways of looking at much the same thing. Daniel Dennett, in his resistance to the Cartesian legacy, has similarly supported the idea of a social self that is a flexible controller of one's own behavior (see a discussion in Franks 2010, 138).

In the pragmatist tradition, this developmental congruence of executive control and management of social situations is no mystery. George Herbert Mead offered two sociological paths to understanding the social nature of voluntary action and agency. One is through his concept of role-taking, and the other is his theory of the act. Role-taking offers a unique theory of self-control which is also social control. Here persons respond to their own emerging actions as they imagine the other will, and use this co-responding to guide the future course of their actions. This presupposes a universe of discourse wherein persons share the meanings of their symbols, making the process both individualistic and social. Mead's theory of the act involves four stages. First, action as some impulse to behave. Second is perception, which is selective, as we perceive most clearly that which answers to, or facilitates our interest and intentions to act. Third is manipulation, which involves doing something with objects, or with people such as role-taking. Fourth is consummation, where the last stage of the act is present in the first stage, as we experience some fulfillment of our intention. This is a teleological account of social behavior, since our conceptions of our intentions literally pull us along the stages of action through to completion, rather than just supply some initial push as if action was a deterministic conditioning process. Another aspect of Mead's voluntarism is therefore that we can modify our intentions at any stage, in light of how that action is going or in light of new information about the situation, in order to change our course of action. Mead's theory of agency is nothing like the Cartesian model of an isolated mind steered by an internal command center issuing orders to the rest of the brain.

Brains and Minds Grow Together

No Cartesian mind is developing within infancy and childhood. Children do acquire facility with the intentions, practices, and attitudes of others through social interaction. If the possession of a "mind" consists of advanced cognitive capacities for focused, flexible, and successful social practices, infants already detect and rely on the minds of older people – infants do notice and take advantage of the more mature cognitions occurring around them. Infants develop understandings of the mindings of others, for they closely attend to the mindful conduct around them, and learn to match them with their own.

It is crucial at this stage of the discussion to prevent Cartesianism from erupting again. Even to ask the typical question, "When does the child leap from observations of mindful behavior to inferences about the inner mental states of others?" is a mistaken and misguided curiosity. The appreciation and utilization of another's mindful conduct is precisely the appreciation of another's mind; there simply is nothing more to discover. Not even the acquisition of full language and the understanding of another's utterance "I believe P" is anything but the appreciation of another's mindful conduct. Either there is replete mindfulness in one's conduct, or there is no mind anywhere. Cartesians, of course, point to adult mental states and events that can consciously occur without any overt action, but these are highly refined capacities not characteristic of the business of ordinary life, and surely not of childhood.

To presuppose Cartesianism, that one's appreciation for another's inner mental life is a matter separable from the appreciation for another's mindful conduct, is to assume that Theory of Mind is not merely distinguishable, but quite separable, from Social Conduct. This notion that the acquisition of theory of mind is just as grounded in observations of behavior yet results from distinct cognitive mechanisms is a demonstrably inadequate notion. Every developmental stage exposes that theoretical inadequacy. The classic example of infants grasping how others can have false beliefs proves to always be a capacity that develops within the wider context of an infant's strengthening ability to anticipate the practical goals of others by observing their conduct (Buttelmann et al. 2009; Caron 2009). Estimating the false beliefs of others especially depends on the child perceiving what others can see and do, and hence what they can know about, while reasoning to judgments about others' true beliefs develops at a later age (Fabricius et al. 2010). Judging the intentions of others is tied to situational context; for example, 3-year-olds not only follow the intentions of others to engage in play, but they figure out normative rules for conducting pretend play, and they enforce rules of play only on those who have willingly entered the field of pretense as actors (Sobel 2007; Williamson et al. 2010; Wyman et al. 2009; see also Mead 1934 on play as preceding role-taking).

A young child's development of executive control is another kind of practical functioning that grows right together with that child's practical ability to appreciate others' mental processes, such as attitudes, emotions, or beliefs. Indeed, developing executive functioning in social conduct has priority, as training in executive function enhances both executive function and theory of mind, yet just training in theory of mind alone won't enhance either executive function or theory of mind (Moses and Tahiroglu 2010). Better than any training is the child's participation in social activities such as playing games, especially games involving a co-player sharing common goals, in which executive function is efficiently enhanced (Qu 2011). Children with more siblings, and hence greater participation in group role-playing dynamics, develop theory of mind faster (McAlister and Peterson 2007). The mutual development of executive function and theory of mind remains consistent from infancy through young childhood (Blaye and Chevalier 2011; Henning et al. 2011). An older child's capacity for reliably identifying others' diverse mental states such as intentions and beliefs is thoroughly dependent on developed skills at conversational language in conjunction with social activities, and typically emerges by age 7 or later (Garfield et al. 2001; Mull and Margaret Evans 2010; Wang et al. 2011).

The heavy reliance on social cognition for the development of one's cognitive capacities is not a temporary matter that diminishes in adulthood. The young social mind only develops further into a mature social mind. Adult learning and practice remains almost entirely social matter, from conversation and reading to participation in scientific inquiry. Common processes underlie theory of mind capacity and participation with coordinated joint action in adulthood (Humphreys and Bedford 2011; Pezzulo and Dindo 2011), and the brain's tendency to prioritize and preserve the impact of social interaction and coordination remains a pervasive adult phenomenon (Oullier et al. 2008; Iani et al. 2011). The dynamics of group intentionality remains a controversial topic, as to be expected when ordinary language, folk psychology, and Cartesian individualism all conspire to lend presumptive credibility to a nominalistic status for intentions, and an instinctive revulsion toward "group mind" lingers across analytic philosophy. Robust efforts at accounting for group or collective intentionality (such as Gilbert 1989; Bratman 1999; Miller 2001; Meggle 2002; Tuomela 2002; Pettit 2003) are increasingly able to take advantage of the experimental and neuroscientific evidence for the mutual development of shared intentionality (see the recent work in Schmid et al. 2008, 2011). Because participation in shared practices with common goals is so tightly interfused with individual goal pursuit (Shteynberg and Galinsky 2011), individual intentions and personal agency cannot be neatly distinguished when empirically studying socially synchronous conduct. Taking the social situatedness of adult cognition seriously, it should further be expected that enculturalization into adult roles powerfully shapes both one's modes of agency and one's self-conception of one's own agency (Hannover and Kühnen 2009).

Socialized Brains Remain Social Minds

In summary, Cartesian-minded psychologists and philosophers have little empirical support for any presumption that knowledge of any mind, one's own mind or minds of others, is developmentally prior to or independent of mindful participation in social activities. We do not first develop internally aware minds and then inferentially project mental states on others. More generally, we do not infer internal mental states that have any radically different ontological status from the traits of external observable behaviors. What we observe from infancy to adulthood is simply mindful behaviors of both ourselves and others; intentionality and mentality is fully in these behaviors, or they are nowhere. The logical point that a person's observably mindful behaviors underdetermines any estimate of that person's ongoing mental states cannot help prove that we go far beyond information-poor behaviors for hypothesizing information-rich mental entities. Such behavioral underdetermination only shows how our judgments about attitudes, intentions, and beliefs are somewhat vague and fallible, not how we must postulate such things so that they transcend all behavior (and any neural events as well).

How we developmentally come to conceive ourselves as individualized mental selves is not to be identified with the strict individualism and proto-dualism of Cartesianism. Put another way, radical Cartesianism has quite a different origin than humanity's long journey of socialization into modes of mentality. The idealistic view that there really are inner mental states only contingently exemplified in behavior must be a view inculcated quite apart from the natural reality that we acquire contingent notions about intentions, beliefs, and the like only through engaging in intentional and thoughtful conduct with others. How the radical dualism and individualism of Cartesianism originated as a cultural construct is a tale best left to intellectual historians; it no longer need concern experimental psychologists or social neuroscientists.

An adequate naturalistic theory of the development of social cognition, partially sketched above, constitutes the solution, or rather the dissolution, of the philosophical "problem" of other minds. This solution shows how understanding group activities and one's social roles has developmental leadership guiding the twin abilities to understand others' minds alongside developing one's own mind. Searle has it backwards – the reality is that the mentality of individuals is dependent on and derived from the collective mentality of organized societies. Searle used the term "mental phenomena" in his own phrasing, and deliberately so, for heightened Cartesian effect. However, naturalism should avoid metaphysical arguments that conflate phenomenality with cognition – such as dismissive diversions about how only individuals have experiences while societies don't – so that empirical studies of actual human cognition can proceed unimpeded by lingering Cartesianism, even Cartesian materialism. Searle supposes that his kind of brain-localized individualism is the only way to stay true to naturalism when he says that "all human intentionality exists only in individual human brains," as quoted above, yet empirical research cannot lend its support here, either.

Cognition is what brains do, and when brains do cognition jointly, then human intentionality and agency exists in multiple brains necessarily functioning together, not in any aggregate of separate brains contingently functioning side by side. From a neurological perspective, why should the Hebbian process of "neurons firing together are wiring together" get arbitrarily halted at the skull or skin? Brains learning together are literally growing together. Put another way, higher level cognitions adequate to social practices require simultaneous and synchronized operations across multiple brains that have grown together through shared experiences, communications, and practices. People can do so many things by themselves because they grew up doing so much with others. Localizing the "real" cognition or the "genuine" agency in just brains taken singly is inadequate both to the natural development of brains and the social psychology of cognition.

References

Adolfs, R. (2003). Cognitive neuroscience of human social behavior. *Nature Reviews Neuroscience, 4*, 165–178.

Adolfs, R. (2009). The social brain: Neural basis of social knowledge. *Annual Review of Psychology, 60*, 693–716.

Aureli, T., Perucchini, P., & Genco, M. (2009). Children's understanding of communicative intentions in the middle of the second year of life. *Cognitive Development, 24*, 1–12.

Bandura, A. (1986). *The social foundations of thought and action: A social cognitive theory*. Englewood Cliffs: Prentice-Hall.

Barr, R. (2010). Transfer of learning between 2D and 3D sources during infancy: Informing theory and practice. *Developmental Review, 30*, 128–154.

Blaye, A., & Chevalier, N. (2011). The role of goal representation in preschoolers' flexibility and inhibition. *Journal of Experimental Child Psychology, 108*, 469–483.

Bogdan, R. (1994). *Grounds of cognition: How goal-guided behavior shapes the mind.* Hillsdale: Erlbaum.

Bogdan, R. (2000). *Minding minds: Evolving a reflexive mind by interpreting others*. Cambridge, MA: MIT Press.

Bogdan, R. (2009). *Our own minds: Executive and social grounds of self-consciousness and self-understanding*. Cambridge, MA: MIT Press.

Bratman, M. (1999). *Faces of intention*. Cambridge, MA: Cambridge University Press.

Brothers, L. (2001). *Mistaken identity: The mind-brain problem reconsidered*. Albany: State University of New York Press.

Buttelmann, D., Carpenter, M., & Tomasello, M. (2009). Eighteen-month-old infants show false belief understanding in an active helping paradigm. *Cognition, 112*, 337–342.

Cacioppo, J., Visser, P., & Pickett, C. (Eds.). (2006). *Social neuroscience: People thinking about thinking people*. Cambridge, MA: MIT Press.

Caron, A. J. (2009). Comprehension of the representational mind in infancy. *Developmental Review, 29*, 69–95.

Chomsky, N. (1968). *Language and mind*. New York: Harcourt, Brace & World.

Csibra, G., & Gergely, G. (2009). Natural pedagogy. *Trends in Cognitive Sciences, 13*, 148–153.

De Jaegher, H., & Froese, T. (2009). On the role of social interaction in individual agency. *Adaptive Behavior, 17*, 444–460.

De Jaegher, H., Di Paolo, E., & Gallagher, S. (2010). Can social interaction constitute social cognition? *Trends in Cognitive Sciences, 14*, 441–447.

Dennett, D. (1991). *Consciousness explained*. Boston: Little, Brown.

Dewey, J. (1922). *Human nature and conduct: An introduction to social psychology*. New York: Henry Holt.

Dewey, J. (1938). *Experience and education*. New York: Macmillan.

Easton, A., & Emery, N. (Eds.). (2005). *The cognitive neuroscience of social behavior*. New York: Psychology Press.

Ellis, B., & Bjorklund, D. (Eds.). (2005). *Origins of the social mind: Evolutionary psychology and child development*. New York: Guilford Press.

Fabricius, W. V., Boyer, T. W., Weimer, A. A., & Carroll, K. (2010). True or false: Do 5-year-olds understand belief? *Developmental Psychology, 46*, 1402–1416.

Fiske, S. T. (2010). *Social beings: A core motives approach to social psychology*. Hoboken: Wiley.

Flinn, M. V., & Ward, C. V. (2005). Ontogeny and evolution of the social child. In B. Ellis & D. Bjorklund (Eds.), *Origins of the social mind: Evolutionary psychology and child development* (pp. 19–44). New York: Guilford Press.

Fodor, J. A. (1987). *Psychosemantics: The problem of meaning in the philosophy of mind*. Cambridge, MA: MIT Press.

Franks, D. D. (2010). *Neurosociology: The nexus between neuroscience and social psychology*. New York: Springer.

Frith, C., & Wolpert, D. (Eds.). (2003). *The neuroscience of social interaction*. Oxford: Oxford University Press.

Garfield, J. L., Peterson, C. C., & Perry, T. (2001). Social cognition, language acquisition and the development of the theory of mind. *Mind & Language, 16*, 494–541.

Gazzaniga, M. (1985). *The social brain*. New York: Basic Books.

Gilbert, M. (1989). *On social facts*. London/New York: Routledge.

Graybiel, A. M. (2008). Habits, rituals, and the evaluative brain. *Annual Review of Neuroscience, 31*, 359–387.

Gredebäck, G., & Melinder, A. (2010). Infants' understanding of everyday social interactions: A dual process account. *Cognition, 114*, 197–206.

Grosse, G., Behne, T., Carpenter, M., & Malinda, M. (2010). Infants communicate in order to be understood. *Developmental Psychology, 46*, 1710–1722.

Hamlin, J. K., & Wynn, K. (2011). Young infants prefer prosocial to antisocial others. *Cognitive Development, 26*, 30–39.

Hannover, B., & Kühnen, U. (2009). Culture and social cognition in human interaction. In F. Strack & J. Förster (Eds.), *Social cognition: The basis of human interaction* (pp. 291–309). Mahwah: Lawrence Erlbaum.

Henning, A., Spinath, F. M., & Aschersleben, G. (2011). The link between preschoolers' executive function and theory of mind and the role of epistemic states. *Journal of Experimental Child Psychology, 108*, 513–531.

Humphreys, G., & Bedford, J. (2011). The relations between joint action and theory of mind: A neuropsychological analysis. *Experimental Brain Research, 211*, 357–369.

Iani, C., Anelli, F., Nicoletti, R., Arcuri, L., & Rubichi, S. (2011). The role of group membership on the modulation of joint action. *Experimental Brain Research, 211*, 439–445.

Jacquette, D. (2011). Intentionality as a conceptually primitive relation. *Acta Analytica, 26*, 15–35.

Jacob, P., & Jeannerod, M. (2005). The motor theory of social cognition: A critique. *Trends in Cognitive Science, 9*, 21–25.

James, W. (1890). *The principles of psychology* (Vol. 2). New York: Henry Holt.

Kitayama, S., & Park, J. (2010). Cultural neuroscience of the self: Understanding the social grounding of the brain. *Social Cognitive and Affective Neuroscience, 5*, 111–129.

Krueger, J. (2011). Extended cognition and the space of social interaction. *Consciousness and Cognition, 20*, 643–657.

Krueger, F., Barbey, A., & Grafman, J. (2009). The medial prefrontal cortex mediates social event knowledge. *Trends in Cognitive Sciences, 13*, 103–109.

McAlister, A., & Peterson, C. (2007). A longitudinal study of child siblings and theory of mind development. *Cognitive Development, 22*, 258–270.

Mead, G. H. (1934). *Mind, self, and society from the standpoint of a social behaviorist.* (C. W. Morris Ed.). Chicago: University of Chicago Press.

Meggle, G. (Ed.). (2002). *Social facts and collective intentionality*. Frankfurt am Main: Hansel-Hohenhausen.

Meltzoff, A. N. (2011). Social cognition and the origins of imitation, empathy, and theory of mind. In U. Goswami (Ed.), *The Wiley-Blackwell handbook of childhood cognitive development* (2nd ed., pp. 49–75). Malden: Wiley-Blackwell.

Miller, S. (2001). *Social action*. Cambridge: Cambridge University Press.

Moses, L. J., & Tahiroglu, D. (2010). Clarifying the relation between executive function and children's theories of mind. In B. W. Sokol et al. (Eds.), *Self and social regulation: Social interaction and the development of social understanding and executive functions* (pp. 218–233). New York: Oxford University Press.

Mull, M. S., & Margaret Evans, E. (2010). Did she mean to do it? Acquiring a folk theory of intentionality. *Journal of Experimental Child Psychology, 107*, 207–228.

Nummenmaa, L., & Calder, A. J. (2009). Neural mechanisms of social attention. *Trends in Cognitive Sciences, 13*, 135–143.

Oullier, O., de Guzman, G. C., Jantzen, K. J., Lagarde, J., & Scott Kelso, J. A. (2008). Social coordination dynamics: Measuring human bonding. *Social Neuroscience, 3*, 178–192.

Payne, C., & Bachevalier, J. (2009). Neuroanatomy of the developing social brain. In M. de Haan & M. R. Gunnar (Eds.), *Handbook of developmental social neuroscience* (pp. 38–59). New York: Guilford Press.

Pettit, P. (2003). Groups with minds of their own. In F. Schmitt (Ed.), *Socializing metaphysics* (pp. 167–193). Lanham: Rowman & Littlefield.

Pezzulo, G., & Dindo, H. (2011). What should I do next? Using shared representations to solve interaction problems. *Experimental Brain Research, 211*, 613–630.

Piaget, J. (1954). *The construction of reality in the child* (M. Cook, Trans.). New York: Basic Books.

Qu, L. (2011). Two is better than one, but mine is better than ours: Preschoolers' executive function during co-play. *Journal of Experimental Child Psychology, 108*, 549–566.

Rakison, D. H. (2005). Infant perception and cognition: An evolutionary perspective on early learning. In B. Ellis & D. Bjorklund (Eds.), *Origins of the social mind: Evolutionary psychology and child development* (pp. 317–353). New York: Guilford Press.

Robbins, P. (2008). Consciousness and the social mind. *Cognitive Systems Research, 9*, 15–23.

Rochat, P. (2009). *Others in mind: Social origins of self-consciousness*. Cambridge: Cambridge University Press.

Russell, B. (1905). On denoting. *Mind, n.s. 14*, 479–493.

Schmid, H., Schulte-Ostermann, K., & Psarros, N. (Eds.). (2008). *Concepts of sharedness: Essays on collective intentionality*. Frankfurt: Ontos.

Schmid, H., Ziv, A., Lehrer, K., & Bernhard, H. (Eds.). (2011). *Self-evaluation: Affective and social grounds of intentionality*. New York: Springer.

Schmitt, F. (Ed.). (2003). *Socializing metaphysics: The nature of social reality*. Lanham: Rowman & Littlefield.

Schulkin, J. (2000). *Roots of social sensibility and neural function*. Cambridge, MA: MIT Press.

Schulkin, J. (2011). *Adaptation and well-being: Social allostasis*. Cambridge: Cambridge University Press.

Searle, J. R. (2010). *Making the social world: The structure of human civilization*. New York: Oxford University Press.

Shteynberg, G., & Galinsky, A. (2011). Implicit coordination: Sharing goals with similar others intensifies goal pursuit. *Journal of Experimental Social Psychology, 47*, 1291–1294.

Simpson, A., & Riggs, K. (2011). Three- and 4-year-olds encode modeled actions in two ways leading to immediate imitation and delayed emulation. *Developmental Psychology, 47*, 834–840.

Sinigaglia, C., & Rizzolatti, G. (2011). Through the looking glass: Self and others. *Consciousness and Cognition, 20*, 64–74.

Skinner, B. F. (1938). *The behavior of organisms: An experimental analysis*. Englewood Cliffs: Prentice-Hall.

Skinner, B. F. (1953). *Science and human behavior*. New York: Macmillan.

Skinner, B. F. (1983). *A matter of consequences*. New York: Alfred A. Knopf.

Smith, E. R. (2008). Social relationships and groups: New insights on embodied and distributed cognition. *Cognitive Systems Research, 9*, 24–32.

Smith, E. R., & Conrey, F. (2009). The social context of cognition. In P. Robbins & M. Aydede (Eds.), *Cambridge handbook of situated cognition* (pp. 454–466). Cambridge: Cambridge University Press.

Sobel, D. M. (2007). Children's knowledge of the relation between intentional action and pretending. *Cognitive Development, 22*, 130–141.

Strouse, G. A., & Troseth, G. L. (2008). 'Don't try this at home': Toddlers' imitation of new skills from people on video. *Journal of Experimental Child Psychology, 101*, 262–280.

Tasker, S. L., & Schmidt, L. A. (2008). The 'dual usage problem' in the explanations of 'joint attention' and children's socioemotional development: A reconceptualization. *Developmental Review, 28*, 263–288.

Theiner, G., Allen, C., & Goldstone, R. (2010). Recognizing group cognition. *Cognitive Systems Research, 11*, 378–395.

Tuomela, R. (2002). *The philosophy of social practices*. Cambridge: Cambridge University Press.

Vogeley, K., & Roepstorff, A. (2009). Contextualising culture and social cognition. *Trends in Cognitive Sciences, 13*, 511–516.

Vygotsky, L. S. (1978). *Mind in society: The development of higher psychological processes*. Cambridge, MA: Harvard University Press.

Wang, F., Zhu, L., & Shi, K. (2011). How do children coordinate information about mental states with social norms? *Cognitive Development, 26*, 72–81.

Wertsch, J. (1991). *Voices of the mind: A sociocultural approach to mediated action*. Cambridge, MA: Harvard University Press.

Wiebe, S., Sheffield, T., Nelson, J. M., Caron, A. C., Clark, N. C., & Espy, K. A. (2011). The structure of executive function in 3-year-olds. *Journal of Experimental Child Psychology, 108*, 436–452.

Williamson, R., Jaswal, V., & Meltzoff, A. (2010). Learning the rules: Observation and imitation of a sorting strategy by 36-month old children. *Developmental Psychology, 46*, 57–65.

Wyman, E., Rakoczy, H., & Tomasello, M. (2009). Normativity and context in young children's pretend play. *Cognitive Development, 24*, 146–155.

Yang, D., Sidman, J., & Bushnell, E. (2010). Beyond the information given: Infants' transfer of actions learned through imitation. *Journal of Experimental Child Psychology, 106*, 62–81.

Zlatev, J., Brinck, I., & Andrén, M. (2008). Stages in the development of perceptual intersubjectivity. In F. Morganti, A. Carassa, & G. Riva (Eds.), *Enacting intersubjectivity: A cognitive and social perspective on the study of interactions* (pp. 117–132). Amsterdam: IOS.

Zmyj, N., Buttelmann, D., Carpenter, M., & Daum, M. (2010). The reliability of a model influences 14-month-olds' imitation. *Journal of Experimental Child Psychology, 106*, 208–220.

Chapter 5
Genetic, Hormonal, and Neural Underpinnings of Human Aggressive Behavior

Pranjal H. Mehta, Stefan M. Goetz, and Justin M. Carré

In 2010, there were 1,246,248 documented cases of violent crimes committed in the United States, and violence is estimated to kill approximately 1.6 million people per year worldwide (www.fbi.gov, Mercy et al. 2002). Although evolutionary theory suggests that aggression and violence were adaptive behaviors that promoted survival and reproduction among our ancestors (e.g., the acquisition of valued resources such as food, shelter, and mates), aggressive behaviors in modern societies have significant social and economic costs (Buss and Shackelford 1997). These include social stigma, job loss, and negative legal consequences for perpetrators as well as substantial monetary and social costs for society (Archer and Southhall 2009).

Research has shown that multiple social and biological factors are implicated in the expression of aggression, but only recently have researchers begun to understand how these factors work together to regulate human aggressive behavior. In this chapter, we review recent studies on the social neuroscience of aggression, including research in the areas of molecular genetics, neuroendocrinology, neuroimaging, and social psychology. Our goal is not to provide an exhaustive review but rather to summarize the main findings from these fields and to highlight recent studies that integrate theories and approaches from disparate areas of research (for a recent comprehensive review, see Siever 2008). We begin by defining aggression and its subtypes. We then selectively review research on the social neuroscience of human aggression with a focus on recent studies. We cover research in neuroimaging, behavioral pharmacology, molecular genetics, neuroendocrinology, and social psychology. We end the chapter by suggesting new directions for future research on aggressive behavior.

What Is Aggression?

Aggression has been defined as "any form of behavior directed toward the goal of harming or injuring another living being who is motivated to avoid such treatment" (Baron and Richardson 1994, p. 7). Although aggression can be intended to cause physical harm (e.g., physical injury or death), not all aggressive behaviors are physical. Nonphysical aggression includes behaviors designed to cause

P.H. Mehta (✉)
Department of Psychology, University of Oregon, Eugene, OR 97403, USA
mehta@uoregon.edu

S.M. Goetz • J.M. Carré
Department of Psychology, Wayne State University, Detroit, MI, USA

D.D. Franks and J.H. Turner (eds.), *Handbook of Neurosociology*, Handbooks of Sociology and Social Research, DOI 10.1007/978-94-007-4473-8_5, © Springer Science+Business Media B.V. 2013

psychological harm (direct insults, psychological abuse), social harm (e.g., spreading rumors to damage someone's reputation, social exclusion), or economic harm (e.g., firing a subordinate or decreasing his or her pay). Researchers typically classify aggression as either reactive or proactive. Reactive aggression, also referred to as impulsive aggression, is a behavioral response to perceived or actual provocation and involves retaliation (Dodge and Coie 1987). Commonly referred to as "hot-blooded," reactive aggression is characterized by anger and impulsivity and is often accompanied by disinhibition and affective instability. In contrast, proactive aggression occurs in the absence of direct provocation and is a goal-oriented behavior aimed at the acquisition of a valued resource (Dodge and Coie 1987). Although the proactive form receives widespread media attention (e.g., serial killings, assassinations, genocide), the reactive form likely accounts for most societal problems associated with aggression (Nelson and Trainor 2007). We focus our literature review below on reactive aggression because it has received greater attention in neuroscience research, and its social and biological causes are better understood.

The Social Neuroscience of Human Aggressive Behavior

Empirical studies indicate that human aggressive behavior is influenced by specific genes, hormones, neural systems, and environmental factors. In this section, we review the main findings from these disparate areas of research with a focus on recent integrative studies. First we discuss the neural systems implicated in aggression with a focus on two specific regions: the amygdala and the orbitofrontal cortex. Next, we discuss neurotransmitters and hormones associated with aggression, including serotonin, testosterone, and cortisol. Third, we discuss two environmental factors linked to aggression: violent media exposure and social rejection. Fourth, we discuss two psychological interventions that can reduce aggression: cognitive reappraisal and self-control. Finally, we end the chapter by discussing directions for future research.

Amygdala–Orbitofrontal Cortex Interactions as a Mechanism for Aggressive Behavior

Animal research indicates that an extensive network of cortical and subcortical regions is involved in the expression of aggressive behavior (Newman 1999; Nelson and Trainor 2007; Siegel et al. 2007). Two regions that have received extensive empirical attention in human research are the amygdala and the orbitofrontal cortex (OFC). According to recent models of human reactive aggression, the amygdala plays a critical role in the affective and motivational drive to respond aggressively to social provocation, while the OFC is thought to be a self-regulatory region that inhibits aggressive impulses (see Fig. 5.1). The findings reviewed below are consistent with these models.

Orbitofrontal Cortex

The OFC is located in the prefrontal cortex, a portion of the brain that appeared later in evolutionary history than subcortical regions such as the amygdala (Kringelbach and Rolls 2004). A number of studies suggest that the OFC functions as a self-regulation and impulse control region and is involved in the top-down inhibition of aggressive behavior. Patients with lesions in the OFC exhibit

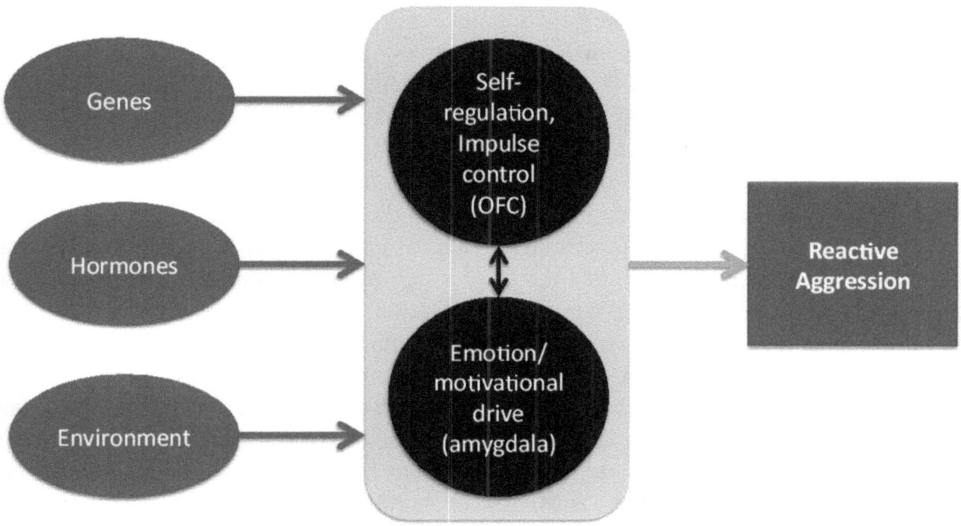

Fig. 5.1 A social neuroscience model of reactive aggression

hyperaggressive behavioral reactions to social provocation (Bufkin and Luttrell 2005; Damasio et al. 1994; Davidson et al. 2000; Koenigs and Tranel 2007; Moretti et al. 2009; Strüber et al. 2008), and human neuroimaging studies indicate that increased OFC activity is associated with low levels of reactive aggression (Bufkin and Luttrell 2005; Damasio et al. 1994; Davidson et al. 2000; Strüber et al. 2008). For example, a recent study examined the relationship between OFC activity and aggressive behavior in the Ultimatum Game, a laboratory model of social decision-making in which people choose between aggression and monetary reward (Mehta and Beer 2010). This game involves two players: a proposer and a responder. The proposer makes an offer as to how to split a sum of money (the stake) with the responder. The responder then decides whether to accept or reject the offer. If the offer is accepted, the stake is split as proposed. However, if the offer is rejected, then both players receive $0. After the responder makes a decision, the game is over. Although responders almost always accept fair offers (e.g., proposer gets 50% and responder gets 50% of stake), responders often reject unfair offers (e.g., proposer gets 80% and responder gets 20% of the stake). Accepting unfair offers guarantees monetary reward, so why do people ever reject them? Psychological evidence indicates that these unfair offer rejections are a form of reactive aggression aimed at retaliating against the other player in the face of perceived social provocation (unfair treatment) (Mehta and Beer 2010).

In this fMRI study, participants were scanned while playing the Ultimatum Game in the role of responder ostensibly with 40 other proposers in one-shot interactions (participants were told they would never play with the same proposers twice). In reality, the offers were experimentally manipulated such that half were relatively fair ($5:$5 split) and the other half were relatively unfair (e.g., an offer of $8 for the proposer and $2 for the responder). The researchers assessed how often participants rejected unfair offers (a behavioral measure of reactive aggression) as well as OFC activity in response to unfair offers compared to fair offers. In support of the hypothesis that OFC is involved in the inhibition of aggressive behavior, the results indicated that bilateral activity in the medial OFC was negatively related to aggressive behavioral reactions to unfair offers. Specifically, individuals who showed decreased activity in the medial OFC after receiving unfair offers tended to reject these offers (high levels of reactive aggression), whereas individuals who showed increased activity in the medial OFC after receiving unfair offers tended to accept these offers (low levels of reactive aggression) (Mehta and Beer 2010).

Other human studies provide convergent support for a relationship between OFC function and the inhibition of impulsive aggression. In a (positron emission tomography) PET study with criminal offenders, Raine and colleagues (1997) reported that affective murderers (i.e., reactively aggressive inmates) demonstrated increased glucose metabolism in subcortical structures (including the amygdala) and decreased glucose metabolism in the prefrontal cortex. Also, psychiatric disorders characterized by high levels of reactive aggression are associated with reduced OFC activity (Coccaro et al. 2007), and lower gray matter volume in the OFC is linked to low impulse control (Matsuo et al. 2009). Although the precise psychological function of OFC in inhibiting aggression is still unclear, it has been theorized that the OFC is part of a self-regulation and impulse control system that integrates emotion, motivation, and cognition to guide context-appropriate behavior (cf. Mehta and Beer 2010). Indeed, not only do patients with OFC lesions show increases in reactive aggression (Blair 2004; Rolls et al. 1994), but they also show increases in impulsive behavior, socially inappropriate behavior, and impaired decision-making (Beer et al. 2003, 2006; Rahman et al. 2001; Bechara et al. 2000; Tucker et al. 1995). These behavioral deficits have been theorized to occur because of a failure to monitor behavior such as failing to consider longer term rewards (Moretti et al. 2009; Beer et al. 2006; De Martino et al. 2006; Bechara et al. 2000). A complementary account of OFC function is that this region is involved in how individuals weigh the costs and benefits of behaving aggressively versus nonaggressively following social provocation, with increased medial OFC activity tipping the cost-benefit analysis toward nonaggression (cf. Mehta and Beer 2010).

Amygdala

The amygdala is a limbic structure that plays a critical role in processing potentially threatening stimuli and mediating various autonomic, neuroendocrine, and behavioral responses that enable an organism to adapt to social and environmental challenges (see Davis and Whalen 2001; LeDoux 2000 for reviews). Animal research indicates that the amygdala is an important component of a neural circuitry that modulates aggressive behavior. Although there is less direct evidence for the amygdala's role in human aggression, indirect evidence suggests that amygdala reactivity may be an important precursor for aggressive behavior in humans. Across a number of studies that used different methodologies, there is robust evidence that amygdala activity increases in response to emotional signals of social provocation (angry faces) and that this amygdala reactivity is stronger in individuals susceptible to aggressive behavior (e.g., Coccaro et al. 2007; Beaver et al. 2008; Lee et al. 2008; Chan et al. 2010; Carré et al. 2012). These findings suggest that hyper-amygdala reactivity to social provocation may be a neural marker for one's propensity to engage in reactive aggression (see Carré et al. 2011, for review)

A recent neuroimaging study more directly linked amygdala function to aggression (Gospic et al. 2011). In this study, participants were scanned while playing the Ultimatum Game in the role of responder using procedures similar to the study described earlier (Mehta and Beer 2010), but the design of this newer study was optimized to detect rapid and slower neural responses to unfair offers. Results showed a rapid amygdala response to unfair offers that was positively related to aggressive behavior (rejecting unfair offers). Interestingly, administration of a benzodiazepine prior to performing the Ultimatum Game effectively reduced amygdala reactivity to unfair offers, and also decreased rejections of unfair offers (Gospic et al. 2011). In line with the study discussed earlier (Mehta and Beer 2010), Gospic and colleagues (2011) also found that prefrontal regions such as the OFC were activated to support the inhibition of unfair offer rejections, but these prefrontal responses came on line later. Together, the findings support a dual-systems model of reactive aggression in line with Fig. 5.1; amygdala activation is associated with a rapid emotional and motivational drive to respond aggressively to social provocation (being treated unfairly), while the OFC is engaged later in the decision-making process to inhibit aggressive impulses.

Amygdala–OFC Connectivity

The studies reviewed above support the view that amygdala is involved in the emotional response to social threat and encourages reactive aggression, whereas the OFC is a self-regulation and impulse control region that inhibits aggression. Recent studies suggest that the functional connectivity between the amygdala and OFC may be another mechanism for aggressive behavioral reactions to social provocation. More specifically, healthy individuals show coupling between amygdala and OFC, but this connectivity is disrupted in psychiatric patients vulnerable to aggressive behavior (cf. Coccaro et al. 2011). Thus, not only do the amygdala and OFC influence aggressive behavior independently, but the neural communication between the two regions seems to play an important role in the inhibition of aggression. This mechanism is supported by neuroanatomical findings, which indicate that the OFC and amygdala have reciprocal connections with one another (Kringelbach and Rolls 2004).

Genetic and Neurochemical Modulators of Human Aggression

Research suggests that various neurotransmitters, genes, and hormones are involved in human aggression. In this section, we synthesize the main findings from these different areas of research. We focus our discussion on a few factors that have received empirical attention in human studies: serotonin, testosterone, and cortisol. We discuss how these factors may regulate human aggressive behavior along with their putative neural mechanisms.

Serotonin

A large correlational literature indicates that enhanced activity in the serotonin system is related to decreases in reactive aggression (Siever 2008; Coccaro et al. 2011). Recent research with pharmacological manipulations provides much needed causal evidence for the role of serotonin in mediating aggression. In one demonstration of this causal relationship, individuals with and without a life history of physical aggression were randomly assigned to receive 40 mg of paroxetine (a drug that acutely augments serotonergic activity) or placebo (Berman et al. 2009). Participants were then placed in the Taylor Aggression Paradigm, a laboratory task that measures physical aggression in response to social provocation. In this task, participants are told they are competing with another participant in a reaction time game, and electric shocks are received and administered. The amount of maximum shock delivered in response to social provocation was the primary measure of aggressive behavior in this study. The findings revealed that augmentation of serotonergic activity via paroxetine significantly reduced physical aggression after social provocation, but only in individuals with a life history of aggression. These findings suggest that enhanced serotonin activity causally reduces aggressive behavior in individuals prone to physical aggression.

If heightened serotonin activity can decrease aggression, might reductions in serotonergic activity increase aggression? Another study provided causal support for this relationship (Crockett et al. 2008). Healthy participants were randomly assigned to receive placebo or tryptophan depletion, a pharmacological manipulation that reduces serotonergic activity. Then participants played the Ultimatum Game in the role of responder, and the fairness of the offers was experimentally manipulated similar to the Ultimatum Game studies reviewed earlier. The results showed that reductions in serotonergic activity via tryptophan depletion causally increased aggressive behavior (rejection of unfair offers) (Crockett et al. 2008).

Serotonergic Gene Polymorphisms

Common variations (polymorphisms) within genes that regulate the serotonergic system can alter human brain function and aggression (Hariri and Weinberger 2003). Two polymorphic genes that have been widely studied in relation to human aggression are monoamine oxydase A (*MAOA u-VNTR*) and the serotonin transporter *(5-HTTLPR)*.

The first evidence in humans for the importance of MAOA in aggression came from the study of large Dutch kindred, whose males were notorious for impulsive aggression (Brunner et al. 1993). Brunner and colleagues (1993) discovered a missense mutation of the *MAOA* gene that resulted in a premature stop codon causing MAOA to be nonfunctional, thus, effectively producing functional MAOA knockouts. Although this finding is informative, the mutation is rare in the population. Nevertheless, within the *MAOA* gene, a more common polymorphism has been described, which is located 1.2 kb upstream of the MAOA coding sequences and consists of a 30-bp repeated sequence present in 3, 3.5, 4, or 5 copies. This variable number of tandem repeats (VNTR) polymorphism is functional: alleles with 3.5 or 4 copies of the repeat sequence are transcribed 2–10 times more efficiently ("high-expression alleles") than those with three or five copies of the repeat ("low-expression alleles") (Sabol et al. 1998). A well-known longitudinal study revealed that the presence of the low-activity allele interacted with a history of childhood maltreatment to predict increased levels of aggression and violence in adults (Caspi et al. 2002). This *MAOA* gene × childhood adversity interaction has conceptually replicated in other studies (e.g., Frazzetto et al. 2007; Reif et al. 2007). Although most studies assessed aggression through self-reported or objective real-world markers of aggression (e.g., violent crimes), one recent study showed an association between the *MAOA* gene and a well-validated behavioral measure of aggression (McDermott et al. 2009). In the study, participants were paid to punish others whom they believed had taken money from them. In reality, participants were playing with a fictitious player whose behavior was experimentally controlled by the researchers. Participants punished their opponents by administering varying amounts of aversive hot sauce, which served as the measure of aggressive behavior. The findings revealed that individuals with the low expression *MAOA* allele behaved more aggressively after social provocation relative to individuals with the high expression allele. That is, low expression allele carriers delivered higher amounts of hot sauce to their opponent, but only after their "opponent" had taken a large amount of money from them.

The serotonin transporter (5HTT) regulates the availability of synaptic serotonin. A widely studied gene within this system is a common functional polymorphism (*5HTTLPR*) (cf. Heils et al. 1996). Individuals with the short allele of this gene have reduced transcriptional activity and therefore reduced reuptake of synaptic serotonin compared to individuals with the long allele. These low activity allele carriers are at greater risk for affective psychiatric disorders such as anxiety and depression, particularly in combination with a life history of stress (e.g., Caspi et al. 2010). Other research has linked low activity allele status to aggressive behavior. Individuals with low activity allele variants in *5HTTLPR* are more likely to show increased childhood aggression (Beitchman et al. 2006), and low activity allele carriers who have adverse childhood environments are more vulnerable to aggressive behavior as adults (Reif et al. 2007). This *5-HTTPLPR* gene effect emerges above and beyond effects of the *MAOA* gene discussed above, suggesting that both of these genes uniquely account for variance in human aggressive behavior (Reif et al. 2007).

Mechanisms for Serotonin-Modulated Aggression

The precise mechanisms for the effects of serotonin activity on human aggression remain unclear, but recent evidence suggests that the OFC, amygdala, and their connectivity are all candidate neural mechanisms. One PET study found increased metabolic glucose response in the left OFC to a serotonergic challenge (meta-chlorophenylpiperazine) in healthy participants, but not among borderline personality disorder (BPD) patients with impulsive aggression (New et al. 2002). Interestingly, this same research group found that administration of fluoxetine (a serotonin reuptake inhibitor) to individuals with BPD was associated with increased glucose metabolic rate in the OFC and an overall decrease in impulsive aggression (New et al. 2004). These findings suggest that serotonergic modulation of the OFC may have an inhibitory effect on impulsive aggression.

In support of heightened amygdala reactivity as a putative mechanism underlying reactive aggression, research suggests that genes that regulate serotonin function are associated with increased amygdala reactivity to facial signals of threat (see Buckholz and Meyer-Lindenberg 2008 and Hariri 2009, for reviews). For instance, Hariri and colleagues (2002) were the first to demonstrate that individuals carrying the "short" allele of the *5HTTLPR* gene demonstrate heightened amygdala reactivity to facial signals of threat, a finding that has been replicated several times (see Munafo et al. 2008, for review). Other research suggests that the *MAOA* gene may bias the socio-emotional circuitry of aggression, including the amygdala (Meyer-Lindenberg et al. 2006). Specifically, individuals with the low expression variant of the *MAOA* gene demonstrated heighted amygdala reactivity to facial signals of threat. Other work indicates that individuals with the low expression variant of the *MAOA* gene scored higher on a trait measure of aggression and interpersonal hypersensitivity and also demonstrated heightened dorsal anterior cingulate cortex (ACC) reactivity to social rejection (Eisenberger et al., 2007). Notably, the positive relationship between interpersonal hypersensitivity and aggression was mediated by heightened dorsal ACC reactivity to social rejection (Eisenberger et al. 2007). Another mechanism may involve connectivity between the amygdala and the prefrontal cortex. Passamonti et al. (2012) found that acute reductions in serotonergic activity via tryptophan depletion reduced functional connectivity between the amygdala and prefrontal cortex in response to angry faces (e.g., connectivity with ventrolateral prefrontal cortex as well as ventral ACC), which may increase one's risk for reactive aggression. Collectively, these findings converge to suggest that serotonergic function may influence aggressive behavior via its interactions with receptors located within a neural circuitry including the amygdala, OFC, and ACC.

Testosterone

Testosterone (T) is a steroid hormone derived from cholesterol. It is produced and released primarily by the testes in men and by the ovaries and adrenal cortex in women. T belongs to a class of hormones called androgens, which are those hormones that are responsible for the development and maintenance of masculine characteristics. In addition to supporting basic physical development, T is also critically involved in regulating social behavior. Naturally occurring and experimentally elevated testosterone levels are positively associated with aggressive behavior in a variety of animal species, especially when the status hierarchy is unstable (Giammanco et al. 2005; Collias et al. 2002; Ruiz-de-la-Torre and Manteca 1999; Oliveira et al. 1996; Sapolsky 1991; Wingfield et al. 1990). In stark contrast to the animal literature, the relationship between individual differences in T and human aggression is relatively weak (see Archer et al. 2005, for review). Even though some studies in humans show that higher circulating T is related to aggression, social dominance, and hyperreactivity to status threats (e.g., Archer et al. 2005; Mehta and Beer 2010; Mehta et al. 2008; Mazur and Booth 1998), other studies have produced inconsistent or null results (Archer et al. 2005). One explanation for these weak effects is that relatively stable levels of T (baseline T) may play less of a crucial role in human aggression than situationally induced fluctuations in T levels (see Carré et al. 2011, for review). It is well-known that T levels rise and fall in competitive social interactions, but only recently have researchers investigated whether dynamic rises in T encourage aggressive and dominant behaviors in humans. In the next section, we review this literature on context-driven T dynamics and human social behavior.

Challenge Hypothesis

John Wingfield and colleagues proposed the *Challenge Hypothesis* to explain how T changes influence social behavior in birds (Wingfield et al. 1990). According to this theory, T levels rise during the breeding season to encourage social competition for mates, and T drops during the non-breeding season to suppress competitive aggression and facilitate care for offspring. Mazur (1985) proposed a similar *Biosocial Model of Status* for T-behavior associations in humans. According to this model, status-relevant social interactions such as competition should cause T levels to fluctuate, and these fluctuations in T should encourage or discourage subsequent status-seeking behaviors such as dominance and aggression.

Although researchers had long known that T levels change during and after competition (Mazur and Booth 1998), researchers had simply assumed that these competition-induced changes in T would influence subsequent status-seeking behaviors. We conducted the first study in humans that explicitly examined the relationship between post-competition fluctuations in T and subsequent social behavior (Mehta and Josephs 2006). We experimentally rigged a competition and collected saliva samples before and after the competition to measure changes in T (Mehta and Josephs 2006). After participants provided the second saliva sample, we measured dominance behavior by asking participants whether they wanted to (a) rechallenge their opponent to a second competition, or (b) complete an alternative noncompetitive task. The results showed that changes in T after losing predicted who wanted to compete again in a second competition. Losers who rose in T were more likely to choose to rechallenge their opponent (73%) than losers who dropped in T (22%). These findings are consistent with the reciprocal model and suggest that a rise in T after a loss of status motivates individuals to reclaim their lost status (choosing to compete again).

We conducted a second study to test whether T responses to competition would also predict subsequent aggressive behavior (Carré et al. 2009). Similar to the previous study, participants provided a saliva sample before and after a rigged competition. After the second saliva sample, participants completed the Point Subtraction Aggression Paradigm (PSAP), a well-validated laboratory task that measures reactive aggression. In this task, participants are paired with a fictitious opponent (actually a computer program) and earn points by pressing Button 1 as quickly as possible or Button 2 to steal points from their opponent. Participants are told their total points will be exchanged for money at the end of the study. During the task, participants have points taken from them by their fictitious opponent, which serves as the experimental manipulation of social provocation. Stealing money from the fictitious competitor by pressing Button 2 is considered aggressive behavior because, like the Taylor Aggression Paradigm and the Ultimatum Game, this behavior represents an intent to cause harm. Consistent with the results of the earlier study (Mehta and Josephs 2006), this study found that changes in T after losing in a competition predicted aggressive behavior in the PSAP. Individuals who lost the competition and rose in T showed more aggressive behavior (stealing more points from their opponents after social provocation) than individuals who lost the competition and dropped in T (Carré et al. 2009). More recent follow-up studies from our labs also show relationships between dynamic T changes and aggressive behavior (Carré et al. 2010; Geniole et al. 2010; Mehta et al. 2010). Together, this recent wave of studies provides strong support for the *Challenge Hypothesis* and *Biosocial Model of Status*, showing that dynamic T responses in status-relevant social interactions have implications for aggression and dominance behaviors. Although all these human studies on dynamic T were correlational, they fit with experimental research in animals, which demonstrates a causal influence of experimentally administered T after competition on aggressive behavior in a second competition (see Gleason et al. 2009 and Oliveira 2009, for reviews).

Neural Mechanisms for Testosterone's Influence on Aggression

Recent studies suggest that T influences human aggression through the OFC and amygdala. In one fMRI study, T levels were measured in saliva and then participants played the Ultimatum Game while being scanned (Mehta and Beer 2010). The findings showed that higher T levels predicted increased aggressive behavior (rejection of unfair offers), and decreases in bilateral medial OFC activity following unfair offers significantly mediated the association between testosterone and aggression. This finding suggests that T increases reactive aggression in part through impairments in the neural circuitry of impulse control and self-regulation (medial OFC). Other recent studies show that T (a) increases amygdala reactivity to angry faces (Hermans et al. 2008; van Wingen et al. 2008), and (b) reduces functional connectivity between OFC and amygdala (van Wingen et al. 2010), providing two additional neural mechanisms for how testosterone may modulate human aggression.

Androgen Receptor Gene

Recently, researchers interested in the genetics of human aggression have turned their attention to a common polymorphism found in the androgen receptor gene. The trinucleotide repeat (*CAG*) has been found to be highly polymorphic (Choong and Wilson 1998) and ranges from 9 to 31 repeats in the human population (e.g., Edwards et al. 1992). *CAG* repeat length is negatively associated with the expression of the androgen gene and androgen receptor (AR) sensitivity (Chamberlain et al. 1994). T exerts its effects primarily through these receptors which are expressed throughout the brain, including regions important in regulating aggression (e.g., amygdala and OFC) (Rubinow and Schmidt 1996; Mehta and Beer 2010, respectively). Thus, AR sensitivity to T may serve as a mechanism to modulate its effects on brain development and subsequent aggressive behavior.

Researchers have found that men who have fewer *CAG* repeats score higher on sexually dimorphic behavioral traits. For example, Simmons and Roney (2011) found that *CAG* length was negatively correlated with prestige and dominance (traits associated with intra-sexual competition) in a sample of men. Other work indicates that rapists and murderers have significantly fewer *CAG* repeats compared to controls in a sample of Indian men (Rajender et al. 2008). Furthermore, a study with adolescent males found that *CAG* repeat length interacted with T to predict a self-report measure of aggressive risk-taking (Vermeersch et al. 2010). Specifically, the authors found that T was positively correlated with aggressive risk-taking, but only among men with relatively fewer *CAG* repeats.

Other work has specifically linked variation in the *CAG* repeat to amygdala reactivity to facial signals of threat. Manuck and colleagues (2010) found an inverse relationship between *CAG* repeats and bilateral ventral amygdala (the principal input region of the amygdala) reactivity when viewing threat cues (angry/fearful faces). On the other hand, reactivity in the dorsal amygdala (principal output region of the amygdala regulating physiological reactivity) was positively correlated with T independent of genotype. These results suggest that the *CAG* polymorphism modulates androgen-sensitive neural circuits associated with aggression.

The Dual-Hormone Hypothesis: Interactions Between Testosterone and Cortisol

Glucocorticoids are a class of hormones that are released by the adrenal glands during physical and psychological stress. The primary glucocorticoid in humans is cortisol (C). Most research on C has focused on the dispositional and situational variables that cause acute changes in C (e.g., Dickerson and Kemeny 2004), but some research indicates that C is negatively associated with aggressive behavior. In one longitudinal study of 314 boys, low basal C levels during preadolescence (age 10–12 years) predicted more aggressive behaviors 5 years later (Shoal et al. 2003). Other studies, however, have shown null effects of C on aggression. These mixed findings suggest that C may interact with other biological systems to modulate human aggression.

We recently proposed the *dual-hormone hypothesis* to reconcile mixed findings on the roles of T and C in human social behavior (Carré and Mehta 2011; Mehta and Josephs 2010). According to the dual-hormone hypothesis, T should have a strong influence on aggression and dominance only when C is low, but T's effect on social behavior should be blocked when C levels are high because C inhibits the neurobiological pathway between T and behavior at multiple levels (see Mehta and Josephs 2010 for a biological rationale). Consistent with the dual-hormone hypothesis, Popma et al. (2007) studied a group of male adolescents and found that T was positively related to physical aggression only in individuals with low C. In individuals with high C, there was no association between T and aggression. Mehta and Josephs (2010) showed a similar pattern of findings in studies of social dominance. A hormone profile of high T and low C was associated with increased dominance across multiple studies. Intriguingly, a profile of high T and high C was associated with submissive behavior after social threat. These dual-hormone effects on social behavior vary across social contexts (threat versus no threat, Mehta and Josephs 2010; social inclusion versus exclusion, Geniole et al. 2010). Together, these findings suggest that T and C jointly modulate human aggression and dominance behavior in a context-dependent fashion. Although the neurobiological mechanisms for dual-hormone modulation of behavior have yet to be studied, the amygdala and OFC are clear candidate regions. Indeed, androgen and glucocorticoid receptors are located in both of these regions, and T and C modulate neural activity in the amygdala and OFC.

Environmental Risk Factors

The research reviewed above provides insights into the biological factors implicated in aggressive behavior. In this section, we review research on environmental risk factors. We focus on two risk environmental factors that have received attention in scientific research – exposure to media violence and interpersonal rejection – and we discuss possible biological mechanisms.

Violent Media Exposure

A number of studies have examined the effects of exposure to violent media on aggressive behavior. In a recent meta-analysis of over 300 studies, the authors found reliable evidence that exposure to violent video games increases aggressive thoughts, feelings, and behavior and decreases empathy and prosocial behaviors (Anderson et al. 2010). Most of the evidence comes from studies of short-term effects (laboratory experiments), but some longitudinal studies also support media violence exposure as a causal risk factor in human aggression. In one recent study of 1,237 German adolescents, media violence exposure at time one predicted a greater propensity toward aggression 12 months later (Krahe and Moeller 2010). Neuroscience studies support the hypothesis that media violence exposure may increase aggression by altering the neural circuitry of aggression. One fMRI study showed that exposure to media violence decreased lateral OFC activity and reduced amygdala–OFC coupling (Kelly et al. 2007), and another study demonstrated that adolescents who reported frequent exposure to violence media had decreased lateral OFC density (Strenziok et al. 2010). These findings suggest that violent media exposure may cause both short-term and long-term changes in aggression by influencing OFC and amygdala function (see Carnagey et al. 2007 for these and related neural mechanisms). Another study suggests that that violent media cues may increase aggression through elevated T levels (Klinesmith et al. 2006). Participants in this study were randomly assigned to interact with a toy gun or a children's toy for 15 min and then could administer various amounts of hot sauce to another person (a measure of aggressive behavior). Saliva samples were collected before and after the experimental manipulation and were analyzed for T levels. The findings showed that people who interacted with the gun administered more hot sauce to the other participant than people who interacted with the child's toy, and this effect of gun exposure on aggression was significantly mediated by increases in T levels after gun exposure. Given previous research linking T to amygdala and OFC, it seems plausible that the effect of increased T levels on aggression following gun exposure may be driven by changes in the amygdala–OFC neural circuit.

Interpersonal Rejection

The act of being rejected or devalued by other people has been shown to be a clear risk factor in aggressive behavior. In fact, a Surgeon General's report concluded that social rejection was the most significant risk factor for violence among adolescents, even more potent than factors such as low socioeconomic status, gang membership, or drug use (cf. Leary et al. 2006; Office of the Surgeon General 2001). A spate of school shootings in the United States illustrates the social isolation–violence relationship. In an analysis of 15 school shootings between 1995 and 2001, 13 out of the 15 perpetrators had a history of being socially rejected – including teasing, bullying, and chronic ostracism (Leary et al. 2003). Experimental evidence also supports a causal effect of interpersonal rejection on aggressive behavior. In many of the experiments, individuals were randomly assigned to receive rejecting or accepting relational feedback from another person (in fact, the feedback is bogus), and factors such as anger, relational aggression (e.g., social derogation), and reactive aggression (e.g., administering varying amounts of aversive hot sauce) were measured after the rejection or acceptance experience. Rejection increased anger and aggression compared to acceptance in many of the studies (see review by Leary et al. 2006). This effect of social rejection on aggression depends on individual differences in social sensitivity and biological differences in serotonergic activity. People high in rejection sensitivity or insecure attachment are more likely to respond to interpersonal rejection with aggression (Leary et al. 2006). Moreover, as described earlier in the chapter, low expression *MAOA* allele carriers show increased activity in the dorsal ACC after social rejection, a region implicated in emotional distress and anger (Eisenberger et al. 2007). These results suggest that the influence of social rejection on aggression may be driven by hyperreactivity in socioemotional neural circuits to rejection experiences (e.g., dorsal ACC, Buckholtz and Meyer-Lindenberg 2008).

Other research indicates that social rejection can increase levels of C, a hormone implicated in psychological stress. In an experiment in which participants were socially rejected or received no rejection, self-esteem moderated cortisol and aggressive behavioral responses to social rejection (Ford and Collins 2010). Compared to individuals high in self-esteem, individuals low in self-esteem showed heightened relational aggression (partner derogation) and increased C in response to social rejection. The association between low self-esteem and relational aggression was mediated by changes in C. Taken together, the findings suggest that social rejection augments aggressive behavior through biological systems associated with stress and socioemotional sensitivity (cortisol, dorsal ACC).

Psychological Interventions

Above, we reviewed some of the biological and social risk factors implicated in human aggressive behavior. In this section, we discuss psychological interventions that can reduce reactive aggression. Although there are many possible interventions, we focus our discussion on two in particular: (1) cognitive reappraisal and (2) self-control training.

Cognitive Reappraisal

Emotion regulation involves cognitive strategies to alter one's emotional response to environmental stimuli. In the context of reactive aggression, cognitive strategies that prolong anger are likely to upregulate aggression, while strategies that reduce anger responses should decrease the propensity toward aggression. In line with this reasoning, rumination – which involves continuing to think about the anger-induced provoking event in a way that prolongs anger – increases anger and reactive aggression (Denson et al. 2011b; Fabiansson et al. 2012). An emotion regulation strategy that may be beneficial for reducing reactive aggression is cognitive reappraisal, which involves reinterpreting an emotional event to reduce its negative emotional impact. For example, an individual who is insulted by another person may try to think about what lessons he or she could learn from the event or think about the event from the perspective of an objective third party (Fabiansson et al. 2012). Recent research supports the hypothesis that cognitive reappraisal can reduce anger and reactive aggression. One study showed that individuals who thought about an anger-inducing event and then engaged in cognitive reappraisal showed less anger compared to participants who thought about an anger-inducing event and then engaged in anger rumination (Fabiansson et al. 2012). A second study used a longitudinal design and found that people who received reappraisal training over the course of a semester showed reduced trait vengeance – an important predictor of aggressive behavior – compared to participants in the control condition (Bartlett and Anderson 2011). Hence, not only can cognitive reappraisal reduce the propensity toward aggression in the short term but also in the longer term. These findings are preliminary, but they suggest that cognitive reappraisal training courses may reduce aggressive behavioral reactions to social provocation in individuals prone to violence.

Neuroscience studies suggest that the aggression-reducing benefits of reappraisal may be due to more effective engagement of prefrontal neural regions including medial and lateral OFC, reduced amygdala activity, and changes in functional connectivity between the frontal cortex and subcortical regions (Fabiansson et al. 2012; Goldin et al. 2008; McRae et al. 2008; Ochsner et al. 2002, 2009). Overall, reappraisal training may reduce the likelihood of reactive aggression through increased top-down neural control and blunted emotional reactivity to social provocation.

Self-control

Theory and research suggest that reduced self-control is a critical psychological mechanism for aggressive behavioral reactions to social provocation. Individuals prone to aggressive behavior are often impulsive (low in self-control), and experimental studies show that reduced self-control mediates the association between social provocation and increased aggressive behavior (Denson et al. 2011b). These studies suggest that self-control training interventions could decrease reactive aggression. A recent study tested this hypothesis by having participants practice motor self-control (using their nondominant hand to do everyday tasks such as brushing teeth) between 8 a.m. and 6 p.m. over a period of 2 weeks (Denson et al. 2011a). The findings revealed that this self-control intervention decreased anger and aggressive behavior following social provocation for individuals high in trait aggression. Although the precise biological mechanisms for the effect of self-control training on aggression remain unknown, it is plausible that the self-control intervention promoted engagement of prefrontal regions implicated in self-regulation and impulse control such as medial OFC, which resulted in the inhibition of aggressive behavior (Mehta and Beer 2010).

Directions for Future Research

There are a number of important directions for future research on the social neuroscience of human aggressive behavior. Here, we discuss some of these directions.

Gene × Hormone Interactions

One area of research that needs more attention is studies that search for theoretically informed gene × hormone interactions. It has been speculated, for example, that T may interact with the serotonergic system to modulate human aggression. Promising new evidence provides initial support for this hypothesis, demonstrating a *MAOA* × T interaction on aggression (Sjoberg et al. 2008). Individuals with the low expression allele who were also high in T levels showed the greatest risk for aggressive behavior. Another study found a *5HTTLPR* × T interaction on stress reactivity. S carriers with high T showed heightened cortisol reactivity to social threat (Josephs et al. 2012), suggesting that these same individuals may be prone to greater emotional reactions to social provocation and reactive aggression. Finally, some studies have shown that hormone receptor genes are related to the neural circuitry of aggression (e.g., androgen receptor genes and amygdala reactivity to angry faces, Manuck et al. 2010), but there has been considerably less work that has tested for biologically relevant hormone receptor gene × hormone interaction on aggression (e.g., androgen receptor gene × T interactions; but see Vermeesch et al. 2010 for a recent example of this fruitful approach). Human research on gene × hormone interactions has only just begun, and we believe that research that takes this approach will greatly improve our understanding of the neurobiology of aggression.

Neuropeptides

Animal research indicates that the neuropeptide vasopressin plays an important role in aggressive behavior in part through interaction with other biological factors such as T, but very little work has examined the influence of vasopressin on human aggression. An earlier study found that individual differences in cerebrospinal fluid arginine vasopressin (AVP) were positively correlated with self-reported aggression (Coccaro et al. 1998). A more recent paper administered AVP and found some initial evidence that it alters psychophysiological correlates of aggressive behavior (Thompson et al. 2004, 2006), but clearly much more research on vasopressin and aggression in humans is needed. Oxytocin is another neuropeptide that also influences social behavior, and its effects on human behavior have been much more well-studied (Bartz et al. 2011). Although most biological theories suggest that oxytocin should encourage prosocial behavior (trust, cooperation), recent human studies suggest that this view is overly simplistic. This research shows that oxytocin administration has divergent effects on human social behavior directed toward ingroup versus outgroup members. More specifically, oxytocin increased aggressive motivation toward outgroup members (outgroup hate) even though oxytocin increased prosocial motivation toward ingroup members (ingroup love) (De Dreu et al. 2010). These results indicate that the effects of oxytocin are context-dependent, which fits with animal models suggesting that oxytocin can promote defensive maternal aggression.

Longitudinal Studies

There is a need for more longitudinal studies that measure a host of biological factors and psychological factors along with aggressive behavior at multiple time points. Such longitudinal studies can illuminate how changes in biological systems (e.g., T levels, OFC function) may track changes in aggressive behavior over time. Such longitudinal studies can inform theories of the psychobiological mechanisms through which environmental risk factors (e.g., media violence) and protective factors (e.g., parental training in cognitive reappraisal) early in life can influence the expression of aggressive behavior in adulthood.

Gender Similarities and Differences

More theoretical and empirical attention is needed to understand how males and females are similar and how they are different in the psychobiological mechanisms of aggressive behavior (Josephs et al. 2011). For example, most research on T and social behavior has focused on males, but a series of recent studies suggest that basal T may also predict social behaviors in females (e.g., social dominance, Mehta et al. 2008, 2009). Other research, however, suggests that acute fluctuations in T predict aggression and dominance only in men (Carré et al. 2009; Mehta and Josephs 2006). Moreover, a greater understanding is needed for how men's and women's aggressive behavior may be expressed differently. Recent research suggests that boys are more likely to show direct forms of confrontation (physical aggression, direct name calling) compared to girls, whereas boys and girls are equally likely to show indirect aggression (sabotaging friendships or romantic relationships, spreading gossip, social exclusion) (Card et al. 2008). Greater attention to biological and cultural issues surrounding gender is required to build more accurate theoretical models of human aggression.

Conclusion

Aggressive and violent behaviors affect millions of people worldwide every year (Mercy et al. 2002). This chapter reviewed the research on the social neuroscience of human reactive aggression, including research on the genes, hormones, neural systems, and environmental factors implicated in aggressive behavior. Researchers have only begun to integrate these perspectives to build comprehensive models of human aggression. Promising new directions for research include longitudinal studies that better delineate the social and biological mechanisms that increase risk for adulthood violence as well as studies that attempt to reduce aggression in at-risk populations through novel pharmacological and psychosocial interventions.

References

Anderson, C. A., Shibuya, A., Ihori, N., et al. (2010). Violent video game effects on aggression, empathy, and prosocial behavior in eastern and western countries: A meta-analytic review. *Psychological Bulletin, 136*, 151–173.

Archer, J., & Southhall, N. (2009). Does cost-benefit analysis or self-control predict involvement in bullying behavior by male prisoners? *Aggressive Behavior, 35*, 31–40.

Archer, J., Graham-Kevan, N., & Davies, M. (2005). Testosterone and aggression: A reanalysis of Book, Starzyk, and Quinsey's (2001) study. *Aggression and Violent Behavior, 10*, 241–261.

Baron, R. A., & Richardson, D. R. (1994). *Human aggression* (2nd ed.). New York: Plenum.

Bartlett, C. P., & Anderson, C. A. (2011). Re-appraising the situation and its impact on aggressive behavior. *Personality and Social Psychology Bulletin, 37*(12), 1564–1573.

Bartz, J. A., Zaki, J., Bolger, N., et al. (2011). Social effects of oxytocin in humans: Context and person matter. *Trends in Cognitive Sciences, 15*, 301–309.

Beaver, J. D., Lawrence, A. D., Passamonti, L., & Calder, A. J. (2008). Appetitive motivation predicts the neural response to facial signals of aggression. *The Journal of Neuroscience, 28*, 2719–2725.

Bechara, A., Damasio, H., & Damasio, A. R. (2000). Emotion, decision making and the orbitofrontal cortex. *Cerebral Cortex, 10*, 295–307.

Beer, J. S., Heerey, E. A., Keltner, D., Scabini, D., & Knight, R. T. (2003). The regulatory function of self-conscious emotion: Insights from patients with orbitofrontal damage. *Journal of Personality and Social Psychology, 85*, 594–604.

Beer, J. S., John, O. P., Scabini, D., & Knight, R. T. (2006). Orbitofrontal cortex and social behavior: Integrating self-monitoring and emotion-cognition interactions. *Journal of Cognitive Neuroscience, 18*, 871–879.

Beitchman, J. H., Baldassarra, L., Mik, H., et al. (2006). Serotonin transporter polymorphisms and persistent, pervasive childhood aggression. *The American Journal of Psychiatry, 163*, 1103–1105.

Berman, M. E., McCloskey, M. S., Fanning, J. R., Schumacher, J. A., & Coccaro, E. F. (2009). Serotonin augmentation reduces response to attack in aggressive individuals. *Psychological Science, 20*, 714–720.

Blair, R. J. R. (2004). The roles of orbital frontal cortex in the modulation of antisocial behavior. *Brain and Cognition, 55*, 198–208.

Brunner, H. G., Nelen, M., Breakefield, X. O., et al. (1993). Abnormal-behavior associated with a point mutation in the structural gene for monoamine oxidase-A. *Science, 262*, 578–580.

Buckholz, J. W., & Meyer-Lindenberg, A. (2008). MAOA and the neurogenetic architecture of human aggression. *Trends in Neurosciences, 31*, 120–129.

Bufkin, J. L., & Luttrell, V. R. (2005). Neuroimaging studies of aggressive and violent behavior: Current findings and implications for criminology and criminal justice. *Trauma, Violence & Abuse, 6*, 176–191.

Buss, D. M., & Shackelford, T. K. (1997). Human aggression in evolutionary psychological perspective. *Clinical Psychology Review, 17*, 605–619.

Card, N. A., Stucky, B. D., Sawalani, G. M., & Little, T. D. (2008). Direct and indirect aggression during childhood and adolescence: A meta-analytic review of gender differences, intercorrelations, and relations to maladjustment. *Child Development, 79*, 1185–1229.

Carnagey, N. L., Anderson, C. A., & Bartholow, B. D. (2007). Media violence and social neuroscience: New questions and new opportunities. *Current Directions in Psychological Science, 16*, 178–182.

Carré, J. M., & Mehta, P. H. (2011). Importance of considering testosterone–cortisol interactions in predicting human aggression and dominance. *Aggressive Behavior, 37*, 1–3.

Carré, J. M., Putnam, S. K., & McCormick, C. M. (2009). Testosterone responses to competition predict future aggressive behavior at a cost to reward in men. *Psychoneuroendocrinology, 34*, 561–570.

Carré, J. M., Gilchrist, J. D., Morrissey, M. D., et al. (2010). Motivational and situational factors and the relationship between testosterone dynamics and human aggression during competition. *Biological Psychology, 84*, 346–353.

Carré, J. M., McCormick, C. M., & Hariri, A. R. (2011). The social neuroendocrinology of human aggression. *Psychoneuroendocrinology, 36*, 935–944.

Carré, J. M., Fisher, P. M., Manuck, S. B., & Hariri, A. R. (2012). Interaction between trait anxiety and trait anger predict amygdala reactivity to angry facial expressions in men but not women. *Social Cognitive and Affective Neurosciences, 7*(2), 213–221.

Caspi, A., McClay, J., Moffitt, T. E., et al. (2002). Role of genotype in the cycle of violence in maltreated children. *Science, 297*, 851–854.

Caspi, A., Hariri, A. R., Holmes, A., et al. (2010). Genetic sensitivity to the environment: the case of the serotonin transporter gene and its implications for studying complex diseases and traits. *The American Journal of Psychiatry, 167*, 509–527.

Chamberlain, N. L., Driver, E. D., & Miesfeld, R. L. (1994). The length and location of CAG trinucleotide repeats in the androgen receptor N-terminal domain affect transactivation function. *Nucleic Acids Research, 22*, 3181–3186.

Chan, S. C., Raine, A., & Lee, T. M. (2010). Attentional bias toward negative affective stimuli and reactive aggression in male batterers. *Psychiatry Research, 176*, 246–249.

Choong, C. S., & Wilson, E. M. (1998). Trinucleotide repeats in the human androgen receptor: A molecular basis for disease. *Journal of Molecular Endocrinology, 21*, 235–257.

Coccaro, E. F., Kavoussi, R. J., Hauger, R. L., Cooper, T. B., & Ferris, C. F. (1998). Cerebrospinal fluid vasopressin levels: Correlates with aggression and serotonin function in personality-disordered subjects. *Archives of General Psychiatry, 55*, 708–714.

Coccaro, E. F., McCloskey, M. S., Fitzgerald, D. A., & Phan, K. L. (2007). Amygdala and orbitofrontal reactivity to social threat in individuals with impulsive aggression. *Biological Psychiatry, 62*, 168–178.

Coccaro, E. F., Sripada, C. S., Yanowitch, R. N., & Phan, K. L. (2011). Corticolimbic function in impulsive aggressive behavior. *Biological Psychiatry, 69*, 1153–1159.

Collias, N. E., Barfield, R. J., & Tarvyd, E. S. (2002). Testosterone versus psychological castration in the expression of dominance, territoriality, and breeding behavior by male village weavers (Ploceus cucullatus). *Behaviour, 139*, 801–824.

Crockett, M. J., Clark, L., Tabibnia, G., Lieberman, M. D., & Robbins, T. W. (2008). Serotonin modulates behavioral reactions to unfairness. *Science, 320*, 1739.

Damasio, H., Grabowski, T., Frank, R., Galaburda, A. M., & Damasio, A. R. (1994). The return of Phineas Gage: Clues about the brain from the skull of a famous patient. *Science, 264*, 1102–1105.

Davidson, R. J., Putnam, K. M., & Larson, C. L. (2000). Dysfunction in the neural circuitry of emotion regulation: A possible prelude to violence. *Science, 289*, 591–594.

Davis, M., & Whalen, P. J. (2001). The amygdala: Vigilance and emotion. *Molecular Psychiatry, 6*, 13–34.

De Dreu, C. K. W., Greer, L. L., Handgraaf, M. J., et al. (2010). The neuropeptide oxytocin regulates parochial altruism in intergroup conflict among humans. *Science, 328*, 1408–1411.

De Martino, B., Kumaran, D., Seymour, B., & Dolan, R. J. (2006). Frames, biases, and rational decision-making in the human brain. *Science, 313*, 684–687.

Denson, T. F., Capper, M. M., Oaten, M., et al. (2011a). Self-control training decreases aggression in response to provocation in aggressive individuals. *Journal of Research in Personality, 45*, 252–256.

Denson, T. F., Pedersen, W. C., Friese, M., et al. (2011b). Understanding impulsive aggression: Angry rumination and reduced self-control capacity are mechanisms underlying the provocation-aggression relationship. *Personality and Social Psychology Bulletin, 37*, 850–862.

Dickerson, S. S., & Kemeny, M. E. (2004). Acute stressors and cortisol responses: A theoretical integration and synthesis of laboratory research. *Psychological Bulletin, 130*, 355–391.

Dodge, K. A., & Coie, J. D. (1987). Social-information-processing factors in reactive and proactive aggression in children's peer groups. *Journal of Personality and Social Psychology, 53*, 1146–1158.

Edwards, A., Hammond, H. A., Jin, L., Caskey, C. T., & Chakraborty, R. (1992). Genetic variation at 5-trimeric and tetrameric tandem repeat loci in 4 human population groups. *Genomics, 12*, 241–253.

Eisenberger, N. I., Way, B. M., Taylor, S. E., et al. (2007). Understanding genetic risk for aggression: Clues from the brain's response to social exclusion. *Biological Psychiatry, 61*, 1100–1108.

Fabiansson, E. C., Denson, T. F., Grisham, J. R., Moulds, M. L., & Schira, M. M. (2012). Don't look back in anger: Neural correlates of reappraisal, analytical rumination, and angry rumination during recall of an anger-inducing autobiographical memory. *NeuroImage, 59*(3), 2974–2981.

Ford, M. B., & Collins, N. L. (2010). Self-esteem moderates neuroendocrine and psychological responses to interpersonal rejection. *Journal of Personality and Social Psychology, 98*, 405–419.

Frazzetto, G., Lorenzo, D., Valeria, C., et al. (2007). Early trauma and increased risk for physical aggression during adulthood: The moderating role of MAOA genotype. *PloS One, 2*, e486.

Geniole, S. N., Carré, J. M., & McCormick, C. M. (2010). State, not trait, neuroendocrine function predicts costly reactive aggression in men after social exclusion and inclusion. *Biological Psychology, 87*, 137–145.

Giammanco, M., Tabacchi, G., Giammanco, S., Di Majo, D., & La Guardia, M. (2005). Testosterone and aggressiveness. *Medical Science Monitor, 11*, 136–145.

Gleason, E. D., Fuxjager, M. J., Oyegbile, T. O., & Marler, C. A. (2009). Testosterone release and social context: When it occurs and why. *Frontiers in Neuroendocrinology, 30*, 460–469.

Goldin, P. R., McRae, K., Ramel, W., & Gross, J. J. (2008). The neural bases of emotion regulation during reappraisal and suppression of negative emotion. *Biological Psychiatry, 63*, 577–586.

Gospic, K., Mohlin, E., Fransson, P., et al. (2011). Limbic justice – Amygdala involvement in immediate rejection in the ultimatum game. *PLoS Biology, 9*, e1001054.

Hariri, A. R. (2009). The neurobiology of individual differences in complex behavioral traits. *Annual Reviews in Neuroscience, 32*, 247–255.

Hariri, A. R., & Weinberger, D. R. (2003). Functional neuroimaging of genetic variation in serotonergic neurotransmission. *Genes, Brain, and Behavior, 2*, 341–349.

Hariri, A. R., Mattay, V. S., Tessitore, A., et al. (2002). Serotonin transporter genetic variation and the response of the human amygdala. *Science, 297*, 400–403.

Heils, A., Teufel, A., Susanne, P., et al. (1996). Allelic variation of human serotonin transporter gene expression. *Journal of Neurochemistry, 66*, 2621–2624.

Hermans, E. J., Ramsey, N. F., & van Honk, J. (2008). Exogenous testosterone enhances responsiveness to social threat in the neural circuitry of social aggression in humans. *Biological Psychiatry, 63*, 263–270.

Josephs, R. A., Mehta, P. H., & Carré, J. M. (2011). Gender and social environment modulate the effects of testosterone on social behavior: Comment on Eisenegger et al. *Trends in Cognitive Sciences, 15*, 509–510.

Josephs, R. A., Telch, M. J., Hixon, J. G., Evans, J. J., Lee, H., Knopik, V. S., McGeary, J. E., Hariri, A. R., & Beevers, C. G. (2012). Genetic and hormonal sensitivity to threat: Testing a serotonin transporter genotype x testosterone interaction. *Psychoneuroendocrinology, 37*, 752–761.

Kelly, C. R., Grinband, J., & Hirsch, J. (2007). Repeated exposure to media violence is associated with diminished response in an inhibitory frontolimbic network. *PloS One, 2*, e1268.

Klinesmith, J., Kasser, T., & McAndrew, F. T. (2006). Guns, testosterone, and aggression: An experimental test of a meditational hypothesis. *Psychological Science, 17*, 568–571.

Koenigs, M., & Tranel, D. (2007). Irrational economic decision-making after ventromedial prefrontal damage: Evidence from the ultimatum game. *The Journal of Neuroscience, 27*, 951–956.

Krahe, B., & Moeller, I. (2010). Longitudinal effects of media violence on aggression and empathy among German adolescents. *Journal of Applied Developmental Psychology, 31*, 401–409.

Kringelbach, M. L., & Rolls, E. T. (2004). The functional neuroanatomy of the human orbitofrontal cortex: Evidence from neuroimaging and neuropsychology. *Progress in Neurobiology, 72*, 341–372.

Leary, M. R., Kowalski, R. M., Smith, L., et al. (2003). Teasing, rejection, and violence: Case studies of the school shootings. *Aggressive Behavior, 29*, 202–214.

Leary, M. R., Twenge, J. M., & Guinlivan, E. (2006). Interpersonal rejection as a determinant of anger and aggression. *Personality and Social Psychology Review, 10*, 111–132.

LeDoux, J. E. (2000). Emotion circuits in the brain. *Annual Reviews of Neuroscience, 23*, 155–184.

Lee, T. M., Chan, S. C., & Raine, A. (2008). Strong limbic and weak frontal activation to aggressive stimuli in spouse abusers. *Molecular Psychiatry, 13*, 655–656.

Manuck, S. B., Marsland, A. L., Flory, J. D., et al. (2010). Salivary testosterone and a trinucleotide (CAG) length polymorphism in the androgen receptor gene predict amygdala reactivity in men. *Psychoneuroendocrinology, 35*, 94–104.

Matsuo, K., Nicoletti, M., Nemoto, K., et al. (2009). A voxel-based morphometry study of frontal gray matter correlates of impulsivity. *Human Brain Mapping, 30*, 1188–1195.

Mazur, A. (1985). A biosocial model of status in face-to-face primate groups. *Social Forces, 64*, 377–402.

Mazur, A., & Booth, A. (1998). Testosterone and dominance in men. *The Behavioral and Brain Sciences, 21*, 353–397.

McDermott, R., Tingley, D., Cowden, J., et al. (2009). Monoamine oxidase A gene (MAOA) predicts behavioral aggression following provocation. *Proceedings of the National Academy of Sciences of the USA, 106*, 2118–2123.

McRae, K., Reiman, E. M., Fort, C. L., Chen, K., & Lane, R. D. (2008). Association between trait emotional awareness and dorsal anterior cingulate activity during emotion is arousal-dependent. *NeuroImage, 41*, 648–655.
Mehta, P. H., & Beer, J. S. (2010). Neural mechanisms of the testosterone-aggression relation: The role of orbitofrontal cortex. *Journal of Cognitive Neuroscience, 22*, 2357–2368.
Mehta, P. H., & Josephs, R. A. (2006). Testosterone change after losing predicts the decision to compete again. *Hormones and Behavior, 50*, 684–692.
Mehta, P. H., & Josephs, R. A. (2010). Testosterone and cortisol jointly regulate dominance: Evidence for a dual-hormone hypothesis. *Hormones and Behavior, 58*, 898–906.
Mehta, P. H., Jones, A. C., & Josephs, R. A. (2008). The social endocrinology of dominance: Basal testosterone predicts cortisol changes and behavior following victory and defeat. *Journal of Personality and Social Psychology, 94*, 1078–1093.
Mehta, P. H., Wuerrhman, E., & Josephs, R. A. (2009). When are low testosterone levels advantageous?: The moderating role of individual versus intergroup competition. *Hormones and Behavior, 56*, 158–162.
Mehta, P. H., Yap, A., & Mor, S. (2010, October). *The biology of bargaining: Dynamic hormone changes during negotiation predict economic profit*. Talk presented at the conference for the Social and Affective Neuroscience Society, Chicago, IL.
Mercy, J., Butchart, A., Farrington, D., & Cerda, M. (2002). Youth violence. In E. G. Krug, L. L. Dahlberg, J. A. Mercy, A. B. Zwi, & R. Lozano (Eds.), *World report on violence and health*. Geneva: World Health Organization.
Meyer-Lindenberg, A., Buckholtz, J. W., Kolachana, B., Hariri, A. R., Pezawas, L., et al. (2006). Neural mechanisms of genetic risk for impulsivity and violence in humans. *Proceedings of the National Academy of Sciences of USA, 103*, 6269–6274.
Moretti, L., Dragone, D., & di Pellegrino, G. (2009). Reward and social valuation deficits following ventromedial prefrontal damage. *Journal of Cognitive Neuroscience, 21*, 128–140.
Munafo, M. R., Brown, S. M., & Hariri, A. R. (2008). Serotonin transporter (5-HTTLPR) genotype and amygdala activation: A meta-analysis. *Biological Psychiatry, 63*, 852–857.
Nelson, R. J., & Trainor, B. C. (2007). Neural mechanisms of aggression. *Nature Reviews Neuroscience, 8*, 536–546.
New, A. S., Hazlett, E., Buchsbaum, M. S., Goodman, M., Reynolds, D., et al. (2002). Blunted prefrontal cortical 18fluorodeoxyglucose positron emission tomography response to meta-chloropiperazine in impulsive aggression. *Archives of General Psychiatry, 59*, 621–629.
New, A. S., Buchsbaum, M. S., Hazlett, E. A., et al. (2004). Fluoxetine increases relative metabolic rate in prefrontal cortex in impulsive aggression. *Psychopharmacology, 176*, 451–458.
Newman, S. W. (1999). The medial extended amygdala in male reproductive behavior: A node in the mammalian social behavior network. *Annals of New York Academy of Sciences, 877*, 242–257.
Ochsner, K. N., Bunge, S. A., Gross, J. J., et al. (2002). Rethinking feelings: An fMRI study of the cognitive regulation of emotion. *Journal of Cognitive Neuroscience, 14*, 1215–1229.
Ochsner, K. N., Ray, R. D., Hughes, B., et al. (2009). Bottom-up and top-down processes in emotion generation: Common and distinct neural mechanisms. *Psychological Science, 20*, 1322–1331.
Office of the Surgeon General. (2001). *Youth violence: A report of the Surgeon General. U.S. Department of Health and Human Services.* Retrieved October, 2011, from http://www.mentalhealth.org/youthviolence/default.asp
Oliveira, R. F. (2009). Social behavior in context: Hormonal modulation of behavioral plasticity and social competence. *Integrative and Comparative Biology, 49*, 423–440.
Oliveira, R. F., Almada, V. C., & Canario, A. V. M. (1996). Social modulation of sex steroid concentrations in the urine of male cichlid fish Oreochromis mossambicus. *Hormones and Behavior, 30*, 2–12.
Passamonti, L., Crockett, M. J., Apergis-Schoute, A. M., et al. (2012). Effects of acute tryptophan depletion on prefrontal-amygdala connectivity while viewing facial signals of aggression. *Biological Psychiatry, 71*(1), 36–43.
Popma, A., Vermeiren, R., Geluk, C. A. M. L., et al. (2007). Cortisol moderates the relationship between testosterone and aggression in delinquent male adolescents. *Biological Psychiatry, 61*, 405–411.
Rahman, S., Sahakian, B. J., Cardinal, R. N., Rogers, R. D., & Robbins, T. W. (2001). Decision making and neuropsychiatry. *Trends in Cognitive Sciences, 5*, 271–277.
Raine, A., Buchsbaum, M., & LaCasse, L. (1997). Brain abnormalities in murderers indicated by positron emission tomography. *Biological Psychiatry, 42*, 495–508.
Rajender, S., Pandu, G., Sharma, J. D., Gandhi, K. P. C., Singh, L., & Thangaraj, K. (2008). Reduced CAG repeats length in androgen receptor gene is associated with violent criminal behavior. *International Journal of Legal Medicine, 122*, 367–372.
Reif, A., Rösler, M., Freitag, C. M., et al. (2007). Nature and nurture predispose to violent behavior: Serotonergic genes and adverse childhood environment. *Neuropsychopharmacology, 32*, 2375–2383.

Rolls, E. T., et al. (1994). Emotion-related learning in patients with social and emotional changes associated with frontal-lobe damage. *Journal of Neurology, Neurosurgery, and Psychiatry, 57*, 1518–1524.

Rubinow, D. R., & Schmidt, P. J. (1996). Androgens, brain, and behavior. *The American Journal of Psychiatry, 153*, 974–984.

Ruiz-de-la-Torre, J. L., & Manteca, X. (1999). Effects of testosterone on aggressive behaviour after social mixing in male lambs. *Physiology & Behavior, 68*, 109–113.

Sabol, S., Hu, S., & Hamer, D. (1998). A functional polymorphism in the monamine oxidase A gene promoter. *Human Genetics, 103*, 273–279.

Sapolsky, R. M. (1991). Testicular function, social rank and personality among wild baboons. *Psychoneuroendocrinology, 16*, 281–293.

Shoal, G. D., Giancola, P. R., & Kirillova, G. P. (2003). Salivary cortisol, personality, and aggressive behavior in adolescent boys: A 5-year longitudinal study. *Journal of the American Academy of Child and Adolescent Psychiatry, 42*, 1101–1107.

Siegel, A., Bhatt, S., Bhatt, R., et al. (2007). The neurobiological bases for development of pharmacological treatments of aggressive disorders. *Current Neuropharmacology, 5*, 135–147.

Siever, L. J. (2008). Neurobiology of aggression and violence. *The American Journal of Psychiatry, 165*, 429–442.

Simmons, Z. L., & Roney, J. R. (2011). Variation in CAG repeat length of the androgen receptor gene predicts variables associated with intrasexual competitiveness in human males. *Hormones and Behavior, 60*, 306–312.

Sjoberg, R. L., Ducci, F., Barr, C. S., et al. (2008). A non-additive interaction of a functional MAO-A VNTR and testosterone predicts antisocial behavior. *Neuropsychopharmacology, 33*, 425–430.

Strenziok, M., et al. (2010). Lower lateral orbitofrontal cortex density associated with more frequent exposure to television and movie violence in male adolescents. *The Journal of Adolescent Health, 46*, 607–609.

Strüber, D., Lück, M., & Roth, G. (2008). Sex, aggression and impulse control: An integrative account. *Neurocase, 14*, 93–121.

Thompson, R., Gupta, S., Miller, K., Mills, S., & Orr, S. (2004). The effects of vasopressin on human facial responses related to social communication. *Psychoneuroendocrinology, 29*, 35–48.

Thompson, R. R., George, K., Walton, J. C., Orr, S. P., & Benson, J. (2006). Sex-specific influences of vasopressin on human social communication. *Proceedings of the National Academy of Sciences of USA, 103*, 7889–7894.

Tucker, D. M., Luu, P., & Pribram, K. H. (1995). Social and emotional self-regulation. *Annals of the New York Academy of Sciences, 769*, 213–239.

van Wingen, G. A., Zylick, S. A., Pieters, S., Mattern, C., Verkes, R. J., Buitelaar, J. K., & Fernandez, G. (2008). Testosterone increases amygdala reactivity in middle-aged women to a young adulthood level. *Neuropsychopharmacology, 34*, 539–547.

van Wingen, G., Mattern, C., Verkes, R. J., Buitelaar, J., & Fernández, G. (2010). Testosterone reduces amygdala-orbitofrontal cortex coupling. *Psychoneuroendocrinology, 35*, 105–113.

Vermeersch, H., T'Sjoen, G., Kaufman, J. M., Vincke, J., & Van Houtte, M. (2010). Testosterone, androgen receptor gene CAG repeat length, mood and behaviour in adolescent males. *European Journal of Endocrinology, 163*, 319–328.

Wingfield, J. C., Hegner, R. E., Dufty, A. M., Jr., & Ball, G. F. (1990). The 'challenge hypothesis': Theoretical implications for patterns of testosterone secretion, mating systems, and breeding strategies. *The American Naturalist, 136*, 829–846.

Chapter 6
Social Neuroscience and the Modern Synthesis of Social and Biological Levels of Analysis

Greg J. Norman, Louise C. Hawkley, Maike Luhmann, John T. Cacioppo, and Gary G. Berntson

The brains of social organisms have neurobiological circuits that recognize, compute, and manipulate socially relevant information. In some cases, these networks appear to have evolved to co-opt older neurobehavioral mechanisms (Eisenberger et al. 2003), in others to have evolved as domain-specific neurobiological mechanisms (Ho et al. 2009), and in still others to have evolved to deal more generally with the complex information processing demands of social species (Dunbar and Shultz 2007). Indeed, the immense processing demands necessary for functioning within a complex social group have been proposed to be one of the primary sources responsible for a neurobiological arms race that has resulted in the development of the relatively large brain size observed in primates (Brothers 1990; Dunbar 2009). Consistent with this hypothesis, average social group size increases with relative neocortex volume in anthropoid primates (Dunbar and Shultz 2007). The processing power of this "social brain" permits the neurobiological mining of the most inconspicuous of socially relevant signals (e.g., variations in body language, eye gaze, tone of voice) from the nearly infinite amount of sensory data in nearly all environmental contexts (Adolphs 2009). Interestingly, the computational capacity of the brain is such that the majority of this processing occurs outside awareness and only comes to our attention when things go wrong. For example, localized lesions to specific bilateral neuroanatomical structures can induce rather striking alterations in social information processing. Frontotemporal dementias and lesions in the ventromedial prefrontal cortex are associated with normal social recognition and memory but increased antisocial behavior, whereas damage to the amygdala diminishes people's ability to infer other's emotional states and causes diminished emotional activation to negative or threatening stimuli (Adolphs et al. 1994; Berntson et al. 2007). Thus, the brains of various social animals are tuned to the relevant frequencies important for social communication and behavior leading some to describe the central nervous system of various social mammals, including humans, as "the social brain" (Adolphs 2009; Brothers 1990; Cacioppo et al. 2004; Dunbar 2009).

Social neuroscience is an interdisciplinary approach to the study of complex social structures, processes, and behaviors. In particular, this field includes the study of social networks, the individuals that create them, and the neural, hormonal, and genetic mechanisms that allow for their existence.

G.J. Norman (✉) • L.C. Hawkley • M. Luhmann • J.T. Cacioppo
Chicago Center for Cognitive and Social Neuroscience, The University of Chicago, Chicago, IL, USA
e-mail: gnorman@uchicago.edu

G.G. Berntson
Chicago Center for Cognitive and Social Neuroscience, The University of Chicago, Chicago, IL, USA

Department of Psychology, The Ohio State University, Columbus, OH, USA

D.D. Franks and J.H. Turner (eds.), *Handbook of Neurosociology*, Handbooks of Sociology and Social Research, DOI 10.1007/978-94-007-4473-8_6, © Springer Science+Business Media B.V. 2013

The broad range of disciplines encompassed by social neuroscience presents a challenge for its practitioners. The multilevel interdisciplinary approach, however, offers an opportunity to refine and calibrate the concepts at one level of analysis (e.g., social psychology) through the application of knowledge from another (e.g., neuroscience) and can also foster conceptual insights as well as experimental tests of otherwise indistinguishable theoretical explanations, while simultaneously increasing the comprehensiveness and relevance of the work.

The ability and inclination to implement such interdisciplinary research is a relatively recent development. Historically, biological and social levels of analysis have been treated as alternative, even mutually exclusive, paths to knowledge. As the field of psychology began to grow in the early twentieth century, distinct schools of thought on how to best describe and explain human nature began to coalesce. The biopsychological perspective was primarily aimed at understanding the physiological substrates underlying behavior, a strategy that often tended to focus attention away from the effects of environmental context. Social psychological views, on the other hand, emphasized contextual influences on multivariate systems in the study of human behavior, a strategy that tends to minimize biological perspectives. Such distinctions are not surprising as human biology is anchored in concrete anatomy and genetics, whereas the social environment is characterized by abstractions used to explain individual relationships, groups, societies, and cultures. Such differences in the implementation of varying levels of analysis have resulted in distinct histories, research traditions, methodologies, and technical demands, yielding what some regard both as an impassable abyss between social and biological approaches and as evidence of the impending demise of psychology as a discipline (Scott 1991).

Although there has been a traditional debate over the contributions of environmental vs. genetic factors in psychological and behavioral processes, it is increasingly apparent that this dichotomy is specious. It is not merely the fact that both genetic and environmental factors are powerful co-determinants of behavior—that has been known for over a century. What is more important is that these determinants are not invariably additive and emergent effects can occur. A given genotype may yield divergent phenotypes as a result of such interactions, and distinct phenotypes may respond to a given environment entirely differently.

Similar to the constrained view of biological sciences throughout most of the twentieth century, social scientists uniformly ignored biological levels of analysis. Two world wars, a great depression, and civil injustices made it clear that social and cultural forces were sufficiently imperative that social scientists were unable to wait for full biological explication of such phenomena. Despite this historical independence of biological and social sciences, during the late twentieth century, sufficient interdisciplinary evidence had accrued that both biological and social scientists began investigating the "social brain" in nonhuman primates (Brothers 1990; Dunbar and Shultz 2007) and humans (Cacioppo and Berntson 1992). Individuals with damage to the amygdala display decreased emotional activation to negative or threatening stimuli and increased ratings of the trustworthiness of social stimuli (Adolphs et al. 1998; Berntson et al. 2007; Tranel et al. 2006). Bilateral damage to specific regions of the temporal lobe leave individuals with the inability to recognize previously familiar faces (Pedersen et al. 1982). Thus, localized damage to particular brain regions is capable of producing rather isolated deficits in the processing of social stimuli, suggesting that understanding such structures may prove useful to understanding particular aspects of social interaction.

It is now evident that integrative multilevel analysis can contribute to the development of more comprehensive models of complex social behavior. The multilevel analyses prescribed by social neuroscience represent a subset of interdisciplinary approaches where the measures, constructs, and theories extend across levels of organization. Efforts to integrate information across levels of analyses are especially challenging given the inherent complexity of biological and social systems, but it is only through such research that a complete understanding of the human mind and behavior will be possible.

Levels of Organization or the Nervous System

Levels of Evaluative Function: Lower Levels and Spinal Reflexes

Spinal reflexes are among the lowest levels of organization in the central nervous system, and their relative simplicity allows for fast and efficient adaptive responding to environmental stimuli. Although capable of operating independently of higher levels, spinal reflexes also provide critical functional support for higher-level functions, an issue to which we will return below.

In his treatise *The Integrative Action of the Nervous System* (1906), Sir Charles Sherrington details spinal organizations that contribute to postural regulation and provide the basic neurological support for locomotion. He also describes spinal substrates for basic, low-level evaluative reactions. Among the most salient of spinal reflexes is the flexor (pain) withdrawal reflex, which represents a primitive but effective evaluative mechanisms for protection against noxious or injurious stimuli. Nociceptive signals, carried by somatosensory afferents, activate flexor neuron pools via interneuron circuits within the spinal cord, resulting in flexor withdrawal responses (Sandrini et al. 2005; Schouenborg et al. 1992; Lundberg 1979).

Although appetitive reflexes may be less obvious than aversive reflexes at the level of the spinal cord, primitive approach/engagement dispositions are also apparent in spinal extensor reflexes. Sherrington (1906) described extensor thrust reflexes to palmer contact that represent low-level reflexive dispositions that promote contact and engagement with the external environment. These approach/engagement reflexes are importantly supplemented by suckling and ingestive reflexes of brainstem origin, which will be considered below. At a trivial level, flexor and extensor reflexes promote diametrically opposing motoric dispositions. The spinal circuits for these reflexes are distinct and separately organized, including differences in peripheral sensory receptors, afferent axonal populations, central interneuronal pathways, and motoneuron output pools. This is not to say that flexor/extensor reflexes are entirely independent. Although the primary neural circuits underlying flexor and extensor reflexes are parallel and distinct, there are rich interactions among these networks—an organizational pattern that Sherrington referred to as the alliance of reflexes. Examples include the crossed-extension reflex, in which activation of the flexor reflex in one limb is associated with a reflex extension of the opposite limb. Interactions among networks for opponent flexor and extensor reflexes for a given limb were also described by Sherrington as a pattern of reciprocal innervation. Reciprocal innervation is the property by which spinal reflex networks that activate a specific outcome (e.g., limb flexion) also tend to inhibit opponent (e.g., extensor) muscles, which synergistically promotes the target response. These organizational patterns are not unique to spinal circuits but represent general neuroarchitectural features that may inform the operations of higher-level systems as well. Behavioral manifestations of the principle of reciprocal innervation, for example, can be seen even at a cognitive level. One example comes from the literature on cognitive dissonance, where the mere selection of an item from among several choices results in increased cognitive valuation of the chosen item and concurrent devaluation of the nonselected items (Egan et al. 2007).

The integrative outputs of spinal approach/withdrawal circuits may provide a basic model for understanding higher-level evaluative processes. For example, flexor withdrawal and extensor approach reflexes are not symmetrical in strength, as flexor withdrawal reflexes are significantly more potent than their antagonistic extensor (approach) reflexes and recover more rapidly than do extensor reflexes after spinal transection. As will be considered below, asymmetric strength of evaluative systems is also apparent at higher levels of the neuraxis where avoidance reactions (anxiety, fear) tend to have a stronger hold on affect when compared to approach reactions (incentive, reward). This makes adaptive sense, as a single failure of the avoidance system can lead to subsequent injury or death, so natural selection may have tuned the avoidance system for preferential control of behavior. The bias toward avoidance reactions represents a reoccurring theme at all levels of the neuraxis and has been termed the negativity bias (Cacioppo and Berntson 1999; Cacioppo et al. 2004, in press).

Despite this negativity bias, flexor/withdrawal reflexes are not always dominant over their opponent processes as extensor/approach reflexes can take precedence over withdrawal processes at lower levels of stimulation or activation. This disposition toward approach behaviors in the context of low levels of activation has been termed the positivity offset (Cacioppo and Berntson 1999; Cacioppo et al. 2004) and characterizes the operations of evaluative processes at multiple levels of the neuraxis. As we will consider below, the asymmetry of neurobehavioral dispositions can lead to a context-dependent outcome in that approach dispositions may predominate at lower levels of evaluative activation but can be trumped by avoidance or withdrawal (negativity bias) at higher levels of evaluative activation.

Spinal flexor and extensor reflexes have separate, although interacting, circuitries and thus can operate in parallel, within the constraints of those neural interactions. Despite this underlying bivalence, the behavioral output of opponent extensor/flexor networks may lie along a bipolar continuum from flexion to extension, the output being constrained by the mechanical coupling of the extensor and flexor muscles around a specific point of articulation at a joint.

Neural Hierarchies

Multilevel perspectives of neuronal organization have been emphasized by scientists and philosophers alike, among the more influential of whom was the nineteenth-century neurologist John Hughlings Jackson. In his essay *Evolution and Dissolution of the Nervous System*, Jackson (1884) laid groundwork for multilevel characterization of neuronal organization. Jackson argued that the evolutionary emergence of higher levels of neuronal organizations does not involve a replacement or displacement of lower levels. Rather, evolutionary development entails a re-representation and elaboration of functions at progressively higher levels of the nervous system. Although rostral levels were thought to be characterized by elaborated networks capable of more sophisticated functions, they were not seen to replace lower levels and in fact remain highly dependent on lower neuraxial substrates. For example, the critical spinal networks for stepping and related locomotor reflexes constitute essential lower processing circuits that support outputs from higher motor systems. In Jackson's view, the proper interpretation of the consequences of brain injuries is not optimally defined by the functions that are lost but rather in the reversion (dissolution) of those functions to lower levels of neural organization.

It is now apparent that the neuraxis is replete with hierarchical organizations comprised of simple reflex-like circuits at the lowest levels, such as the brainstem and spinal cord, and neural networks for more integrative computations at higher levels (for reviews, see Berntson et al. 1993; Berntson and Cacioppo 2000; Berridge 2004). The relatively simple neural circuitry characteristic of lower levels of the neuraxis is essential for survival as it allows for rapid computations and subsequent motor outputs. The adaptive function of such circuits is obvious as it may be more important in some circumstances to perform a rapid but imperfect response rather than a more elaborate and protracted performance that may produce a more elaborate outcome, as the additional time consumed by such processes could lead to a negative outcome. As environmental challenges grow increasingly complex, more integrated neuronal processing may be more adaptive, and higher-level analytical and response mechanisms may come into play. Moreover, learned anticipatory processes may promote more strategic avoidance of adaptive challenges prior to their occurrence. The increasing amount of information that must be processed and integrated by progressively higher-level systems may lead to neurocomputational bottlenecks which require a slower and more serial mode of processing. Based on hierarchical interconnections, higher-level systems may depend heavily on lower-level systems for the transmission and preliminary processing and filtering of afferent sensory and perceptual data and for implementing sensory motor subroutines that support executive outputs. The advantages and disadvantages associated with higher-level (integrative, flexible, but capacity limited) and lower-level (rapid, efficient, but rigid) processing were a likely source of evolutionary pressure for the preservation of lower-level substrates, despite higher-level elaborations and re-representations (Berntson and Cacioppo 2012). Together these interacting hierarchical structures allow neural systems to rapidly respond through low-level processing (e.g., pain-withdrawal reflexes) while more rostral neural substrates permit a more elaborated response over time and allow evaluation of future strategies and subsequent consequences. Hierarchical representations do not merely reflect theoretical models or cognitive curiosities but are empirically documented by neuroanatomical and functional analyses of neural systems throughout the brain (Berntson et al. 1993).

Neural Heterarchies

Additional neuroarchitectural complexities exist beyond strict hierarchical organization patterns, as long descending pathways also exist which bypass intermediate levels and directly synapse onto lower levels of the neuraxis (Wakana et al. 2004). This type of organization is documented by the presence of direct, long descending projections from higher neuraxial systems to lower motor neurons, effectively bypassing intermediate levels. In addition to the well-known anatomy of somatomotor systems (Wakana et al. 2004), this pattern of organization is also apparent in the autonomic nervous system (Berntson and Cacioppo 2000). For example, the baroreflex is a tightly organized brainstem-mediated reflex system that serves to maintain blood pressure homeostasis. Increases in blood pressure activate specialized cardiovascular mechanoreceptors, which then feed back into brainstem reflex circuitry leading to reciprocal increases in vagal cardiac output and decreases in sympathetic cardiac and vascular tone. These responses collectively lead to decreases in heart rate, cardiac output, and vascular tone, which synergistically compensate for the blood pressure perturbation. In contrast to this lower-level, homeostatic reflex regulation, higher-level systems, for example, with even mild psychological stress, are capable of overriding the baroreflex and yielding concurrent increases in blood pressure and heart rate. This non-homeostatic modulation of cardiovascular processes may arise in part from descending inhibition of brainstem baroreflex networks. It also likely reflects the actions of long descending projections from higher neurobehavioral substrates that bypass intermediate reflex circuits and project monosynpatically to lower autonomic source nuclei. In this fashion, cortical and limbic structures are able to bypass intermediate hierarchical elements and directly control lower levels (see Berntson et al. 1994).

The presence of long ascending and descending pathways in neural organizational patterns, combined with lateral interconnections between levels, has previously been described as a neural heterarchy (see Berntson et al. 1993; Berntson and Cacioppo 2000). Heterarchical organization patterns have the components of hierarchical systems, as higher levels are in continuous communication with lower-level systems via intermediate levels, but have the additional capacity to interact over widely separated levels via direct connections. Direct neuronal projections from higher brain systems to lower-level systems allow for manifestations of higher computational re-representative networks that are not constrained by intermediate-level organizations. This affords cognitive and behavioral flexibility when needed but also allows for intermediate-level processing when necessary. The multiple levels of organization and associated functional flexibility comes with a disadvantage, as a heterarchical organization opens the possibility for functional conflicts between distinct levels of processing, such as when an organism must inhibit pain withdrawal in order to achieve a higher order goal. We will return to this issue below.

Levels of Evaluative Function: Intermediate Levels—Decerebration

Although primitive approach/withdrawal dispositions are represented at spinal levels, they are substantially developed and elaborated at brainstem levels. Classical demonstrations of the functional capacity of brainstem networks come from studies of experimental isolation of the brainstem and spinal cord (decerebration) and from tragic cases of human decerebration (Berntson and Micco 1976; Tuber et al. 1980; Berntson et al. 1983; Ronca et al. 1986, Yates et al. 1993; Harris et al. 2006). Although acute postsurgical somatomotor rigidity historically obscured the behavioral capacities of the experimental decerebrate, with longer survival times and the resolution of this rigidity, a great deal of organizational capacity is apparent at brainstem levels (Bard and Macht 1958; Berntson and Micco 1976; Norman et al. 1977). Decerebrate animals, for example, can right themselves and locomote, eat and drink on encountering appropriate goal objects, groom, and display aggressive, defensive, and escape behaviors to noxious stimuli (see Berntson and Micco 1976; Norman et al. 1977; Adams 1979).

Considerable functional capacity is also apparent in tragic cases of human decerebration (anencephaly and hydranencephaly), generally resulting from a failure of cell migration early in neurodevelopment. Although these infants generally do not survive for more than a few weeks after birth, they show a relatively intact array of infantile reflexes, including flexor and extensor reflexes, stepping reflexes, and a wide range of brainstem reflexes including tonic neck reflexes and suckling reflexes, among others.

It is worthy to note that brainstem neurobehavioral substrates do not entail a mere assemblage of rigidly regulated and tightly organized reflex networks, as both decerebrate animals (Norman, et al. 1977; Mauk and Thompson 1987) and humans (Tuber et al. 1980; Berntson et al. 1983) have been shown to display neural plasticity and associative learning.

Among the more thoroughly studied of brainstem evaluative processes are those supporting approach–avoidance action dispositions related to taste hedonics. Similar to the organization of the spinal cord, the neuroarchitecture underlying approach and avoidance dispositions appears to be relatively independent and under separate control in brainstem circuitry (Berntson et al. 1993; Berridge and Grill 1984; Steiner et al. 2001). Taste hedonics and associated intake/rejection responses offer a prime example of brainstem evaluative systems. Orofacial displays to taste, represented by stereotyped, reflex-like negative rejection/ejection responses to aversive stimuli (gaping, tongue protrusion) and positive intake responses (smiling, licking, swallowing), are well conserved in mammals. Such responses can be seen early in development and are readily apparent in decerebrate organisms. The positive and negative responses to gustatory stimuli mirror the evaluative reflexes of the spinal cord in that they reflect opposing patterns of approach/avoidance dispositions. Similar to spinal reflexes, the behavioral output of these systems cannot be interpreted as lying along a single bipolar continuum extending from approach (highly positive) to avoidance (highly negative). Although this depiction can be useful, it belies the underlying complexity of hedonic processes as experimental evidence suggests that gustatory approach/withdrawal systems are partially independent and do not converge on a single hedonic integrator (Berridge and Grill 1984).

Just as one can tighten extensor and flexor muscles simultaneously, intake and rejection responses are not incompatible and can become coactive. For example, although the probability of rejection responses to a glucose solution increases following the addition of a bitter compound, this can occur without a reciprocal reduction in probability of intake responses. Similarly, increasing both bitter and sweet concurrently leads to increases in both intake and rejection responses (Berridge and Grill 1984). Thus, it is clear that taste preference, as measured by behavioral consumption and represented on a bipolar scale, does not always represent the underlying bivariate hedonic state. This does not rule out interaction between the approach/avoidance responses, of course, but suggests that the mixing positive and negative valences of hedonic stimuli do not simply yield a null average of the two or a state of indifference (Berridge and Grill 1984).

Gustatory approach/avoidance responses are represented by distinct positive and negative hedonic dimensions which conform to the positivity offset and negativity bias as described above. Gustatory evaluative processes mediated by brainstem systems are more complex than their behavioral output (total intake continuum), and knowledge of this fact facilitates a more accurate description of evaluative processes based on the underlying bivariate substrates.

Levels of Function: Higher-Level Representations

As one moves to the highest levels of the neuraxis, the re-representation and elaboration of evaluative processes becomes evermore apparent, and neuron-organizational complexity expands dramatically. The brainstem and spinal cord are highly sensitive to aversive and hedonic stimuli and can yield appropriate behavioral responses, but this so-called reptilian brain (MacLean 1985) lacks much of the behavioral flexibility and adaptability characteristic of intact organisms. Although decerebrates may ingest palatable foods, they do not display typical goal-seeking behavior in the absence of a food stimulus but rather are prisoners of the momentary stimulus or environmental context (see Berntson et al. 1993; Berntson and Micco 1976). Decerebrate's lack of the flexibility and variety of behavior seen in intact animals results from the devolution of the nervous system to its more primitive representations. It is not until the development of the paleomammalian brain (limbic system and archicortex) and the neomammalian brain (neocortex) that we see the full evolution and elaboration of evaluative processes (MacLean 1985). It is with the development of rostral brain structures that one begins to see the emergence of goal-directed behaviors that reflect anticipatory processes and expectancies that liberate the organism from the immediate exigencies of this stimulus or that.

In view of the expanding complexity of rostral evaluative substrates, it seems unlikely that these networks would simplify from the basic bivariate evaluative structure of lower substrates to become a single bipolar hedonic integrator. In contrast, with the expanding cognitive and computational complexity of evaluative processes at higher neuraxial levels, there is a parallel expansion of the complexity of the underlying mediating neural systems. Higher evaluative processes entail planning, strategizing, and anticipatory processes that can require access to associative networks, attentional and computational resources, etc. Moreover, while lower evaluative substrates may entail simple approach/withdrawal dispositions, higher motivational processes become further differentiated and nuanced. Berridge (1996) characterizes the "liking" aspects of motivation as those which entail the hedonic and response eliciting properties of a stimulus or motivational context. These are apparent in the orofacial intake/ingestive responses to positive hedonic tastes as described above for the decerebrate organism. The decerebrate, however, largely lacks what Berridge terms the "wanting" aspects of motivation, which entail an attentional focus on and goal-seeking behaviors directed toward a desired stimuli, state, or context. This latter aspect of evaluative processes is heavily dependent on the increased computational capacity of higher levels of the neuraxis and is mediated by more elaborate neural circuitry.

It should not be surprising that the neuroarchitecture of higher evaluative processes entails more complex and distributed networks, which are not as readily dichotomized into positive and negative substrates as is the case with lower-level representations. Indeed, many computational, attentional, and memorial processes may be commonly deployed whether for positive and negative evaluative processing. Moreover, the further development and elaboration of evaluative systems, such as that between "liking" and "wanting," may entail added neuroanatomical complexity. Historically, the nucleus accumbens (nACC) has been depicted as a neural integrator of reward and positive hedonic states (Hoebel et al. 1999; Koob 1992; Berridge and Grill 1984). In the 1940s, Robert Heath, working on psychiatric patients with indwelling electrical brain stimulators, showed that patients would report pleasurable states and would self-administer stimulation to various brain regions, especially areas in and around the nACC (Heath 1972). More recently, electrical stimulation of the nACC has been reported to elicit a smile, associated with euphoric responses (Okun et al. 2004). It is now clear that nearly all rewarding stimuli or positive hedonic states are associated with dopamine release in the nACC, and lesions or blockade of dopamine receptors in the nACC reduces rewards and positive hedonics (Hoebel et al. 1999; Robinson and Berridge 2003; Wise 2006). In this regard, the nACC contrasts with the amygdala, which has generally been implicated in fear conditioning, negative affect, and aversive states (see Bush et al. 2009), a topic to which we will return below.

Although these findings are consistent with a differentiation of positive and negative neural substrates at higher levels of the neuraxis, similar to that seen at lower levels, there are added complexities in higher substrates. The nACC, in fact, may not be a simple monolithic reward integrator. Recent work has suggested important phenomenological and computational distinctions within the nACC. For example, it has been shown that the "liking" (positive hedonic effect, reward) and "wanting" (incentive salience, goal-striving) aspects of hedonic states are mediated by distinct anatomical regions of the nACC (Berridge 1996; Pecina et al. 2006). Moreover, negative stimuli may also activate the nACC, and other distinct areas may be involved in suppression of negative evaluative processing (Pecina et al. 2006). These complexities caution against the overly simplistic ascription of discrete neural loci to the mediation complex neuropsychological phenomena. Nevertheless, there remain clear differentiations between higher neural substrates mediating positive and negative evaluative processes.

A hemispheric lateralization of positive and negative evaluative processes has been reported, with the right hemisphere implicated more in negative affective processing or avoidance dispositions and the left hemisphere more involved in positive affect or approach dispositions (Cacioppo and Gardner 1999; Davidson 1990; Harmon-Jones et al. 2004). For example, positive affective stimuli induce greater activation in the left hemisphere (Canli et al. 1998; Davidson 1998, 2004; Lee et al. 2004; Nitschke et al. 2006; Pizzagalli et al. 2005), and patients with damage to the left hemisphere have a higher probability of experiencing depression and overall negative affect (Davidson 1998). Similarly, facial expression and reaction time data suggest a left hemisphere predominance for positive affect and a greater right hemisphere representation for negative affect (Davidson 2004; Root et al. 2006). The relative right hemispheric bias for withdrawal/avoidance reactions may be related to the right lateralization of visceral/nociceptive afferent projections (Craig 2005) and is consistent with the finding that stimulation of the left insula gives rise to parasympathetic cardiac activation whereas right insula stimulation induces sympathetic activation (Oppenheimer 1993, 2006).

Further, within-hemisphere differentiation is also apparent in cortical representations. Pleasantness rating of odors, for example, is related to the degree of medial orbitofrontal activation as measured by fMRI, whereas unpleasantness was more related to activation of the dorsal anterior cingulate (Grabenhorst et al. 2007). Similarly, deciding on the lesser of two punishments yielded greater activation in the dorsal anterior cingulate, whereas deciding between the larger of two rewards yielded greater activation in the ventromedial prefrontal cortex (Blair et al. 2006).

The amygdala is a structure that has been especially implicated in fear and negative affect, since the classic studies of Walter Rudolf Hess on brain stimulation in the waking animal (Hess 1954). The amygdala appears to be a critical nodal point in subcortical circuits that allow for rapid detection and response to threat and for the learning of fear-related cues (Ledoux 1996; Öhman and Mineka 2001). These circuits allow more elaborated processing of threat-related cues than do lower-level brainstem substrates but remain highly efficient, as they can operate without the need for extensive cortical processing (Tooby and Cosmides 1990; Larson et al. 2006; Ledoux 1996; Öhman and Mineka 2001). Although the amygdala may also participate in classical thalamo-cortical-limbic circuits, direct thalamo-amygdaloid pathways are a sufficient substrate for fear reactions and simple fear conditioning, providing for a "quick and dirty transmission route" (LeDoux 2000). The thalamo-amygdaloid subcortical circuit may support simple fear conditioning and fear reactions in the absence of awareness ("blindsight") following visual cortical injuries (see Weiskrantz 1986; De Gelder et al. 1999; Pegna et al. 2005). In contrast, relational learning (e.g., contextual conditioning) and the processing of more complex threat-related cues may be more dependent on higher-level cortical processing (Berntson et al. 1998; see Bush, and LeDoux 2009). Recent research supports this heterarchical organization as auditory fear conditioning induces plasticity in amygdala neurons prior to apparent changes in cortical areas suggesting that early plasticity in amygdala neurons results from direct thalamoamygdala projections (Quirk et al. 1995; Öhman and Mineka 2001). The more direct, efficient, but relatively limited direct thalamo-amygdaloid and the more elaborated, integrative, and flexible thalamo-cortical-amygdaloid circuits represent distinct heterarchical levels of processing.

Biological Influences on Social Processes: Examples from Oxytocin and Loneliness Research

Just as the neural representations of social interaction do not reside in any single neural structure, the neurochemistry mediating such properties is similarly diverse. A variety of endogenous compounds, for example, play an important role in regulating social behavior. The ability to recognize familiar individuals is a necessity for functioning in society and requires an additional level of processing for social cues. The nonapeptide oxytocin has been consistently implicated in species typical social and reproductive behaviors (Carter et al. 1995; Insel and Young 2000). Oxytocin and the structurally similar vasopressin are necessary for social recognition within social rodents (Dantzer et al. 1987). Additionally, central administration of oxytocin can induce maternal behavior in virgin rats (Pedersen et al. 1982) and increases social interaction between adults (Witt et al. 1992). In fact, the demonstrated role of oxytocin and vasopressin in the social bonds of Microtine rodents (voles) has become a particularly useful model into the nature of social behavior. The highly social prairie vole forms enduring bonds that can last for the lifetime of the individual (Getz et al. 1993). In contrast, the behavior of the closely related Montane Vole is characterized by a relatively asocial existence, forming no long-lasting social bonds. Pharmacological studies have demonstrated that the neuropeptides oxytocin and vasopressin play critical roles in establishing the social bonds between mates of the prairie vole (Winslow et al. 1993; Williams et al. 1994). Subsequent studies suggest that the species differences in behavior are related to individual differences in the expression pattern of oxytocin and vasopressin receptors within the brain (Insel and Shapiro 1992; Ross et al. 2009). Specifically, oxytocin and vasopressin receptors are concentrated within components of the mesolimbic dopamine reward circuitry of the pair-bonding prairie vole, but not of the solitary voles. It is this difference that appears to underlie in part the species differences in behavior.

Over the past decade, the role of oxytocin in human social behavior has received increasing attention, as it has been shown that the oxytocinergic system is remarkably sensitive to social interactions. Early childhood neglect impairs the increase in peripheral OT levels typically triggered by mother–infant interaction (Fries et al. 2005), and early parental separation may alter central OT receptor sensitivity (Meinlschmidt and Heim 2007). Conversely, positive social interactions, such as social support, lead to increased levels of OT (Heinrichs et al. 2003), lending weight to the suggestion that OT may mediate the well-known beneficial effects of social support on stress responsivity and health (Uvnas-Moberg 1998). From experimental work in animals, the effects of OT on affect are thought to be a consequence of its effects within particular limbic structures including the hypothalamus, bed nucleus of the stria terminalis, and amygdala (Lim et al. 2005). Within humans, intranasal OT attenuates amygdala activity and reduces amygdala-brainstem coupling in response to threatening social stimuli (Kirsch et al. 2005). Similarly, oxytocin modulates the evaluation of socially relevant faces by influencing the activity of the amygdala as well as the fusiform gyrus (Domes et al. 2007b; Petrovic et al. 2008). The biological effects of OT on neural structures translate into observable changes in behavior. Intranasal oxytocin increases trust, independent of risk taking

(Kosfeld et al. 2005), and improves the ability to correctly infer the mental state of others (Domes et al. 2007a, b). Additionally, exogenous OT increases the number and duration of gazes toward the eye region of neutral faces (Guastella et al. 2008), an important behavior for the recognition of emotion in social interactions (Adolphs and Tranel 2003). Oxytocin receptor gene (OXTR) variants have also been associated with loneliness (Lucht et al. 2009). Taken together, the decades of research on the role of OT in rodents coupled with recent development in human studies suggest that OT modulates social perception, social cognition, and social behavior and thereby possibly promoting social approach and affiliation.

Ancestral humans who were inclined to form social connections, communicate and work together, share food and defense, and retaliate in the face of violations of reciprocity norms had a selective advantage to survive and pass on their genes (Cacioppo et al. 2006; Cacioppo and Patrick 2008). Indeed, it is only by operating within highly complex social networks that humans have become such a formidable species. This has likely conferred a powerful evolutionary pressure for social bonding and affiliation.

When an individual perceives that their particular social needs are not adequately met, a complex set of feelings (e.g., loneliness) are enacted which serve to drive that individual to seek the fulfillment of these needs (Russell et al. 1980; Weiss 1973). In line with Jackson's view on evolutionary re-representation, the motivational and affective aspects of loneliness may operate, in part, through the co-option and activation of more primitive aversive (pain) and appetitive (reward) systems (Cacioppo and Hawkley 2003; Cacioppo et al. 2006). By modulating systems already highly adept at promoting or preventing particular behaviors, loneliness may motivate individuals to avoid the phylogenetically and ontogenetically dangerous state of social isolation while the social reward of connecting with others promotes individuals to repair and maintain social connections. Thus, it is not surprising that social rejection, or even the threat of such, is perceived as a highly aversive event and activates localized portions of the anterior cingulate gyrus, a structure that is also implicated pain (Vogt and Sikes 2008). Additionally, feelings of loneliness tend to activate primate survival mechanisms that serve to heighten the sensitivity to diverse threats. Thus, the perception of social isolation may lead to more than transient variations in hedonic states—it appears to promote a generalized increase in the perception of other threats (Cacioppo and Hawkley 2003). In order to protect against an increasingly threatening world, lonely individuals tend to display defensive behaviors such as preventative rejection of others that may temporarily serve to fend off treachery, rejections, or attacks (Cacioppo and Hawkley 2009). Indeed, humans are particularly sensitive to threats originating from the social world (e.g., ostracism), as they can have rather large influences on ones survival and thus reproductive fitness. Although perhaps understandable from a broad evolutionary perspective, in contemporary society, these reactions may be maladaptive. An important goal of social neuroscience is to enhance our understanding of the origins, mechanisms, and processes underlying social behavior. This is important both from a basic science perspective and for its insights and applications in optimizing human potential.

Social Psychological Influences on Biological Processes

Although much of the above discussion has emphasized biological influences on social processes, the latter can also powerfully impact the former. A central tenant of social neuroscience is that reference to any single level of analysis is insufficient to describe phenomena that operate across multiple levels. While all human behavior derives from biological processes, unidirectional reductionism often fails to provide a simple, singular, or satisfactory explanation for complex interactions at molecular levels, let alone the behavior of complex social organisms. Indeed, molar constructs, such as those routinely employed by social and behavioral scientists, provide a means of organizing and structuring highly complex activity that not only permits but informs and directs investigations into its biological underpinnings. Moreover, even after biological processes are understood, concepts derived from higher levels of analysis (e.g., psychological) may continue to provide a more efficient description and understanding of social phenomena than the biological. The added value of the multilevel analysis is not achieved through simple unidirectional reductionism but through the refining and calibration of constructs operating at one level of analysis (e.g., social cognition) using knowledge derived from different levels (e.g., genetic contribution to social cognition, societal influences on the cognitive representation of the social world). This is what has been referred to as reciprocal or calibrative reductionism (Berntson et al. 2012). In contrast to a process of reduction to progressively lower levels of organization, reciprocal reductionism seeks a mutual calibration and refinement in concepts and understandings across levels. Through this process, biological concepts and mechanisms will not likely look like the biology of today.

In everyday life, the employment of psychological constructs to explain and predict behavior is fundamental to our existence in a complex social environment. These very constructs are what allow us to perceive the emotions and intentions of others without having to decipher the countless neuronal interactions occurring within their brain. Our recent ability to study and deconstruct these very neuronal processes has allowed for the discovery of numerous important findings on how social organisms perceive the social environment they inhabit. However, this new found ability is far from a sign of the impending demise of psychology as a field as some have suggested, rather, it is an expansion of the tools and development of novel theoretical modeling now available to psychologists. Psychological constructs are composed of countless component processes that interact dynamically to form emergent representations that are subsequently given linguistic labels. Emergent properties appear from combinations or interactions among the elements that are not readily predictable from the known properties of the elements. From an epistemological perspective, the inability to predict the behavior of emergent systems means that in no meaningful sense could the lower-level properties (neuronal interactions) be said to "explain" or account for emergent properties. Thus, knowledge of underlying neuronal processes of social cognitive and affective states serves as a compliment, not a substitute, to the isolated study of social psychological processes.

The interaction between the social world and physical processes can be characterized by dynamic reciprocal interplay where all levels, form the molecular to the psychological, interact to form the emergent properties responsible for behavioral and physiological processes of everyday life.

Conclusion

Recent developments in biological and social levels of analysis have made it increasingly clear that multilevel, cross-disciplinary research contributes to our understanding of the inherent complexities of human social processes. Utilization of this broad conceptual perspective has already facilitated the dismantling of the artificial academic constructions that have historically interfered with a multimethod, integrative study of human behavior and fostered the aggressively separate development of social and biological levels of analysis. Although the task of integrating fields as seemingly disparate as molecular biology and social psychology seems rather daunting, it is a task that must be accomplished if we are to gain a complete understanding of the human mind and behavior. The ultimate goal of social neuroscience is to promote meaningful reductionism and extensionism so that knowledge and constructs at multiple levels of organization and analysis can mutually inform, elucidate, and constrain theory and research at other levels. Importantly, this multilevel research agenda does not require that all research must be conducted across levels of analysis. There remains much to be discovered within the social and biological realms themselves. Rather, social neuroscience represents first and foremost a multidisciplinary perspective that guides the conceptual development and understandings of multilevel phenomena and offers the theoretical framework within which to integrate and bridge concepts and findings across levels of organization and analysis.

References

Adams, D. B. (1979). Brain mechanisms for offense, defense, and submission. *The Behavioral and Brain Sciences, 2*, 201–241.

Adolphs, R. (2009). The social brain: Neural basis of social knowledge. *Annual Review of Psychology, 60*, 693–716.

Adolphs, R., & Tranel, D. (2003). Amygdala damage impairs emotion recognition from scenes only when they contain facial expressions. *Neuropsychologia, 41*(10), 1281–1289.

Adolphs, R., Tranel, D., Damasio, H., & Damasio, A. (1994). Impaired recognition of emotion in facial expressions following bilateral damage to the human amygdala. *Nature, 372*(6507), 669–672.

Adolphs, R., Tranel, D., & Damasio, A. R. (1998). The human amygdala in social judgment. *Nature, 393*(6684), 470–474.

Bard, P., & Macht, M. B. (1958). The behavior of chronically decerebrate cats. In G. E. W. Wolsten-holme & C. M. O'Connor (Eds.), *Neurological basis of behavior* (pp. 55–75). London: Churchill.

Berntson, G. G., & Cacioppo, J. T. (2012). Reductionism. In L. L'Abate (Ed.), *The role of paradigms in theory construction*. New York: Springer-Science.

Berntson, G. G., & Micco, D. J. (1976). Organization of brainstem behavioral systems. *Brain Research Bulletin, 1*, 471–483.

Berntson, G. G., Boysen, S. T., & Cacioppo, J. T. (1993). Neurobehavioral organization and the cardinal principle of evaluative bivalence. *Annals of the New York Academy of Sciences, 702*, 75–102.

Berntson, G. G., Bechara, A., Damasio, H., Tranel, D., & Cacioppo, J. T. (2007). Amygdala contribution to selective dimensions of emotion. *Social Cognitive and Affective Neuroscience, 2*(2), 123–129.

Berntson, G. G., Norman, G. J., & Cacioppo, J. T. (2012). Evolution of neuroarchitecture, multilevel analyses and calibrative reductionism. *Interface Focus, 2*(1), 65–73.

Berntson, G. G., Sarter, M., & Cacioppo, J. T. (1998). Anxiety and cardiovascular reactivity: The basal forebrain cholinergic link. *Behavioural Brain Research, 94*, 225–248.

Berntson, G. G., Tuber, D. S., Ronca, A. E., & Bachman, D. S. (1983). The decerebrate human: Associative learning. *Experimental Neurology, 81*, 77–88.

Berridge, K. C. (1996). Food reward: Brain substrates of wanting and liking. *Neuroscience Biobehavioral Review, 20*, 1–25.

Berridge, K. C. (2004). Motivation concepts in behavioral neuroscience. *Physiology & Behavior, 81*(2), 179–209.

Berridge, K. C., & Grill, H. J. (1984). Idohedonic tastes support a two-dimensional hypothesis of palatability. *Appetite, 5*, 221–231.

Blair, K., Marsh, A. A., Morton, J., Vythilingam, M., Jones, M., Mondillo, K., Pine, D. C., Drevets, W. C., & Blair, J. R. (2006). Choosing the lesser of two evils, the better of two goods: Specifying the roles of ventromedial prefrontal cortex and dorsal anterior cingulate in object choice. *Journal of Neuroscience, 26*(44), 11379–11386.

Brothers, L. (1990). The social brain: A project for integrating primate behaviour and neurophysiology in a new domain. *Concepts in Neuroscience, 1*, 27–51.

Bush, D. E. A., Schafe, G. E., & LeDoux, J. E. (2009). Neural basis of fear conditioning. In G. G. Berntson & J. T. Cacioppo (Eds.), *Handbook of neuroscience for the behavioral sciences* (Vol. 2, pp. 762–764). Hoboken: Wiley.

Cacioppo, J. T., & Berntson, G. G. (1992). Social psychological contributions to the decade of the brain. Doctrine of multilevel analysis. *The American Psychologist, 47*(8), 1019–1028.

Cacioppo, J. T., & Berntson, G. G. (1999). The affect system: Architecture and operating characteristics. *Current Directions in Psychological Science, 8*, 133–137.

Cacioppo, J. T., & Gardner, W. L. (1999). Emotion. *Annual Review of Psychology, 50*, 191–214.

Cacioppo, J. T., & Hawkley, L. C. (2003). Social isolation and health, with an emphasis on underlying mechanisms. *Perspectives in Biology and Medicine, 46*(3 Suppl), S39–52.

Cacioppo, J. T., & Hawkley, L. C. (2009). Perceived social isolation and cognition. *Trends in Cognitive Sciences, 13*, 447–454.

Cacioppo, J. T., & Patrick, B. (2008). *Loneliness: Human nature and the need for social connection*. New York: W. W. Norton and Company.

Cacioppo, J. T., Lorig, T. S., Nusbaum, H. C., & Berntson, G. G. (2004). Social neuroscience: Bridging social and biological systems. In C. C. Morf, T. Panter, & C. Sansone (Eds.), *The sage handbook of methods in social psychology*. Thousand Oaks: Sage.

Cacioppo, J. T., Hawkley, L. C., Ernst, J. M., Burleson, M., Berntson, G. G., Nouriani, B., & Spiegel, D. (2006). Loneliness within a nomological net: An evolutionary perspective. *Journal of Research in Personality, 40*, 1054–1085.

Cacioppo, J. T., Larsen, J. T., Smith, N. K., & Berntson, G. G. (in press). The affect system: What lurks below the surface of feelings? In A. S. R. Manstead, N. H. Frijda, & A. H. Fischer (Eds.), *Feelings and emotions: The Amsterdam conference*. New York: Cambridge University Press.

Canli, T., Desmond, J. E., Zhao, Z., Glover, G., & Gabrieli, J. D. (1998). Hemispheric asymmetry for emotional stimuli detected with fMRI. *Neuroreport, 9*, 3233–3239.

Carter, C. S., DeVries, A. C., & Getz, L. L. (1995). Physiological substrates of mammalian monogamy: The prairie vole model. *Neuroscience and Biobehavioral Reviews, 19*(2), 303–314.

Craig, A. D. (2005). Forebrain emotional asymmetry: A neuroanatomical basis? *Trends in Cognitive Science, 9*(12), 566–571.

Dantzer, R., Bluthe, R. M., Koob, G. F., & Le Moal, M. (1987). Modulation of social memory in male rats by neurohypophyseal peptides. *Psychopharmacology, 91*(3), 363–368.

Davidson, R. J. (1990). Approach-withdrawal and cerebral asymmetry: Emotional expression and brain physiology. *Journal of Personality and Social Psychology, 58*(2), 330–341.

Davidson, R. J. (1998). Anterior electrophysiological asymmetries, emotion, and depression: Conceptual and methodological conundrums. *Psychophysiology, 35*, 607–614.

Davidson, R. J. (2004). What does the prefrontal cortex "do" in affect: Perspectives on frontal EEG asymmetry research. *Biological Psychology, 67*, 219–233.

De Gelder, B., Vroomen, J., Pourtois, G., & Weiskrantz, L. (1999). Non-conscious recognition of affect in the absence of striate cortex. *Neuroreport, 10*, 3759–3763.

Domes, G., Heinrichs, M., Glascher, J., Buchel, C., Braus, D. F., & Herpertz, S. C. (2007a). Oxytocin attenuates amygdala responses to emotional faces regardless of valence. *Biological Psychiatry, 62*(10), 1187–1190.

Domes, G., Heinrichs, M., Michel, A., Berger, C., & Herpertz, S. C. (2007b). Oxytocin improves "mind-reading" in humans. *Biological Psychiatry, 61*(6), 731–733.

Dunbar, R. I. (2009). The social brain hypothesis and its implications for social evolution. *Annals of Human Biology, 36*(5), 562–572.

Dunbar, R. I., & Shultz, S. (2007). Evolution in the social brain. *Science, 317*(5843), 1344–1347.

Eisenberger, N. I., Lieberman, M. D., & Williams, K. D. (2003). Does rejection hurt? An FMRI study of social exclusion. *Science, 302*(5643), 290–292.

Egan, L. C., Santos, L. R., & Bloom, P. (2007). The origins of cognitive dissonance: Evidence from children and monkeys. *Psychological Science, 18*, 978–983.

Fries, A. B., Ziegler, T. E., Kurian, J. R., Jacoris, S., & Pollak, S. D. (2005). Early experience in humans is associated with changes in neuropeptides critical for regulating social behavior. *Proceedings of the National Academy of Sciences of the United States of America, 102*(47), 17237–17240.

Getz, L. L., McGuire, B., Pizzuto, T., Hofmann, J. E., & Frase, B. (1993). Social organization of the prairie vole (Microtus ochrogaster). *Journal of Mammalogy, 74*, 44–58.

Grabenhorst, F., Rolls, E. T., Margot, C., da Silva, M. A., & Velazco, M. I. (2007). How pleasant and unpleasant stimuli combine in different brain regions: Odor mixtures. *Journal of Neuroscience, 27*, 13532–13540.

Guastella, A. J., Mitchell, P. B., & Dadds, M. R. (2008). Oxytocin increases gaze to the eye region of human faces. *Biological Psychiatry, 63*, 3–5.

Harmon-Jones, E., Vaughn, K., Mohr, S., Sigelman, J., & Harmon-Jones, C. (2004). The effect of manipulated sympathy and anger on left and right frontal cortical activity. *Emotion, 4*, 95–101.

Harris, R. B., Kelso, E. W., Flatt, W. P., Bartness, T. J., & Grill, H. J. (2006). Energy expenditure and body composition of chronically maintained decerebrate rats in the fed and fasted condition. *American Journal of Physiology, 147*(3), 1365–1376.

Heath, R. G. (1972). Pleasure and brain activity in man. Deep and surface electroencephalograms during orgasm. *The Journal of Nervous and Mental Disease, 154*(1), 3–18.

Heinrichs, M., Baumgartner, T., Kirschbaum, C., & Ehlert, U. (2003). Social support and oxytocin interact to suppress cortisol and subjective responses to psychosocial stress. *Biological Psychiatry, 54*(12), 1389–1398.

Hess, W. R. (1954). *Diencephalon: Autonomic and extrapyramidal functions* (Monographs in biology and medicine, Vol. III). New York: Grune and Stratton.

Ho, T. C., Brown, S., & Serences, J. T. (2009). Domain general mechanisms of perceptual decision making in human cortex. *Journal of Neuroscience, 29*, 8675–8687.

Hoebel, B. G., Rada, P. V., Mark, G. P., & Pothos, E. N. (1999). Neural systems for reinforcement and inhibition of behavior: Relevance to eating, addiction, and depression. In D. Kahneman, E. Diener, & N. Schwarz (Eds.), *Well-being: The foundations of hedonic psychology*. New York: Russell Sage Foundation.

Insel, T. R., & Shapiro, L. E. (1992). Oxytocin receptor distribution reflects social organization in monogamous and polygamous voles. *Proceedings of the National Academy of Sciences of the United States of America, 89*(13), 5981–5985.

Insel, T. R., & Young, L. J. (2000). Neuropeptides and the evolution of social behavior. *Current Opinion in Neurobiology, 10*(6), 784–789.

Jackson, J. H. (Ed.). (1884). *Evolution and dissolution of the nervous system (Croonian lectures)*. New York: Basic Books.

Kirsch, P., Esslinger, C., Chen, Q., Mier, D., Lis, S., Siddhanti, S., et al. (2005). Oxytocin modulates neural circuitry for social cognition and fear in humans. *The Journal of Neuroscience, 25*(49), 11489–11493.

Koob, G. F. (1992). Drugs of abuse: Anatomy, pharmacology, and function of reward pathways. *Trends in Pharmacological Sciences, 13*, 177–184.

Kosfeld, M., Heinrichs, M., Zak, P. J., Fischbacher, U., & Fehr, E. (2005). Oxytocin increases trust in humans. *Nature, 435*(7042), 673–676.

Larson, C. L., Schaefer, H. S., Siegle, G. J., Jackson, C. A., Anderle, M. J., & Davidson, R. J. (2006). Fear is fast in phobic individuals: Amygdala activation in response to fear-relevant stimuli. *Biological Psychiatry, 60*(4), 410–417.

LeDoux, J. E. (1996). *The emotional brain: The mysterious underpinnings of emotional life*. New York: Simon and Schuster.

LeDoux, J. (2000). Emotion circuits in the brain. *Annual Review of Neuroscience, 23*(1), 155–184.

Lee, G. P., Meador, K. J., Loring, D. W., Allison, J. D., Brown, W. S., Paul, L. K., Pillai, J. J., & Lavin, T. B. (2004). Neural substrates of emotion as revealed by functional magnetic resonance imaging. *Cognitive and Behavioral Neuroscience, 17*, 9–17.

Lim, M. M., Bielsky, I. F., & Young, L. J. (2005). Neuropeptides and the social brain: Potential rodent models of autism. *International Journal of Developmental Neuroscience, 23*(2–3), 235–243.

Lucht, M. J., Barnow, S., Sonnenfeld, C., Rosenberger, A., et al. (2009). Associations between the oxytocin receptor gene (OXTR) and affect, loneliness, and intelligence in normal subjects. *Progress in Neuro-Psychopharmacology & Biological Psychiatry, 33*(5), 860–866.

Lundberg, A. (1979). Multisensory control of spinal reflex pathways. In R. Granit & O. Pomeiano (Eds.), *Reflex control of posture and movement* (pp. 11–28). Amsterdam: Elsevier.

Mauk, M., & Thompson, R. F. (1987). Retention of classically conditioned eyelid responses following acute decerebration. *Brain Research, 403*, 89–95.

Meinlschmidt, G., & Heim, C. (2007). Sensitivity to intranasal oxytocin in adult men with early parental separation. *Biological Psychiatry, 61*(9), 1109–1111.

Nitschke, J. B., Sarinopoulos, I., Mackiewicz, K. L., Schaefer, H. S., & Davidson, R. J. (2006). Functional neuroanatomy of aversion and its anticipation. *NeuroImage, 29*, 106–116.

Norman, R. J., Buchwald, J. S., & Villablanca, V. J. (1977). Classical conditioning with auditory discrimination of the eye blink in decerebrate cats. *Science, 196*, 551–553.

Öhman, A., & Mineka, S. (2001). Fears, phobias, and preparedness: Toward an evolved module of fear and fear learning. *Psychological Review, 108*, 438–522.

Okun, M. S., Bowers, D., Springer, U., Shapira, N. A., Malone, D., & Rezai, A. R. (2004). What's in a 'smile?' Intra-operative observations of contralateral smiles induced by deep brain stimulation. *Neurocase, 10*, 271–279.

Oppenheimer, S. (1993). The anatomy and physiology of cortical mechanisms of cardiac control. *Stroke, 24*, 13–15.

Oppenheimer, S. M. (2006). Cerebrogenic cardiac arrhythmias: Cortical lateralization and clinical significance. *Clinical Autonomic Research, 16*, 6–11.

Pecina, S., Smith, K. S., & Berridge, K. C. (2006). Hedonic hot spots in the brain. *The Neuroscientist, 12*, 500–511.

Pedersen, C. A., Ascher, J. A., Monroe, Y. L., & Prange, A. J., Jr. (1982). Oxytocin induces maternal behavior in virgin female rats. *Science, 216*(4546), 648–650.

Pegna, A. J., Khateb, A., Lazeyras, F., & Seghier, M. L. (2005). Discriminating emotional faces without primary visual cortices involves the right amygdala. *Nature Neuroscience, 8*(1), 24–25.

Petrovic, P., Kalisch, R., Singer, T., & Dolan, R. J. (2008). Oxytocin attenuates affective evaluations of conditioned faces and amygdala activity. *The Journal of Neuroscience, 28*(26), 6607–6615.

Pizzagalli, D. A., Sherwood, R. J., Henriques, J. B., & Davidson, R. J. (2005). Frontal brain asymmetry and reward responsiveness: A source-localization study. *Psychological Science, 16*, 805–813.

Quirk, G. J., Repa, C., & LeDoux, J. E. (1995). Fear conditioning enhances short-latency auditory responses of lateral amygdala neurons: Parallel recordings in the freely behaving rat. *Neuron, 15*(5), 1029–1039.

Robinson, T. E., & Berridge, K. C. (2003). Addiction. *Annual Review of Psychology, 54*, 25–53.

Ronca, A. E., Berntson, G. G., & Tuber, D. A. (1986). Cardiac orienting and habituation to auditory and vibrotactile stimuli in the infant decerebrate rat. *Developmental Psychobiology, 18*, 79–83.

Root, J. C., Wong, P. S., & Kinsbourne, M. (2006). Left hemisphere specialization for response to positive emotional expressions: A divided output methodology. *Emotion, 6*, 473–483.

Ross, H. E., Freeman, S. M., Spiegel, L. L., Ren, X., Terwilliger, E. F., & Young, L. J. (2009). Variation in oxytocin receptor density in the nucleus accumbens has differential effects on affiliative behaviors in monogamous and polygamous voles. *The Journal of Neuroscience, 29*(5), 1312–1318.

Russell, D., Peplau, L. A., & Cutrona, C. E. (1980). The revised UCLA loneliness scale: Concurrent and discriminant validity evidence. *Journal of Personality and Social Psychology, 39*, 472–480.

Sandrini, G., Serrao, M., Rossi, P., Romaniello, A., Cruccu, G., & Willer, J. C. (2005). The lower limb flexion reflex in humans. *Progress in Neurobiology, 77*, 353–395.

Schiller, L., Jahkel, M., Kretzschmar, M., Brust, P., & Oehler, J. (2003). Autoradiographic analyses of 5-HT1A and 5-HT2A receptors after social isolation in mice. *Brain Research, 980*(2), 169–178.

Schouenborg, J., Holmberg, H., & Weng, H. R. (1992). Functional organization of the nociceptive withdrawal reflexes. II. Changes of excitability and receptive fields after spinalization in the rat. *Experimental Brain Research, 90*, 469–478.

Scott, T. R. (1991). A personal view of the future of psychology departments. *The American Psychologist, 46*, 975–976.

Sherrington, C. S. (1906). *The integrative action of the nervous system*. New Haven: Yale University Press.

Steiner, J. E., Glaser, D., Hawilo, M. E., & Berridge, K. C. (2001). Comparative expression of hedonic impact: Affective reactions to taste by human infants and other primates. *Neuroscience and Biobehavioral Reviews, 25*, 53–74.

Tooby, J., & Cosmides, L. (1990). The past explains the present: Emotional adaptations and the structure of ancestral environment. *Ethology and Sociobiology, 11*, 375–424.

Tuber, D. S., Berntson, G. G., Bachman, D. S., & Allen, J. N. (1980). Associative learning in premature hydranencephalic and normal twins. *Science, 210*(4473), 1035–1037.

Tranel, D., Gullickson, G., Koch, M., & Adolphs, R. (2006). Altered experience of emotion following bilateral amygdala damage. *Cognitive Neuropsychiatry, 11*(3), 219–232.

Uvnas-Moberg, K. (1998). Oxytocin may mediate the benefits of positive social interaction and emotions. *Psychoneuroendocrinology, 23*(8), 819–835.

Vogt, B. A., & Sikes, R. W. (2008). Cingulate nociceptive circuitry and roles in pain processing: The cingulate premotor pain model. In B. A. Vogt (Ed.), *Cingulate neurobiology & disease* (Infrastructure, diagnosis, treatment, Vol. 1). Oxford: Oxford University Press.

Wakana, S., Jiang, H., Nagae-Poetscher, L. M., van Zijl, P. C., & Mori, S. (2004). Fiber tract-based atlas of human white matter anatomy. *Radiology, 230*(1), 77–87.

Weiskrantz, L. (1986). *Blindsight: A case study and implications*. Oxford: Oxford University Press.

Weiss, R. S. (1973). *Loneliness: The experience of emotional and social isolation*. Cambridge, MA: MIT Press.

Williams, J. R., Insel, T. R., Harbaugh, C. R., & Carter, C. S. (1994). Oxytocin administered centrally facilitates formation of a partner preference in female prairie voles (Microtus ochrogaster). *Journal of Neuroendocrinology, 6*(3), 247–250.

Winslow, J. T., Hastings, N., Carter, C. S., Harbaugh, C. R., & Insel, T. R. (1993). A role for central vasopressin in pair bonding in monogamous prairie voles. *Nature, 365*(6446), 545–548.

Wise, R. A. (2006). Role of brain dopamine in food reward and reinforcement. *Philosophical Transactions of the Royal Society of Biological Sciences, 361*, 1149–1158.

Witt, D. M., Winslow, J. T., & Insel, T. R. (1992). Enhanced social interactions in rats following chronic, centrally infused oxytocin. *Pharmacology, Biochemistry, and Behavior, 43*(3), 855–861.

Yates, B. J., Jakus, J., & Miller, A. D. (1993). Vestibular effects on respiratory outflow in the decerebrate cat. *Brain Research, 3*(2), 209–217.

Chapter 7
Can the Two Cultures Reconcile? Reconstruction and Neuropragmatism

Tibor Solymoski

> *That character of everyday experience which has been most systematically ignored by philosophy is the extent to which it is saturated with the results of social intercourse and communication. Because this factor has been denied, meanings have either been denied all objective validity, or have been treated as miraculous extra-natural intrusions. If, however, language, for example, is recognized as the instrument of social cooperation and mutual participation, continuity is established between natural events (animal sounds, cries, etc.) and the origin and development of meanings. Mind is seen to be a function of social interactions, and to be a genuine character of natural events when these attain the stage of widest and most complex interaction with one another. Ability to respond to meanings and to employ them, instead of reacting merely to physical contacts, makes the difference between man and other animals; it is the agency for elevating man into the realm of what is usually called the ideal and spiritual. In other words, the social participation affected by communication, through language and other tools, is the naturalistic link which does away with the often alleged necessity of dividing the objects of experience into two worlds, one physical and one ideal.*
>
> – John Dewey, *Experience and Nature*, 1929 (LW1: 6–7)

> *Over-specialization and division of interests, occupations and goods create the need for a generalized medium of intercommunication, of mutual criticism through all-around translation from one separated region of experience into another. Thus philosophy as a critical organ becomes in effect a messenger, a liaison officer, making reciprocally intelligible voices speaking provincial tongues, and thereby enlarging as well as rectifying the meanings with which they are charged.*
>
> – John Dewey, *Experience and Nature*, 1925 (LW1: 306)

Neuroscience and the Humanities

As the word itself indicates, neurosociology permeates the traditional boundary between natural science and social science. This permeation is not too surprising, given the biological fact that humans are social animals and that our bodily makeup, including but not limited to the neural, manifests this

T. Solymoski (✉)
Department of Philosophy, University of Southern Illinois, Carbondale, IL 62901, USA
e-mail: tibor@neuropragmatism.com

D.D. Franks and J.H. Turner (eds.), *Handbook of Neurosociology*, Handbooks of Sociology and Social Research, DOI 10.1007/978-94-007-4473-8_7, © Springer Science+Business Media B.V. 2013

social orientation. The pattern of permeation appears to move along a spectrum, from the most general subject matter of physics to more specialized subsets, from chemistry then to biology, of which neuroscience is, in some sense, a further specialization. From particles to molecules to tissues and organ systems, onward to organisms and species, and finally to the sciences of a specific and relatively odd species, *H. sapiens sapiens* – a science of themselves performed by themselves – the sciences have progressed from the natural to the social. As the progress of permeation continues, the next logical step seems to be toward the individual, to the existential, and to the humanistic. Indeed, as popular neuroscience, from the writings of Sacks and Damasio to those of Ramachandran and Eagleman, illustrates, neuroscience has something to say about the human condition.

To the neurosociologist, such permeation is unsurprising; after all, societies are constituted by as much as they constitute individual persons – persons who have their very own nervous systems to boot. From the perspective of the sciences, the territory of the humanities is the next and indeed final area targeted for the expansion of science. Yet the relationship between these two cultures is so seemingly incompatible that expansion of one into the other can only be seen as an act of hostility: indeed, science's encroachment is invasive and often derogatory; the response of the humanities has been proactively defensive, as it has made countless attempts to assassinate its perceived conqueror – perhaps succeeding, at least in the American context, at fatally wounding the authority (or, more optimistically, *perceived* authority) of science.

Neurosociologists have a vested interest in how the cultures of science and the humanities relate to one another. Not only do they stand at a key position in what many conceive as a war, neurosociologists also hold in this position many promising tools for reaching rapprochement. These scientists, however, cannot do this work alone. Fortunately, they have allies in the humanities. These allies share both a contemporary interest in the exciting developments in neuroscience and a historical heritage in their philosophy.

As David Franks has shown, the American pragmatist and sociologist, George Herbert Mead has provided a strong philosophical framework for recent developments in the intersection of neuroscience and sociology (Franks 2010). Among Mead's close allies at the University of Chicago is the preeminent American pragmatist of the first half of the twentieth century, John Dewey. Dewey's conception of philosophy is both naturalistic and interdisciplinary. These two aspects of his thought provide a sound anchor point for addressing this significant cultural problem of the conflict between the humanities and the sciences. This problem rears its philosophical face in many ways. From the apparent limits of science to account for qualitative experience to normative concerns over the present course of civilization, the conflict between the sciences and the humanities is at root a social and educational problem over communication. This conflict is entailed by a larger philosophical question of how to reconcile the image of humans in the world put forth by science and the image put forth by the humanities (Sellars 1963). This question of reconciliation, however, has been treated as a primarily linguistic one, with little regard to the experiential contexts in which specialized language is developed, used, and modified. When the experiential context is considered, this conflict between science and the humanities is well seen as a cultural or social one. The problem of reconciliation is a consequence of inadequate representationalist theories of truth. Once experiential context is incorporated, as it is in the pragmatic theory of truth, the problem of reconciliation is overcome through the project of reconstruction.

To this day, C. P. Snow's classic work, *The Two Cultures*, continues to be the most cogent and popular articulation of the conflict. Jerome Kagan has revisited Snow's ideas in his recent book, *The Three Cultures: Natural Sciences, Social Sciences, and the Humanities in the 21st Century*. As the title suggests, little progress seems to have been made on reconciling the two cultures: indeed, the emergence of a third hardly denotes movement in the right direction. A novel approach to the relationship that does not share Kagan's pessimism is presented in Edward Slingerland's *What Science Offers the Humanities*. In a welcome manner, Slingerland advocates a pragmatist account of truth. However, his focus is on a general pragmatic theory that does not directly embark on the project of reconstruction.

I contend that, despite Kagan's pessimism for reconciliation, there is a plausible means of addressing the conflict between science and the humanities that comes from the work of John Dewey. Dewey's philosophical insights provide an underappreciated but nevertheless productive platform for addressing myriad problems based on unwarranted dualisms. It is from such a platform that I put forth a distinction between reconciliation and reconstruction. This distinction corresponds to the different notions of truth put forth by Kagan and Slingerland. So long as this distinction is neglected, rapprochement between specialized disciplines will remain out of reach. However, once a pragmatist framework, especially one informed by neuroscience, is established and recognized, the social conflict articulated by Snow as the two cultures may very well be resolved thereby opening up greater possibilities for human amelioration.

My argument proceeds as follows. First, I introduce Snow's conflict between the two cultures, from which Kagan's modifications are also introduced. By way of Slingerland, I turn to Dewey's distinction between reconciliation and reconstruction, in which the role of the philosopher as a liaison officer is further elucidated. I then consider recent developments in the neuroscience of mirror neuron systems and cognitive/motor systems more generally. I argue that these systems can serve as a common medium for science and the humanities for addressing questions about shared meanings, especially with regard to rapprochement between science and the humanities. I close by advocating a new metaphor for a science of consciousness (consciousness as cooking) that aims at bringing together both scientists and humanists.

Snow's Two Cultures

C. P. Snow is often recognized as the one who coined the term "two cultures," but we should not mistake him for discovering the tension within Western civilization that dates back to Plato and the sophists and poets. Writing at the middle of the twentieth century, Snow draws on his life experience as both a working scientist and a relatively successful novelist. He not only popularized the opposition between the intellectual literary folk and the scientists and engineers, he did so by abstracting from his own lived experience (Snow 1959: 2). Most striking to Snow was that despite immense similarities in race, class, income, and origin of members of both cultures, they were nevertheless living in vocational worlds entirely closed off from each other (ibid., 2). Snow saw the two cultures as not only speaking different languages but as living in "two galaxies," so extreme were the misperceptions each culture had for the other (ibid., 16). In his essay, Snow gave a telling anecdote of how immense the barrier between the humanities and the sciences had become, writing "Once or twice I have been provoked and have asked the company of [humanist intellectuals] how many of them could describe the Second Law of Thermodynamics. The response was cold: it was also negative. Yet I was asking something which is about the scientific equivalent of: *Have you read a work of Shakespeare?*" (ibid., 15–16).

The scientist sees himself, on Snow's view, as a realist motivated to investigate the natural world for the sake of his fellow man. The scientist sees the literary intellectual as self-absorbed, uncaring, and unconcerned with the practical difficulties of human life. The humanist, on Snow's view, sees himself as the champion of tradition, as the guardian of the intellectual, that is, the literary triumphs of civilization. Without such literary men, this side of the divide asserts, life is hollow, meaningless, and not worth living. Snow admonished members of both cultures for missing the creative moment of this clash of cultures to address the shared existential concerns over living well – a creative opportunity that no doubt would be mutually beneficial to each enterprise to boot (ibid., 16).

Snow speculated that the origin of the divide between the two cultures is multifarious. One integral reason he did consider is the nature of education as both poison and cure. The hyper-specialization of British education then and much education today has led to and continues to generate these separate cultures, separate languages, and separate worlds – indeed separate experiences. The remedy to this insularity, for Snow, resides in changing education so that it is both full of breadth and full of rigor (ibid., 50–51). Despite his reflective insight, Snow's articulation of the problems and its solutions are made in broad strokes. More detail and careful analysis is needed before any serious headway can be made. Unfortunately, since Snow's day, the conflict has only worsened in scope, as Jerome Kagan argues.

Kagan's Three Cultures

Kagan begins his book by emphasizing the consequences of Snow's work. Snow was controversial because he emphasized the ability of science to solve the human problems of his day, such as world hunger. Kagan describes how he was struck by the failure of science not only to complete such a task but of how radically changed the two cultures are since Snow's time. In the past 50 years, science has become a big enterprise that both consumes and generates significant resources. Because of this productivity, scientists have become confident, if not arrogant, especially in comparison with their humanist counterparts. This difference is reflected in the respect given by scientists both inside and outside the academy. The changes are particularly significant with the rise of the social sciences thanks to the development of methods distinct from those of the natural sciences. These significant differences in focus, practice, and stature motivated Kagan to reexamine Snow's two cultures for today.

The differences are illustrated in at least two ways in Kagan's book. First, there is simply the number of pages devoted to each culture. The chapter on "The Natural Sciences" receives 52 pages of attention, whereas there are two chapters of nearly equal length on the social sciences, totaling 117 pages. Only 22 pages are devoted to the humanities. This is especially telling in light of the concluding chapter that takes a rather pessimistic view of the current state of affairs for human civilization.

The second important means of distinguishing the three cultures is Kagan's nine dimensions of primary difference. These differences reflect various conceptions of truth along a spectrum from a valueless and objective realism to a value-laden, subjective idealism. They are as follows, in Kagan's own words (Kagan 2009: 2–3):

1. The primary questions asked, including the degree to which prediction, explanation, or description of a phenomenon is the major product of inquiry
2. The sources of evidence on which inferences are based and the degree of control over the conditions in which the evidence is gathered
3. The vocabulary used to present observations, concepts, and conclusions, including the balance between continuous properties and categories and the degree to which a functional relation was presumed to generalize across settings or was restricted to the context of observation
4. The degree to which social conditions, produced by historical events, influence the questions asked
5. The degree to which ethical values penetrate the questions asked and the conclusions inferred or deduced
6. The degree of dependence on external financial support from government or industry
7. The probability that the scholar works alone, with one or two others, or as a member of a large team
8. The contribution to the national economy
9. The criteria members of each group use when they judge a body of work as elegant or beautiful

I now take each of these in turn to summarize Kagan's judgments on how each culture matches up. Here I follow his extremely helpful (ibid., 4–5).

Primary interests: for natural scientists, the aim is the most general, the "[p]rediction and explanation of all natural phenomena." Social scientists bridge the gap between natural science and the humanities by specifying the general aim of prediction and explanation to "human behaviors and psychological states." Humanists become even more specific in the shift away from an explanation to "[a]n understanding of human reactions to events and the meanings humans impose on experience as a function of culture, historical era, and life history."

Primary sources of evidence and control of conditions: natural scientists focus on "[e]xperimentally controlled observations of material entities." Whereas social scientists are concerned with "[b]ehaviors, verbal statements, and less often biological measures, gathered under conditions in which the contexts cannot always be controlled." The emphasis on control is further diminished with humanists whose primary sources are "[w]ritten texts and human behaviors gathered under conditions of minimal control."

Primary vocabulary: "Semantic and mathematical concepts whose referents are the material entities of physics, chemistry, biology…, [which are] assumed to transcend particular settings" are the tools of the natural scientist. The assumed realism of natural science is tempered by the constructivism of the social scientists, whose tools are primarily "[c]onstructs referring to psychological features, states, and behaviors of individuals or groups, with an acceptance of the constraints that the context of observation imposes on generality." This restriction on generality, as associated with a realism granting primacy to natural science, is strengthened by the idealism of the humanities, whose tools include "[c]oncepts referring to human behavior, and the events that provoke them with serious contextual restrictions on inferences."

The influence of historical conditions: as should be becoming apparent, the differences between the sciences and the humanities are on a spectrum, from generality to specificity, from realism to idealism, and from maximum controllability to minimal controllability. It is no different with regard to the perceived or valued influence of historical conditions. The natural scientist sees such influence as minimal, the social scientist is more modest, and the humanist takes very seriously the import of history.

Ethical influence: heretofore, the bridging done by the social sciences may seem to suggest a preference for the natural sciences' way of doing things. The preference sways toward the humanists with regard to ethical influence. The natural scientist believes a minimum degree of ethical influence is essential to the scientific enterprise, whereas both social scientists and humanists see a major ethical influence on their respective enterprises.

Dependence on outside support: natural scientists are heavily dependent on support beyond their personal or institutional means, often relying on private and public grants. Social scientists are less dependent yet still require greater resources than the humanist who is "[r]elatively independent."

Work conditions: again, the spectrum is helpful. Natural scientists collaborate in groups both big and small, social scientists work in small groups and often alone, while the work of the humanist is "solitary."

Contribution to the national economy: natural scientists provide a major contribution to the economy through industrial and medical developments. Social scientists are able to contribute to medical and other social problems. Humanists, however, make a minimal contribution.

Criteria for beauty: here the spectrum is relatively ambiguous but can be expressed in terms of correspondence versus coherence. Natural scientists are most concerned with theories that are simple both in formulation and in parsimony, namely, theories ought to correspond to the fewest and most general parts: "Conclusions that involve the most fundamental material components in nature inferred from evidence produced by machines and amenable to mathematical descriptions." Social scientists share with the natural scientists a preference for broad or general theories that correspond to the world, yet they, like the humanists, have a preference for the human: "Conclusions that support a broad theoretical view of human behavior." Humanists become less concerned with the material facts or correspondence with reality. Rather they value "[s]emantically coherent arguments described in elegant prose" (never mind that this description does beg the question: after all, what counts as elegant if not beauty, which is just what we seek to define).

All of these differences can be expressed on a continuum between realism and idealism, of fact and value. This deep dichotomy is a pernicious one at the heart of the Western mind and, as I will show below, is the main target of dissolution for pragmatic reconstruction. One of the roots of this dichotomy is a neglect of the various conceptions of truth assumed to be at work in various enterprises. Kagan offers a useful view of four varieties of truth at work within the sciences and the humanities. He, however, neglects the pragmatic alternative. I now turn to Kagan's discussion of truth, to which I offer a direct challenge from a pragmatist perspective.

Theories of Truth

In his characterization of the three cultures, Kagan recognizes that there are varying conceptions of "*truth* and the related notions *correct, valid, coherent*, and *right*" (ibid., 40, emphasis in original). He asks "what do natural scientists, social scientists, and humanists point to when they claim they are communicating a true idea?" (ibid., 40). He presents four common answers:

1. *Correspondence*: an idea is true when it corresponds to a real thing or event that exists outside of and independently of any observers. Kagan's example: "the moon is or is not present in the sky" (ibid., 40).
2. *Logical consistency*: an idea is true when it does not contradict and/or offers logical support to other ideas known (believed) to be true – often seen as foundational or a priori. Kagan's example: "if velocity equals the ratio of distance over time then distance equal[s] the product of velocity and time" (ibid., 40).
3. *Semantic coherence*: an idea is true when it fits with or seems highly plausible in light of the larger contextual narrative woven to interpret or understand the facts of a case. Kagan's example: "a historian's suggestion at the end of a narrative of World War II that Churchill did not attend Roosevelt's funeral because of his lingering anger over being embarrassed by Roosevelt during their meetings with Stalin has a claim to truth if this idea strikes most readers as coherent with the complete text" (ibid., 40).
4. *Compelling feeling*: an idea is true when it does effect specific emotions that are either pleasing or upsetting to the believer in his/her relation to the idea. Kagan's example: "the feeling accompanying the thought that parental sacrifice for a child is right and abuse is abhorrent" (ibid., 40).

Even though Kagan recognizes that all four of these notions of truth can be found or are at work in each of the three cultures, he finds that natural scientists tend to "trust only the first two; social scientists the first and third; [and that] humanists rely on the last two" (ibid., 40).

While these four theories of truth are undoubtedly common, Kagan's discussion of truth ends in a rather bizarre fashion. His section on truth closes with a brief consideration of the German philosopher, Jürgen Habermas, whose views have been widely discussed in many circles. Kagan states that "Habermas has been criticized for trying to unite the two distinct traditions of American pragmatism with Wittgenstein's emphasis on the importance of consensual understanding" (ibid., 42). This statement is both bizarre and problematic. Bizarre because it so clearly flies in the face of all he has said so far about truth and therein lies the problem, which I now elaborate by following the four notions of truth Kagan himself expounds. First, Kagan's discussion here fails to meet the evidential standards of correspondence as he provides no citation of the relevant experts or publications. Second, it is not clear, from what Kagan says, if there is a significant logical inconsistency between American pragmatism and the language games of the later Wittgenstein – of which there has been substantial debate (cf. Hickman 2007a). Third, Kagan's brevity suggests a lack of coherent understanding of American pragmatism or Wittgenstein. Finally, Kagan's mentioning of American pragmatism effects in me both a feeling of disappointment and a feeling of opportunity. I am disappointed because Kagan misses an opportunity for dealing with these conflicting notions of truth and the consequences – not just for the three cultures but for the current state and future of civilization as well. As I argue throughout this essay, the missed opportunity is a fifth notion of truth that productively entails the other four.

The last pages of Kagan's book are a meditation on whether human life today is an improvement over 200 years ago (ibid., 267–275). Kagan expresses significant doubt over whether life since the beginning of industrialization has improved as much as we like to think it has. While there are certainly reasons to suggest that life has improved for many, it is not clear that human well-being has improved for all humans or that the cost of the human modifications to the rest of the planet are commendable. Kagan goes so far as to draw, speculatively in the penultimate paragraph of the book, a parallel between how an intelligent alien species observing earth may treat humans, given our activity, and how humans do treat an infestation of poisonous snakes: in either case, once the species had become "such a serious threat to all forms of life [it became clear that] it was time to have this animal restrained, culled, or, perhaps, eliminated" (ibid., 274). Kagan concludes with a little bit of hope that some benefit may result from greater humility across the three cultures, through which "each group [recognizes that it] is potent in its own territory but impotent in the territory of the other" (ibid., 275).

Kagan's hope results from his understanding that each culture is up to different things, so far as most people – the participants of each culture and the observers of the participants – are concerned. Kagan's pessimism results from his recognition that nobody seems to actually understand the territory of the other yet act as if the territory is just like one's own or clearly visible and understandable from their particular vantage point. To return to my feelings about Kagan's position, my disappointment is not just that Kagan does not talk more about American pragmatism or that he is pessimistic. Rather, it is that there already is important work on the relationship of the humanities and the sciences that advocates a pragmatist theory of truth – a theory that recognizes both the differences and the similarities between the three cultural territories and thereby provides the means of communication across borders. Edward Slingerland's *What Science Offers the Humanities: Integrating Body and Culture* is a fresh perspective on the relationship between the sciences and the humanities that recognizes the promise of a pragmatic approach to both truth and the rapprochement of science and the humanities. Unfortunately, Kagan does not seem to be aware of Slingerland, nor does Slingerland give much discussion to Snow. This is unfortunate doubly so because Slingerland provides an account of truth that entails the four presented by Kagan, yet Slingerland's account neglects John Dewey's contribution to pragmatic inquiry and the means of reconstruction that give us greater melioristic hope than Kagan seems able to do.

Pragmatism, Inquiry, and Truth

Edward Slingerland takes an explicitly pragmatist and naturalist approach to the relationship between the sciences and the humanities. For Slingerland, it is the humanities that have more to learn from the sciences than vice versa. Slingerland defends this claim by noting that scientists, like humanists, are humans embedded and participating in cultures. Humanists are just as much human as scientists are, but, unlike scientists, most humanists are entirely ignorant of how actual science works. Nevertheless, humanists are happy to enjoy the fruits of science – especially, as Slingerland notes, to criticize science despite their ignorance (Slingerland 2008, 299–302).

Slingerland does recognize that the humanities do have something to contribute to the sciences. However, he seems to believe that such a contribution will come once the humanists embrace the sciences. Slingerland uses the pragmatist account of truth as the alternate means of rapprochement to the representational objectivism advocated in Kagan's account of natural science. Upon reviewing Slingerland's account of truth, I use it as a platform for articulating Dewey's theory of inquiry as the philosophical means of reconstructing the conflict between the sciences and the humanities.

Slingerland demonstrates an impressive breadth of knowledge across the humanities and the sciences. Central to his attack on the philosophical biases that keep humanists and scientists from mutually benefiting each other is the rejection of objectivism. Objectivism, in brief, holds that there is an objective reality independent of human activity and that this reality is discovered or revealed to humans through the activity of scientific inquiry. Postmodernists, by and large, have attacked objectivism to the point that any authority granted to any method claiming to produce knowledge is nothing more than a power game. Slingerland notes, however, that once we give up on the objectivist's goal of a God's eye view of reality, a better alternative to objectivism and the correspondence theory of truth awaits (ibid., 238). Central to his presentation of the pragmatic account of truth is the notion that truth is the successful achievement of goals and that active engagement with one's environment instead of passive representation of the world better accounts for our experience as embodied organisms embedded in biocultural environments (ibid., 238–240).

Following classical pragmatists Charles Sanders Peirce and William James, and noting contemporary thinkers Susan Haack and Ian Hacking, Slingerland presents an account of truth that entails Kagan's four variant conceptions of truth discussed above. Slingerland's presentation, however, does not go into the detail that Peirce, James, or Dewey does in their accounts. While similar, the accounts given by these three pragmatists are not identical. However, for my present purposes, I limit myself to Dewey's general pattern of inquiry. It captures the general traits of most if not all pragmatist accounts of inquiry and truth. Dewey's pattern of inquiry is presented in the following five steps that an organism undergoes (from Dewey 1910, MW6: 236–237):

1. A felt difficulty
2. Its location and definition
3. Suggestion of a possible solution
4. Development of reasoning of the bearings of the suggestion
5. Further observation and experiment leading to its acceptance or rejection, that is, the conclusion of belief or disbelief

An organism engaged in its environment is an active organism whose activities are dynamically entangled and transacting with its active environment. This transaction, it is worth noting, is how Dewey reconstructed the concept of *experience* (Dewey 1925, LW1: 12–13). When this interaction is disturbed – when the equilibrium is upset – there is a felt difficulty or, as Dewey elsewhere describes it, a problematic situation. In order to restore equilibrium, the difficulty must be identified. The difficulty is conceived in terms of doubt. That is, an organism's beliefs are its habits of actions that allow it to effectively and efficiently get about in its environment. When such activity is disrupted, one or some of those beliefs are no longer doing their work. This cessation of activity, of work, is the experience of doubt – an experience of irritation, of *dis-ease*. In order to resolve the doubt, a modification needs to be made either to the organism, including but not limited to its belief system, or to the environment itself. This experience of doubt and of its resolution through inquiry and the fixation of belief is Kagan's fourth variety of truth. The location and definition of the difficulty, of the doubt, coincides with Kagan's first and second varieties of truth, correspondence, and logical consistency. Kagan's third variety, semantic coherence, can also be at work in Dewey's step of location and definition; regardless, it is surely at play in the third step in which a possible solution is imagined. From there, any of the first three of Kagan's varieties of truth could be at work in Dewey's fourth step in which critical consideration is given to the possible solutions imagined. This entanglement of the first three conceptions of Kagan's are integrated with the fourth conception in the final step of Dewey's pattern. If the feeling about the proposed solution or belief is unsatisfactory, then it is back to the drawing board: truth is not yet attained by the organism. But once the feeling is satisfactory, truth is not only established, the organism may now take ameliorative action.

What Slingerland's account provides is not only the recognition that there is an alternative to the accounts of truth presented by Kagan but also that the pragmatist account is productive in the organized activity of inquiring organisms – namely, human beings with regard to the relationship of the humanities and the sciences. In other words, despite perceived differences in the territories of these cultures, as Kagan puts it, there is a common general pattern of inquiry that applies equally to each culture. Moreover, and back to Slingerland's preference for humanists' attending to science rather than vice versa, the origin of the pragmatic conception of inquiry and of truth is in the experimentalism cultivated in the sciences. Central to this experimental conception is the emphasis on practice and activity. The four notions of truth Kagan presents neglect the primacy of action in human experience. Once the primacy is acknowledged, then pragmatists can start doing the work of reconstruction of the dichotomies that permeate the Western mind.

Experimentalism, as the classical pragmatists understood it, does not fit into the categorization Kagan provides. Unlike Kagan's view and the modern conception of experience on which it is based, the experimental method rejects the dichotomies between mind/world and value/fact. The spectrum from the subjectivist, value-laden humanities to the more objective but still value-laden social sciences to the objective but value-free natural sciences is outright rejected from the perspective of pragmatism. As my treatment of truth should indicate, there is a reorientation of the issues at play that permits further productive activity instead of becoming dragged down in theoretical yet impractical logic chopping. Furthermore, as Slingerland notes, once truth is conceived as having to do with achieving goals, the truth remains objective while becoming valuable: Kagan's spectrum collapses. As William James noted, the "truth is *one species of good*" (James 1907, 75). Once truth is humanized and instrumentalized, the pragmatic project of reconstruction, as opposed to reconciliation, not only becomes possible but tractable as well. To this opposition, I now turn.

Reconstruction and Reconciliation

> The problem of reconciliation arises and persists for one reason only. As long as the notions persist that knowledge is a disclosure of reality, of reality prior to and independent of knowing, and that knowing is independent of a purpose to control the quality of experienced objects, the failure of natural science to disclose significant values in its objects will come as a shock.
>
> – John Dewey, *The Quest for Certainty*, 1929 (LW4: 35)

> In short the problem of reconstruction in philosophy, from whatever angle it is approached, turns out to have its inception in the endeavor to discover how the new movements in science and in the industrial and political human conditions which have issued from it, that are as yet only inchoate and confused, shall be carried to completion. For a fulfillment which is consonant with their own, their proper direction and momentum of movement can be achieved only in terms of ends and standards so distinctively human as to constitute a new moral order.
>
> – John Dewey, 1948 (MW12: 275)

If the pragmatic conception of truth as the expedient amelioration of exigencies is taken as an alternative to the four varieties of truth that Kagan presents, then we get over the problem of reconciling the sciences and the humanities simply by rejecting the premises on which the problem is based. The social activities of scientists and humanists do indeed suggest a serious conflict with regard to fact and value. However, the project of reconstruction reorients our conceptions of both fact and value so they are not diametrically opposed but thoroughly entangled (Dewey 1920/MW12). Facts are value-laden, and values are anchored in facts.

Truths effect new productivities in an organism's engagement with its environment. Humans are social organisms interacting with their biocultural environments, which include other humans. Instead of trying to determine whether the realism of science or the idealism of humanism is superior or prior, pragmatists argue that the very idea of disclosing reality to a subject divorced from the world is wrongheaded and doomed to failure. The problem of reconciling the cultures is a problem based on this failing conception of experience. Once we take the Darwinian perspective as pragmatists must, then we conceive an evolutionary and developmental account of how experience evolves. This continuity between the physical and natural world (the territory of natural science) and the personal experience of the individual (the territory of the humanities) must pass through the interactions of a social group (the territory of social science). In short, the individualism and atomism of modernity is rejected in light of Darwin and in favor of a social and developmental account of the individual.

Such an account has been developed in part by pragmatists like Dewey and Mead. However, they lacked the experimental techniques and the results of experimentation that we now have at our disposal in an array of disciplines. Nevertheless, there was recognition of growing specialization among the sciences and the humanities in Dewey's day. This led Dewey to develop a largely neglected account of inquiry, which he simply called logic, that aimed to guide the project of reconstruction according to the method of intelligence (Cf. Dewey 1929/LW4; Dewey 1938/LW12).

Larry Hickman has concisely and usefully described Dewey's approach to the conflict between realism and idealism, of which the conflict between science and the humanities is a social variety. In Hickman's words:

> Dewey simply bypassed the chasm this debate has opened. He proposed that the two sides – the one that emphasizes facts and the one that emphasizes values – are at bottom connected as *phases* or *moments* within inquiry. Whereas both facts (as facts-of-a-case) and values (as ideals, or ends-in-view) are essential components of problem-solving activities, realism errs when it attempts to make a fact into something independent that exists outside of and apart from the ideals (ends-in-view) that arise from active discrimination of the features of lived experience (Hickman 2007b, 158).

When truths are understood to be the products of problem-solving activities that are then applicable to other human problems, the fruits of inquiry from both science and the humanities can be mutually beneficial. This is especially so when we consider the differences between the subject matters of science, commonsense, and philosophy.

As Dewey argued in his 1938 *Logic: The Theory of Inquiry* (LW12: 71–72), commonsense inquiries begin with practical affairs of everyday life. These inquiries become refined into literature and tradition as each seeks to pass down lessons of experience to the next generation often but not always in a self-critical manner. Science begins with the same subject matter of ordinary affairs. It differs from the humanities in its concern for creating artificial situations through which to isolate and control specific variables. In doing so, scientists establish new relations in the world that permit scientists in particular and humans in general to interact with their environments in more stable and often novel ways. Ideally, the products of scientific inquiry, Dewey argued, feedback into our everyday commonsensical experience. Unfortunately, this feedback and its subsequent modification of both further humanistic and scientific activity have not occurred to the degree that would be amelioratively transformative of human social life. This lack of sufficient feedback (and subsequently the impossibility of feeding forward, of productive anticipation) is at the heart of the conflict between the cultures of the sciences and of the humanities.

This failure of navigating between the two cultures in such a way that their territories are not opposed or traversed with difficulty is a failure of philosophy. For Dewey, the proper aim of philosophy is not only to be a liaison officer between special disciplines, making them mutually intelligible, but also to develop a general method of inquiry that grows out of the specific methods of both the humanities and the sciences (Dewey 1938/LW12). This is the general pattern of inquiry introduced in the previous section and the method of intelligence described by Hickman above.

The reasons for this failure of philosophy are multifaceted (McCumber 2001), have been addressed elsewhere (Shook and Solymoski 2012), and are beyond the scope of this chapter. My aim here, instead of rehearsing these failures, is to suggest a new way forward by bridging pragmatist philosophy and recent advances in neuroscience as a plausible means of *reconstructing* – as opposed to reconciling – the sciences and the humanities.

Neuropragmatism and Mirror Neurons

Neurophilosophical pragmatism is a return to the method of intelligence advocated by Dewey with the benefit of the developments in the sciences, particularly of life and mind, since Dewey's day that are now at our disposal for the project of reconstruction. As I have argued elsewhere (2011b), the conflict of reconciliation is endemic to contemporary pragmatism, explicitly in neopragmatism and implicitly in neurophilosophy. Among the virtues of Edward Slingerland's work is his recognition that neuroscience offers a bridge between the sciences and the humanities (Slingerland 2008, 298 and 308). In this section, I pursue the possibilities of such a bridge by first suggesting that the nervous system does not identify or correspond to mentation but rather coordinates meaningful activity within a body and with a body and its environment. Secondly, in light of this reconstructed conception of experience as the nexus of brain-body world, I focus specifically on recent work on mirror neurons. I suggest that mirror systems are part of the underlying biological operations from which cultural activities like scientific and humanistic inquiry grow. I then turn to a new metaphor for a science of consciousness, consciousness as cooking, to serve as a guide in this plausible project for scientists and humanists to tackle together.

With rejecting representational truth and objectivism, we pragmatists are no longer concerned with questions such as the relationship between mind and brain, mind and body, or mind and world. As organisms already embedded in environments with which we interact, questions about the exterior world are atavistic nonstarters. Experience is no longer a matter of a spectator passively watching sense data go about on a veiled screen (Dewey 1925/1927/LW1: 235 and 259). Instead, it is, as Dewey argued, reconstructed as the interaction of an organism and its environment (ibid., 12–13). If experience amounts to organic-environmental transaction, if such a transaction is prior to any distinction between organism and environment, then experience evolves as does the rest of nature. Among the products and ongoing contributors to evolution in this organic-environmental transaction are nervous systems.

In studying nervous systems, humans are understandably interested in how brains and nerves contribute to mental life. If we are to avoid the hazards of dualism, our inquiries must not seek answers to questions having to do with representation or correspondence between a physical and a mental world. Rather our questions become functional. As Larry Hickman has noted, following both Dewey and Richard Feynman, scientists are interested in understanding how nature works (Hickman 2001, 25–35) not what nature is, independently of function and context.

Following the pattern of inquiry and the focus on function, instead of representation, one of the questions neuroscientists can ask is "what role or function does the nervous system have in experience?" Another way of putting the question is "what work does the nervous system do in the transaction of an organism with its environment?" Even though there is no overwhelming consensus on the answer to such a question, there is a growing body of work coming out of dynamical systems theory that suggests that a major role the nervous system plays is that of coordinating the systems of the body with one another as well as the body (a system of systems) with the larger environment (Cf. Rockwell 2005; Chemero 2009; Solymoski 2011a).

Given the problem of the relationship between the sciences and the humanities, developing an understanding of the means by which scientists and humanists as communicators are able to relate to one another would benefit the reconstructive project. Recent work on mirror neuron systems provides such a promising platform.

Mirror neurons are clusters of nerve cells in the premotor areas of the frontal lobes that innervate throughout the brain, including "the temporal, parietal, and frontal lobes as well as… the insula, amygdala, basal ganglia, and cerebellum" (Cozolino 2006, 186, 193). Perhaps best known for their role in imitation between at least two organisms, mirror neurons play a vital role in goal-directed behavior (ibid., 187). When a child is observing an adult open a can of soda pop, there are patterns of neural activity going on in the adult and in the child in the same parts of their brains. When the child goes on to open the can of soda pop herself, the same parts of her brain that were active when she observed the adult are equally active and at work. These patterns of activity are not the same as when the person (child or adult) sees just the can or even the holding or sipping of the can. It is the specific purpose of opening a can that is registered by the mirror neuron system in the relatively specific pattern produced through observation, expressly for the purpose of learning to *perform successfully* that specific act (ibid., 187). Beyond imitative learning of technical behavior, mirror neurons are also involved in the process of empathy and the development of a theory of mind (ibid., 195–198).

The importance of mirror neurons for the project of reconstruction that I am presenting goes beyond the simple learning through imitation that mirror neurons underlie. Louis Cozolino summarizes the work in which mirror neurons are involved by relating it to their functional organization within the larger cortical context:

> The structures of mirror neurons are not special in and of themselves; they serve this mirroring function due to their location. They reside in association areas of the frontal cortex where networks converge to process high-level information. Mirror neurons lie at the crossroads of the processing of inner and outer experience, where multiple networks of visual, motor, and emotional processing converge… It is because of their privileged position that mirror neurons are able to bridge observation and action. Mirror systems have helped us to understand how our brains link together in synchronization of such group behaviors as hunting, dancing, and emotional attunement… They are most likely involved in the learning of manual skills, the evolution of gestural communication, spoken language, group cohesion, and empathy (ibid., 187).

Cozolino raises a number of central issues for a neuropragmatic reconstruction of the sciences and the humanities. First, there is the disregard for structure in favor of integrated function. Second, despite the somewhat Cartesian phrasing of inner versus external experience, Cozolino recognizes the role of mirror neurons in coordinating what goes on beyond the boundary of the skin with what goes on within it. Third, the convergence of myriad systems with the mirror neuron system provides us with some of the neural means of experience. That is, we have the possibility of a natural scientific account of qualitative experience at our disposal that does not eliminate such experience. Rather it provides the means of effecting such an end in view. (To be clear, by "neural means," I do not mean the *only* means: without the rest of the organism-environment unity, there would be no experience, for there would be no (inter)activity, purposeful, or otherwise.) Fourth, mirror systems are integrally social. Not only do they coordinate the body with itself and its environment, mirror systems are emphatically social in their operation. Finally, this social integration that mirror systems underlie runs the gamut of social activities that humans partake in as well as that in which social scientists and humanists take interest.

Furthermore, this entanglement of abilities, this integration of systems, resembles the varieties of truth that Kagan saw operating separately that I argued are synthesized in the pragmatic account of truth. The integrated innervation of emotion, action, and *re*-presentation – the environment is not represented but presented anew thus affording novel activity – fits the pragmatic conception of truth as that which affords its holder a probabilistically greater chance of achieving one's goals. Not only do we have with mirror neurons a liaison between various accounts of human experience and activity, we also have a neural illustration of one of the biological bases of inquiry.

To be sure, my advocating mirror neuron systems as a plausible means for a neuropragmatic reconstruction of truth is not intended to discourage other neural means. In fact, others have already suggested to various degrees pragmatic approaches to truth and action that are informed by the relevant neuroscience. Teed Rockwell, for example, takes work from computational neuroscience to push the neurophilosopher Paul Churchland toward embracing a pragmatist theory of truth (see Rockwell 2011). More extensively, the neuroscientist Jay Schulkin takes to heart pragmatism's emphasis on action in a series of books that provide neurobiological backing to many of the central insights of classical pragmatism (see Schulkin 2000, 2004, 2006, 2009). For instance, Schulkin persuasively argues that central to the evolution of mentation is the activity of problem solving. Problem solving, as Dewey's pattern of inquiry suggests, involves significant cognitive, affective, and motor activities that are not clearly differentiated experimentally or conceptually (Schulkin 2004, 2009). The role of central dopamine in motivation and perseverance toward achieving one's goals (Schulkin 2006) is integral to a pragmatist conception of truth, where truths are what afford us successful action in achieving our aims in life. Beyond the role of dopamine, Schulkin discusses the heavy innervation of cognitive and motor systems, arguing that the evidence is better understood when we do not distinguish the cognitive from the motor (Schulkin 2000, 2009). This is just another reiteration of the pragmatist claim that theory and practice are inextricably linked.

In short, the mirror neuron system is an integral component of an informed account of the neural roots of the pragmatist conception of truth as it is directly tied to successful action. Mirror neurons, however, are not the whole neural story, as work in computational and behavioral neuroscience suggests. As Dewey and Mead both recognized, any attempt to locate one aspect of mentation to one and only one part of the brain is dangerous folly. Such Cartesian materialism – the position that holds that mind or consciousness is so narrowly confined in a part of the brain or the whole brain itself and thus not distributed throughout brain, body, and world (see Rockwell 2005) – is the inevitable consequence of representationalist theories of truth. Such objectivism demands without warrant that there must be a direct correspondence between mental and physical activities. Pragmatism is at odds with such a view because it asserts that the nervous system is a coordinative circuit of activity (see Solymoski 2011a).

This coordinative integration that bridges various activities within the body and the environment does, however, allow for a notion of correspondence that is not the representationalism that pragmatists reject in objectivism. Another form of correspondence is the type found between a key and a lock. This work is done synaptically with neuromodulators at a degree of complexity several of orders of magnitude greater than a typical key and lock system. The innervation of mirror systems with motor systems provides the means of connecting correspondence with activity. This activity ties into the logical consistency and semantic coherence theories of truth that Kagan presents insofar as we take heed of the pragmatic emphasis on the instrumentality of language and on the desire to evade performative contradictions. Just as bodily actions are true insofar as they achieve the ends in view of the organism, propositions are true insofar as they affect productive activity (Dewey 1938/LW12). Finally, the compelling feeling variety of truth is no longer tied to an atomistic subjectivism that stinks of nihilistic relativism. Instead, the feelings of doubt are modulated in part by mirror systems, suggesting that such feelings are objectively shareable.

These connections are admittedly skeletal and likely to bring about derision from both the sciences and the humanities. Nevertheless, the consideration by members of both cultures would be doing the project of reconstruction a service simply by considering these bold conjectures, especially if scientists and humanists do so together. Since I recognize that these reflections on truth and mirror neurons may strike some as fanciful if not outright specious, I close with a more plausible and surely more grandiose project for both cultures.

Within contemporary neurophilosophy, there is debate over the appropriate scope of mentation that has been usefully expressed in metaphor. The orthodoxy is that consciousness (or mentation more generally) is like digestion: it happens within the body, specifically a part of the body (the gut for digestion, the brain or a cortex for mentation). The main challenger is that consciousness is like dancing. That is, it is something a person does with his/her body, in an environment. The orthodoxy fits nicely with the culture of natural science, whereas the dancing metaphor pushes the social sciences to a greater extent. These approaches both rightly reject the old humanist method of introspection. Yet neither approach, based as they are on non-pragmatic conceptions of truth, is able to account for the qualitative feel of experience. I have suggested that a better metaphor that emphasizes the continuity across brain, body, and world is that human consciousness is like cooking (Solymoski 2011a). This metaphor combines the digestion metaphor by extending consciousness beyond the body just as cooking is an extension of digestion. In doing so, it brings about the bodily aspects of dancing as well as the communal proclivities of producing a meal with and for many persons that not only meets the basic metabolic needs of life but does so in an aesthetically pleasing manner.

Such a metaphor draws out the pragmatic approach to experience, inquiry, and truth. It is my hope that pursuit of a science of consciousness that is motivated by such a metaphor could bring together – indeed, its success requires it – scientists and humanists in novel, exciting, and productive ways. Such a science is a long ways off, but rapprochement between the sciences and the humanities is not necessarily as distant. Indeed, such reconstructive rapprochement, from my neuropragmatic standpoint, with its natural allies in neurosociology, is the greatest hurdle to a satisfactory and productive theory of conscious life. Such a theory, no doubt, will have consequences beyond our having greater effective control over experience: it will effect a new moral order, just as Dewey imagined the end in view of reconstruction to be.

References

Chemero, A. (2009). *Radical embodied cognitive science*. Cambridge: MIT Press.

Cozolino, L. (2006). *The neuroscience of human relationships: Attachment and the developing social brain*. New York: W. W. Norton and Company.

Dewey, J. 1910 (1988). How we think. In J. A. Boydston (Ed.), *The middle works of John Dewey* (Vol. 6). Carbondale/Edwardsville: Southern Illinois University Press.

Dewey, J. 1920/1948 (1988). Reconstruction in philosophy. In J. A. Boydston (Ed.), *The middle works of John Dewey* (Vol. 12). Carbondale/Edwardsville: Southern Illinois University Press.

Dewey, J. 1925/1929 (1988). Experience and nature. In J. A. Boydston (Ed.), *The later works of John Dewey* (Vol. 1). Carbondale/Edwardsville: Southern Illinois University Press.

Dewey, J. 1929 (1988). The quest for certainty. In J. A. Boydston (Ed.), *The later works of John Dewey* (Vol. 4). Carbondale/Edwardsville: Southern Illinois University Press.

Dewey, J. 1938 (1988). Logic: The theory of inquiry. In J. A. Boydston (Ed.), *The later works of John Dewey* (Vol. 12). Carbondale/Edwardsville: Southern Illinois University Press.

Dewey, J. 1948 (1988). Introduction: Reconstruction as seen Twenty-Five Years Later in reconstruction in philosophy. In J. A. Boydston (Ed.), *The middle works of John Dewey* (Vol. 12). Carbondale/Edwardsville: Southern Illinois University Press.

Franks, D. (2010). *Neurosociology: The nexus between neuroscience and social psychology*. New York: Springer.

Hickman, L. (2001). Philosophical tools for technological culture. In S. Rice (Ed.), *Philosophy of education 2001*. Urbana: Philosophy of Education Society.

Hickman, L. (2007a). Some strange things they say about pragmatism: Robert Brandom on the pragmatists' semantic 'mistake'. *Cognition, 8*(1), 105–113.

Hickman, L. (2007b). *Pragmatism as post-postmodernism: Lessons from John Dewey*. New York: Fordham University Press.

James, W. (1907). *Pragmatism*. New York: Longmans, Green, and Co.

Kagan, J. (2009). *The three cultures: Natural sciences, social sciences, and the humanities in the 21st century*. New York: Cambridge University Press.

McCumber, J. (2001). *Time in the ditch: American philosophy and the McCarthy Era*. Evanston: Northwestern University Press.

Rockwell, W. T. (2005). *Neither brain nor ghost: A nondualist alternative to mind-brain identity theory*. Cambridge: MIT Press.

Rockwell, W. T. (2011). Beyond eliminative materialism: Some unnoticed implications of Churchland's pragmatic pluralism. *Contemporary Pragmatism, 8*(1), 173–190.

Schulkin, J. (2000). *Roots of social sensibility and neural function*. Cambridge: MIT Press.
Schulkin, J. (2004). *Bodily sensibility: Intelligent action*. Oxford/New York: Oxford University Press.
Schulkin, J. (2006). *Effort: A behavioral neuroscience perspective on the will*. Mahwah: Lawrence Erlbaum.
Schulkin, J. (2009). *Cognitive adaptation: A pragmatist perspective*. Cambridge/New York: Cambridge University Press.
Sellars, W. (1963). Philosophy and the scientific image of man. In W. Sellars (Ed.), *Science, perception and reality* (pp. 1–40). Atascadero: Ridgeview Publishing Company, 1991.
Shook, J., & Solymoski, T. (2012). Neuropragmatism: A way forward. In J. Giordano (Ed.), *Neuroscience, neurotechnology, and neuroethics*. New York/Cambridge: Cambridge University Press.
Slingerland, E. (2008). *What science offers the humanities: Integrating body and culture*. New York: Cambridge University Press.
Snow, C. P. 1959 (1998). *The two cultures*. New York: Cambridge University Press.
Solymoski, T. (2011a). Neuropragmatism, old and new. *Phenomenology and the Cognitive Sciences, 10*(3), 347–368. (published online April 2011).
Solymoski, T. (2011b). A reconstruction of freedom in the age of neuroscience: A view from neuropragmatism. *Contemporary Pragmatism, 8*(1), 153–171.

Chapter 8
Notes Toward a Neuroethics

David D. Franks

Back in the mid-1970s, Amitai Etzioni shocked most of sociology by arguing that a culture could be harmful to human nature and thus to its people. The notion that humans had no nature and were infinitely flexible had become axiomatic ever since the popular writings of anthropologists like Clyde Kluckhohn (1944) and Ashley Montagu (1967). To make such negative judgments about a culture was thought to be nothing but parochial ethnocentrism. Behind this assumption was the old enlightenment idea that the human mind was a blank tablet writ on by experience which differed fundamentally from culture to culture.

With the recent rise of neuroscience, we now know differently, and many have accepted Etzioni's challenge to sociology. In fact, according to numerous leading figures in anthropology and sociology, the individualized assumptions of our own westernized culture are alien to neuroscience's findings about the social nature of the human brain (Geertz 1974; Sampson 1981, 1988; Westen 1985; Scheff 1990). More recently, Berrios and Markova (2003: 9) have put it starkly:

> The concept of self is a construct. It is not a "natural kind" sited somewhere in the human brain. The western concept of self emphasizes individualism and autonomy but this view is cultural and no more scientific or truthful or advanced than the ... collective view of self developed in other cultures and which revolves around family or clan rather than the individual.

Currently, evolutionary theory views the cooperative and social capacities of early Homo sapiens to be fully as important to the development of modern humans as tool use.

Connections Between the Social and Ethical

Gazzaniga (2005) argues that the human brain is by nature moralistic and that these ethics derive from the social skills that enabled us to survive. He says that the human brain is a decision-making device and social decisions must constantly be made with the same lightening speed as the social interactions they support. Innate tendencies aid such quickness. Churchland (2011:59) agrees. She sees social and moral behavior as part of a single continuum noting that the same regions of the prefrontal cortex increase in activity whether the subject witnesses a purely social event or a conventionally moral one. Somewhat different from Gazzaniga, but still social, is her position that attachment may be the platform for morality.

D.D. Franks (✉)
Department of Sociology, Virginia Commonwealth University, Richmond, VA 23284, USA
e-mail: daviddfranks@comcast.net

D.D. Franks and J.H. Turner (eds.), *Handbook of Neurosociology*, Handbooks of Sociology and Social Research, DOI 10.1007/978-94-007-4473-8_8, © Springer Science+Business Media B.V. 2013

Universal ethical issues focus on what forms of killing are considered murder and what forms of sexual intercourse are considered incest. Other universal issues deal with the obligations of child-care, what constitutes telling lies, and what it means to break promises and the elements of family loyalty. Regardless of the variety of interpretations any one culture can make, every culture deals with them.

Three Major Categories of Research into the Brain's Involvement with Ethics

According to Gazzaniga (2005:169), effective social decisions are made possible by the large cerebral cortex and orbital cortex in humans. Information produced by scanning devices that constitute the source of our knowledge about the ethical brain can be organized into three topics: (1) the emotions which create the necessity for ethical decisions, (2) ToM, or theories of other people's minds that we all must have in order to interact normally, and (3) mirror neurons that make us literally feel the pain of others. For example, without the discomfort that is a part of compassion and empathy, we would have no need for altruistic cultural rules. These rules lead to their avoidance. We then rationalize these rules into abstract moral systems.

Findings

fMRIs and other scanning techniques have found regions of the brain that increase their activity with one kind of moral emotion but not others. However, these activations only occur when we are preparing for real actions motivated by particular ethical and emotional preferences. If we are pondering these ethical issues merely intellectually, such brain activation does not occur (Gazzaniga 2005: 167).

According to the purely correlational information produced by these scanning devices, the moral emotions are associated with activity in the brain stem and more enthusiastically in the limbic areas. The brain correlates of theory of mind are the orbital frontal lobe and the medial part of the amygdala. Abstract moral reasoning involves many brain systems.

Decision making and moral reasoning are made particularly difficult when we have to make choices about saving the lives of a *few* people with whom we identify personally or saving *many* people whom we know only as distanced abstractions. There are various ways in which subjects can be confronted with such decisions, but in all of them, there is a strong tendency to choose saving one person whom they know or even who is merely present over a larger number of people who are not present. A popular technique in such studies is to confront subjects with the "trolley problem": a trolley is going briskly down a track headed for five people you cannot see. Should you throw the person in front of you on the tracks killing one person and stopping the trolley or kill the five down the tracks? Subjects inevitably chose to let the train run over the unseen five rather than sacrifice the person present. The dilemma is a choice between the personal and the abstract with the personal winning out. (No option is given to sacrifice the subject.)

Interestingly, these particular dilemmas create an increase in brain activity associated with emotion and with moral cognition. Gazzaniga suggests that through evolutionary time, neural structures, which bound altruistic instincts to emotion, had survival value for the social group. In fact, he sees this morality as "the glue that over the long haul keeps our species from destroying itself."

Confirmation of the brain as ethical and thus social in its very nature comes from Marc Hauser's (2006) findings[1] that regardless of age, sex, and culture, most subjects made similar moral choices but none could give reasons for them. They are spontaneous and intuitive reflecting a subconscious mechanism that responds to moral challenges, in particular brain-driven ways. Like any sociological "account" (Scott and Lyman 1968), we act first and make rationalizations for those reactions later. Again, for a social animal, it is critical that we anticipate each other's behavior, especially considering the speed of social interaction. Churchland (2011: 60) is interested in morality in all mammals as well as humans and thus stresses the fact that cross-cultural similarities are common from land ownership to suitable forms of humor and insults. Shunning, for example, is a common form of punishment for primates as is forms of reconciliation. Both are associated with "neural circuitry that causally implicate distress or comfort respectively."

Simulation and Folk Theories of How We Know the Minds of Others

Two theories about how such anticipations are enabled predominate in the social neuroscience literature. One is known as simulation theory (ST) and borrows strongly from the literature on mirror neurons. Essential here is that we do not just "put ourselves in other's shoes" in an imaginative manner consistent with the role-taking processes of G. H. Mead but in the context of emotions and mirror neurons we literally feel what others feel albeit unconsciously in our motor context. We understand others because actually we are like them (Gazzaniga 2005: 173). "The brain is built to feel not only our own experiences but those of others," he writes.

The other theory for how we understand, and thus anticipate the minds of others, is folk theory known as TT for theory-theory since it comes from our common popular culture. This is nothing more than common sense, and to give it a special label is superfluous. These "theorists" disagree on whether ethical intuitions are brain-driven or culture-driven and therefore do not play an important part in the issues of interest herein. Obviously both are a factor.

Work on mirror neurons is supportive of ST theorists, but research on such neurons is indirect and dependent on differences in blood flow from areas rather than directly on examining single neurons or synapses in their own location—or so goes the criticism. But Gazzaniga points to EEG studies as early as 1954 and continually confirmed thereafter, showing that participants have a particular brain wave response not only when performing a behavior but also when observing the same behavior in others. (See Churchland 2011: 154 for possible qualifications.) Assuming that the EEG studies are tapping in to the same processes as those creating empathy, we have another route as to how we understand the world generally, including how we understand others like ourselves. Critical here is that the mirror neurons with their role in understanding other people may be the precursors to true language. The common notion is that we went from understanding gestures to understanding the abstractions involved in language. Gazzaniga says that hand and speech are linked in humans, but as interesting as this is, he leaves the connection between this and ethics vague. My surmise at this point is that hand gestures may be precursors to language. What does this have to do with morality? The common notion is that language is essential for abstract moral reasoning, but according to de Waal (2009), empathy is common among nonlinguistic primates, especially bonobos.

Gazzaniga and Churchland suggest that we should not look for universal ethics producing hard and fast (absolute) truths but for a universal ethic that is clearly contextualized or situational as well as influenced by emotions. These ethics contribute to our social survival. Issues here are abundant and complex. For example, soldiers can kill the enemy and legal institutions of the vast majority of societies can make this a legal act, but many soldiers have misgivings nonetheless, and empathy for

[1] Though Hauser has been criticized from misrepresenting his data, these allegations are not aimed at the finding reported here.

the enemy is not altogether rare. A certain give and take between culture, brain, and its genetics must be part of the picture.

Coming back to human nature, it does seem that its existence is confirmed in numerous ways by the developing knowledge about our social brains and the strong evidence that we are ethical animals to the core. It is obvious that cultures exist that diminish these brain-driven ethical impulses and one significant example of such a culture currently is our own. In light of the widely published reactionary activists who promote abandoning social protections and the rising imbalance between the wealthy and the other classes, Carl Sagan's warning (Gazzaniga 1998) takes on even more dire consequences:

> It is a foreboding I have—maybe ill placed—of an America in my children's generation ... our critical functions in steep decline, unable to distinguish what's true and what feels good, we slide almost without noticing, into superstition and darkness.

However, this must be balanced by the fact that our social and thus ethical nature is a potential even when our culture does not support it. Current lack of justice in our country—be it the federal government's withholding of bona fide justice for prisoners of Guantanamo Bay or in our economy—produces a pain that testifies to our ethical sides and its potential for drawing us closer together. More evidence for this is presented below.

Frans de Waal and Innate Empathy

For a more positive approach, we can turn to Frans de Waal and his arguments in The Age of Empathy: Nature's Lessons for a Kinder Society (2009). Here we find more other-directed tendencies, not only in us but also in mammals and primates generally. He and Gazzaniga share the conviction that we have a definable nature and that certain cultures (like ours) can be harmful to it. In de Waal's case, the ethical part of our empathetic nature emerges not just from the fact that we are social animals to the core but because we are mammals and in particular primates. He is careful to balance empathy, however, with the dark side of animal and human behavior. This balance leaves potential room for Jonathan Turner's (2000) argument that we come from a surly and relatively isolated ape that had weak ties with others.

De Waal's critique of western society focuses on American implementations of enlightenment notions of self-interest. These became fully expressed by the arguments of our leading economist Milton Freedman who insisted "few trends could so very undermine the foundations of our free society as the acceptance by cooperative officials of a social responsibility other than to make as much money for their stock holders as possible." If those who live in poverty were disadvantaged politically by such an ideology, they were even more endangered by our cultural belief in Social Darwinism that further justified such lack of concern. On top of this came Richard Dawkin's metaphor of the Selfish Gene that most readers took literally. One of these was Enron's now incarcerated Jeff Shilling who thought he was mimicking "Nature Red in Tooth and Claw" with the cutthroat competition he fostered in his own company. Enron even had a long and detailed "code of ethics" to this effect.

According to de Waal, the flaw in the business beliefs that discourage any feeling for others is that empathy abounds in nature as much as teeth and claws and that this is illustrated not only by primate behavior but by other large-brain mammals such as whales, dolphins, and elephants. Gazzaniga claims that the ethical anatomical part of the brain is the prefrontal lobes and the abstractions thereof, but for de Waal, it is the limbic system and its emotions. Moreover, the ethical aspect of mammalian and primate behavior is not just empathy but a widespread tendency to react negatively to unfairness.

Read Montague's Findings on Innate Ethics

Both Montague (2007) and de Waal make use of the "ultimatum game" to seal their cases for an innate, brain-driven theory of fairness which is part and partial of our evolutionary adaptation via our cooperative social natures. De Waal does not imply that all is sweetness and light; he reminds us that "power and hierarchy are such a central part of primate society that conflict is just around the corner." His focus, however, is on what one might refer to as the "humaneness" of large-brained mammals and primates.

Both use a research technique called the ultimatum game. It consists on a "proposer" and a "responder." The proposers are given 100 dollars by the experimenter, and they are instructed to make an offer to split it into any amount they want with the responder. The responder can accept the amount of the offer or reject it. The base assumption is that rationally, the responder would accept any small amount because "something is better than nothing." But with strong regularity, the norm of fairness makes this not the case. An offer of only 20 dollars is far more often rejected with an attitude that conveys something to the effect of "I don't need your precious money, just keep it!" Fairness works another way also. Responders do not want to "hog" too much money because of possible resentments. The findings show that they simply want equality.

Both authors describe how primitive tribes have strict norms about fairly sharing the results of the hunt. Most of us have witnessed this reaction in our children and pets to others getting more than they are getting, but de Wall gives us systematic studies of these judgments starting with a pair of monkeys that were trading available pebbles for the handlers' cucumber slices. When one monkey received the more desirable grapes however, the other monkey became agitated and threw both pebbles and cucumbers away. In order to demonstrate further the importance of social comparisons in this process, the handlers waved grapes around before the trading began to show they were available. As long as each was trading for less desirable cucumbers, there were no problems. In terms of rejecting positive but unfair offers, such was not the case—fairness was out the window. De Wall sees the "fairness" exhibited in these last experiments as the "most egocentric kind that is also found in human children." As we may suspect, when we move up to chimpanzees, the fairness norm became more in evidence. Their protests to either chimps or other humans who are seen as perpetuators can be horrific.

Montague (2007: 184) points out that these experiments with the ultimatum game show the same results cross-culturally, giving testimony to the innate character of norms of fairness. De Wall (2009:185) also cites anthropological findings that 15 small-scale cultures differ in their notions of fairness with most being close to equal. Once again, we see evidence of players rejecting large portions of the 100 dollars so as not to feel greedy.

More brain-related evidence comes from John Cohen's neuroimaging lab in Princeton (Montague 2007:185). He uses correlational data from the ultimatum game showing how brain responses increase with the *degree* of fairness made by the proposer. When responders rejected an offer, many brain regions were activated, but the most active region was the anterior insular. This area is associated with visceral feelings of disgust—an appropriate metaphor in expressing rejection of an inappropriate offer.

Montague goes further than de Waal by making the sociological point that the "one shot" character of the ultimatum game is not the way human beings interact. We interact in a social context of a future-oriented reciprocity where tit for tat is the norm and a priority is given to taking the role of the other. It was in this ongoing interactional context that our norms for fairness arose.

Ernest Fehr cited in Montague (2007:93) is an economist who studies what he refers to as "inequality aversion." He and de Waal use this term, but for Montague it refers to the notion that one is uncomfortable with splits that favor themselves in the ultimatum game but is even more uncomfortable with splits favoring the other. This is essential if we see inequality aversion in the context of a long-term process. Fehr adds findings that we tend to punish those that break the norms of fairness even when there is no gain for the self for doing so. All of Fehr's findings lead to the conclusion that humans have a built-in need to cooperate and will do so at a cost with no direct gain to themselves.

While many will see this as a case for innate altruism, others may argue that altruism is really self-interest in disguise (Montague 2007:189). As long as self-interest cannot be defined in such a way that allows it to be falsifiable, it is a personal belief or an ideology not testable by scientific methods. For example, a martyr on a cross can be seen as being self-interested in making his point, changing people's minds, or obtaining rewards in the afterworld.

Conclusions

How can we square the above with the often horrific capacity of human beings for torture, rape, mutilation of private parts, and other forms of mayhem inflicted on others of their kind? Aside from such acts of violence, how do we square innate ethics with the recurring unfairness of taxation and the unequal distribution of wealth and medical services in our country (de Waal 2009: 197)? Answers to such questions are important because however strong the evidence is for our ethical natures, we are left sensing a lack of coherence without it.

Specifically we can ask, what keeps the victims of the overuse of power from protesting on the streets and demanding fairness? Part of the answer is that impoverished persons are so debilitated by their social environments and so outside of the political system that they do not know about the possible vehicles for voicing their feelings. In terms of voting, they may think "that's just for other folks." It is axiomatic in sociology that revolutions depend on an articulate, organized middle class with at least some small access to power. But the most serious answers (as if this were not enough) have to do with unconscious forces described by systems justification theory (Jost et al. 2004). They point to a general, implicit tendency among minority group members to identify with members of the dominant group. This tendency is often even stronger among those who are most disadvantaged (Franks 2010: 81). One is reminded here of Marx's methodologically questionable notion of false consciousness. The place of ideologies that convince minority group members that their economic and educational disadvantages are fair (like pie in the sky) should also be appreciated.

The study of brain-driven ethics has been established by the evidence and has a bright future, but it has multiple causes including a combination of genetics that can only be understood in relation to culture. The innately ethical human brain and the genes that support it are a consequence of our social and cooperative natures. This places it solidly under the umbrella of neurosociology. As suggested here, it would be naive of us not to balance our ethical natures with the blatantly immoral, non-empathetic tendencies that are all too common in the human animal.

References

Berrios, G. E., & Markova, L. S. (2002). The self and psychiatry: A conceptual psychiatry. In T. Kircher & A. David (Eds.), *The self in neuroscience and psychiatry* (pp. 9–39). Cambridge, UK: Cambridge University Press.

Churchland, P. S. (2011). *Braintrust: What neuroscience tells us about morality*. Princeton: Princeton University Press.

De Waal, F. (2009). *The age of empathy: Nature's Lessons for a kinder society*. New York: Crown Bubl.

Franks, D. D. (2010). *Neurosociology: The nexus between neuroscience and social psychology*. New York: Springer Press.

Gazzaniga, M. S. (1998). *The mind's past*. Berkeley: University of California Press.

Gazzaniga, M. S. (2005). *The ethical brain: The science of our moral dilemmas*. New York: Harper.

Geertz, C. (Ed.). (1974). *Myth, symbol and culture*. New York: Norton.

Hauser, M. (2006). *Moral minds: How nature designed our universal sense of right and wrong*. New York: Harpercollins.
Jost, J. T., Banaji, M. R., & Nosek, B. (2004). A decade of system justifications theory: Accumulated evidence of conscious and unconscious bolstering of the status quo. *Political Psychology, 25*(6), 881–919.
Kluckhohn, C. (1944). *Mirror for man*. New York: Fawcett.
Montagu, A. (1967). *On being human*. New York: Hawthorn.
Montague, R. (2007). *Your brain is (almost) perfect: How we make decisions*. New York: Penguin Group.
Sampson, E. (1981). Cognitive psychology as ideology. *American Psychologist, 36*(7), 730.
Sampson, E. (1988). The debate on individualism: Indigenous psychologies of the individual and their role in personal and societal functioning. *American Psychologist, 43*(1), 15.
Scheff, T. J. (1990). *Microsociology: Discourse, emotion, and social structure*. Chicago: University of Chicago Press.
Scott, M., & Lyman, S. (1968). Accounts. *American Sociological Review, 31*, 46.
Turner, J. H. (2000). *On the origins of human emotions: A sociological inquiry into the evolution of human affect*. Stanford: Stanford University Press.
Westen, D. (1985). *Self and society: Narcissism, collectivism, and the development of morals*. Cambridge: Cambridge University Press.

Chapter 9
Emergence and Reductionism in Sociology and Neuroscience

David D. Franks

In this chapter, I discuss the place of emergence and reduction in sociology and neuroscience along with a critique of both types of causal analysis. Paramount in this discussion is Roger Sperry's early thesis of mind as an emergent from the synaptic circuitry of the brain and also as a causal force that can, in certain circumstances, actually change the brain's synaptic structures. Views from leading figures in sociology and neuroscience on both emergence and reduction are presented. I argue that both emergence and reduction are acceptable ways of conducting sociological and neuroscientific research. This chapter ends with an argument for the validity of Roger Sperry's thesis and its applications to certain disorders.

A Short History of Early Emergence and Reductionism

The history of emergence starts with the early reductionism of neuroscience. Sharon Begley (2007: 134) begins this history with the seventeenth-century English philosopher Henry More, who was convinced that it was ridiculous to consider that the brain had anything to do with such things as human aspirations, love, faith, and consciousness. That changed when his contemporary, Thomas Willis, the leader of a group of alchemists, philosophers, and physicians, became convinced that the brain was an important subject of investigation. However, from the perspective of neurosociology, they got started on the wrong track. While they were convinced that the brain was necessary for everything the mind does, their new perspective took on a belief that mind was *really nothing but* the surges of chemicals and activities of material. There is nothing more to a conscious state than its neural correlates (Begley 2007: 135). They did away with Descartes' absolute dualism between mind and matter by creating an equally disastrous reductionism that there was nothing *but* matter. Even today how matter can be necessary to producing intangible thought is an unsolved mystery. Begley quotes Robert Doty who argued in 1998 that neuroscientific work fails to explain feelings. He says, "I could give you the most detailed neurophysiological account of what the brain is doing when you feel sad, and if you have never felt sad this exhaustive explanation would fall short of enabling you to understand sadness (Begley 2007: 135)." For Doty, if the mind is seen as the result of physics and chemistry, then it is "…but the babbling of a robot, chained ineluctably to crude causality." On the other hand, neuroscientific philosopher John Searl in 1990 asked how a world of seemingly disembodied subjective experiences like the feeling described above could

D.D. Franks (✉)
Department of Sociology, Virginia Commonwealth University, Richmond, VA 23284, USA
e-mail: daviddfranks@comcast.net

D.D. Franks and J.H. Turner (eds.), *Handbook of Neurosociology*, Handbooks of Sociology and Social Research, DOI 10.1007/978-94-007-4473-8_9, © Springer Science+Business Media B.V. 2013

dove-tail with an objective world of tangible matter. As we shall see, ideas about mind and brain vary widely in neuroscience. Leslie Brothers (2001: 8) sees the apparent division through the eyes of Wittgenstein as being merely two separate language games or social practices, mind being one language and brain being another.

Types of Reductionism

In any discussion of emergence, reductionism is just around the corner since each one is what the other is not. However, there are many uses and meanings of reductionism which are hidden perhaps by its frequent association with the clarity and precision sought in science. As Murphy (2003: 61) says, any technical critique of reductionism necessitates differentiating among at least five separate meanings. Some of these meanings are problematic, and some are not.

1. *Methodological* reductionism is a research strategy that separates the object into parts. This is unproblematic because that is exactly what analysis means: the separation of the object into parts that then allow for the tracing of the relationship between these parts.
2. *Epistemological* reductionism: The theories and laws of the higher levels of science should be traced in causal terms to arise from the lower levels' laws and ultimately from those of physics.
3. *Logical* or definitional reductionism: language referring to one type of entity can be translated without any loss to the language about another type of entity.
4. *Causal* reductionism: all causation is bottom up. Ultimately herein the parts of subatomic parts of the system determine all the parts on a higher level.
5. *Ontological* reductionism: Higher-level entities are seen as nothing but the sum of its parts. In other places, this is called "nothing buttery" by its critics. Since Murphy believes that this is ambiguous, we have two subtypes:

   (a) As we go up the levels, we need no new metaphysical entities added to higher levels from these lower ones. For example, we need no immaterial mind to get consciousness.
   (b) Atomic reductionism: This is a stronger version of the above. Only the entities at the very lowest level are *really* real.

Murphy sees causal reductionism as the most problematic while ontological reductionism is the least problematic. It is basically "physicalism." Atomic reductionism expresses more of an attitude than a philosophical thesis. It is difficult to state it without employing the nonsense phrase "really real," so it is not clear how it could be refuted.

Quale

The sensory "feel" of something is referred to as *quale*, and from an experiential point of view, it involves a subjective and idiosyncratic world which is qualitatively different from the physical one where objects are ultimately the same to me as they are to you. Some vegetables like kale taste good to my wife but not to me. It would be ridiculous to say that one of us is right and the other is wrong; it just is. To believe that the other person's subjective experience is actually in *error* would obliterate that person's experience in this regard. Nor can these experiences of qualia be reduced to words. Experience, like emotion, is ineffable. The actual experience of lust is certainly of a different order from the word lust. Quale is simply irreducible and incapable of reduction to something else. As de Sousa (1987) describes it, "a property is *emergent* if it could not be deduced from the lower level properties on the basis of which it could be explained." The feeling of pain then is more than its neural correlates, and its irreducibility is what Begley calls the "explanatory gap" in neuroscience. The notion of quale then is the opposite of reductionism, but both are implicit in each other since quale is that which cannot be reduced, in this case to biological processes.

Emergence in Sociology

Emergence has an important place in sociology because it gives us a distinctive unit of analysis just as social interaction does, and gives us a clear way of setting the boundaries of our field. Solid cross-disciplinary research depends on the knowledge of such boundaries.

Groupthink

Emergence occurs when the whole is more than the sum of its parts taken separately. The last phrase "taken separately" is important. Otherwise one can ask, what is this mysterious, undefined force that makes it more than its parts? One common notion of emergence in sociology is groupthink (See Janis 1972). Such thinking is recognized widely enough that many corporations take explicit steps to avoid it. A common example found in introductory textbooks is the thinking that took place in 1941 before the Japanese air attack when a group of high ranking officers were called together to decide whether to prepare for such an attack or not. Each person taken separately thought such an attack was likely, but they feared the reactions of the group and the higher ranking officer in charge of deliberations. As it turned out, a decision emerged with which no one privately agreed. In many cases, this reluctance to voice an opinion would be unconscious.

Other examples include the Bay of Pigs decision by President John F. Kennedy's advisors not to give air cover to the land forces because the Cuban population was expected to rise up and physically support the invaders. Loyalty to Castro by the Cubans was seen as contrary to Kennedy's beliefs as well as that of the American people. Only a few members of the Cuban expatriate invaders survived the invasion, and an awakened Kennedy was reluctant to help them, which made political enemies for him on the part of many Cuban-Americans.

Durkheim and Emergence

Long before groupthink was an accepted concept in our field, Durkheim argued that society was composed of social facts existing *sui generis*—external and constraining to the individual. A critical stipulation was that one social fact could only be explained by another social fact. This, he argued, created the very boundaries of the discipline and separated it clearly from individualistic, psychological approaches at that time which explained society as the sum of the separate members within it. For example, Thomas Hobbs argued that each societal member, being blessed with an inherent rationality, made a contract with the state in order to avoid a war of all against all—nature red in tooth and claw as he mistakenly put it. (See Franz de Waal's 2009 *Age of Empathy* which challenges this view of nature and animals.) While Durkheim did not use the term "emergent," he was nonetheless the epitome of an emergent thinker. Sawyer (2002: 227) argues that the central premise unifying all of Durkheim's work is the attempt to account for both the emergence of the social upward from individuals *in interaction with each other* and downward causation from the social to the individual interactions.

George Herbert Mead on Emergence

In George Herbert Mead's pragmatic social psychology, the concept of emergence is used to explain the generation of novelty. Complete determinism is not capable of explaining this. To his cohort of pragmatists, a listing of all possible causes of a phenomenon would mean that the dependent variable would have nothing left to it that had not already been explained. It would actually be nothing but its past causes. Following Mead, Miller (1973) says that grass in the environment, once eaten by a deer, becomes food. Grass is simply grass without a digestive track to which it can accommodate. Without this digestive tract, grass would forever remain grass. This is not to say that its potential for food does not exist. But this possibility is a human creation; it is not "there" in nature independent of human thought about such possibilities. While this at first sight may appear unrelated to the versions of emergence that appear above and below, it does reveal certain consistencies in the concept, namely that emergence is involved in something new and different from its parts taken separately and that a completely deterministic, causal approach leaves nothing of our subject matter.

Emergence and Reductionism in Neuroscience

Not surprisingly emergence holds sway more in sociology, and reductionism holds sway in neuroscience. However, the vast complexity of the brain tends toward an acceptance of holist approaches that certainly imply emergent features. Below we will look at the opinions of leading neuroscientists about these matters. Actually, I find a significant amount of both in leading neuroscientists who write for the educated public.

Antonio Damasio

Damasio is very clear about his nonreductive stance, at least when dealing with societal issues (1994: 124):

> I am not attempting to reduce social phenomena to biological phenomena but rather to discuss the powerful connection between them. It should be clear that although culture and civilization could not have arisen from single individuals the behavior was generated in collectives of individuals interacting in specific environments. Culture and civilization could not have arisen from single individuals and thus, cannot be reduced to biological mechanisms and even less can be reduced to a subset of genetic specifications the behavior and thus even less, can they be reduced to a subset of genetic specifications.

Their comprehension demands not just general biology but the methodologies of the social sciences as well. Following this up-down statement, however, Damasio makes clear how important the down-up causal relationship is. Talking about certain social rules and regulations, he suspects "that the neural representations of the wisdom they embody and of the means to implement that wisdom are inextricably related to the neural representation of innate regulatory biological processes" (p 125).

Furthermore, Damasio respects the nonreducibility of quale though he does not use the word… "the magnitude of the feeling, and the beauty of the feeling are not endangered by realizing that survival, brain and proper education have a lot to do with the reasons why we experience such feelings."

Joseph LeDoux

Another leading neuroscientist who warns against the exclusive acceptance of reductionism is Joseph LeDoux (2002: 328). He says that "reduction has a bad name because carried to its extreme it would require that we … describe poetry in terms of subatomic particles. This is the so-called absurd type of reduction that we have to avoid." He is looking for nonabsurd reductions that make sense consistent with those suggested by Smith and Franks (1999) below.

Gerald Edelman

In similar fashion, Edelman, a Nobel Prize winner and leading philosopher of science, does not mince words (2004: 166):

> To reduce a theory of one's behavior to a theory of molecular reactions is simply silly, a point made clear when one considers how many different levels of physical biological and social interactions must be put in place before high-order consciousness emerges.

V. S. Ramachandran

V. S. Ramachandran (1998: 264), another leading neuroscientist, warns against reduction becoming an end in itself to the extent that it becomes a fetish. He says that unfortunately, because reductionism is used so often in solving problems, it is therefore also believed to be sufficient for explaining them. Rather than reductionism by itself, he says what is really needed are attempts to bridge different levels of discourse.

Francis Crick

The most famous neuroscientist who embraces extreme ontological reductionism is the Nobel Prize winner Sir Francis Crick in his 1994 book *The Astonishing Hypotheses*. But he has drawn numerous criticisms from his neuroscientific colleagues. His famous description of reductionism is that "'You', your joys and your sorrows, your memories and your ambitions, your sense of personal identity and free will are in fact no more than the behavior of a vast number of nerve cells and their assorted molecules." The quotes around "you" are because to Crick there is no "you" per se. It is mere epiphenomena. In Murphy's terms, this is causal as well as ontological reductionism and is thus the most problematic.

Smith and Franks (1999: 4) agree with Crick's critics. Exclusive acceptance of only one side of the up-down emergent model or the down-up reductionist model in regard to the brain forces an unnecessary choice between the two. This one-sided choice assumes that biological forces like brains and their genes are unworthy of our attention. Actually, the establishment of both the social and the biological is an *empirical* matter. In the examples above of "groupthink," one can accept the fact of emergence and also be interested in studying the factors involved in making each individual feel more constrained or less constrained by the situation and/or leadership. Many sociologists have been so alarmed by the extreme ontological reduction claimed in E.O. Wilson's *Sociobiology: A Synthesis* (1975) that they have repudiated any hint of reductionism simply on principle. According to them, joining emergence and reductionism will seem strange, but a part of the scientist's assumptive order was that mind could not move matter. As it was put, "no physical action awaits on anything than another physical action" (Sperry 1993). In opposition to this, Sperry argued that the causal potency of an idea becomes just as real as that of a neuronal synapse. Mental forces direct electrochemical traffic between neurons at the cellular level (see Sperry in Franks 2010: 8). This is to say that the emergent whole works back to exert influence on the parts that gave it life. The emergent character of mind does not mean that it is absolutely free of its parts but that it overrides the physical and chemical elements giving it birth, and in turn can exert downward control over neural activity (Henninger 1994: 23). The causal chains in the brain are twofold and cybernetic. There is no "nothing buttery" in Sperry's scheme. The potency of an idea is just as real as that of a molecule, cell, or nerve impulse. First, we have the upward chain of causation going from the parts to the emergent mind. Second, we have the downward control by the mind of the parts from which it originally arose. For Sperry, says Schwartz (2002: 42), the myriad of conscious experiences cannot exist apart from the brain. Sperry did not posit a disembodied mind or consciousness as classical Descartian dualists do. As we will see below, the causal potential of mind in controlling its parts has been used to combat severe cases of obsessive-compulsive disorder and even depression.

Other Antireductionists in Neuroscience

According to Schwartz, most of the philosophers of neuroscience adhere to some form of reductive materialism. An exception, he says, is David Chalmers, professor of philosophy and director of the Center of Consciousness at the Australian National University. He complains about the prevailing don't-have-a-clue materialism and disagrees with the reduction of philosophers like Daniel Dennett and Patricia and Paul Churchland who see mind as epiphenomenal and thus without Sperry's causal forces of mind. Briefly, "easy problems" are those that can be resolved by traditional scientific methods like the difference between wakefulness and sleep or how we discriminate or categorize environmental stimuli. The "hard problem of consciousness" is that of explaining *experience*. It is widely agreed, Chalmers says, that we have no good explanation of how or why a given physical basis gives rise to a rich subjective life and its vast qualia. It seems objectively unreasonable that it should and yet it does (Chalmers 1995). Schwartz (2002: 49) says "that more and more scholars are concluding that our deep inner sense of a mental life not fully encompassed by the electrical interactions of neural circuits is not delusional." He goes on to quote others who say that the elements of consciousness "transcend the reach of reductionists' neurobiological explanations" and that the brain "has an ambiguous relationship to the mind." Chalmers says that many people including himself started out thinking that they could simultaneously take consciousness seriously and remain a materialist, but he now realizes this is not possible. Schwartz ends by citing Chalmers again. "There is no prior principle that says that all natural laws will be physical laws; to deny materialism is not to deny naturalism."

Evidence for Sperry's Thesis that Emergent Mind Can Move Matter

An understanding of distinctions in mind and matter in neurosociology hinges on what we mean by the tangible and the intangible, the abstract and the concrete. According to the American pragmatists, the real is that which resists one's push. It has an existence in a particular time and space. As jello-like as the brain is it also has this quality. In contrast according to Mead, the term mind as a thing is a reification. It has no tangible substance and is the process of using intangible symbols to communicate with one's self and others. The universal, as the ancient Greeks knew, has no external existence in objective time and space. When I look out of my window, I do not see an entity called space. I see the trees and my cat down the hill. Space is my intangible idea of how long it would take for me to get there. This "airy" character of emergent mind is obviously qualitatively different from the cellular and tangible nature of the sources out of which mind emerged. Rather than immersing us in the quale of smells and tastes given by immediacy, mind gives us control over such experiences only by distancing us from them—by allowing us to stand away from them. One of the most distanced products of mind is the "generalized other" that transcends the viewpoint of any particular person and gives us the notion of the objective order which is the same for me as it is for you. Given this imposing contrast between the real and the mental, it was not surprising that Sperry was charge with espousing a irreconcilable dualism, but such was not the case. His position of emergent mentalism was closely tied to, and absolutely dependent on, the tangible brain (Sperry 1993: 16). The crux of the matter is this: If it can be shown that by using the generalized other one can change the neuronal circuits of the brain, this would confirm the emergent theory espoused by Sperry. Important in making this possible is well-known plasticity of the human brain. It was the brazen insight of an animal researcher that this plasticity could indeed be brought under human control. By the time these findings were used on humans, an important ingredient became the utilization of the distance granted from what we know as the generalized other. The generalized other is distanced in the sense that it is impersonal.

Michael Gazzaniga

Professor Gazzaniga is another leading neuroscientist who was a student of Sperry's and finally asked the kind of research questions that demonstrated once and for all the differences in the left verbal hemisphere and the right mute side communicating by electricity (see Chap. 11). In a most recent volume, *Who's in Charge? Free Will and the Science of the Brain* (2011), he says: "Like generosity and pettiness, like love and suspiciousness, responsibility has a 'strongly emergent' property—a property that, though derived from biological mechanisms, is fundamentally distinct and obeys different laws, as do ice and water". Hopefully this statement and the ones above and below will establish once and for all that neuroscience can no longer be characterized by exclusive determinism and exclusively reductionistic tendencies.

Creating Downward Control in the Laboratory

According to Schwartz and Begley 2002 the original work on the minded and willful creation of brain circuits began with experiments on monkeys by a psychologist named Edward Taub. With procedures that will seem appalling to us now, he tested the hypothesis that monkeys who had sensory nerves cut to one of their arms retained the latent capacity for purposeful behavior in these limbs, but this could occur only if the monkeys were forced to use the arms and only if they were routinely and immediately rewarded. He demonstrated that under tightly controlled conditions, monkeys could be made to use their numbed and supposedly incapacitated arm when motivated by near starvation. They were also conditioned to use their numb arm in order to avoid intense electric shock if they failed to try. Under the conditions which Taub imposed, feeling was not necessary for initial success at movement, and with further conditioning, the monkeys almost fully recovered the use of their damaged arms.

These crucial findings had to come after the death of the monkeys. An autopsy showed that new neuronal paths had been created by their extreme motivation and conditioning. These new paths allowed the eventual effective use of their feeling-deprived limbs. The somatosensory cortex controlling feeling from the body to the "cognitive" prefrontal cortex had been literally rezoned. The deafferented zone was no longer empty. Instead of receiving sensory input from the arm into the somatosensory cortex, the supposedly empty zone in the cortex had been replenished over the years by neuronal axons from the face! The monkeys' consistent attempts to use their bad arms had rewired their brains in a way legitimately described as "massive cortical reorganization" (Schwartz and Begley 2002: 159). The monkeys had been able to feel again. This allowed the possibility that human neural wiring did not have to stop at early adolescence as neuroscientific dogma insisted at the time. Taub next applied the principles learned with the monkeys to human stroke patients. He put their paralyzed arm in a sling and an oven mitt on their good hand so they were as motivated as possible to use their bad arm. Scans of his rehab patients' brains demonstrated significant cortical remapping as the patients learned to use their arms. Granted, this took a great deal of time and a trainer with them 6 hours a day for 10 days, but at the end, the brain area producing movements of the once paralyzed arm doubled in size. Eventually these principles were applied to OCD patients and even to those suffering from depression (see Franks 2008). The significance of these studies to the problem of emergent mindedness and agency is that Sperry had finally been vindicated empirically and that will was an important ingredient in the process of mind over matter.

The Distanced Generalized Other as a Lever of Control in Human Therapy

There are at least two practices that can change the synaptic structures of the brain. One is the use of "will" as outlined above even though it is artificially induced by laboratory researchers. Another is by means of taking the attitude of the generalized other and the self-awareness this enables. The latter allows a decentering from the self into the perspective of the generalized other that makes possible the well-known sanity within insanity, expressed by some schizophrenics, such as Zelda Fitzgerald. This implies that there is a perspective in consciousness that is free from the immediacy of psychological distress. This ability was used by Schwartz to develop programs treating people with OCD. One advantage that these patients had was that most of them were intelligent and fully aware of the folly of their compulsions, just like the ability of schizophrenics to experience the sanity of insanity however short lived. He relied on the patients, own capacities of reflexivity and ability to see themselves from the generalized other's point of view. One way to do this was to show the patients images from PET scans of the circuits that produced the OCD. This way patients could interpret their malady as a problem of chemistry rather than of their own failed identities. They could then get a detached standpoint apart from their bodies. Schwartz developed a four-step program helping them to identify their impulses quickly and to dismiss their urges as inauthentic illusions. It was the hardest stage which he called "refocusing." Here the patient needed another course of action to take the place of the problematic urges. In biological terms, patients were beginning to create a good circuit from a dysfunctional one. Other details can be found in Schwartz and Begley (2002: 85).

The last step in placing Sperry's emergent mind over matter was revaluing. It was described as a deeper form of relabeling which contained all the other steps and enabled quick recognition of OCD thoughts as senseless and false. Ten weeks of this program on 18 patients resulted in positive behavioral changes and brain scans showing decreased metabolic activity in the particular brain parts that were overactive during the OCD.

Plasticity as a Treatment in Strokes

The same general principles were successful in stroke patients. Taub et al. (2004) felt that patients with stokes demobilizing one region of the brain could make use of another brain region to take up the function of the damaged part. This he referred to as "constraint-induced movement therapy." He put the victim's unparalyzed arm in a sling and the arm's good hand in a big glove so they were useless to the patient. This would give the patient no choice but to try to use their supposedly useless arm for daily functions. Taub also had them extend their arms to reach a distanced peg as well as other hard to reach objects. If at first they had a hard time, which all did, the therapist would help them with small nudges. As a result, after only 10 days, the patients regained significant use of their arms. This was against all the beliefs of the medical field at that time. Many patients recover from strokes spontaneously, so for a larger study, he recruited 41 patients who had been paralyzed for a number of years. Twenty-one received his therapy, and the others were used as controls undergoing strength, balance, and stamina training. Even after 2 years, the experimental group retained their edge over the control group. Taub used transcranial magnetic stimulation with chronic stroke patients who had one of their arms and hands almost rendered useless. In this process, one area of the brain was paralyzed to disable the many spots that controlled the hand in the motor cortex. This allowed him to see which regions were involved in performing the movements of the hand and arm regions they could muster as weak as they were. Taub's therapy caused the area of the motor cortex controlling the hand to grow to over double in size.

Depression

Remarkably, Taub's findings above were also utilized to treat depression. Once a person has had one depression, they are vulnerable to having other depressions in the future. Therefore, much of his research was involved in trying to cut down relapses. Taub's findings above were utilized to treat this tendency. His four-step treatment which was so successful in OCD was also utilized as a major part of his depression therapy. Patients were taught to be aware of their thoughts and to negate them as mere brain events. The program consisted of 8 two hour sessions weekly. Focus was on learning to attend to the moment by concentrating on breathing and bodily sensation. However powerful the negative thoughts, they were not to be seen as reflecting reality. Teasdale et al. (2000), for example, randomly assigned half of 145 patients who had suffered one past episode of depression in the last 5 years to receive mindfulness cognitive therapy and the other half to receive their regular care. These were not easy patients. Three-fourths of the sample were made up of patients who had suffered three or four past depressions. Nonetheless, mindfulness-based therapy with its intensive program nearly halved the rate of relapse. In sum, enough has been said to conclude as Sperry did, that the mind and will are qualitatively different emergents which cannot be explained by the brain's material substance. We have seen that Taub's therapies for gaining such minded control over specific parts of the brain are not easy. But this only confirms Randall Collins, statement in the 2010 symposium on the social brain that will is an important ingredient in social matters.

Conclusions

Implicit in the above is a verification of the idea of agency. The constructionist Stephan Fuchs (2001) sees agency as a residual category resulting from variance in structural processes that are not explained. As such, agency becomes a default explanation. However, the change in synaptic structures documented by Schwartz is a substantial description of self-healing regardless of the programmatic help that is needed to accomplish it. In the final analysis, only the patient can accomplish this dramatic change in brain structure. The other implicit thread in this chapter is that determinism and reductionism no longer have the complete sway in neuroscience that once made E.O. Wilson so popular in that field. I have resurrected Mead's concern that determinism is a belief system—indeed an ideology—that would do away with its subject matter. But Wilson's reductionism is no longer uniformly accepted in neuroscience. We have seen the many reservations held about his "ontological reductionism" by contemporary leading neuroscientists. Most importantly, we have argued that there are times for reduction and time for the recognition of emergent top-bottom analysis in our field.

References

Begley, S. (2007). *Train your mind, change your brain: How a new science reveals our extraordinary potential to transform ourselves*. New York: Ballentine Books.

Brothers, L. (2001). *Mistaken identity: The mind-brain problem reconsidered*. New York: State University of New York Press.

Chalmers, D. (1995). Facing up to the problem of consciousness. *Journal of Consciousness Studies, 2*(3), 200–220.

Collins, R. (2010). *Preliminary remarks: Didactic seminar on the social nature of the brain*. Atlanta: American Sociological Meetings.

Crick, F. J. (1994). *The astonishing hypothesis: The scientific search for the soul*. New York: Scribner's.

Damasio, A. R. (1994). *Descartes' error: Emotion, reason and the human brain*. New York: Avon Books.

De Wall, F. (2009). *The age of empathy: Natures lessons for a kinder society*. New York: Harmony Books.

De Sousa, R. (1987). The rationality of emotion. Cambridge, MA: MIT Press.

Doty, R. (1998). The five mysteries of the mind, and their consequences. *Neuropsychologia, 36*, 1069–1076.

Edelman, G. (2004). Wider than the sky. New Haven: Yale University Press.

Franks, D. (2008). The controversy of mind over matter: Mead's solution and applications from neuroscience. In N. K. Denzin (Ed.), *Part one: Blue ribbon papers in studies of symbolic interaction* (Vol. 31, pp. 32–61). New York: Emerald Group Publishing.

Franks, D. (2010). *Neurosociology: The nexus between social psychology and neuroscience*. New York: The Springer Press.

Fuchs, S. (2001). Beyond agency. *Sociological Theory, 19*(1), 24–40.

Gazzaniga M. (2011). *Who's in charge: Free will and the science of the brain*. New York: Harper Collins Publisher.

Henninger, P. (1994). Roger Sperry's theory of consciousness, its antecedents and consequences. In W. TenHouten (Ed.), *Social Neuroscience Bulletin, 7*(2).

Janis, I. L. (1972). *Victims of groupthink*. New York: Houghton Mifflin.

Janis, I. L. (1982). *Groupthink: Psychological studies of policy decisions and fiascoes* (2nd ed.). New York: Houghton Mifflin.

LeDoux, J. (2002). *Synaptic self: How our brains become who we are*. New York: Penguin Group.

Miller, D. (1973). *George Herbert Mead: Self, language and the world*. Austin: University of Texas Press.

Murphy, N. (2003). What ever happened to the soul?: Theological perspectives on neuroscience and the self. In J. LeDoux, J. Debiec, & H. Moss (Eds.), *The self from soul to brain* (pp. 51–64). New York: New York Academy of Sciences.

Ramachandran V. S., & Blakeslee, S. (1998). *Phantoms in the Brain: Probing the Mysteries of the Human Mind*. New York: Harper Collins Publisher.

Sawyer, R. K. (2002, July 2). Durkheim's dilemma: Toward a sociology of emergency. Sociological Theory, 20, 227–247.

Schwartz, J., & Begley, S. (2002). *The mind and the brain: Neuroplasticity and the power of mental force*. New York: Regan Books.

Smith, T., & Franks, D. (1999). Introduction: Emergence, reduction and levels of analysis in the neurosociological paradigm. In D. D. Franks (Ed.), *Mind, brain and society toward a neurosociology of emotion. Social perspectives on emotion* (pp. 3–17). Stanford: JAI Press.

Sperry, R. (1993). A mentalist view of consciousness. Social Neuroscience Bulletin, W. TenHouten ed. 6(2),15

Taub, E., Uswatte, G., King, D. K., Morris, D., Crago, J. E., & Chatterjee, A. (2006). A placebo-controlled trial of constraint-induced movement therapy for upper extremity after stroke. *Stroke, 37*, 1045–1949.

Teasdale, J., Segal, Z., Williams, J., Ridgeway, V., Soulsby, J., & Chatterjee, A. (2000). Prevention of relapse/occurrence in major depression by mindfulness based cognitive therapy. *Journal of Counseling and Clinical Psychiatry, 68*, 615–623.

Wilson, E. O. (1975). *Sociobiology: A synthesis*. Cambridge: Belknap Press of Harvard University Stand.

Part II
Neurology, Self, Interaction, and Sociality

Chapter 10
Neurology and Interpersonal Behavior: The Basic Challenge for Neurosociology

Jonathan H. Turner

The Need for a More Robust "Theory of Mind"

The notion of "the theory of mind" (ToM) has gained a certain currency in recent years. This label denotes the processes by which one person is able to understand the thoughts, intentions, dispositions, motives, and perspectives of others. In so doing, an individual is in a better position to predict the responses of these others. Empathy, for example, is often viewed as a behavioral capacity critical to a theory of mind. Over 30 years ago, D. G. Premack and G. Woodruff (1978) posted the question: Do chimpanzees have a theory of mind? And, over the ensuing years, neurosciences have sought to isolate the brain mechanisms by which human and nonhuman primates take on the mental perspective of others (Call and Tomasello 2008). Studies have shown, for example, that the medial prefrontal cortex (mPFC), posterior temporal sulcus (pSTS), temporoparietal junction (TPJ), right lateralized (Saxe et al. 2006) are all involved in theory of mind dynamics. The literature continues to expand as efforts are made to link brain systems or modules to particular capacities that enable theory of mind. One problem with this literature, however, is that many of the modules that are seen as elements of the theory of mind are rather generalized processing systems that have many other functions, and thus, it is hard to know if these modules are really the critical structures making the theory of mind work.

What is remarkable to me is that the literatures in philosophy, neuroscience, neuroeconomics, and other fields taking up the question of the theory of mind virtually never mention the work of George Herbert Mead (1934, 1938), who developed a theory of mind some 100 years ago. It might be understandable if Mead had been a sociologist since work in our discipline is almost always ignored by economists, psychologists, neuroscientists, and philosophers, but Mead was a philosopher, and philosophers are at the forefront of much of the debate over the dynamics involved in the theory of mind. (But see Chaps. 8 and 20 of this volume.) Mead's notion of role-taking captures the basic problem of the theory of mind: Individuals read the significant or conventional gestures of others to put themselves in each other's place, and in so doing, they are able to determine their perspectives and dispositions of others to act. And, on the basis of this taking on the role of the other, individuals can predict their behaviors and thereby engage in cooperative interactions. Mead also offered an explanation of how individuals acquire this capacity to role-take.

J.H. Turner (✉)
Department of Sociology, University of California, Riverside, CA 92521, USA
e-mail: jonathan.turner@ucr.edu

D.D. Franks and J.H. Turner (eds.), *Handbook of Neurosociology*, Handbooks of Sociology and Social Research, DOI 10.1007/978-94-007-4473-8_10, © Springer Science+Business Media B.V. 2013

Humans are born into ongoing, organized social contexts, and thus, if they are to survive, they must adapt to these contexts (what Mead termed "society"). The behavioral capacities that make humans unique are acquired through a reinforcement process in which those actions that allow an infant to meet its needs, or "impulses" (Mead 1938), will be retained in its behavioral repertoire. Through a combination of naturalistic conditioning of this sort, coaching and operant conditioning by others in an infant's environment, and biological maturation, infants successively acquire the capacity to (a) read and use conventional gestures or signs marking common meanings between the receiver and sender of signals (vocalizations and body language); (b) role-take or assume the dispositions, motives, perspectives, and likely courses of actions of others in increasingly complex social settings of others and generalized others; (c) imaginatively rehearse alternative courses of actions and select that course of action that will facilitate cooperation with others; (d) see self as an object of evaluation by others as they emit gestures and signals; and then (e), on the basis of (a) through (d), make decisions about those behaviors that will facilitate cooperation with others.

Mead was not concerned about the neurology of these processes, and indeed, he probably would have argued that they are learned by virtue of humans' need to interact in, and thus adapt to, ongoing social contexts. Yet, Mead also assumed that (a) through (e) above are unique to humans, and this assumption implies that there is something unique about human neurology. Mead was wrong in making this assumption because other animals can also engage in many of the behaviors that are seen as essential to the theory of mind. Indeed, as I will argue in this chapter, comparative neuroanatomy of humans with other higher primates (the great apes) can yield insights into just how the brain was rewired during hominin evolution to give humans greater facility in role-taking, use of conventional gestures, deliberative thought, self-reflection and self-evaluation, and other processes driving the theory of mind.

I emphasize Mead not only to correct for the injustice of ignoring his thinking in most discussions of the theory of mind but also because Mead's approach allows for a theory of mind that is based upon *the process of interaction in social contexts* and the neurology of the brain allowing humans to acquire the behavioral capacity for activating these processes. Even though the research on the theory of mind nominally emphasizes the processes by which individuals achieve, to use phenomenological terminology, intersubjectivity (Schutz 1932; Weber 1922), the actual research does not emphasize the process of interaction in a very robust sense, nor does it seem to be very sociological in general and view the production and reproduction of sociocultural formations as built from theory of mind dynamics. Thus, Mead alone adds much to the theory of mind debate, but if we look at the cumulative work of microsociology, we will see that sociology has a great deal to offer work on the theory of mind and the underlying neurology driving the processes that make up minded interactions. Indeed, the sociological work on interaction processes generating intersubjectivity is so far superior to what is evident in other disciplines; I prefer to drop the label of "theory of mind" as hopelessly simplistic and a rather long-winded label for a complex set of basic interaction processes. In its place, I will expand upon Mead's notion of role-taking to include other key interactive processes involved in creating a sense of intersubjectivity among persons in social contexts and, then, address the neursociology of these interactive processes.

Thus, to have an adequate theory of mind, it is first necessary to have an adequate theory of interaction dynamics; then and *only then*, can we ask the neurological question: What is the neurological basis of these interaction dynamics, and which of these neurological modules is evident in present-day mammals and primates and humans and which, if any, is unique to humans? A neurosociology of interaction processes must, therefore, involve not only a concern with the neurological substrates that make interaction dynamics possible but also the evolution of these substrates as they distinguish humans from other higher primates.

Using Mead's Theory to Expand the Conceptualization of Interaction Dynamics

Role-Taking and Role-Making

Long ago, Ralph H. Turner (1962) proposed that the reciprocal of Mead's notion of role-taking should be *role-making*. When individuals role-take, they read the gestures of others, while at the same time, present gestures to others in order to make roles themselves in a social context. With this presentation of gestures—what Goffman (1959) termed the "presentation of self"—individuals provide the information necessary for others to role-take. Sometimes, this information is deliberately orchestrated, and at other times, it simply slips out as individuals interact. In either case, humans can be seen as not only role-taking but also as role-making animals, and the more individuals role-make, the easier will the role-taking efforts of others become.

Role-making and role-taking both depend upon individuals possessing "stocks of knowledge at hand"—to use Schutz's (1967 [1932]) adaptation of Edmund Husserl's vocabulary. Humans carry in their frontal cortex inventories of roles for most situations, and they draw upon these inventories when making a role for themselves and when trying to determine the role of others. Table 10.1 outlines some of the basic types of roles that humans store in their long-term memory banks in the frontal cortex. As an individual role-makes and role-takes, this person implicitly assembles a role for self as well as a role for others. These respective roles will also carry moral evaluations and be valenced with emotions pulled up from subcortical areas of the brain, via the neuronets connecting the prefrontal cortex to emotion modules in the subcortex of the human brain. What is needed from neuroscience is a better understanding of how these inventories of roles are stored, how they are retrieved, how they become emotionally valenced, and how they are assembled along the patterns outlined in Table 10.1.

Table 10.1 Inventories of types of roles stored in stocks of knowledge

1. *Preassembled roles*. These are roles that are widely known, often by virtue of being attached to structural units within institutional domains (e.g., economy, kinship, religion, education). These roles are learned early in life, and even though individuals may never actually play a particular preassembled role, they will have the role stored in their stocks of knowledge and, as a result, be able to identify the role in role-taking. There are also more general preassembled roles that are widely known outside particular institutional domains, such as the roles of host, guest, friend, lover, and companion.
2. *Combinational roles*. These are, as the label suggests, combinations of roles that are well understood. Often preassembled roles are combined to form a combinational role, as would be the case with a host who is also a daughter entertaining her family kin unit. Or, a generalized role like companion can be combined with a work role.
3. *Generalized roles*. These are generalized roles that individuals generally understand and that can be added to almost any other role. The role of being assertive, social, upbeat, gracious, shy, reserved, serious, and diligent, for example, are all stored in people's stocks of knowledge, and they can recognize variants of these role alone or in combination with other roles.
4. *Transituational roles*. These are roles that people carry with them from situation to situation. They are typically associated with membership in particular social categories or categoric units. For instance, gender, class, and ethnicity generate expectations for how individuals will play roles, and typically, individuals carry these roles and play them out attached to another role.

Note: These four types of roles are only examples of how roles are cognitively stored in stocks of knowledge. There are certainly hundreds and perhaps thousands in a complex society of roles stored in these stocks. There are, no doubt, other dimensions along which roles are organized in the frontal cortex, but the key point is that role-taking and role-making are made much easier because individuals already have the building blocks of a role stored in their stocks of knowledge, thereby making role-taking and role-making considerable easier and faster

Some time ago (Turner 1998, 2002), I began to recognize that what Mead and Ralph. H. Turner (1962) produced together was a useful way to conceptualize other key interpersonal dynamics. Contained in Mead's and R. Turner's conceptualization is a much more robust concept of role, beyond their normal sociological definition. In fact, too much is packed into Mead's and Turner's notion of role; it is better to limit their definition to lines of behavioral outputs that are assembled from inventories of elements stored in the frontal cortex than to make them the crucible for *all* interpersonal processes. But, the imagery of "taking" (through reading situational gestures and cues) and "making" (by deliberately and unconsciously orchestrating presentations of gestures) is useful when isolating other interpersonal processes beyond roles, and so, as will become evident, I will conceptualize all interpersonal dynamics through the prism provided by the notions of *taking* and *making*.

Status-Taking and Status-Making

Probably, the best way to make the above point clear is to address the dynamics of *status-taking* and *status-making*, since the notions of role and status are often conflated. But, as the very large literature on status dynamics that has accumulated over the last 50 years documents (especially compared to the literature on roles, which has virtually disappeared in sociology), status dynamics can and should be conceptualized as distinct from, though interconnected with, role dynamics. True, each status position in a social structure generally has a set of expectation states (established by status beliefs and norms) that constrain the roles that individual's play. Moreover, role-taking will often rely upon status cues as a source of useful information in role-taking. The converse can also be true: roles can be used to determine the status location of others when social structures are more fluid and ambiguous. In such cases, roles become cues for status-taking, or the process whereby individuals determine the status of others in a situation, *vis-à-vis* self. Moreover, as researchers such as Joseph Berger and colleagues (e.g., Berger et al. 1972, 1977, 1992; Berger and Zelditch 1985; Berger 1988) have clearly documented in the large literature of status organizing processes, interaction revolves around a process of not only determining the status of others (the status-taking dynamic), but this research as well as other research projects have demonstrated that status dynamics revolve around games of *micropolitics* and *microeconomics* in which individuals seek to assert a given status *vis-à-vis* others and the expectations states on self and others that this status carries (Clark 1987, 1990).

Some neurological data document that the amygdala—the center for *fear* and *anger* in all mammals and reptiles—is involved in status dynamics; this finding alone signals that the neurology of status is somewhat different than that for roles. True, much like roles, people store inventories of status information in their frontal cortex and use these stocks of knowledge in status-taking and status-making, but status dynamics arouse more intense emotions than do role dynamics. The more individuals status-make and seek to establish their superiority over others, the more will *anger* and *fear* be activated in the amygdala, probably because of the micropolitics of rank and power that flow in many interactions. These involve implicit and explicit aggressive and fear responses in the interaction between a person and others. The more successful are persons in establishing their higher rank, the more subcortical areas generating, at a minimum, *satisfaction* or, at a maximum, *happiness* will also be activated—above and beyond what is evident for successful role-making. As Theodore Kemper and Randall Collins (1990) have theorized, individuals experience positive emotional energy when they have successfully established higher status than others, whereas they experience negative emotions when they have failed to make for themselves the status that they had expected to establish in a situation, and the more individuals lose status during the course of interaction, the more will the core centers for *anger*, *fear*, and *sadness* as well as *shame* and *alienation* be activated. To a lesser

Table 10.2 Normatizing through culture-taking and culture-making

Normatization is the process of culture-taking and culture-making in which individuals establish expectations for how individuals should interact during the course of an encounter. These expectations revolve around the following axes:

1. *Microframing of the encounter*: The process of culture-taking and culture-making that imposes expectations for what can be included and, conversely, what is to be excluded as subjects of talk and nonverbal behaviors.
2. *Categorizing the encounter*: The process of culture-taking and culture-making in which individuals typify (a) the categoric-unit memberships of participants in the encounter; (b) the relative amounts of work-practical, social, and ceremonial activity to be conducted in the encounter; (c) the degree of intimacy to be achieved with others along a continuum of treating others as personages (people as only representatives of categoric units or as incumbent in positions of corporate units), persons (with some knowledge of others as individuals), and intimates (with more in-depth knowledge of others); and (d) the relative authority/power of self and others, and on the basis of these nodes of categorization, expectations for behaviors of self and others are developed.
3. *Forming communication in the encounter*: The process of culture-taking and culture-making by which expectations for the proper modes of (a) talk and conversation as well as (b) expressions of body language and demeanor.
4. *Ritualizing the encounter*: The process of culture-taking and culture-making in which expectations are developed for the appropriate rituals to (a) open and close interaction, (b) form and structure the flow of interaction, (c) symbolize the significance of the interaction, and (d) repair breaches to the interaction.
5. *Emotionally energizing the encounter*: The process of culture-taking and culture-making whereby expectations for the nature and valence of (a) emotions to be felt by a person and (b) emotions to be displayed to others are established.

extent, the same dynamics hold for role-making, but I would argue that status dynamics activate more areas of the brain because they are explicitly tied to people's ability to sustain their sense of self-worth and power *vis-à-vis* others in social contexts.

Culture-Taking and Culture-Making

Mead borrowed from W. Wundt's "folk psychology" the notion that individuals not only role-take with specific others but also with "generalized others." The generalized other is a "community of attitudes," and there can be multiple communities at varying levels of generality. If we unpack this notion of role-taking with generalized others, we can reconceptualize the process *as culture-taking* or reading situational cues, scanning stocks of knowledge, and assembling the relevant cultural elements guiding the behaviors of individuals in a situation. The converse of this process of culture-taking is *culture-making* where individuals orchestrate gestures and use situational props to assert what cultural elements they think should be relevant in a situation. From people's mutual efforts at culture-taking and culture-making, a set of situational expectations is generated, and these expectations revolve, I believe, around the dimensions of situational culture summarized in Table 10.2.

Individuals carry cultural codes in their stocks of knowledge, stored in the frontal cortex and tagged with emotions from subcortical areas of the brain. Part of these stocks of knowledge are inventories of understandings as to when and where particular types of codes can be invoked and used and when they cannot. Table 10.2 simply emphasizes the dimensions of a process that I have, at times, termed *normatization* or the assembling of expectations for episodes of interaction (Turner 2002, 2010). These expectations revolve around categorizing (persons and situations) in an appropriate manner, establishing the relevant frames (Goffman 1974), using the appropriate forms of talk and body language, invoking the appropriate rituals (Goffman 1967, 1981, 1983), and experiencing and displaying the proper emotions and feelings (Hochschild 1979, 1983).

At a more general level, these situational expectations are constrained by beliefs and ideologies, institutional norms, and cultural values, and thus, as people scan stocks of knowledge, they first determine the relevant values, beliefs, and broader norms before they fine-tune these during the process of normatization. Individuals generally know that episodes of interaction, or *encounters* to use Goffman's (1983) term, are embedded in larger-scale social structures that evidence an ideology (moral codes about proper conduct within an institutional domain) and broad institutional norms, and they understand the societal value premises on which these ideologies and institutional norms are based. These become, in essence, a *general or macroframe* that delimits the options of individuals as they *culture-take* and *culture-make* within an encounter. They will also seek to set up situational expectations in the encounter with a more microframe that allows them to (a) establish categories for self, others, and situations; (b) settle on forms of talk; (c) enact rituals needed to open, close, track, and repair the interpersonal flow; and (d) emit emotions that are acceptable within the general frame provided by values, ideologies, and institutional norms. Yet, since there are still many options, even within this more delimited microframe, a considerable amount of ongoing negotiation during the processes of culture-taking and culture-making will occur in an encounter.

I have, of course, unpacked the notion of generalized other considerably beyond Mead's formulation, but this is necessary in order to have a more robust conception of culture and how it is assembled on the ground during the course of interaction. Since culture is almost always moral, it reveals both a cognitive and an emotional dimension. Emotions *moralize* cognitions and give them power to constrain interaction by allowing others to sanction those who deviate from culture prescriptions and proscriptions or, even more effectively, by making deviants feel *ashamed* and *guilty* for their conduct (thus leading them to sanction themselves and become motivated to make amends to others).

Thus, the neurology of culture-taking and culture-making revolves around stores of emotionally charged cognitions about how people should behave in situations and the mechanisms by which individuals can scan relevant stores of knowledgeability to assemble mutually acceptable cultural expectations. This process is facilitated, as I emphasized above, by two *levels of framing*, first at the level of society and macrostructure and, second, at the level of the encounter embedded in macrostructures and mesostructures. If the embedding is unclear, then individuals must work very hard to figure out how to normatize an episode of interaction because they do not have the first level frame. As far as I know, there is no great understanding of framing dynamics in the human brain. Clearly, the frontal cortex is involved as the place where cognitions, tagged with emotions, are in long-term storage, while the hippocampus is also involved for shorter term memories (less than 2 years old) that eventually, if reactivated sufficiently with emotional content, will be shipped up to the frontal lobe for longer term storage. The prefrontal cortex is probably involved in both levels of framing, rapidly invoking the relevant macro-level frames that, in turn, direct the process of microframing at the level of the encounter. And, once framed at the microlevel, categorization, talk and body language, feelings, and rituals become clear—thereby allowing the encounter to be normatized.

Motive-Taking and Motive-Making

The theory of mind has always emphasized that individuals need to understand each other's intentions and dispositions to act in particular ways. Mead made much the same point, and I would add that the underlying question is how individuals go about understanding each other's motives to mobilize energy in situations. Like other interpersonal processes, this understanding is achieved through mutual reading of gestures, but in this case, individuals are *motive-taking* and *motive-making*. This process is facilitated, I believe, by the fact that certain motive states are activated in virtually all

Table 10.3 Universal need states driving motive-taking and motive-making

1. *Verification of identities*: Needs to verify one or more of the four basic identities that individuals present in all encounters:
   (a) *Core identity*: The conceptions and emotions that individuals have about themselves as persons that they carry to most encounters.
   (b) *Social identity*: The conception that individuals have of themselves by virtue of their membership in categoric units which, depending upon the situation, will vary in salience to self and others; when salient, individuals seek to have others verify this identity.
   (c) *Group identity*: The conception that individuals have about their incumbency in corporate units (groups, organizations, and communities) and/or their identification with the members, structure, and culture of a corporate unit; when individuals have identity.
   (d) *Role identity*: The conception that individuals have about themselves as role players, particularly roles embedded in corporate units nested in institutional domains; the more a role identity is lodged in a domain, the more likely will individuals seek to have others verify this identity.
2. *Making a profit the exchange of resources*: Needs to feel that the receipt of resources by persons in encounters exceeds their costs and investments in securing these resources and that their shares of resources are "just" and "fair" compared to (a) the shares that others receive in the situation and (b) reference points that are used to establish what is a just share. Reference points can include others in a situation, alternative situations where exchange can occur, norms specifying just rates of exchange, or abstracted distributions of what people in general receive in exchanges.
3. *Group inclusion*: Needs to feel that one is a part of the ongoing flow of interaction in an encounter, and the more focused is the encounter, the more powerful is this need.
4. *Trust*: Needs to feel that others' are predictable, sincere, respective of self, and capable of sustaining rhythmic synchronization through talk and body language.
5. *Facticity*: Needs to feel that, for the purposes of the present interaction, individuals share a common intersubjectivity, that the situation is indeed as it seems, and that the situation has an obdurate character.

episodes of interaction, especially face-to-face interaction in focused encounters (Goffman 1981, 1983). Other motives that are idiosyncratic to a situation or unique to a particular individual will need to be actively asserted by individuals through their gesturing and understood by the motive-taking of others. But, people begin with implicit understandings that all interactions involve the more universal motive states delineated in Table 10.3. The central task of motive-taking is then to learn how salient any of these motive states are in a given situation, and when persons successfully motive-make, they communicate this information to others, thereby increasing the likelihood that they will meet their need states. For I would argue that persons engaged in motive-taking are motivated to allow others to meet their needs if they can without compromising meeting their own needs because the failure to do so will generally breach the interaction. If individuals do not motive-make effectively, even these more universal needs become ambiguous to others, and they must then more actively motive-take because, as just noted, the failure to meet any salient need state will generally breach the interaction or, at a minimum, cause the arousal of negative emotions.

Mead implied and most subsequent symbolic interactionists have emphasized that *verification of self or identity* is the most powerful motive state directing a person's behavior (Stryker 1980; McCall and Simmons 1978; Burke and Stets 2009; Turner 2010). Individuals are constantly motive-taking to determine which of many identities that a person has is most salient in a situation and if it is possible to verify this identity. The second most salient motive state is to *realize a profit in the exchange* of resources and especially a profit that meets normative standards of what would be considered "fair" and "just" in a given culture (as determined by culture-taking). Thus, in individuals motive-taking, they seek information about what resources are being exchanged, at what ratio, and at what level of fair exchange; reciprocally, individuals in their motive-making are signaling through their emotional arousal on a positive to negative scale whether or not the exchange of resources is fair and just.

I see these two motive states as the most important in that individuals will devote most of the motive-taking and motive-making efforts to determining if these needs are being realized. There is clearly some hard-wiring for these motive states, since calculations of fairness and justice run through the primate line and perhaps the higher mammalian line as well (Brosnan and de Waal 2003; Brosnan et al. 2005; deWaal 1991, 1996 and Franks Chap. 15 of this volume.). The amygdala is clearly the key subcortical region of the brain in calculations of justice, providing the emotional (moral) outrage for injustices, whereas the anterior cingulate and neurotransmitter system provide much of the positive emotions for when justice and fairness are realized. The frontal cortex stores stocks of knowledge about justice norms and criteria for calculating justice, whereas the prefrontal cortex connects these stocks of knowledge to the relevant subcortical emotion centers in ways that are not fully understood but critical to understanding human behavior.

Higher primates are one of the few mammals that can identify their reflection in a mirror as their own image, and as I will argue later, this neurological capacity was selected upon and blown into the behavioral ability to possess multiple types of identities, including those listed in Table 10.3. Since identities are emotionally laced cognitions about self, they are a powerful motive force in human interaction. Again, the prefrontal cortex connects the cognitive part of an identity to subcortically generated emotions, but again, it is unclear just how this is done. It may be, as Burke and Stets (2009) have emphasized, that there is an *identity standard* stored in the frontal cortex, and this standard is evoked by the prefrontal cortex in making decisions about whether or not an identity has been verified by others. And, depending upon whether or not the person perceives the identity standard to be met, the prefrontal cortex sends to subcortical emotion centers signals activating an array of positive or negative emotions. Emotions operate as a positive feedback loop when comparisons of behavioral outputs and reactions of others to these outputs to the identity standard are congruent, whereas the arousal of negative emotions signals that there is disjuncture between the identity standard, on the one side, and the behavioral outputs and responses of others to these outputs, on the other. Again, the neurology here is not clear because sociologists have not designed studies using imaging technologies to determine how the brain is operating during the self-verification process. Psychologists and economists, among the social scientists, have been far more active in imaging studies than sociologists, but the questions posed by sociologists tend to be less interesting to researchers in these disciplines.

The process of calculating justice and fairness in exchanges is complicated by the dynamics briefly outlined in Table 10.3 where comparison points are mentioned. Resource payoffs relative to costs and investments in receiving these payoffs are always made relative to a comparison point or points. Depending upon *which* comparison point or set of points listed in Table 10.3 is invoked, the sense of justice or injustice will vary, and so the key to understanding individuals and collective senses of fairness revolves around the neurology of invoking comparison points—clearly, something done in the prefrontal cortex. Just how this is done is unclear, even after a considerable amount of research by neuroeconomics on decision-making dynamics (Loewenstein et al. 2008).

The other motive states listed in Table 10.3 are also important but not as powerful as verification of self or identity and fairness in exchange payoffs. People do need to feel that they are part of the ongoing interpersonal flow, but this need for *group inclusion* does not necessarily mean high solidarity with others. Individuals always want to feel themselves to be part of an interaction, but their need for solidarity or simple inclusion will vary. Individuals also need to feel that others are *trustworthy*, sincere, and respectful of self. And they have needs for *facticity* or the sense that things are as they seem and that, for the purposes of a given interaction, self and others are experiencing the same obdurate world.

When any of the need states listed in Table 10.3 are not realized, negative emotions are aroused to the point of breaching the encounter. The more powerful the need state and the greater the failure to meet these need, the greater will be the negative emotional reaction, and the more likely will the encounter be disrupted.

Table 10.4 Repression, defense, transmutation, and targeting emotions

Repressed emotions	Defense mechanism:	Transmutation to:	Target of:
Anger, sadness, fear, shame, guilt, and alienation	Displacement	Anger	Others, corporate units, and categoric units
Anger, sadness, fear, shame, guilt, and alienation	Projection	Little but some anger	Imputation of anger, sadness, fear, shame, or guilt to dispositional states of others
Anger, sadness, fear, shame, guilt, and alienation	Reaction formation	Positive emotions	Others, corporate units, categoric units
Anger, sadness, fear shame, guilt, and alienation	Sublimation	Positive emotions	Tasks in corporate units
Anger, sadness, fear shame, guilt, and alienation	*Attribution*	*Anger*	*Others, corporate units, or categoric units*

Because of this potential for highly intense negative emotional arousal (see Tables 19.2 to 19.4 in Chapter 19), individuals motive-take to be sure that they understand which needs among others are most salient and whether or not these needs are being met. They rely mostly on emotional cues to make this determination because negative emotions demand attention because of their disruptive potential. As long as emotions of others remain on the positive side, individuals can relax their motive-taking, but part of the reciprocal of motive-taking—that is, motive-*making*—is to signal not only one's needs but whether or not these needs are being realized in an interaction. Again, the emission of negatively valenced emotional cues is critical to keeping others informed during the process of motive-making.

When individuals have more idiosyncratic need states, they will have to be more demonstrative in their motive-making if they expect others to motive-take with them successfully. And the more idiosyncratic are need states of individuals, the more likely will something go wrong in the situation because signaling unique motives is difficult, particularly since individuals may not be fully conscious of their needs until it is too late. They may only come to recognize that they had a need after it has gone unfulfilled, thereby creating *frustration* to go along with *anger*, *fear*, *hurt*, and other negative emotions. For those motive-taking, they will initially focus on the more universal need states listed in Table 10.3 to be sure that these needs are covered, and the result may be a failure to recognize additional need states that others may bring to an encounter. Thus, failure in motive-taking and motive-making is endemic to human social interaction—once again emphasizing the fragility of sociality and group solidarity among humans.

Emotion-Taking and Emotion-Making

Human interaction is not possible without individuals' capacities to signal and read emotional phonemes strung together by syntax. As Chap. 19 argues, the first language of hominins and later humans was the language of emotions. This language is visually based because humans, as an evolved primate, are visually dominant and thus rely primarily on vision for sensory inputs from the environment. *Emotion-taking* is primarily a process of reading gestures visually, particularly of face but also body movements, positioning, and countenance. Individuals' sources of information on emotions can come through voice inflections via the auditory sense modality (Wernicke's area on the posterial temporal lobe near the inferior parietal lobe) or acts of touching through the haptic sense modality (parietal lobe), but emotion-taking will typically involve efforts to visually verify emotional inputs via other sense modalities.

Emotion-making is, to some degree, outside of volitional control because emotions emanate from subcortical areas of the brain—for example, amygdala, hippocampus, peptide, and neurotransmitter systems—and cannot be completely controlled by the neurons from the frontal lobe and prefrontal cortices, by Broca's area in the temporal area controlling speech production, or by the Sylvian fissure (separating the parietal from frontal lobes) regulating many muscles and the musculoskeletal system. Emotions can be partially filtered through the cortical censors, and humans are such adept emotion-takers that they can generally recognize this distorting effect of the filters and make the appropriate assessments of "the real" emotions that others are experiencing. At times, people reveal considerable expressive control and are able to regulate and orchestrate their emotional displays in accordance with cultural expectations, despite their "real" emotions. Still, if emotions are strong and especially if they are negative, this control is always tenuous, even when negative emotions have been repressed and are out of conscious awareness (see discussion below). Nonetheless, because culture-taking and culture-making inevitably pull emotion ideologies, feeling rules, and display rules from culture (Hochschild 1979, 1983), culture-taking and culture-making inevitably force some efforts at control of emotion-making.

Emotion-taking and emotion-making are complicated by the fact that humans are capable of activating *defense mechanisms* to control their emotional experiences and their displays of emotions to others. Furthermore, defense mechanisms among those emotion-taking can distort what they read in the emotional signals of others. Thus, the most important source of information of many interactions can be distorted either at the behavioral output or at the reflected appraisals stage of emotion-taking. Humans have very large neurons extending down from the prefrontal cortex through the amygdala to other emotion-generating subcortical modules that can push emotional memories from conscious awareness, storing them in the hippocampus, it appears. The result is that people are often unaware of powerful emotions that may affect their behavioral outputs. For example, individuals may be shocked when others tell them (from their emotion-taking efforts) that they seem *sad*, *mad*, *anxious*, *fearful*, or *angry*. Or, a person can often be truly surprised by their sudden emotional reactions to certain events because the emotions had been part of a repressed memory which was stimulated by the current situation. Such is the power of repression and other defense mechanisms. At times, the repression is so complete that the repressed emotions are not even evident at the behavioral output end of interactions, with the result that others cannot effectively emotion-take with a person. Still, repression is never complete, at least in the long run, and over time, those who consistently emotion-take with a person will begin to make attributions (see below) about the sources of this person's emotional responses and dispositions.

When emotions are repressed, they will generally reemerge but often in transmuted forms. For example, as many clinicians would argue, repressed *shame* will often come out as aggression and *anger*. Moreover, repressed emotions will generally increase in intensity over time, and as they do so, they will eventually break through the cortical censors and become an inadvertent part of emotion-making that others pick up in their emotion-taking. Because individuals are implicitly aware of repression of emotions, they typically redouble their efforts at emotion-taking to be sure that they understand another's feelings, and in cultures like those in western societies that have an influential psychiatric subculture culture, people are consciously looking for signs of repression. A lot is at stake in the dynamics revolving around repression because if people cannot honestly emotion-make (that is, "give off" their true feelings) and accurately emotion-take, the most important bonding force in human interaction—the arousal of emotions—can remain ambiguous, thus creating tensions and breaches in ongoing interactions.

If an episodes of interaction are embedded in more inclusive sociocultural formations (e.g., groups, organizations, communities, institutional domains), then expectations for emotions will be clearer than when interactions are not embedded because the expectations follow from culture and are activated in roles and status positions that further specify what emotions should be felt and expressed. Thus, the more structured is an encounter, the more individuals can make assumptions about the emotions of others in their emotion-taking efforts, and the more they can orchestrate emotional displays to conform to expectations tied to culture and social structure. As Hochschild (1983) has argued, individuals must often engage in "emotion work" under conditions where they are expected to feel and display particular emotions that they do not feel. And thus, the more emotion work by individuals, the more likely are unacceptable emotions to be held in check, but like repression, it is even more difficult to hold back emotions that a person is consciously feeling but is forbidden to display. At some point, these emotions will come out and be picked up by others in emotion-taking. And so, like repression, emotion-taking by individuals will often implicitly seek to determine if there are other emotions being held back consciously by individuals. And the longer is an episode of interaction, the more likely are these less acceptable emotions to become part of emotion-making by a person and, hence, part of the emotional information gleaned by emotion-taking by others.

The neurology of these dynamics is not well understood, but the large increase in the size of the hippocampus in humans compared to other higher primates (Turner 2000; Stephan 1983; Stephan and Andy 1969, 1977; Stephan and Baron 1981; Stephan et al. 1988; Eccles 1989), controlling for body size which is correlated with brain size, is one indicator that natural selection on humans and their hominin ancestors created a subcortical area to store unconscious memories, but this area is also part of the normal memory dynamics of all mammals. On the one side through the transition cortices, working memory is sustained, with short-term memories being shuttled to the hippocampus for storage. If these memories are activated again, especially with their original emotional tags (what Damasio 1994, termed the "somatic marker hypothesis"), then they are more likely to be remembered and after 2 years shipped up to the frontal lobe for longer term memory and often marked by cultural symbols. From the other side, unpleasant and painful emotions tied to particular situations can be shipped down from the prefrontal lobe to the hippocampus to remove them from conscious awareness, and they can be kept there for long periods of time. Yet, since the hippocampus is designed to ship memories back to the frontal lobe, it is often difficult to hold them from consciousness. Still, if the cortical blockers are powerful and the memory stays in the hippocampus for long periods of time, a person may remain unaware of certain painful memories. Yet, this person may still reveal some of these emotions in emotion-making and, as noted above, may remain unaware that these emotions are "leaking" out and becoming part of others emotion-taking.

Attribution-Taking and Attribution-Making

The dynamics of attribution have been well studied in psychology (e.g., Weiner 1986), but I want to put a more sociological slant to this large literature. People make causal attributions for their experiences, and especially those experiences that arouse emotions, whether positive or negative. Thus, attribution dynamics are activated almost anytime; people's efforts at role-taking and role-making, status-taking and status-making, culture-taking and culture-making, and motive-taking and motive-making become implicated in emotion-taking and emotion-making.

In addressing attribution dynamics, I am revising the terminology of much attribution theory employed in psychology to fit my more sociological purposes. When individuals make *self-attributions*, they are seeing themselves as responsible for their emotional experiences. When they make *internal attributions* in others, they are seeing the internal states and dispositions of others (as determined by role-, status-, culture-, motive-, and emotion-taking) as the cause of their emotional experiences. When they make *external attributions*, they see not only others but the sociocultural context and larger social structures as responsible for their emotional experiences.

I have argued that, from a sociological perspective, attribution is the most important defense mechanism after repression, a point not pursued in the general literature within psychology on attribution (Turner 2008). When individuals make internal or external attributions, they often do so to protect an identity from self-attributions for negative emotional experiences. And they do so for powerful and painful emotions like *shame* and *guilt* as well as emotions such as *fear*, *anger*, and *sadness* that have made them feel ashamed and guilty (see Table 19.4 in Chapter 19 for a model on the structure of guilt and shame). As Edward Lawler (2001) first stressed, negative emotions have a *distal bias*, with individuals blaming others, situation, and social structures for their negative emotional experiences. They often do so through repression, removing self-attributions as a potential option. In contrast, positive emotional experiences have a *proximal bias*, with individuals seeing themselves or others in the immediate situation as responsible for their feelings. These biases—distal for negative and proximal for positive emotions—have large effects on interactions and people's commitments to others and social structures (Turner 2008), and hence, it is important to understand the neurology of what occurs.

Since identity verification is the most important motive state for humans, repression of negative emotions about self is one means for meeting this need state. Once repressed, the emotion will intensify and often transmute to a more intense or entirely different emotions. One way to release the pressure that builds up from repression of negative emotions is to make external attributions and blame others, local situations, and more meso- and macro-level social structures by venting either more intense or transmuted negative emotions on them. It is more likely that individuals will vent these emotions on remote structures because immediate others in an encounter can fight back and negatively sanction a person (thus attacking this person's identity). At the same time, blaming the local situation in which a person must interact on a regular basis only spoils the nest and generates negative responses from others in the situation. Thus, it is safer to blame social structures or categories of others for negative emotions and to dump the intensified and/or transmuted emotions on remote objects that cannot directly attack and sanction self.

Table 10.4 summarizes some of these attribution dynamics by connecting them to negative emotional arousal, repression as the master defense mechanism, and activation of secondary defense mechanism, among which *attribution* is the most sociologically important because it affects people's commitments to meso- and macro-level social structures (this point is emphasized by bold-facing attribution processes in Table 10.4). Thus, attribution allows individuals to "blow steam" outward toward safer objects from the pressures that build up in the pressure cooker that is created by repression of negative emotions, especially emotion directed at self. As the center for fear and anger, the amygdalae, coupled with the neurotransmitter and neuroactive peptide system that can generate *sadness* and *depression*, are all involved. Yet, just how conscious awareness in the prefrontal cortex of the transmutation and targeting of attributions interacts with these subcortical modules in the brain remains unclear, but from a sociological perspective, this is an important arena of potential research.

People are almost always attribution-making; they assert who or what is responsible for particular emotional states, and if these states are salient in the situation (and often, when they are not), they go about asserting their attributions, seeking to make them the attributions of others and the group. The tendency to make one's attributions known to others provides, of course, the fodder for attribution-taking where each participant to an interaction seeks to determine the causal attributions of the others. If these stand in conflict, some negotiation will be necessary, but even then, the interaction may reveal an underlying tension because individuals are not making the same attributions. However, when attributions line up, and especially when they target others or social structures external to the group for negative emotional experiences, attribution dynamics are critical to creating and sustaining local group solidarity, while reducing commitments to targets of negative external attributions. Or, if the attributions are for positive emotional experiences, consensus over attributions can increase local group solidarities, while at the same time, increasing commitments of group members to larger-scale social units such as organizations, communities, institutions, and even whole societies (Lawler 2001; Lawler et al. 2009; Turner 2002, 2008, 2010).

A Distant Mirror: Clues About the Origins of Human Behavioral Capacities

What Did Natural Selection Have to Work on?

The behavioral capacities of humans in interaction are the end points of long-term evolution of the hominin ancestral line. Unlike Mead and many contemporary sociologists' views, these behavioral abilities are not *all* unique to humans. We are evolved apes, and thus, our closest relatives in the primate order—the great apes (chimpanzee, gorilla, and orangutan)—offer clues about what our distant ancestors were like because we shared a common ancestor with extant apes some eight million years ago. By examining the abilities of present-day apes, we can gain insight about the basic materials that natural selection was given as it developed the hominin line culminating in *Homo sapiens sapiens*.

What, then, are the key behavioral capacities of living primates, particularly the great apes but other primates as well? If humans' closest relatives all possess these capacities, it is likely that they were present in the last common ancestor to humans and extant primates. Hence, they were subject to millions of years of selection to hone the fine-grained and complex interpersonal processes summarized above. Let me simply list some of these behavioral capacities of present-day primates (references cited are the data source, particularly on the neurological mechanisms that generate these capacities):

1. Visual dominance over haptic and auditory sense modalities, thereby subordinating other sense modalities to vision (Maryanski and Turner 1992; Ettlinger 1977; Jarvis and Ettlinger 1977; Passingham 1982:51–55).
2. Behavioral propensity to follow the gaze and eye movements of others (Hare et al. 2001, 2006; Povinelli 2000; Itakura 1996; Baizer et al. 2007).
3. Ability of infants to imitate orofacial movements of caretakers within weeks of birth (Emde 1962; Ekman 1984; Sherwood et al. 2004).
4. Propensity for face-monitoring for signs of action of conspecifics, particularly for emotional content (Leslie et al. 2004; Gazzaniga and Smylie 1990).
5. Capacity to communicate meanings and coordinate actions through nonverbal signals, especially by eyes and face (Menzel 1971; Turner and Maryanski 2008).
6. Enhanced propensity to use imitation to learn appropriate signals and behaviors (Tomonaga 1999; Subiaul 2007; Horowitz 2003; Gergely and Csibra 2006).
7. Increased cortical control of subcortical areas through projections of axons (Raghanti et al. 2008; Sherwood 2007; Sherwood et al. 2005).
8. Enlarged decision-making prefrontal cortex (Semendeferi et al. 2002; Rilling and Insel 1999).
9. The ability among the great apes to learn and use language at the level of a 3-year-old human child (Geschwind 1965a, b, c; Damasio and Geschwind 1984; Rumbaugh and Savage-Rumbaugh 1990; Savage-Rumbaugh and Lewin 1994; Savage-Rumbaugh et al. 1988, 1993; Bickerton 2003).
10. The ability to recognize an image in a mirror as a reflection of self as an object in the environment (Gallup 1970, 1979, 1982).
11. Rhythmic synchronization, especially of emotions, via mirror neurons (Rizzolattti et al. 2002).
12. Propensities for reciprocity in the give and take of resources (Cosmides 1989; deWaal 1989, 1991, 1996).
13. Propensity to compare shares of resources with others in making judgments of fairness (Brosnan et al. 2005).
14. Capacity to experience variants among primary emotions, particularly among happiness, fear, anger, and sadness (Darwin 1872; Turner 2000; Parr et al. 2005).
15. Capacity to experience empathy with conspecifics (deWaal 1996, 2009).

By discovering the neurological modules responsible for these behavioral capacities and propensities of current higher-order primates, we will go a long way to understanding *what modules natural selection worked on to produce the more extensive abilities of humans* (Passingham 1973, 1975; Sherwood et al. 2008). Many of the references cited for the fifteen behavioral capacities listed above document at least some of the modules involved. Still there is much more detailed work to be done.

The human brain is about three times as large as that of our closest relative, the common chimpanzee, and so one obvious change in neurology was growth of the brain. The neocortex gets much of the press in highlighting the uniqueness of the human brain, but in fact, subcortical areas of the human brain where emotions are generated are, on average, twice as large as their counterparts in great ape brains. Moreover, I have hypothesized that subcortical areas of our hominin ancestors grew much earlier than subsequent increases in the size of the neocortex, suggesting that natural selection first enhanced hominin's emotional capacities long before going to work on expanding the neocortex and increasing cognitive abilities (Turner 2000). Thus, by taking measurement on, and tracing new neuronets among, the areas of the brain of humans for the fifteen capacities listed above, we can gain considerable insight into the neurology of human interaction.

The Directional Bias of Natural Selection During Hominin Evolution

As Alexandra Maryanski and I outlined in Chap. 19, apes are not highly social in at least one clear sense: They do not form permanent social relations at the group level. Moreover, a cladistic analysis of the last common ancestor to humans and the great apes indicates that this ancestor was individualistic, mobile, and isolated, with virtually no permanent ties to conspecifics beyond what were necessary for infant care by mothers and reproductive acts between adult males and females. This lack of social structures doomed most apes to extinction when the forests of Africa began to recede and forced many primates to live on the predator-ridden savanna (see Chap. 19, as well as Maryanski and Turner 1992; Turner and Maryanski 2008). Apes were not sufficiently organized at the group level to survive, and none of the features of apes listed above would dramatically promote increased fitness of group organization because none leads directly to group-organizing propensities. Rather, some can increase the subtly and complexity of interaction, but they do not represent group-level bio-programmers, and so, it should not be surprising that natural selection worked to enhance groupness among hominins by selecting on the interpersonal capacities of apes in the distant past.

Nor would anatomical features of apes that evolved to live in trees, not on the ground, be of much help in increasing ape fitness in terrestrial habitats. Hence, the number of species of apes declined rapidly over the last ten million years, with just the handful now left and, ironically, facing extinction by the one line of apes—Homo—that somehow beat the odds and developed sufficient levels of organization to survive the savanna and, now, to encroach on the habitats of their closest relatives.

But making a low-sociality animal like an ape, more social is not easy. Mutations as an evolutionary process are too slow. Moreover, mutations on neurology are almost always harmful, as R. A. Fisher (1930) for all of his warts pointed out long ago. Thus, selection worked on the tail ends of Bell curves of existing neurological features and behavioral capacities to move hominins in a particular direction, which enhanced interpersonal sociality that eventually led to tighter-knit social organization at the group level. By looking down the fifteen capacities listed earlier, we could construct a thought experiment how selection went to work to make low-sociality apes more social (for details, see Chap. 14). In this thought experiment, we need to do is ask the question: Could these features of apes enhance sociality that could eventually lead to more cohesive group structures, thereby increasing fitness? We already know the end point: the basic processes involved in interaction outlined in section "Using Mead's Theory to Expand the Conceptualization of Interaction Dynamics", and so, starting with the list above, we can construct a scenario about how selection moved these capacities and, along the way, added capacities to get an animal capable or role-taking and role-making, status-taking and status-making, culture-taking and culture-making, motive-taking and motive-making, emotion-taking and emotion-making, and attribution-taking and attribution-making. Such an animal can develop bonds of sociality and dramatically increase group solidarity, thereby increasing its fitness in a predator-ridden open country savanna.

Because neurology is complex and because it has such far-reaching effects on behavior, the modules of the brain have evolved slowly, building upon the basic structure of all mammalian brains. To understand why particular modules have evolved, it is useful to take a long evolutionary view in which changes to the structure of the brain of the last common ancestor to humans and modern apes are examined with an eye toward what they can tell us about the selection pressures that were rebuilding key modules. Clearly, selection was rewiring the brain to make low-sociality apes more social and capable of forming permanent groupings. Hominin evolution began with selection pushing for increased sociality, which means interactions that forged social bonds.

What Processes Increase Sociality and Bonds of Solidarity?

How do humans form bonds of social solidarity? When copresent, people rely upon visual cues from their dominant sense modality, and as would be expected given the wiring of the brain, they first read body language signaling emotional disposition as revealed by face, eyes, hands, body positioning, and movements. They rely secondarily on talk and voice inflections. They emit greeting rituals that establish the interpersonal mood of an encounter, and they use rituals throughout the encounter to frame and reframe the interaction during the process of culture-taking and culture-making. Once framed, they can readily categorize self and others, use appropriate forms of talk and body language, and display appropriate emotional dispositions. During these dynamics of culture-taking and culture-making, they role-take and role-make, status-take and status-make, and motive-take and motive-make. In so doing, they can understand the dispositions of others and make roles and occupy status locations that will enable them to successfully motive-make and emotion-make, especially with regard to needs for self-verification and fair exchanges of resources.

As the interaction proceeds, talk and body language will fall into a rhythm and synchronization, and as this synchronization ensues, individuals become emotionally entrained which, in turn, escalates the flow of positive emotions (Collins 2004). As positive emotions circulate, individuals make self and internal attributions for their feelings, seeing self and others in the encounter as responsible for these feelings (in accordance with the proximal bias). As positive emotions flow, individuals perceive that they have verified self and identities and received valuable resources from others through role-, status, culture-, motive-, emotion-, and attribution-taking and role-, status-, culture-, motive-, emotion-, and attribution-making. The longer this synchronization and arousal of positive emotional energy ensues, the more likely are the emerging bonds of solidarity to be symbolized in some way with totems, whether real objects, words, and phrases marking group boundaries and signifying the bonds of solidarity that have emerged (Collins 2004). In this way, group relations can be moralized.

When seen in this light, forming stronger bonds of solidarity is not only complicated but also a great deal of work. Humans, as evolved apes, do not have powerful bio-programmers for group formation; instead, group formation *must be achieved indirectly* via the arousal of positive emotions through a series of interpersonal processes. Groups are a remarkable accomplishment for humans, not some automatic response to copresence. We need to *work at* forming social bonds, and we must do so each and every time an encounter is formed and iterated.

Thus, one of neurosociology's goals is to determine which brain modules are involved in these interpersonal dynamics that lead to increased sociality and solidarity among humans. That is, what is the neurology behind role-taking and role-making, status-taking and status-making, culture-taking and culture-making, motive-taking and motive-making, emotion-taking and emotion-making, and, attribution-taking and attribution-making? A theory of mind, in my view, must isolate the modules involved in all of these processes. Research using imaging technologies can perhaps provide some of the needed information on the neurology of interpersonal behavior. Yet another tool is comparative neuroanatomy that compares the brains of the great ape with those of humans; the differences in these brains will, I believe, be primarily differences generated by natural selection to enhance humans capacity for role-taking and role-making, status-taking and status-making, culture-taking and culture-making, motive-taking and motive-making, emotion-taking and emotion-making, and attribution-taking and attribution-making. What sociology offers the theory of mind, then, are some specific dimensions of interpersonal behavior that can, to a very large extent, be caused by the neurology of the human brain.

There may be clues provided by imaging studies on higher primates because many of the dimensions of human interpersonal activity are built from modules in higher-primate brains. For example, if reciprocity, justice calculations, empathy, language production, self-perception, attention to face, rhythmic synchronization, and emotional arousal are all part of the behavioral repertoire of the great apes, the neurology involved should be observable through imaging studies on higher primates, thereby providing clues as to the neurological basis of these same behaviors—perhaps greatly enhanced by alterations of neurology—in humans.

Conclusions

I have tried to recast the theory of mind by extending the early insights of George Herbert Mead (1934, 1938). In so doing, I have incorporated at a very abstract but still fundamental level the interpersonal processes that drive the theory of mind for humans. As an evolved ape, it should not be surprising that many of these processes can also be observed in higher primates, and indeed, we should use imaging research to study these behaviors in primates and compare findings with those on humans. The comparison will give us data on how natural selection reworked the basic brain of great apes to made humans capable of forming stronger social bonds, developing more permanent group structures, and increasing the solidarity of individuals in these group structures. This was a remarkable accomplishment since most apes went extinct over the last ten million years. For a small set of species on the human clade (hominins), however, natural selection was able to do what it had failed to do for virtually all other apes that had ever existed: Make them more social and group-oriented.

We can still see how difficult it is for humans—despite significant rewiring of our neurology—to create and sustain cohesive groups. It is a lot of work for humans to interact, and indeed, one reason that texting is so popular today, I would hypothesize, is that it allows humans to sustain a sense of group inclusion without all of the interpersonal work involved in role-taking and role-making, status-taking and status-making, culture-taking and culture-making, motive-taking and motive-making, emotion-taking and emotion-making, and attribution-taking and attribution-making. Once in place, group structures can be elaborated into larger-scale structures like organizations that, in turn, can reduce the interpersonal burdens on people by formalizing status, roles, culture, motive states, and emotional responses. Indeed, Max Weber's (1922) famous typology of bureaucracy can be view as a sociocultural formation that *reduces* the interpersonal burdens on its incumbents, although as the work of Hochschild (1979, 1983) documents, it can also increase the emotional burdens of individuals.

Social thinkers in general, and sociologists in particular, have tended to perceive of humans as *naturally* social and as *naturally* forming cohesive groups like family. Such arguments make an assumption that humans are like most other mammals: driven by biogrammers in the brain for troop, pack, herd, pod, and group behaviors. But, humans are apes, and apes are weak tie and not very social animals, despite the many efforts at anthropomorphizing them by imputing "families" and other such human social structures to them. In reality, people must work at forming and sustaining groups, and tension, conflict, and volatility are very common in groups because we have no natural bio-programmers to override negatively charged emotions, as do most other mammals. In fact, the popularity of "reality TV" can be partly explained by humans' fascination with how others try and often fail to organize cohesive social bonds.

Humans must rely upon the neurological systems, built up from their primate ancestors, to become well organized, and once this angle of vision is assumed, the task of neurosociology is to discover the brain modules by which this capacity for group organization is made possible. If we simply assume that groups are natural, there is little neurology to explore. If we assume that it is *not natural* or easy but an uneasy balance between our ape ancestry and rewired brains, the challenge to neurosociology and neuroscience in general is clear.

References

Baizer, J. S., Baker, J. F., Haas, K., & Lima, R. (2007). Neurochemical organization of the nucleus *Paramedianus dorsalis* in the Human. *Brain Research, 1176*, 45–52.

Berger, J. (1988). Directions in expectation states research. In M. J. Webster & M. Foschi (Eds.), *Status generalization: New theory and research.* Stanford: Stanford University Press.

Berger, J., & Zelditch, M. (1985). *Status, rewards, and influence.* San Francisco: Josey-Bass.

Berger, J., Cohen, B. P., & Zelditch, M., Jr. (1972). Status characteristics and social interaction. *American Sociological Review, 37*, 241–255.

Berger, J., Hamit Fisek, M., Norman, R. Z., & Zelditch, M., Jr. (1977). *Status characteristics and social interaction: An expectation-states approach.* New York: Elsevier.

Berger, J., Norman, R. Z., Balkwell, J. W., & Smith, R. F. (1992). Status inconsistency in task situations: A test of four status processing principles. *American Sociological Review, 57*, 843–855.

Bickerton, D. (2003). Symbol and structure: A comprehensive framework for language evolution. In M. S. Christiansen & S. Kirby (Eds.), *Language evolution: The states of the art* (pp. 77–93). Oxford: Oxford University Press.

Brosnan, S. F., & de Waal, F. B. M. (2003). Fair refusal by capuchin monkeys. *Nature, 425*, 128–140.

Brosnan, S. F., Schiff, H. C., & de Waal, F. B. M. (2005). Tolerance for inequity may increase social closeness in chimpanzees. *Proceedings of the Royal Society of London, 272*, 253–258.

Burke, P. J., & Stets, J. E. (2009). *Identity theory.* New York: Oxford University Press.

Call, J., & Tomasello, M. (2008). Do chimpanzees have a theory of mind: 30 years later. *Trends in Cognitive Science, 12*, 187–192.

Clark, C. (1987). Sympathy biography and sympathy margin. *American Journal of Sociology, 93*, 290–321.

Clark, C. (1990). *Misery loves company: Sympathy in everyday life.* Chicago: University of Chicago Press.

Collins, R. (2004). *Interaction ritual chains.* Princeton: Princeton University Press.

Cosmides, L. (1989). The logic of social exchange: Has natural selection shaped how humans reason? *Cognition, 31*, 187–276.

Cosmides, L., & Tooby, J. (1992). Cognitive adaptations for social exchange. In J. H. Barkow, L. Cosmides, & J. Tooby (Eds.), *The adapted mind: Evolutionary psychology and the generation of culture* (pp. 19–136). New York: Oxford University Press.

Damasio, A. (1994). *Descartes' error: Emotion, reason, and the human brain.* New York: G. P. Putman.

Damasio, A., & Geschwind, N. (1984). The neural basis of language. *Annual Review of Neuroscience, 7*, 127–147.

Darwin, C. (1872). *The expression of the emotions in man and animals.* London: John Murray.

de Waal, F. B. M. (1989). Food sharing and reciprocal obligations among chimpanzees. *Journal of Human Evolution, 18*, 433–459.

de Waal, F. B. M. (1991). The chimpanzee's sense of social regularity and its relation to the human sense of justice. *American Behavioral Scientist, 34*, 335–349.

de Waal, F. B. M. (1996). *Good natured: The origins of right and wrong in humans and other animals.* Cambridge: Harvard University Press.

de Waal, F. B. M. (2009). *The age of empathy: Nature's lessons for a kinder society.* New York: Three Rivers Press.

Eccles, J. C. (1989). *Evolution of the brain: Creation of self.* London: Routledge.

Ekman, P. (1984). Expression and the nature of emotion. In K. Scherer & P. Edman (Eds.), *Approaches to emotion* (pp. 319–343). Hillsdale: Lawrence Erlbaum.
Emde, R. N. (1962). Level of meaning for infant emotions: A biosocial view. In W. A. Collins (Ed.), *Development of cognition, affect and social relations* (pp. 1–37). Hillsdale: Lawrence Erlbaum.
Ettlinger, G. (1977). Cross-modal equivalence in non-human primates. In A. M. Schriver (Ed.), *Behavioral primatology* (Vol. 1). Hillsdale: Erlbaum.
Fisher, R. A. (1930). *The genetical theory of natural selection*. Oxford: Clarendon.
Gallup, G. G., Jr. (1970). Chimpanzees: Self-recognition. *Science, 167*, 86–87.
Gallup, G. G., Jr. (1979). *Self-recognition in chimpanzees and man: A developmental and comparative perspective*. New York: Plenum Press.
Gallup, G. G., Jr. (1982). Self-awareness and the emergence of mind in primates. *American Journal of Primatology, 2*, 237–248.
Gazzaniga, M. S., & Smylie, C. S. (1990). Hemisphere mechanisms controlling voluntary and spontaneous mechanisms. *Annual Review of Neurology, 13*, 536–540.
Gergely, G., & Csibra, G. (2006). Sylvia's recipe: The role of imitation and pedagogy. In N. J. Enfield & S. C. Levinson (Eds.), *The transmission of cultural knowledge* (pp. 229–255). Oxford: Berg Press.
Geschwind, N. (1965a). Disconnection syndromes in animals and man, part I. *Brain, 88*, 237–294.
Geschwind, N. (1965b). Disconnection syndromes in animals and man, part II. *Brain, 88*, 585–644.
Geschwind, N. (1965c). Disconnection syndromes in animals and man. *Brain, 88*, 237–285.
Goffman, E. (1959). *The presentation of self in everyday life*. New York: Penguin.
Goffman, E. (1967). *Interaction ritual*. Garden City: Anchor Books.
Goffman, E. (1974). *Frame analysis: An essay on the organization of experience*. New York: Harper and Row.
Goffman, E. (1981). *Forms of talk*. Philadelphia: University of Pennsylvania Press.
Goffman, E. (1983). The interaction order. *American Sociological Review, 48*, 1–17.
Hare, B., Call, J., & Tomasello, M. (2001). Do chimpanzees know what conspecifics know? *Animal Behavior, 61*, 139–159.
Hare, B., Call, J., & Tomasello, M. (2006). Chimpanzees deceive a human competitor by hiding. *Cognition, 101*, 495–514.
Hochschild, A. R. (1979). Emotion work, feeling rules and social structure. *American Journal of Sociology, 85*, 551–575.
Hochschild, A. R. (1983). *The managed heart: Commercialization of human feeling*. Berkeley: University of California Press.
Horowitz, A. C. (2003). Do chimps ape? Or apes human? Imitation and intension in humans (*Homo sapiens*) and other animals. *Journal of Comparative Psychology, 117*, 325–336.
Itakura, S. (1996). An exploratory study of gaze-monitoring in non-human primates. *Japanese Psychological Research, 38*, 174–180.
Jarvis, M. J., & Ettlinger, G. (1977). Cross-modal recognition in chimpanzees and monkeys. *Neuropsychologia, 15*, 499–506.
Kemper, T. D., & Collins, R. (1990). Dimensions of microinteraction. *American Journal of Sociology, 96*, 32–68.
Lawler, E. J. (2001). An affect theory of social exchange. *American Journal of Sociology, 107*, 321–352.
Lawler, E. J., Thye, S., & Yoon, J. (2009). *Social commitments in a depersonalized world*. New York: Russell Sage.
Leslie, K. R., Johnson-Frey, S. H., & Grafton, S. T. (2004). Functional imaging of face and hand imitation: Towards a motor theory of empathy. *NeuroImage, 21*, 601–607.
Loewenstein, G., Rick, S., & Cohen, J. (2008). Neuroeconomics. *Annual Reviews, 59*, 647–672.
Maryanski, A., & Turner, J. H. (1992). *The social cage: Human nature and the evolution of society*. Stanford: Stanford University Press.
McCall, G. P., & Simmons, J. L. (1978). *Identities and interactions*. New York: Basic Books.
Mead, G. H. (1934). *Mind, self, and society*. Chicago: University of Chicago Press.
Mead, G. H. (1938). *Philosophy of the act*. Chicago: University of Chicago Press.
Menzel, E. W. (1971). Communication about the environment in a group of young chimpanzees. *Folia Primatologica, 15*, 220–232.
Parr, L. A., Waller, B. M., & Fugate, J. (2005). Emotional communication in primates: Implications for neurobiology. *Current Opinion in Neurobiology, 15*, 716–720.
Passingham, R. E. (1973). Anatomical differences between the Neo-cortex of man and the other primates. *Brain Behavioral Evolution, 7*, 337–359.
Passingham, R. E. (1975). Changes in the size and organisation of the brain in man and his ancestors. *Brain and Behavior Evolution, 11*, 73–90.
Passingham, R. E. (1982). *The human primate*. Oxford: Freeman.
Povinelli, D. J. (2000). *Folk physics for apes: The chimpanzee's theory of how the world works*. Oxford, UK: Oxford University Press.

Premack, D., & Woodruff, G. (1978). Does the chimpanzee have a theory of mind. *Behavior and Brain Science, 1*, 515–526.
Raghanti, M. A., Stimpson, C. D., Marchiewicz, J. L., et al. (2008). Differences in cortical serotonergic innervation among humans, chimpanzees, and macaque monkeys: A comparative study. *Cerebral Cortex, 18*, 584–597.
Rilling, J. K., & Insel, T. R. (1999). The primate neocortex in comparative perspective using magnetic resonance imaging. *Journal of Human Evolution, 42*, 505–533.
Rizzolattti, G., Fadiga, L., Fogassi, L., & Gallese, V. (2002). From mirror neurons to imitation: Facts and speculations. In W. Prinz & A. N. Meltzoff (Eds.), *The imitative mind: Development, evolution and brain bases* (pp. 247–266). Cambridge: Cambridge University Press.
Rumbaugh, D., & Savage-Rumbaugh, E. S. (1990). Chimpanzees: Competencies for language and numbers. In W. Stebbins & M. Berkley (Eds.), *Comparative perception* (Vol. 2). New York: Wiley.
Savage-Rumbaugh, S., & Lewin, R. (1994). *Kanzi: The ape at the brink of the human mind*. New York: Wiley.
Savage-Rumbaugh, S., Seveik, R., & Hopkins, W. (1988). Symbolic cross-model transfer in two species. *Child Development, 59*, 617–625.
Savage-Rumbaugh, S., Murphy, J., Seveik, J., Brakke, K., Williams, S. L., & Rumbaugh, D. (1993). *Language comprehension in the ape and child* (Monographs of the Society for Research in Child Development, 58). Chicago: University of Chicago Press.
Saxe, R., et al. (2006). Reading minds versus following rules: Dissociating theory of mind and executive control in the brain. *Social Neuroscience, 1*, 284–298.
Schutz, A. (1967 [1932]). *The phenomenology of the social world*. Evanston: Northwestern University Press.
Semendeferi, K., Lu, A., Schenker, N., & Damasio, H. (2002). Humans and great apes share a large frontal cortex. *Nature Neuroscience, 5*, 272–276.
Sherwood, C. C. (2007). The evolution of neuron types and cortical histology in apes and humans. In T. M. Preuss & J. H. Kaas (Eds.), *Evolution of nervous systems 4: The evolution of primate nervous systems* (pp. 355–378). Oxford: Academic.
Sherwood, C. C., Holloway, R. L., Erwin, J. M., & Hoff, P. R. (2004). Cortical orofacial motor representation in old world monkeys, great apes and humans. *Brain Behavior and Evolution, 63*, 82–106.
Sherwood, C. C., Holloway, R. L., Semendeferi, K., & Hoff, P. R. (2005). Is prefrontal white matter enlargement a human evolutionary specialization? *Nature Neuroscience, 8*, 537–538.
Sherwood, C. C., Subiaul, F., Tadeusz, H., & Zawidzki, W. (2008). A natural history of the human mind: Tracing evolutionary changes in brain and cognition. *Journal of Anatomy, 212*, 426–454.
Stephan, H. (1983). Evolutionary trends in limbic structures. *Neuroscience and Biobehavioral Review, 7*, 367–374.
Stephan, H., & Andy, O. J. (1969). Quantitative comparative neuroanatomy of primates: An attempt at phylogenetic interpretation. *Annals of the New York Academy of Science, 167*, 370–387.
Stephan, H., & Andy, O. J. (1977). Quantitative comparison of the amygdala in insectivores and primates. *Acta Antomica, 98*, 130–153.
Stephan, H. F., & Baron, B. (1981). New and revised data on volumes of brain structures in insectivores and primates. *Folia Primatoligica, 35*, 1–29.
Stephan, H., Baron, G., & Frahm, H. (1988). Comparative size of brains and brain components. In H. Steklis & J. Erwin (Eds.), *Neurosciences* (Vol. 4). New York: Alan Liss.
Stryker, S. (1980). *Symbolic interactionism: A structural version*. Menlo Park: Benjamin-Cummings.
Subiaul, F. (2007). The imitation faculty in monkeys: Evaluating its features, distribution, and evolution. *Journal of Anthropological Science, 85*, 35–62.
Tomonaga, M. (1999). Attending to the others' attention in Macaques' joint attention or not? *Primate Research, 15*, 425.
Turner, R. H. (1962). Role-taking: Processes versus conformity. In A. Rose (Ed.), *Human behavior and social processes* (pp. 20–40). Boston: Houghton Mifflin.
Turner, J. H. (2000). *On the origins of human emotions: A sociological inquiry into the evolution of human affect*. Stanford: Stanford University Press.
Turner, J. H. (2002). *Face to face: Toward a theory of interpersonal behavior*. Palo Alto: Stanford University Press.
Turner, J. H. (2008). *Human emotions: A sociological theory*. Oxford: Routledge.
Turner, J. H. (2010). *Theoretical principles of sociology* (Microdynamics, Vol. 2). New York: Springer.
Turner, J. H., & Maryanski, A. (2008). *On the origins of societies by natural selection*. Boulder: Paradigm Press.
Weber, M. (1968 [1922]). *Economy and society*. Berkeley: University of California Press.
Weiner, B. (1986). *An attribution theory of motivation and emotion*. New York: Springer.

Chapter 11
Relationships Between Neurosociology, Foundational Social Behaviorism, and Currents in Symbolic Interaction

David D. Franks

Of the three categories above, the most difficult to define is symbolic interaction. Easiest in a sense is social behaviorism in which George Herbert Mead did the defining for those willing to read his various lectures as in *Mind, Self and Society* (1934), *The Philosophy of the Act* (1938), as well as *The Philosophy of the Present* (1959), which contains the Carus Lectures presented at Berkeley the year before his death in 1931. Even though it was not highly developed in his time, he insisted that the physiological mechanisms underlying the social were indispensable to understanding minds, and in several places he made this point. For example,

> …This experienced world does not appear except when the various excitements reach certain points in the central nervous system; it is also true that if you cut off any of those channels you wipe out so much of that world.
>
> What the behaviorist… ought to do is to take the complete act… the whole process of conduct…as his unit in his analysis. In doing that he has to take into account, not simply the nervous system, but also the rest of the organism, for the nervous system is only a specialized part of the entire organism. (1934:111)

For further examples, on pages 111–115 in *Mind Self and Society*, he mentions the central nervous system at least 17 times. As Clark McPhail[1] points out, Mead worked in what is now called neuroscience before joining Dewey at the University of Chicago. This is an important point because the vast majority of current symbolic interactionists today are not aware of this and even may or may not be interested in Mead's works.

As this author remembers, symbolic interaction (SI) was promulgated as a society in the late 1960s and was born in large part as a reaction to Talcott Parsons and traditional structural sociology. For a significant part of the membership since the 1980s, contemplating what Mead said was a matter of unproductive interpretation. They rejected the third person, scientific voices espousing "truth statements" like Mead's.

Instead there was an emphasis on the new and the untraditional. At times, annual sessions took the form of theatrical performances and especially ethnographies. David Maines had to urge his colleagues away from the extremes of the postmodernist "linguistic turn" wherein "everything was within the text." "New readings," wrote Denzin (1992: 513), "are justified not because they yield more truth, but because they are new." Maines (1995) countered that "… if the new is good because it's new and if the old is bad because it's boring, then what do we do if we come up with something

[1] Private communication. Prof. McPhail was influential in SI and was its distinguished lecturer in 2005. He brought Mead's initial experience as a neurological researcher at the University of Chicago to my attention.

D.D. Franks (✉)
Department of Sociology, Virginia Commonwealth University, Richmond, VA 23284, USA
e-mail: daviddfranks@comcast.net

D.D. Franks and J.H. Turner (eds.), *Handbook of Neurosociology*, Handbooks of Sociology and Social Research, DOI 10.1007/978-94-007-4473-8_11, © Springer Science+Business Media B.V. 2013

old that works?" As we shall see, this is especially true in regard to the pragmatic tradition of Dewey and Mead.

Herbert Blumer, Mead's teaching assistant, became known as his "spokesman" and coined the term symbolic interaction. In doing so, he moved SI significantly far from the social behaviorism of Mead and the early American pragmatism. (See McPhail and Rexcoat 1979.) Because of Blumer's influence, the interpretive symbol as an intangible word became the emphasis, and the pragmatic notion that words subserve action was lost.

The diversity within the SI society should not be ignored though that is not the point of this essay. A few were even interested in neuroscience. There was always a friendly tension between Denzin espousing ethnographic interpretations of movies and Carl Couch, who was committed to developing general principles of group processes and traditional hypothesis testing. Sheldon Stryker was another adherent of Mead's who diverged from the members of the society by espousing a "social structuralist position" (see Stryker 1985). For many SI members, the objectivity of neuroscience, if they attended to it at all, was really a construction of male domination. Auto-ethnologies came into vogue, and for many an exclusive reliance on qualitative methods replaced the impersonal character of traditional science. There was a decided move away from Mead's behaviorist pragmatism and toward a complete relativization of Mead. This was in spite of his clear insistence on maintaining the "possibility of error" which extreme relativism does not allow because it implies a truth from which that error diverges. The possibility of error is also an *a prior* assumption of hypothesis-testing. The new view implied that traditional science was just another story.

It is important to keep these two versions of symbolic interaction separate so one does not confuse Mead's foundational efforts with current tendencies. There is also significant variation among current symbolic interactionists. For example, Clark McPhail was another SSSI member who took a more traditional methodological position in studying collective behavior and was very interested in Mead's interest in the neuroscience of his day. A third, less visible group teach and use foundational concepts but are not members of the association. Examples of these are Victor Gecas and Andrew Weigert as well as Darwin Thomas.

Symbolic interaction would seem to be as different from neuroscience as any field in sociology; if brain science can be seen as playing a relevant role in symbolic interaction, it has passed the most difficult test of its relevancy to all of sociology.

This chapter will demonstrate that neurosociology can confirm important concepts in foundational symbolic interaction and more recent SI and also broaden its range. I will discuss neurological insights and interests by early symbolic interactionists and their recognition of correlations between thinking and movements of bodily parts as used in speech. The very name symbolic interaction implies a stress on the intangible and symbolic nature of its referent and public constructions. The tangible resists our push. Abstractions like space, time, and grammar do not; to think of holding them physically is an absurdity. In this sense, the symbol is extrasensory—a perfect illustration of what we call mind as an emergent from the physical brain. Since the symbol's link to its referent is customary and not fixed causally, it follows that there will be a major stress on the interpretive nature of symbolic life. We shall see how brain science is largely consistent with this because "the brain actively selects, interprets, edits and changes the very quality of incoming information to fit its own requirements and limitations." The brain projects its own character on the world just as symbols do (Franks 2007).

Another issue given attention in this chapter deals with the unconscious and recent neuroscience research. This material broadens the range of foundational symbolic interaction. Neuroscientists and cognitive psychologists have collected a large amount of very credible evidence for the importance of the unconscious and symbolic interaction, with its emphasis on the self-conscious control of behavior, ignores the unconscious at its own peril. Gazzaniga writes that we are only conscious of two percent of what the brain does. Thought and language, as important as they are, do not characterize all of the brain's thought. Rather than thinking in words, the right side of the brain thinks in some type of gestalt-based mode (Gazzaniga 1989: 21). According to Jonathan Turner, SI tends to overemphasize thought and language as important as they are. "If all of thought was internalized speech we would think very slowly because speech is lineal and must draw out one phoneme at a time. Our brains work much faster than that, and this is why we have Broca and Wernicke's area to down load and up load speech to the brain's way of thinking." Scheff (1990: 51) also emphasizes the impressive speed of social interaction. He tells us that "Because of the ambiguity of expression it seems necessary that internal searches must be so rapid that they must occur as many different parallel processes." He quotes Anderson (1983) that internal problem-solving events take only 1 ms (one thousandth of a second).

In spite of their differences, social behaviorism and research on mirror neurons are intimately based on a similar epistemology—that of Mead's pragmatism including as it does the metatheory of *transaction.*

After a presentation of the social construction of memory, this chapter deals with the curbs that pragmatism puts on social construction, and Gazzaniga's findings from his split-brain research will be described to show how it gives robust empirical evidence to Lyman and Scott's theory of accounts.

Early Symbolic Interactionists' Interests in the Neurological Processes

Aphasia

According to the notable SI leaders Lindesmith et al. (1988), the work on aphasia by neurologists Henry Head (1926), Kurt Goldstein (1948), and A.R. Luria (1976) was of great interest because it gave strong evidence for the important causal relationship between human speech and thought. To symbolic interactionists as well as neuroscientists among others, these two processes are critical for being human. Clearly these cognitive abilities were seriously disabled in the syndrome of aphasia that attacked Broca's area and Wernicke's area of the temporal lobes. The findings of these neurologists provided important evidence for the influence of language on all things human. For Lindesmith et al., human thought and language are intimately related, and they chided the early neuroscientists above for reifying them into separate processes. Such reification is now justified by current neuroscience and Jonathan Turner's remarks above. It followed that thinking and speaking were also intricately related since making "symbolic indications" to one's self was a key process in this perspective. In severe aphasia, intelligence and personality are disorganized, as well as memory of words. They add that "The scaffolding on which speech is developed is built up in relation to hearing, vision, and the sensory motor skills that were involved in uttering words" (Lindesmith et al. 1988: 137).

The loss of capacities in aphasia may include an inability to read, write, or name familiar objects (Lindesmith et al. 107). Out of sight, out of mind characterizes their mind-frames, and they are stuck in the concrete, having difficulty with abstractions. These neurologists also note that the focus on the concrete meant that the continuous flow of experience is disrupted and the patient's images of things not present were unstable and disorganized.

The Revisionist Nature of Perception

Another early agreement between neuroscience and Meadian theory is the revisionist nature of perception. In terms of its connection with the world, the source of perception can be seen as coming from two directions. One is the sensed world of stimuli that indeed, we may interpret, but which we do not initiate. A loud noise may be interpreted any number of ways, but the fact that it happened in the first place is not such a matter. Such stimuli come to us as passive observers and can be seen as "read out *from*" the world. The old Enlightenment British empiricists saw this as the main avenue of knowledge since to them it was the major link between the knower and the known. This epistemology is variously known as "copy" or "correspondence theory" since the human brain was seen as a blank, impressionable *Tabula Rasa* filled in by sensed experience much like the impression on malleable dough being formed by the "cookie cutter" of experience. Learning theory is consistent with this since the person is seen as passive and knowledge is mainly "read out *from*" the world separate from our will. This goes against the voluntarism so important to Mead and symbolic interaction theorists.

More compatible with SI theory and its focus on conscious interpretation underlying social constructionism is what we project *onto* the world. Here the focus of causation comes from us with the help of the interpretations and definitions of the situation stemming from our social groups. This theory of perception was typical of British idealists. A prime example of this position can be seen clearly in our perceptions of gestalt pictures that lend themselves to equally plausible interpretations even though the sensory drawings and dots originally drawn and comprising the possible perceptions remain stable. The fact that sensation by itself gives only sensation and nothing else was not recognized by the British empiricists.

> As Albert Einstein …clearly recognized, the belief in an objective world with scientific objects in it the same for all observers, is a theoretically inferred world, not a purely empirically given knowledge. (N.S.F. Northrop 1948:43)

The Chicago pragmatists saw these positions as dualistic and the source of never-ending debates that assured a lack of intellectual progress. In place of thinking of the link between the knower and the known world as the rationalist's reason or the empiricist's bodily-given senses, Mead and Dewey saw the link as stemming from human earthly *behavio*r. We become conscious of the world through the way it responds to our actions on it. In neurosociological terms, we only act successfully to that which the object allows. Researchers working with mirror neurons use the concept of "affordances" that an object makes possible. It is difficult, indeed, to use the stem of a daisy as a nail or to bounce a solid lead square.

In order to avoid the pitfall of dualism which places the mind and body, the individual and society, the knower and the known, and the organism and environment in antithetical and contradictory realms, pragmatism looks at these contrasts in a *transactional* manner.

> If interaction assumes the organism and the environmental objects to be present separate forms of existence… prior to their entry into joint as investigation, then transaction assumes no pre-knowledge of either organism or environment alone as common system… (Dewey and Bentley 1949: 123)

In neurosociological terms, pertaining to the dualism between concrete body and intangible mind, how do we get from the electrochemical, synaptic connections within the brain to the direct but intangible experience we are conscious of here and now?

Neuroscience has produced strong evidence that we cannot copy the world the way it is. First, our five senses cannot possibly register all of the multitudinous possible stimuli in the "obdurate world" to use the phrase of symbolic interactionist's major theorist, Herbert Blumer. Buckminster Fuller warned that what humans can sense is "less than a millionth of reality" (Christian 1977).

But more than that, even if we could sense these things, our brain-given senses are *transducers*. In terms of sight, Blumer's obdurate world turns out to be made up of electronic waves of different lengths traveling at a speed of 186,000 miles per hour (Christian 1977: 193). The eyes of the brain change electronic waves into a form to which the brain can accommodate. For example, the computer program that changed WordPerfect to Microsoft Word was a transducer. The brain is highly revisionist because we can never see the world simply as light waves, i.e., as the world is. What we can do is accurately perceive the obdurate world as the consequence of this brain/environment relationship. Thus, while we may have a *relational*, transactional view, we avoid the kind of relativism where, as

Mead says, "anything goes." Arendt (1958: 236) stresses this revisionism in her quotation of Eddington that "whatever we perceive is as different from the incoming data as a telephone number is to its subscriber." Likewise, Mead spoke of this as the "objective reality of perspectives." Here the perspective is made up of what an animal brings to the environment in terms of sensory transducers (read *into*) and what the environment offers in terms of what can be successfully transduced (read out *from*). Since this "transactional slab" is objectively there in nature, and although it is relational, it still has a relational deterministic character that is not relative. This places important limits on constructionism that some interactionists ignore. They often see color as solely determined by the fact that different societies socialize their members to see different colors and to ignore others. But color is not completely socially constructed. We can readily learn to see other colors that our linguistic categories ignore. (See also Lakoff and Johnson 1999: 23–26.)

The revisionist character of the perception of color may be clearly seen when we consider that the electric spikes which different senses send to the brain are all the same. The experiential difference for the individual is the brain area to which the stimulus is sent (Franks 2003). A similar argument can be made for all the senses other than color. When I smell an odor, gaseous molecules are sent to the lining of the olfactory membranes of my nose. Here the same kind of argument must be made as that of color. There is no smell in the rose per se; fragrances are a transaction of odor messages and how they stimulate the lining of my nose. To say it is in the rose is a "semantic fallacy."

Another reason that a non-revisionist, correspondence theory fails is that each of our senses must provide an extreme simplification of what is "out there" independent of our will. The limited size of the brain only allows reception of a small portion of what is "out there." Each sensed image must be reduced by a factor of 100. Gazzaniga (1985) tells us that rather than incorporating each microscopic flash, the eye constructs a gestalt or a pattern of the "shape of things."

An excellent statement on the revisionist statement of perception is written by Wentworth and Ryan (1992: 29) about the vast number of processes the brain goes through in constructing the sight of a doll:

> Once in the brain, "feature extraction" takes place. Edges, horizontal and vertical lines, mappings of textures and shadings are abstracted from the impulse-coded doll and scattered across the visual cortex. This twice shattered doll has connections among several specialized areas and seconds, with the experience of the doll antedated so that no processing lag time is noticeable.

Consistent with the above, they state that "the brain is neither the passive mirror of the world nor its camera.... Like a master carpenter, the brain takes the blue print, tools and materials and brings them together into a dwelling." Then they add, "What a wonderful physiological foundation for social constructionism" (Wentworth and Ryan 1992: 30).

The Revisionist Nature of Memory

Daniel Schacter's revisionist characterization of memory is set forth in his classic, The Seven Sins of Memory (2001). We do not remember the actual act in long-term memory. What we remember is the last time we remembered something. Furthermore, every long-term past is a reconstructed past. Therefore we do not have direct memories of the original acts. Lindesmith et al. (1988: 137). have focused on the social construction of collective memory following Mead's theory of the symbolic nature of the past.

Our memory is very often tied to the self and/or our own interests. Again, Turner states that memory emotions tagged to cognition are stored in the hippocampus for 2 years and then shipped up to the frontal lobes where they are activated. He says that self is a proxy for emotions since cognitions involving self will almost always arouse them.

In principle, collective memories are no different when they construct self-serving memories of the group. Individually we remember those events that confirm or threaten the self or our groups. We remember events that produce emotion, and one of the things we are most emotional about is ourselves. We remember best those things that interest us. Thus, the seven sins are really about self-maintenance rather than sins. The past is remembered in self-enhancing ways.

In the late 1990s, considerable attention was drawn to the possibility of making persons believe they had been abused when they were not, or even thinking they had been perpetrators of abuse when they were not. These instances were caused by the inadvertent suggestions of therapists or other interrogators (See Loftus 1999). However one looks at memory, it is interpretive to the core.

Split-Brain Research as Evidence for the Validity of the Symbolic Interaction Theory of Accounts

This section explores one way that neuroscience can give empirical confirmation of SI hypotheses that would otherwise be impossible in that framework. The theory of accounts was developed in 1968 by Scott and Lyman. Here they diverged from the psychological notion of a motive as the "well-spring" of action and shifted emphasis to the more sociological notion of motives as words that we direct to others when we have done something untoward and need to "save face." Motives here are *imputed* by one's audience and *avowed* by the actor. They have little to do with the truth about one's private motives. Actors can, or cannot, believe their own accounts; they are part of facework to make their audiences accept our behavior.

They depend on the social context. Men will talk to other men about their girl friends in ways they would never do when talking to their girls. They are presenting different selves, one that is "on the make" and tough and one that is "in love, thoughtful, and tender".

Closely associated with accounts is *Techniques of Neutralization* by Sykes and Matza (1957). These techniques allow delinquents to "neutralize or justify their activities, which reduces conventional social controls and can allow their behavior to continue". Once again, they have little to do with the truth.

Neuroscientific Supports for Accounts

Michael Gazzaniga (1985) began his split-brain research by operating on patients suffering from extreme epilepsy. This involved severing the corpus callosum in half. This is a mass of 200 million cables in the middle of the brain that enables the two halves of the brain to communicate with each other. Such communication is needed because the two halves are capable of different things and thus complement each other. The left brain communicates linguistically in right-handed people, and the right brain is mute, communicating only by electronic means. The differences in the two hemispheres have been exaggerated in the past, but the fact remains that both are needed. When a written message is presented to the right hemisphere, the severed right side can no longer inform the conscious left side what it is doing. What is remarkable is that when patients whose corpus callosums have been severed are asked to draw a horse, they will start drawing one without any recognition of what they are doing until the final form of the horse emerges.

When patients were asked why they drew what they did, they always came up with a quick answer that made sense to them and that they thoroughly believed. Here the action occurred first, and then the verbal explanation followed. Of course, they were the only one who is deceived. Even more relevant to Scott and Lyman's sociology of accounts are two other experiments. In the first, the experimenters sent a message to the right brain of the patients that they should rise and leave the room. When they did so, they were asked why they left. The answers came quickly that they were going to get a coke or get some fresh air. They never said that they did not know why they left. In the second case, the right brain was told to laugh. When the subjects did so, they were asked why they were laughing, and again, they never said, "I don't know," but a reason is forthcoming that only they believe. The verbalizations the sociologists term "accounts" or "vocabulary of motives" are to neuroscientists "confabulations." The only recourse the sociologist has available to get evidence about the true reasons for this behavior would be to ask the subject, but this would be patently absurd since the subjects believed their own accounts. In the cases above, it was clear to the researchers that they did these things because they were asked. In this research, neuroscience provided empirical evidence for the existence of accounts that sociologists could not have obtained by their own methods.

Mirror Neurons and a Confirmation of the Pragmatic Priority of Action

Because the brain is the organ by which we know, anyone studying its implications unwittingly becomes an epistemologist. This is especially true if we seek, in Lakoff and Johnson's words, "an empirically responsible philosophy." In 1999 in their *Philosophy and the Flesh*, they argued that human language was largely metaphorical and that these metaphors involved our experience with bodily actions…. "Do you grasp what I mean?" "I hammered that home to them." "You are talking over my head or beneath me"—I won't bore you with other thousands of examples. But they also insisted that philosophy be informed by what we know of the mind and its embodiment through neuroscience. Because we all share bodily experiences, this increases intersubjectivity. Society is dependent on this for its very existence.

Mirror Neurons

As discussed in Chap. 14 by Liza Azziz-Zadeh and Sook-lei Liew, a mirror neuron fires when we watch others perform a certain action. But it does more than that. Unconsciously it actively simulates, or copies, other's actions on the observer's motor cortex. In an important sense, we do behaviorally what we watch. These neurons are inherently social and directed toward the actions of others; they also actively combine motor activity and perception. Furthermore, and most important, mirror neurons fire when the action observed can be seen as *intentional on the part of the observed actor*, thus preparing us for the lightening quick responses which are made necessary by the speed of social interaction. (Some reservations of how this reflects intentionality can be found in Churchland 2011: 139–143.)

For researchers in mirror neurons, the relation between our motor intentions and the way we perceive the world is an "affordance." An affordance is something that the environment offers to an animal's behavior such as surfaces that provide support, objects to be manipulated, and substances that can be eaten. Interestingly, the old Chicago pragmatists used the very same word. Researchers Rizzolatti and Sinigaglia (2008) argue that the visual perception of an object implies the immediate and automatic selection of those properties that facilitate our behavioral interaction with it. For these neuroscientists, we perceive the world not so much as "it is" in some final sense but in terms of how it answers to or facilitates our actions.

This means once again that what we sense and reason about is limited. The environment becomes constructed and objectified only in relation to the individual's limited motor capacities, senses, and brains. The German term *Umwelt* captures this organism/environment relation very well. It refers to the world carved out for our attention by our own capacities, sensitivities, and motor repertoires (Becker 1964).

The research on mirror neurons implies the existence and relevance of just this social behaviorist view. The mirror neurons of dancers responded most strongly to observations of familiar dance steps in line with the style of dancing the observers routinely used. More pointedly, when human subjects were shown videos of motor behaviors of different animals communicating, like monkeys lip-smacking, dogs barking, and humans talking, mirror neurons became engaged only at the observation of humans talking, even though visually it was obvious that these all were communications with moving mouths. Since we have no spaces in our motor cortex for barking and lip-smacking, we cannot simulate them.

Mirror Neurons and Speech

Here again we find supportive evidence for the early pragmatist's program. As we have seen, the *products* of speech and human symbolic thought are *extra*sensory, like time and space, or the notion of redness in and of itself, independent of any actual red object. But the *process* of speech is behavioral. An important tenet of social behaviorism, and thus foundational symbolic interaction, is that as humans solve problems they are really talking to themselves, making subliminal movements with those muscles associated with speech like the tongue, lips, and larynx. Deaf children privately thinking to themselves likewise make subliminal hand and body gestures of sign language that underlie their symbolic thought. The congenitally blind, who have never seen gestures involved in sign language, nonetheless make motor movements when they talk. Gestures are not ancillary to talk but an inherent part of it just as grammar is. We even gesture on the phone with no one to see us.

How do mirror neurons enter the picture at this point? Once again, we do not just passively hear talk—we actively *do* the talk, in this case, the talk of other people. This is because talk is behavioral. We perceive talk not so much as sounds directly through our ears, but we *comprehend* it more indirectly through making the same speech movements ourselves in the motor cortex (Iacoboni 2008: 103). We simultaneously enact the same speech movements others are making. Their talk literally becomes our talk with the provision that it is automatic and unconscious.

As seen in this volume, more than several social neuroscientists see mirror neurons as capturing the social intent of others. This position holds that concepts are represented in the brain by the same sensory–motor circuitry on which the concept relies. The general *concept* of grasping would be represented by increased activity in the sensory motor areas devoted to grasping *actions*. Some of these neuroscientists include Azziz-Zadeh and Damasio (2008), and Gibbs (2006). They also describe the last 20 years of research that gives evidence that language is biologically linked to action through mirror neurons in a way congruent with the priority given to action by the pragmatist social behaviorists. They insist that language is not separate from motor behavior but rather inherently grounded in actions. This is an important part of Mead's social behaviorism. Neuroscientific research provides convergent validity here and is a powerful tool to strengthen evidence for the validity of the perspective of social behaviorism or the "embodied semantic hypothesis."

Mead's Four Stages of the Act

Mead's four stages of the act are more in line with mirror neuron research. First, comes action as some impulse to behave toward an object, and second, what we perceive most clearly is what answers to or facilitates our interests and intentions to act. As Damasio wrote in his *Descartes'*

Error (1994), "Perception is more involved with action than we think. Perceiving is as much acting on the environment as it is about receiving signals from it." All perception is selective and our behavioral intentions do the selecting. Books on the shelf that we have ignored for years come immediately and clearly into sight as we reach for them.

The next stage of the act is manipulation—*doing* something with the object or, if the act involves people, taking the role of the other in social interaction. Last, we have consummations. Here consummation reaches back teleologically to pull the different stages into being. This is teleological because the later stages are present in the first stages and control the others. The whole framework is voluntaristic because in the first stage, the image of consummation is symbolic and open to the actor's change. This also contrasts with a version of reinforcement theory wherein the stimulus comes first, we then perceive it, we next think (at least sometimes), and last, we act in a way dictated by our conscious thought.

Conclusion

In this chapter, I have shown how neuroscience can be helpful to perspectives or fields that some might (wrongly) consider diametrically opposed to it—foundational social behaviorism and even symbolic interaction in some cases. I have also shown how the neuroscience of mirror neurons confirms the pragmatic metatheories of Dewey and Mead. This is a prime example of David Maines' rebuttal to Denzin that "if the new is good because it's new and if the old is bad because its boring, then what do we do if we come up with something old that works?" Hopefully I have given enough evidence to illustrate that pragmatism works as far as mirror neurons are concerned. Neuroscience additionally can provide helpful critiques of the perspective of symbolic interaction, in this case, creating a balanced approach to social constructionism and relativism.

References

Anderson, J. R. (1983). *The architecture of cognition*. Cambridge, MA: Harvard University Press.

Arendt, H. (1958). *The human condition*. New York: Double Day Anchor Books.

Aziz-Zadeh, L., & Damasio, A. (2008). Embodied semantics for actions: Findings from functional Bain imaging. *The Journal of Physiology, 102*(1–3), 335–376.

Becker, E. (1964). *The revolution in psychiatry: The new understanding of man*. New York: The Free Press of Glencoe.

Christian, J. (1977). *Philosophy an introduction to the art of wondering* (2nd ed.). New York: Holt, Rhineheart and Winston.

Churchland, P. (2011). *Braintrust: What neuroscience tells us about morality*. Princeton: Princeton University Press.

Damasio, A. (1994). *Descartes' error: Emotion, reason and the human brain*. New York: Avon books NY.

Denzin, N. (1992). *Symbolic interaction and cultural studies: The politics of interpretation*. Cambridge, MA: Blackwell.

Dewey, J., & Bentley, A. (1949). *The knower and the known*. Boston: Beacon.

Franks, D. (2003). Mutual interests, different lenses: Current neuroscience and symbolic interaction. *Symbolic Interaction, 26*(4), 613–630.

Franks, D. (2007). Neurosociology. In G. Ritzer (Ed.), *Blackwell encyclopedia of sociology* (p. 3185). England: Blackwell Publishing.

Gazzaniga, M. (1985). *The social brain: Discovering the networks of the mind*. New York: Basic Books.

Gazzaniga, M. (1998) *The Mind's Past*. University of California Press: Berkeley

Gibbs, R. (2006). *Embodiment and cognitive science*. New York: Cambridge University Press.

Goldstein, K. (1948). *Language and language disturbances: Aphasic symptom complexes and their significance for medicine and theory of language*. New York: Grune and Sutton.

Head, H. (1926). *Aphasia and kindred disorders of speech* (Vol. 1). New York: Macmillian.

Iacoboni.M. (2008) *Mirroring people: The new science of how we connect with others*. Farrar, Straus and Giroux: New York
Lakoff, G., & Johnson, M. (1999). *The philosophy of the flesh: The embodied mind and its challenge to western thought*. New York: Basic Books.
Lindesmith, A., Strauss, A., & Denzin, N. (1988). *Social Psychology* (7th ed.). Englewood Cliffs: Prentice Hall. Inc.
Loftus, E. (1999). Creating false memories. In The Scientific American (Ed.), *The scientific American book of the brain*. Guilford Conn: Lyons Press.
Luria, A. R. (1976). *Cognitive development: Its culture and social foundations* (M. Martin-Morillas & L. Solotaroff, Trans., M. Cole Ed.). Cambridge, MA.: Harvard University Press.
Maines, D. (1995). *Some Comments on postmodernism: The so-called new interpretive turn*. Paper presented at the annual meeting of the Midwest Sociological Society, Chicago.
Matza, D., & Sykes, G. (1961). Juvenile delinquency and subterranean values. *American Sociological Review, 26*(5), 712–719.
McPhail, C., & Rexroat, C. (1979). Mead VS Blumer: The divergent methodological perspectives of social behaviorism and symbolic interactionism. *American Sociological Review, 44*(June), 449–467.
Mead, G. H. (1934). *Mind, self and society: From the standpoint of a social behaviorist*. Chicago: University of Chicago Press.
Mead, G. H. (1938). *The philosophy of the act*. Chicago: University of Chicago Press.
Mead, G. H. (1959). *The philosophy of the present: The Paul Carus lectures*. La Salle: The Open Court Publishing Co.
Northrop, F. S. C. (1948). *The logic of the sciences and the humanities*. New York: Macmillan.
Rizzolatti, G., & Singigalia, C. (2008). *Mirrors in the brain: How our minds share actions and emotions*. New York: Oxford University Press.
Schacter, D. L. (2001). *The seven sins of memory: How the mind forgets and remembers*. New York: Houghton Mifflin.
Scheff, T. J. (1990). *Microsociology: Discourse, emotion and social structure*. Chicago: The University of Chicago Press.
Scott, M., & Lyman, S. (1968). Accounts. *American Sociological Review, 31*, 46.
Sook-lie, L., & Aziz-Zadeh, L. (2011). The neuroscience of language and action in occupations: A review of findings from brain and behavioral sciences. *The Journal of Occupational Science Incorporated, 18*(2), 1–18.
Stryker, S. (1985). Symbolic interaction and role theory. In G. Lindzey & E. Aronson (Eds.), *The handbook of social psychology* (pp. 311–378). New York: Random House.
Sykes, G., and Matza D. (1957). "Techniques of Neutralization: A Theory of Delinquency." *American Sociological Review. 22*, 664–70
Wentworth, W. M., & Ryan, J. (1992). Balancing mind body and culture: The place of emotion in social life. In D. D. Franks & V. Gecas (Eds.), *Social perspectives on emotion* (pp. 25–46). Stamford: JAI Press.

Chapter 12
What Are the Neurological Foundations of Identities and Identity-Related Processes?

An Examination of How the Default Mode Network Relates to Identity Theory

Richard E. Niemeyer

Introduction

With increasing frequency, social scientists are investigating how an individual's biology, social networks, and culture reciprocally interact to cumulatively influence their personality, cognitive-behavioral patterns, and psychophysical health (Fowler et al. 2011; Franks 2010; Freese et al. 2003; Guo et al. 2010; Machalek and Martin 2004; Shanahan et al. 2008; Turner and Maryanski 2005; Udry 1995). In this chapter, we extend this line of research by examining the neurological foundations of an important part of sociological social psychology – identity theory (Burke and Stets 2009).

For the last decade, neuroscientists have been identifying and describing the structural and functional characteristics of what are commonly referred to as intrinsic connectivity networks (ICN) (Biswal et al. 2010; Corbetta and Shulman 2002; Fox et al. 2006; Gusnard et al. 2001; Hagmann et al. 2008; Raichle and Gusnard 2005; Raichle et al. 2001). Generally speaking, an ICN is a collection of functionally interconnected brain regions that dynamically coordinate their respective neural activities to generate specific mental states or to perform specific behavioral and cognitive acts (Deco et al. 2011; Greicius et al. 2009; Jiao et al. 2011; van den Heuvel et al. 2009).

According to current research, a specific ICN called the *default mode network* appears to be responsible for generating several of the cognitive processes that identity theory argues are an essential part of social interaction (Andrews-Hanna et al. 2010a, b; Buckner et al. 2008; Leech et al. 2011; Spreng and Grady 2010). Specifically, these include the possession of a self; the capacity to engage in self-reflection; the ability to develop, maintain, manipulate, and utilize stocks of knowledge to interpret the meaning of physical and social phenomena; the ability to ascribe meanings to oneself; the ability to take the role of the other; and the ability to covertly organize and rehearse potential lines of action (Burke and Stets 2009). Given the significant overlap between the cognitive processes generated by the default mode network and those discussed by identity theory, the goal of this chapter is to illustrate the applicability of research related to the default mode network to the identity theory research program.

In the following sections, we will argue that the patterns of neuronal activity generated by the default mode network represent the concrete referents of identity theory's abstract concepts; as such, empirical insights regarding the functional properties of the default mode network can be incorporated into identity theory research in the same manner as any other empirical referent. Further, we

R.E. Niemeyer (✉)
Department of Sociology, University of California,
Riverside, CA 92521, USA
e-mail: richard.niemeyer@email.ucr.edu

D.D. Franks and J.H. Turner (eds.), *Handbook of Neurosociology*, Handbooks of Sociology and Social Research, DOI 10.1007/978-94-007-4473-8_12, © Springer Science+Business Media B.V. 2013

will argue that research on the default mode network can supplement identity theory by offering the research program the ability to physiologically validate the existence of its concepts and processes, as well as elaborate in neurological terms their underlying causal mechanisms. Finally, we will briefly examine how an understanding of the neurological foundations of identity-related processes can expand the explanatory power of identity theory into new domains of research.

Overview of Identity Theory

Identity theory is a microsociological research program that examines how social structures, individuals' identities, and individuals' social interactions mutually influence each other (Burke and Stets 2009).[1] The goal of identity theorists is to explain how an individual's identities initially form; once formed, how identities organize an individual's behavior in predictable ways; and finally, how the meaning of an identity and the associated cognitive and behavioral processes potentially change in response to repeated interactions with others (Burke and Stets 2009).

According to identity theory, there are three bases of identities: personal identities, role identities, and social identities (Burke and Stets 2009). These identities consist of the set of meanings that define the characteristics of a person as an individual; within the context of the roles they perform; and as a member of a social group, respectively. Regardless of type (e.g., personal, role, or social), an identity has two dimensions. The first is called the *conventional dimension*, which reflects the sociocultural definitions and behavioral expectations associated with an identity. The second dimension is called the *idiosyncratic dimension*; it consists of the individual's distinctive interpretation of the meanings and responsibilities associated with an identity, as well as their definition of the kinds of behaviors that one can appropriately engage in (Burke and Stets 2009:39).

For example, Burke and Stets (2009) note that the conventional dimension that defines the "role-responsibilities" of a professor includes teaching courses, grading papers, and holding office hours. But, an individual may inject into their professor role-performance an idiosyncratic dimension that defines the role's responsibilities to include mentor, confidant, or friend. Similarly, the conventional dimension of an individual's moral identity may state that "a good person does not steal" (Stets 2010). But, the idiosyncratic dimension of the identity may qualify the conventional definition with "a good person does not steal *from members of their social network*; outsiders, though, are fair game."

Identity theorists argue that individuals learn the conventional and idiosyncratic dimensions of their respective identities through a combination of socialization and the reflected appraisals they receive from others (Burke and Stets 2009). Following the structural symbolic interaction tradition (Stryker 2002), identity theorists define socialization as the process by which individuals are formally taught by the members of their society the meanings of the signs, symbols, and behaviors that characterize their culture. Generally speaking, it is through this socialization process that individuals learn the conventional dimensions of their identities. The set of meanings that eventually come to define an individual's identity is called the identity standard (Burke and Stets 2009). Following Cooley (1902), a reflected appraisal is the information that people convey to one another during a social

[1] Like most sociological research programs, identity theory has evolved over time. In the past, it was possible to distinguish three separate lines of research carried out by George McCall and J.L. Simmons; Sheldon Stryker and colleagues; and Peter J. Burke, Jan E. Stets, and colleagues. Recently, Burke and Stets (2009) articulated a version of identity theory that incorporates these distinct research programs into a single framework. Following this specification, this chapter will treat the concepts and theories offered by each of the three programs as elements of a single framework.

interaction that indicates how each is defining and interpreting the meaning of the other. For example, Christy's approving look during Conrad's lecture may indicate to Conrad that he is being perceived as a competent professor. If Conrad continues to experience this identity verification from Christy and his other students, the standard for his professor identity will be "competent."

The identity control process is a theoretical model of how an individual's identity standard dynamically influences his or her behavior as he or she goes about his or her daily tasks (Burke and Stets 2009). To briefly summarize, the model postulates that the meanings stored in the identity standard guide an individual's behavior (e.g., a competent professor lectures *this* way and not *that* way). Others then interpret the meaning of this behavior and provide feedback in the form of reflected appraisals. Once perceived, a person compares the meaning of this feedback with the meanings stored in his or her identity standard. An individual's identity is said to be verified if the meaning of the reflected appraisal matches the meaning stored in his or her identity standard. Generally speaking, positive emotions are the phenomenological consequence of identity verification. If the meaning of the reflected appraisal does not match the identity standard, then an error signal is generated; generally speaking, this error signal is experienced as a negative emotion. This negative emotion, in turn, motivates the individual to engage with the stimulus until the meanings being perceived match those stored in the identity standard. Importantly, identity theory argues that this process can occur either consciously or subconsciously (Burke and Stets 2009). In short, this means that the perception and interpretation of the meaning of self-relevant stimuli, the comparison of this meaning with the meanings stored in the identity standard, and the motivation to respond to the failure to verify an identity can occur as an automatic or deliberately controlled cognitive process.

Finally, identity theorists argue that the identity control process can occur even when other individuals are not directly present to provide a reflected appraisal. For example, when *anticipating* a social interaction, an individual will draw upon the meanings stored in his or her identity standard to construct a course of action that reflects these meanings. Identity theorists maintain that this is done by the individual mentally rehearsing potential lines of behavior and imagining how others may react (Burke and Stets 2009). Importantly, identity theorists further claim that these imagined reflected appraisals generate the same positive and negative emotions as do those experienced during face-to-face interaction. Beyond serving to organize future behaviors into a coherent plan of action, the anticipatory feelings of approval serve to partially legitimate an individual's cultural understanding and personal definitions of an identity; in short, this is because emotions in general act to reinforce or dissuade behaviors regardless of their source (e.g., imagined vs. real). Once this set of behaviors is performed in front of others, the emotions generated by the *actual* reflected appraisals act to fully legitimize or delegitimize these understandings (Burke and Stets 2009:39–41); this is because experienced reflected appraisals deliver a larger emotional impact than their imagined counterpart.

To illustrate more clearly how the identity control process operates, consider the following example. Imagine an individual who defines the meaning of his student identity using terms such as "hardworking," "disciplined," and "responsible" and who is unable to turn in an assignment on time due to car trouble. When the student eventually gives the assignment to the professor, the professor expresses disapproval and recommends taking the class more seriously. Here, both the professor's look and verbal statement are stimuli that indicate to the student that the professor is interpreting turning in the assignment late to mean that the student is not hardworking, disciplined, and responsible.

Upon perceiving the meaning of these stimuli and contrasting them with the meanings stored in the student's identity standard, the student experiences a negative emotion and is thus motivated into action. In this case, the student "takes the role of the other" and views the situation from the perspective of the professor; upon doing so, the student thinks, "Oh! The reason the professor said that is because she is not aware of my car trouble." Realizing this, the student explains to the professor his situation and then apologizes for turning his assignment in late. Upon considering the student's explanation, the professor smiles and tells the student that she is sorry for misjudging his devotion to the class. Importantly, the professor's second response represents a new set of stimuli that are perceived by the student and compared to his "student identity standard." Because these new perceived meanings match the meanings stored in the standard, no error signal is generated; instead, the student experiences the positive emotions that occur when an identity is verified.

The Neurological Foundations of an Identity

Recall from above that identity theorists define an identity as the conventional and idiosyncratic meanings that are self- and socially ascribed to an individual. As noted by Burke and Stets (2009), identity theory's definition of "meaning" is significantly influenced by the work of the classical sociologist George H. Mead (1934). Interestingly, this connection to Mead allows us to identify the neurological equivalent of identity theory's concept of meaning.

According to Mead (1934), the meaning of an object is determined by how an individual is prepared to act toward it. In regard to the meaning of a telescope, Mead says:

> If we want to trace the responses of the astronomer, we have to go back into his *central nervous system*, back to a whole series of *neurons*; and we find something there that answers to the exact way in which the astronomer approaches the instrument under certain conditions....the values which we say the instrument has are values through the relationship of the object to the person who has that sort of attitude. If a person did not have that particular *nervous system*, the instrument would be of no value. *It would not be a telescope*. (Mead 1934:29; emphasis added)

Notice here the degree to which Mead binds this concept of meaning to the existence of some neurological substrate; indeed, in the last line of the quote, he explicitly notes that the neurological processes underlying the behavioral attitude toward the telescope is exactly what makes it a telescope. Without this biological substrate, the telescope as such would not exist. Clearly then, the biological substrate is an important dimension of Mead's definition of meaning.

In the neuroscience literature, Mead's "biological substrate" is called a cell assembly (Hebb 2002). A cell assembly is defined as a collection of neurons that coordinate their individual patterns of neural activity to represent information about the environment. To borrow Mead's example, this means that a discrete collection of neurons responds in a deliberate and coordinated fashion whenever an individual perceives a telescope, while another cell assembly responds whenever he or she perceives a horse. Interestingly, the same cell assembly that is involved in the perception of a specific aspect of the environment is also activated whenever the object is mentally recalled (Squire and Kandel 2009). This means that a telescope is represented in the brain by the same collections of neurons regardless of whether it is "directly perceived" or "remembered." Given this, we argue that conventional and idiosyncratic meanings constituting an individual's identity standard are represented in his or her brain by cell assemblies.

According to neuroscience research, knowledge about the environment can be divided into distinct types (Squire and Kandel 2009). Knowledge that an individual can consciously recollect is called declarative knowledge or explicit memory.[2] Semantic memory is a subtype of declarative memory and is defined as an individual's knowledge of general facts, i.e., the capitals of major countries, the names of popular sport teams, the grammar and vocabulary of their language, etc. Episodic memory is a second subtype of declarative memory that is defined as the knowledge an individual has about

[2] In the following sections, we will use the terms information, knowledge, and memory interchangeably.

specific events that they experienced from the first-person perspective, i.e., their memory of the time they traveled to Paris after graduating from high school or the memory of the time they watched their favorite sports team win the championship game.

A third type of declarative memory is sometimes referred to as semantic autobiographical memory (D'Argembeau and Salmon 2011; Levine et al. 2004). Like episodic memory, semantic autobiographical memory is a form of self-relevant knowledge; like semantic memory, though, this knowledge about the self is not tied to a specific "episode" in an individual's autobiographical history. In other words, semantic autobiographical memory is knowledge about oneself that has been abstracted from specific experiences. For example, an individual's memory of their mood yesterday during lunch is episodic knowledge; an individual's knowledge about their general disposition is a semantic autobiographical memory. Importantly, then, the key difference between semantic, episodic, and semantic autobiographical memory is not so much the content of the knowledge per se, but the degree to which it is tied to a specific moment or abstracted from numerous individual events.

Although the identity theory concepts of conventional and idiosyncratic dimensions are defined in terms of "sociocultural" versus "personal meanings," and the neuroscience concepts of semantic and episodic memory are defined in terms of "general" versus "particular" experiences, we argue nonetheless that these two sets of concepts are interchangeable. The reason for this can be illustrated by the following example. Generally speaking, a professor is defined as an individual that teaches at the university or college level; the primary responsibilities of a professor include designing and teaching courses, performing research, and grading assignments. This general knowledge about the meaning of the word professor is semantic knowledge. In the above summary of identity theory, I noted that Burke and Stets (2009) cite teaching courses, grading papers, and holding office hours as examples of the conventional dimension of the role-responsibilities of a professor. As can be seen here, the semantic and conventional meanings of a professor are essentially the same.

Burke and Stets (2009) further note that the idiosyncratic dimension of the professor identity can include being a friend or mentor. It is reasonable to assume that an individual could develop this idiosyncratic definition after *specific interactions* with "mentor-type" or "friend-type" professors during her time in graduate school. It is also possible that a professor could learn these meanings after experiencing *specific instances* of positive reinforcement from her students as she engaged in mentor-like or friend-like behaviors. Because an individual can deliberately recall from the first-person perspective specific examples of these kinds of experiences, they are episodic memories. The general idiosyncratic meaning of "professor as friend" that is abstracted from these specific episodic instances, though, is semantic autobiographical knowledge. Based upon these similarities, the neurological concepts of episodic and semantic autobiographical memory are interchangeable with the identity theory concept of idiosyncratic dimension of meaning.

Relevant Measures of Neural Activity

As a consequence of the correspondences discussed above, we argue that neuroscience research on declarative knowledge can be used to supplement sociological research on identity-related processes. Researchers have recently begun to investigate the neurological foundations of how individuals recall semantic, episodic, and semantic autobiographical knowledge using a noninvasive neuroimaging technology called functional magnetic resonance imaging (fMRI) (Buckner et al. 2008; Burianova et al. 2010; D'Argembeau and Salmon 2011; Kompus et al. 2009; Levine et al. 2004; Spreng and Grady 2010; Spreng et al. 2009). Given the importance of fMRI to this line of research, it is necessary to briefly review how this technology works.

fMRI measures the neural activity underlying a cognitive process by taking advantage of what is called the *blood oxygen level-dependent effect* (BOLD effect) (Cacioppo and Berntson 2005). Like all cells in the body, neurons must draw oxygen from the blood in order to function. Because of this, the body constantly delivers to the brain a steady flow of blood in order to maintain its resting state or baseline levels of cellular activity, e.g., level of neural activity that occurs when an individual is not actively engaged in a cognitive or behavioral act. Also, like all other cells in the body, increasingly larger levels of cellular activity require increasingly larger amounts of oxygen. In order to meet this larger demand, the body increases the amount of blood delivered to a brain region whenever it is "activated" to perform some cognitive or behavioral function.

Because the oxygen demands of neurons are relatively low when a brain region is "at rest," the ratio of the amount of oxygenated to deoxygenated hemoglobin in the blood is high. This is because the body is able to deliver new oxygenated blood to the region faster than the associated neurons are able to consume it. When neurons suddenly increase their level of activity at the onset of a cognitive or behavioral act, they briefly begin to consume oxygen from the blood at a faster rate than the body can replace it. As a result of this, the ratio of oxygenated to deoxygenated hemoglobin decreases; this transient state is immediately followed by an increase in the amount of blood sent to the active brain region, which, in turn, increases the ratio of oxygenated to deoxygenated hemoglobin. To summarize, fMRI is able to measure changes in neural activity by measuring changes in the ratio of oxygenated to deoxygenated hemoglobin. Further, fMRI allows researchers to determine where in the brain a specific cognitive act occurs by identifying which brain regions demonstrate a change in blood oxygen level at both the onset and during the entire performance of the cognitive activity.

In light of how fMRI operates, neuroscientists examining the neurological foundations of declarative knowledge generally employ two types of research designs (Cacioppo and Berntson 2005; D'Argembeau and Salmon 2011). The first of these involves using fMRI to record changes in neural activity while an individual deliberately recalls a semantic, episodic, or semantic autobiographical memory. For example, Wiggs and colleagues (Wiggs et al. 1998) examined the neural correlates of semantic memory by showing participants a black-and-white line drawing of common objects (such as a thimble, an American football, or an igloo) and then asking them to either recall the object's name or specify the color typically associated with the object (e.g., thimbles are generally silver and American footballs are generally brown). Similarly, the neurological correlates of episodic memory are generally examined by recording changes in brain activity as individuals recall very specific moments in their life (i.e., what they did on their last vacation, etc.) (D'Argembeau and Salmon 2011). Semantic autobiographical memory, in turn, is examined by asking individuals to evaluate if a series of descriptive statements accurately describes them; specifically, such statements include the following: I am a shy person. I am a father. I am a good person. etc. (D'Argembeau and Salmon 2011; Ochsner et al. 2005; Rameson et al. 2010). The recorded levels of "task-induced" neural activity are then compared to the levels of activity recorded before the onset of the task in order to determine which regions demonstrated increased levels of oxygen consumption. Areas of the brain that demonstrate an increase in oxygen consumption are then identified as playing a functional role in enabling the memory recall.

The second research design involves a series of tasks that sequentially require the recollection of semantic, episodic, and semantic autobiographical knowledge, and the patterns of increased neural activity are then compared across tasks. Brain regions that only show increased levels of activation during specific kinds of memory recall are said to be functionally discrete; brain regions that show increased activation across multiple tasks are said to serve as a common foundation for multiple types of recollection.

The Neurological Foundations of Declarative Knowledge

More than a decade of extensive research, employing the above empirical designs has consistently identified the following brain regions as responsible for enabling the recollection of semantic, episodic, and semantic autobiographical knowledge (see Fig. 12.1). First, portions of the lateral temporal and the posterior inferior parietal cortices have routinely been demonstrated to be functionally involved in the recollection of semantic memory (Binder et al. 2009; Prince et al. 2007; Wiggs et al. 1998). Second, portions of the medial prefrontal cortex, posterior cingulate cortex/precuneus, hippocampus, and anterior cingulate cortex have all been reliably shown to play a functional role in recalling episodic memory (see Fig. 12.1) (Addis et al. 2009; D'Argembeau and Salmon 2011; Kwan et al. 2010; Levine et al. 2004; Leyhe et al. 2009; Prince et al. 2007). Third, a comparison of the brain regions activated to generate semantic autobiographical knowledge reveal a significant overlap with those that generate general semantic knowledge; e.g., both semantic and semantic autobiographical knowledge are generated by regions of the brain that are responsible for recalling knowledge that is not tied to a specific event. The key difference though is that the recollection of semantic autobiographical knowledge involves "episodic" areas of the brain that process self-related information and enable deliberate cognitive control (Levine et al. 2004). This means semantic autobiographical knowledge is neurologically distinct from general semantic knowledge to the degree that the former requires a greater amount of cognitive effort to recall, as well as by the fact that it contains a self-relevant dimension.

To summarize our argument thus far, the identity theory concepts of conventional and idiosyncratic meanings of an identity are conceptually identical to semantic, episodic, and semantic autobiographical memory. Further, because an individual's identity standard is defined as the complete stock of meanings a person has for a given identity, the concept of an identity standard is co-terminate with an individual's complete stock of episodic, semantic, and semantic autobiographical knowledge. Recognizing this correspondence is important for three reasons. First, it provides an important clue for identifying which neuroscience research programs may be applicable to identity theory, e.g., those research programs that are related to the three forms of memory. Second, the fact that all three forms of memory are generated by physiologically distinct mechanisms indicates that the difference

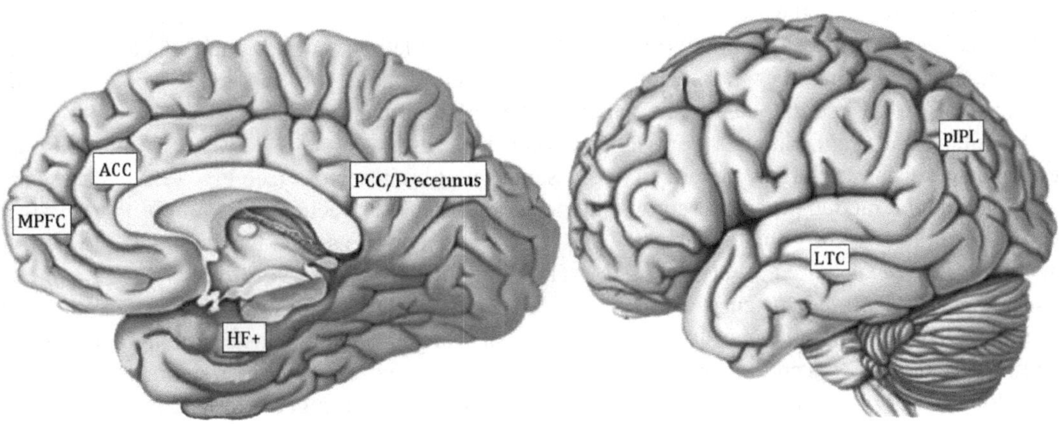

Fig. 12.1 Brain regions responsible for generating episodic memory (*left*) involve the medial prefrontal cortex (*MPFC*), anterior cingulate cortex (*ACC*), hippocampal formation (HF+), and posterior cingulate cortex/preceunus (*PCC/preceunus*). Brain regions responsible for generating semantic memory (*right*) include lateral temporal cortex (*LTC*) and posterior inferior parietal lobe (PIPL). Views of the brain are taken from midsagittal, dissected down the midline of the brain (*Left*) and an exterior view (*Right*)

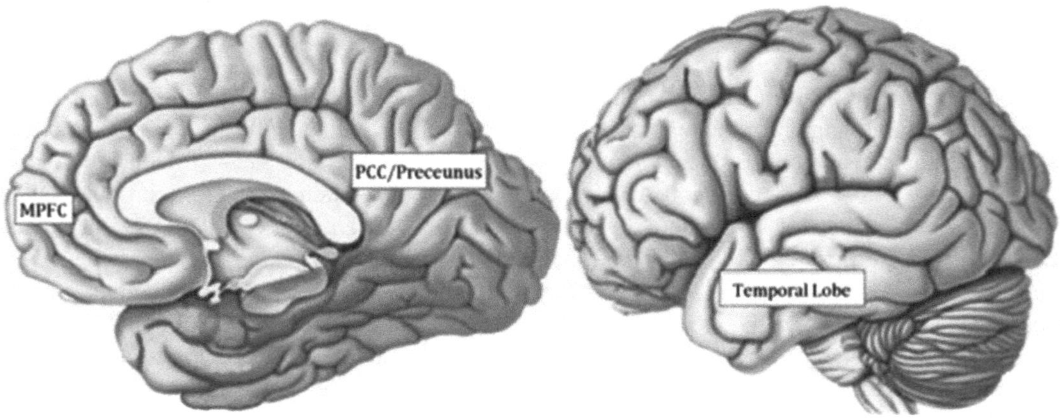

Fig. 12.2 Brain regions involved in self- and reflected-appraisals. Medial prefrontal cortex (*MPFC*) and posterior cingulate cortex (*PCC*) as seen from the midsagittal view (*Left*) and temporal lobe as seen from the exterior view of the brain (*Right*). Locations of brain regions are approximate

between idiosyncratic and conventional meanings is more than simply a heuristic distinction; instead, it is an accurate representation of empirical reality. And finally, the fact that fMRI can be used to observe the neurological processes underlying declarative knowledge means that this same neuroimaging technology can be used to study how the conventional and idiosyncratic meanings of an identity may or may not affect an individual's behavior. The larger implications of this last point will be discussed in greater detail below.

The Neurological Foundations of Self- and Reflected Appraisals

As was noted in the overview above, an important assumption of identity theory is that individuals can engage in self- and reflected appraisals (Burke and Stets 2009). Using fMRI in the manner discussed above, neuroscientists have investigated three research questions related to these two processes: first, what are the neurological processes responsible for generating self- and reflected appraisals; second, to what degree do the neurological foundations of self- and reflected appraisals overlap; and third, what does the existence of a shared neurological substrate tell us about how individuals are able to accurately assess how they are being interpreted by others.

A remarkable finding that is consistently confirmed across studies shows that both self- and reflected appraisals are largely generated by the same set of brain regions (see Fig. 12.2); specifically, these include the medial prefrontal cortex, the posterior cingulate/precuneus, and multiple regions of the temporal lobe (for original evidence of this correlation, as well as an extensive summary of similar findings, see Ochsner et al. 2005). At the same time, differences between the two cognitive processes can be identified in terms of the strength of neural activity in specific brain regions. For example, Ochsner and colleagues found that the medial prefrontal and the posterior cingulate cortices are more strongly activated during self-appraisals than they are during reflected appraisals (Ochsner et al. 2005). Likewise, reflected appraisals were shown to employ stronger neural activity in many of the temporal brain regions that were noted above to play a role in the generation of episodic and semantic memory. Interestingly, this and other studies have found that reflected appraisals from the perspective of close others tend to demonstrate stronger activity in the brain regions associated with episodic memory and emotion, while reflected appraisals from the perspective of unfamiliar others tend to show stronger neural activity in the brain regions responsible for generating semantic knowledge (Ochsner et al. 2005).

Generally speaking, these patterns of neural activation relate to identity theory in the following ways. First, the fact that each form of appraisal can be characterized by a distinct pattern of neural activity empirically validates identity theory's taken-for-assumption that self- and reflected appraisals are distinct cognitive processes. Second, the involvement of brain regions associated with declarative knowledge during the performance of self- and reflected appraisals empirically validates identity theory's argument that individuals draw upon their identity standard when defining themselves or interpreting how they are being defined by others. At the same time though, the indication that reflected appraisals from the perspective of close others are neurologically distinct from those performed from the perspective of unfamiliar others may suggest the need for identity theory to better specify how a person's level of familiarity with others may modify identity-related processes.

The Neurological Foundations of the Implicit Perception of Self-relevant Meanings

A distinguishing characteristic of the self- and reflected appraisals discussed above is the degree to which an individual explicitly decides whether or not a stimulus is self-relevant; reflecting this, neuroscience researchers often refer to these processes as explicit appraisals. As Rameson et al. (2010) note though, people implicitly detect the self-relevance of a stimulus more often than they do so explicitly; specifically, Rameson and colleagues comment:

> …when flipping through television channels, one usually does not explicitly ask oneself: "Am I a funny person who likes comedies or a serious person who prefers documentaries?" Instead, one's implicit self-knowledge guides one's behavior to make an appropriate selection in a way that is adaptive, automatic and below the level of awareness. (Rameson et al. 2010:701)

As was noted in the overview of identity theory, the identity control process also assumes that individuals can both consciously and subconsciously perceive self-relevant stimuli. Unlike neuroscience research though, identity theorists have never empirically specified how the implicit detection of self-relevant stimuli is possible. Or, in other words, how is a person able to automatically distinguish the small subset of stimuli that represents a reflected appraisal from totality of stimuli that constantly bombards their sensory system.

Recently, Rameson and colleagues used fMRI to compare and contrast the patterns of neural activity associated with the implicit and explicit perception of self-relevant stimuli (Rameson et al. 2010). In this study, a group of participants were asked to rate how well the adjectives "athletic" and "scientific" applied to their self-concept, as well as rate how important each concept was to their sense of self. Participants that indicated that "athletic" was both an accurate description of themselves and an important part of their self-concept were coded by the researchers as possessing an athletic identity. Likewise, participants that indicated that "scientific" was both an accurate and important description of their self-concept were coded as possessing a scientific identity.

Following this, each participant was shown a stream of 88 emotionally neutral images of athletic- and scientific-related situations and asked to identify for each one whether or not a human being was present in the scene. The requirement to detect a person in the image was used to divert the participant's conscious attention away from the "athletic" or "scientific" theme that underlined each scene. The goal of this task was to determine if nonconsciously processed stimuli are "tagged" by the brain as being self-relevant and, if so, which areas of the brain are responsible. After performing this task, the participants were then shown 40 athletic- and 40 scientific-related adjectives and asked to quickly respond with either "me" or "not-me." Here, the goal of the task was to determine which brain regions were involved in the explicit evaluation of the self-relevance of a stimulus.

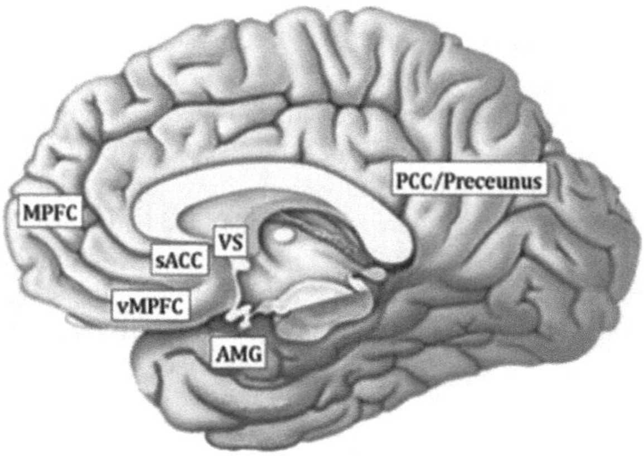

Fig. 12.3 Brain regions involved in implicit and explicit detection of self-relevant stimuli. Locations of brain regions are approximate as seen in the midsagittal view. Brian regions include medial prefrontal cortex (*MPFC*), ventral medial prefrontal cortex (*vMPFC*), amygdala (*AMG*), ventral striatum (*VS*), posterior cingulate cortex (*PCC*), and subgenual anterior cingulate cortex (*sACC*)

According to Rameson et al.'s (2010) results, the implicit and explicit detection of self-relevant stimuli are largely processed by the same brain regions. Specifically, these regions include the medial prefrontal and ventral medial prefrontal cortices, the amygdala, the subgenual anterior cingulate cortex, the posterior cingulate/precuneus, and the ventral striatum (see Fig. 12.3). Further, the strength of neural activity generated by a brain region during the explicit performance task is significantly correlated with the strength of the neural activity generated during the performance of the implicit task. For this reason, Rameson et al. (2010) conclude that processing self-relevant information engages roughly the same neural structures regardless of whether it is done in an implicit or explicit manner.

Importantly though, Rameson and colleagues are careful to note that a brain region called the dorsal medial prefrontal cortex was observed to be uniquely involved in the explicit perception of self-relevant stimuli (Rameson et al. 2010). Following a theoretical model proposed by Northoff and Bermpohl (2004), Rameson and colleagues interpret this pattern of neural activity to mean that the ventral medial prefrontal cortex is functionally responsible for tagging and representing incoming stimuli as self-relevant at the preconscious level; the dorsal medial prefrontal cortex, in turn, is functionally responsible for consciously evaluating this relevance. Accordingly, this is why the ventral medial prefrontal cortex is involved in both implicit and explicit tasks; e.g., regardless of the task type, incoming stimuli must be tagged and internally represented before they are consciously reflected upon. Interestingly, Northoff and Bermpohl (2004) reference a study by Kawasaki and colleagues in which the ventral medial prefrontal cortex began to process self-relevant stimuli 200 ms after it was presented, while increased activity in the dorsal medial prefrontal cortex did not occur until 300–800 ms later (Kawasaki et al. 2001). They suggest that this difference in activation time is indicative of functional specialization of the ventral and dorsal medial prefrontal cortices.

Beyond empirically validating identity theory's assumption that the mind is able to consciously and subconsciously perceive self-relevant stimuli, the above studies suggest that altered functionality in the dorsal and ventral medial prefrontal cortices commonly associated with mental illness may act to modulate identity-related processes in significant ways (Northoff and Bermpohl 2004). For example, an altered pattern of neural activity in the dorsal medial prefrontal cortex of patients diagnosed with either severe depression or anxiety is significantly related to several cognitive symptoms that are indicative of these illnesses (Kim et al. 2011; Lemogne et al. 2009; Northoff 2007; Sheline et al. 2010). Specifically, these symptoms include the tendency to overly attend to negative stimuli currently in the environment; the tendency to overestimate the potential for negative consequences to occur during the planning of future actions; the tendency to incorrectly attribute negative meanings to positive or neutral stimuli; the inability to experience positive emotions in response to pleasurable or rewarding stimuli; the tendency to define oneself in negative terms (i.e., worthless, stupid, ugly, not good enough, etc.); as well as the tendency to uncontrollably ruminate on negative thoughts (Grimm et al. 2009; Mor and Winquist 2002).

In light of the role that the dorsal medial prefrontal cortex plays in the various identity processes discussed thus far, the relationship between this brain area and these cognitive symptoms suggests that depression and anxiety negatively influence the self- and reflected appraisal, perception, and role-rehearsal processes, as well as negatively modulating the positive emotions experienced when an identity is confirmed. Given this, identity theory would predict that individuals diagnosed with depression or anxiety should have difficulty establishing and maintaining a positive identity. Interestingly, qualitative interviews with young adults diagnosed with bipolar disorder report experiencing this exact problem (Inder et al. 2008).

The Neurological Foundations of the Identity Control Process

An important argument made by identity theorists is that the conventional and idiosyncratic meanings of an identity, the ability to engage in self- and reflected appraisals and the ability to perceive and interpret self-relevant information, are not individual processes that occur in isolation; instead, they are all a part of the larger, continuously ongoing, identity control process (Burke and Stets 2009). Up until now, we have largely investigated the neurological foundations of the individual components of the identity control process; in this section, we will illustrate how research on the default mode network can be used to integrate these different patterns into a single neurological system.

The default mode network is comprised of 11 brain regions that are structurally organized into 2 subnetworks connected by 2 central hubs (see Fig. 12.4) (Andrews-Hanna et al. 2010b; Buckner and

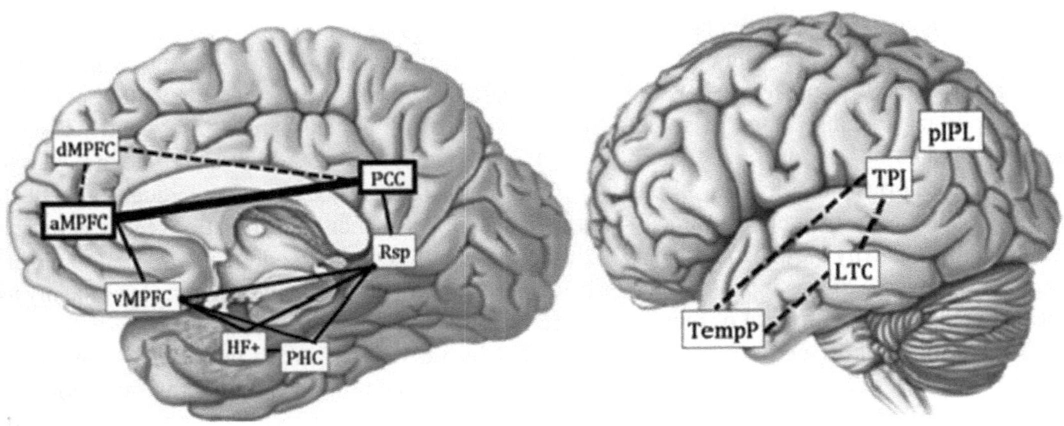

Fig. 12.4 Brain regions comprising the default mode network. The thick *solid line* represents the connection of the two main hubs of the default mode network: the posterior cingulate cortex (*PCC*) and anterior medial prefrontal cortex (*aMPFC*). The posterior inferior parietal lobe (*pIPL*) and the brain regions connected by thin solid lines represent the medial temporal lobe (*MTL*) subsystem which includes ventral medial prefrontal cortex (*vMPFC*), retrosplenial cortex (*Rsp*), hippocampal formation (*HF+*), and the parahippocampal cortex (*PHC*). Brain regions connected by thin dashed lines represent the dorsal medial prefrontal cortex (*dMPFC*) subsystem and include temporal pole (*TempP*), lateral temporal cortex (*LTC*), and temporal parietal junction (*TPJ*). Locations of brain regions are approximate as seen in the midsagittal view (*Left*) and exterior view (*Right*) of the brain

Carroll 2007; Buckner et al. 2009; Hagmann et al. 2008; Sporns 2011). One of the central hubs is defined by an area of the brain that is commonly called the anterior medial prefrontal cortex; the other hub is the posterior cingulate cortex discussed previously. The first of the subnetworks is called the dorsal medial prefrontal cortex system. As its name suggests, this subsystem contains the dorsal medial prefrontal cortex that was discussed above, as well as a small number of other brain regions. The second subnetwork is called the medial temporal system; this network consists of the medial temporal lobe, the ventral medial prefrontal cortex, and the hippocampus (all of which were already discussed above), as well as a few other brain regions (Leech et al. 2011).

The default mode network *as a whole* has been empirically demonstrated to be functionally responsible for generating and controlling three broadly defined cognitive states and processes (Stawarczyk et al. 2011). The first of these cognitive processes is called endogenously focused attention (Buckner et al. 2008; Buckner and Carroll 2007; Fox et al. 2005). This pattern of neural activity is associated with the deliberate generation and effortful control of internal cognitions; phenomenologically, these internal cognitions are experienced whenever an individual covertly thinks, deliberates, ruminates, recalls, considers, or fantasizes about something in their "mind."

The internal cognitions generated by the default mode network can be divided into two subtypes based upon their content and structural-functional connectivity patterns (Leech et al. 2011). The first subtype of internal cognitions is largely performed by the medial temporal subsystem and consists of episodic/autobiographical memory, imaginary scenario construction, and prospection (Leech et al. 2011). Episodic/autobiographical memory was already defined above. Imaginary scenario construction, in turn, is defined as the ability to imagine with great detail fictitious or hypothetical events (Buckner 2010; Buckner and Carroll 2007). Finally, prospection is defined as the ability to view oneself from a future perspective, e.g., thinking about themselves making dinner later that evening or taking a trip the following weekend (Buckner 2010; Buckner and Carroll 2007). According to identity theory, these cognitive processes are nearly identical with the mental activity that they argue an individual engages in when they rehearse potential lines of future action. The second set of internal cognitions is predominantly generated by the dorsal medial prefrontal cortex subsystem and involves the processes of self-reflection, as well as self- and reflected appraisals that were also discussed above (Corbetta et al. 2008; Fox et al. 2005; Leech et al. 2011).

The fact that the totality of the individual brain regions discussed above is a part of a single structural-functional network explains why the same brain regions are routinely activated in so many different cognitive processes; in short, each individual mental act is simply a specific instance of a more general cognitive process (e.g., deliberately generated and controlled internal mentations). This is an important insight because it indicates that the various component parts of the identity control process can only be *fully* understood as a coherent and dynamic whole. At the same time though, the fact that the various subprocesses all possess clearly defined empirical referents indicates that each dimension of the identity control process can be isolated for explicit consideration. Finally, understanding how and why the subcomponents of the default mode network dynamically interact to generate the larger whole makes it possible for researchers to empirically move between levels of analysis without committing ecological or reductionist errors.

The second type of cognitive activity generated by the default mode network is commonly referred to as the default state (Raichle et al. 2001). Stated in sociological terms, this state consists of what Alfred Schutz (1967) called duration: a "continuous coming-to-be and passing-away" of undirected thoughts, which is phenomenologically experienced as the constant transition from a "now-thus" to a "new now-thus." The reason that this mental state is referred to as "default" is because the associated pattern of neural activity is what the brain "defaults" to whenever not engaged in a cognitively effortful task (Preminger et al. 2011; Raichle and Gusnard 2005). Finally, the third process related to the default mode network is sometimes referred to as the sentinel hypothesis (Buckner et al. 2008; Stawarczyk et al. 2011). This specific pattern of activity is believed to represent a state of "watchfulness" (Gilbert et al. 2007), whereby the brain engages in the general and unfocused monitoring of an individual's internal and external environment for salient or potentially relevant stimuli.

Researchers have repeatedly shown that the cognitive and neurological content of default and sentinel processes is heavily imbued with self-relevant information (Buckner et al. 2008; Christoff et al. 2009; Smallwood and Schooler 2006; Spreng et al. 2010). This means that when the mind wanders, it tends to think about autobiographical memories, future plans and goals, and other self-related thoughts. Likewise, the sentinel state of activity involves a constant cycling through episodic and semantic knowledge that may need to be recalled to address a situation in the environment. In a certain sense, the sentinel state of activity represents a constant simulation of the external environment based on the brain's previous experiences.

In light of a theoretical model recently proposed by Northoff (2011), these two dimensions of the default mode network may relate to the identity control process in the following way. Recall from above that the ventral medial prefrontal cortex acts as a filter that tags incoming stimuli as self-relevant. Because the neurological representation of a perceived stimulus and the memory of said stimulus are identical (Iacoboni et al. 2005; Squire and Kandel 2009), and because the ventral medial prefrontal cortex is a part of a larger neural system that is constantly cycling through memories of past stimuli, it may be the case that the ventral prefrontal cortex is able to detect self-relevant stimuli by matching the incoming signal to the corresponding pattern of neural activity stored in memory. If this hypothesis is correct, the model would not only explain how the perception component of identity control process occurs but it would also empirically verify that the identity control process is constant and ongoing, as well as occurring at both the subconscious and conscious level.

To summarize, the default mode network is a large network of structurally and functionally connected brain regions that engages in three types of neural activity. These three processes represent the identity control process as it occurs at both the conscious and subconscious levels.

Summary and Conclusion: The Importance of Neuroscience Research to Identity Theory

The goal of this chapter was to illustrate the empirical and theoretical relevance of neuroscience research on the default mode network to the identity theory research program. Toward this end, we examined how specific identity theory concepts and processes can be empirically described in terms of their neurological correlates. We then concluded our argument by demonstrating how the totality of these individual brain regions are in fact component parts of a larger dynamic system of ongoing neurological activity. This realization is significant because it physiologically validates identity theory's argument that the individual dimensions of identities and identity-related processes are in fact a coherent complex process.

In light of the above arguments, we conclude that neuroscience can aid identity theory in four ways. First, it provides identity theory the capacity to define its concepts with greater precision. Because concepts are merely symbolic constructs created by the researcher, their ability to accurately reflect the phenomena they represent is limited by the observational power of the research technologies available (Collins 1994). As such, a single concept can erroneously be used to refer to two sets of unrelated phenomena simply because the characteristics that distinguish them occur at an inaccessible level of analysis. A similar problem can occur with hypothesized theoretical statements when a single process is argued to describe two empirically distinct phenomena.

For example, the neuroscience research discussed above clearly indicates that the concept of the idiographic dimension of an identity indiscriminately refers to two distinct empirical processes, e.g., semantic autobiographical and episodic memory. Thus, the ability to specify the neurological correlates of an identity enables identity theorists to formulate a more precise conceptual definition of the idiographic dimension.

Second, neuroscience can identify elements of an identity process that may be occurring below the social, behavioral, or cognitive level of analysis. This is an important contribution because a failure to account for these influences will increase the probability that a theory will fail to correctly predict or accurately explain the outcomes of social interactions. To illustrate this point, consider the following proposition: *A change in cognitive variable X is positively related to a change in environmental variable Y.* According to this proposition, a researcher should expect to see an increase in *variable X* whenever they observe an increase in the value of *variable Y*.

Let us now assume though that in reality *environmental variable Y* is only positively related to *cognitive variable X* if and only if *neurological variable Z* is above some threshold *k*. Under these circumstances, the proposition would only appear to be correct if *neurological variable Z* happened by chance to be above threshold *k* at the time the theory was applied. In short, a failure to consider the input of the neurological level of analysis may result in an accurate theory *by chance*, but its rightful incorporation can result in a correct theory *by design*. Because variations in hormone levels and functional-connectivity patterns within and between individual have been shown to impact the neurological mechanisms underlying identity-related processes, it is essential to understand how the psychological dimensions of identity theory are influenced by their biological substrate (Burnett and Blakemore 2009; Cacioppo and Patrick 2008; Fair et al. 2008).

Third, an understanding of the biological foundations of social behavior can help identity theorists specify the scope conditions of their theories with greater accuracy. Simply stated, scope conditions are defined as the set of circumstances under which a theoretical claim is expected to hold (Cohen 1989). Accurately specified scope conditions are important because they can aid researchers in diagnosing why a proposition failed to explain or predict a phenomenon of interest. On the one hand, said failure can stem from the fact that the relationships between the variables specified in the model do not accurately reflect the "real" nature of the processes they are meant to represent. On the other hand, it could also be that the "correct" theoretical explanation was applied to the "correct" set of phenomena, but not under the "correct" conditions. Knowing the relevant scope conditions of a theory can help discern between these two possibilities by allowing researchers to determine if the observed deviations from the expected outcome are attributable to conditions exogenous to the model. For this reason, then more precisely stated scope conditions are better than their less precise counterparts because they provide a greater degree of accuracy when diagnosing the sources of "theory failure."

Finally, neuroscience research can help "grow" identity theory in new directions. According to Berger, Zelditch, and Wagner (Berger et al. 1989; Berger and Zelditch 1993; Wagner 1984, 2007; Wagner and Berger 1985), sociological theories grow in at least five ways; one of these ways is called proliferation. Formally stated, theoretical proliferation is defined as the following: "We say of two theories, T1 and T2, that T2 is a proliferation of T1 if T2 enlarges the range of application of the concepts and principles in T1 to social phenomena beyond the original domain or the original set of problems within a domain addressed by T1 (Berger and Zelditch 1993)." As was noted above, the recognition of the neurological foundations of identity-related processes provides important insights into how mental illness can negatively impact the development and maintenance of identities. Given this, the proliferation of identity theory into either the sociology of mental illness or clinical psychology may provide important insights into how mental illness, the identity control process, and social structures interact to influence the quality of life of pathological populations. These insights, in turn, may suggest new holistic interventions that take into consideration a larger domain of potential solutions for assisting the mentally ill in terms of identity-related processes.

References

Addis, D. R., Sacchetti, D. C., Ally, B. A., Budson, A. E., & Schacter, D. L. (2009). Episodic simulation of future events is impaired in mild Alzheimer's disease. *Neuropsychologia, 47*(12), 2660–2671.

Andrews-Hanna, J. R., Reidler, J. S., Huang, C., & Buckner, R. L. (2010a). Evidence for the default network's role in spontaneous cognition. *Journal of Neurophysiology, 104*(1), 322–335.

Andrews-Hanna, J. R., Reidler, J. S., Sepulcre, J., Poulin, R., & Buckner, R. L. (2010b). Functional-anatomic fractionation of the brain's default network. *Neuron, 65*(4), 550–562.

Berger, J., & Zelditch, M. (1993). *Theoretical research programs: Studies in the growth of theory*. Stanford: Stanford University Press.

Berger, J., Wagner D. G., & Zelditch, M. Jr. (1985). Theory growth, social processes, and metatheory. In J. H. Turner (Ed.), *Theory building in sociology: Assessing theoretical accumulation* (pp. 19–42). Newbury Park: Sage.

Binder, J. R., Desai, R. H., Graves, W. W., & Conant, L. L. (2009). Where is the semantic system? A critical review and meta-analysis of 120 functional neuroimaging studies. *Cerebral Cortex, 19*(12), 2767.

Biswal, B. B., Mennes, M., Zuo, X. N., Gohel, S., Kelly, C., Smith, S. M., Beckmann, C. F., Adelstein, J. S., Buckner, R. L., Colcombe, S., Dogonowski, A. M., Ernst, M., Fair, D., Hampson, M., Hoptman, M. J., Hyde, J. S., Kiviniemi, V. J., Kotter, R., Li, S. J., Lin, C. P., Lowe, M. J., Mackay, C., Madden, D. J., Madsen, K. H., Margulies, D. S., Mayberg, H. S., McMahon, K., Monk, C. S., Mostofsky, S. H., Nagel, B. J., Pekar, J. J., Peltier, S. J., Petersen, S. E., Riedl, V., Rombouts, S., Rypma, B., Schlaggar, B. L., Schmidt, S., Seidler, R. D., Siegle, G. J., Sorg, C., Teng, G. J., Veijola, J., Villringer, A., Walter, M., Wang, L. H., Weng, X. C., Whitfield-Gabrieli, S., Williamson, P., Windischberger, C., Zang, Y. F., Zhang, H. Y., Castellanos, F. X., & Milham, M. P. (2010). Toward discovery science of human brain function. *Proceedings of the National Academy of Sciences of the United States of America, 107*(10), 4734–4739.

Buckner, R. L. (2010). The role of the hippocampus in prediction and imagination. *Annual Review of Psychology, 61*, 27–48.

Buckner, R. L., & Carroll, D. C. (2007). Self-projection and the brain. *Trends in Cognitive Sciences, 11*(2), 49–57.

Buckner, R. L., Andrews-Hanna, J. R., & Schacter, D. L. (2008). The brain's default network – Anatomy, function, and relevance to disease. In *Year in cognitive neuroscience 2008* (pp. 1–38). Oxford: Blackwell Publishing.

Buckner, R. L., Sepulcre, J., Talukdar, T., Krienen, F. M., Liu, H. S., Hedden, T., Andrews-Hanna, J. R., Sperling, R. A., & Johnson, K. A. (2009). Cortical hubs revealed by intrinsic functional connectivity: Mapping, assessment of stability, and relation to Alzheimer's disease. *Journal of Neuroscience, 29*(6), 1860–1873.

Burianova, H., McIntosh, A. R., & Grady, C. L. (2010). A common functional brain network for autobiographical, episodic, and semantic memory retrieval. *NeuroImage, 49*(1), 865–874.

Burke, P. J., & Stets, J. E. (2009). *Identity theory*. New York: Oxford University Press.

Burnett, S., & Blakemore, S. J. (2009). Functional connectivity during a social emotion task in adolescents and in adults. *European Journal of Neuroscience, 29*(6), 1294–1301.

Cacioppo, J. T., & Berntson, G. G. (2005). *Social neuroscience: Key readings*. New York: Psychology Press.

Cacioppo, J. T., & Patrick, W. (2008). *Loneliness: Human nature and the need for social connection*. New York: WW Norton & Company.

Christoff, K., Gordon, A. M., Smallwood, J., Smith, R., & Schooler, J. W. (2009). Experience sampling during fMRI reveals default network and executive system contributions to mind wandering. *Proceedings of the National Academy of Sciences, 106*(21), 8719.

Cohen, B. (1989). *Developing sociological knowledge*. Chicago: Nelson-Hall.

Collins, R. (1994). Why the social sciences won't become high-consensus, rapid-discovery science. *Sociological Forum, 9*(2), 155–177.

Cooley, C. H. (1902). *Human nature and the social order*. New York: Charles Scribner's Sons.

Corbetta, M., & Shulman, G. L. (2002). Control of goal-directed and stimulus-driven attention in the brain. *Nature Reviews Neuroscience, 3*(3), 201–215.

Corbetta, M., Patel, G., & Shulman, G. L. (2008). The reorienting system of the human brain: From environment to theory of mind. *Neuron, 58*(3), 306–324.

D'Argembeau, A., & Salmon, E. (2011). The neural basis of semantic and episodic forms of self-knowledge: Insights from functional neuroimaging. In C. López-Larrea (Ed.), *Sensing systems in nature*. Austin: Landes Biosciences.

Deco, G., Jirsa, V. K., & McIntosh, A. R. (2011). Emerging concepts for the dynamical organization of resting-state activity in the brain. *Nature Reviews Neuroscience, 12*(1), 43–56.

Fair, D. A., Cohen, A. L., Dosenbach, N. U. F., Church, J. A., Miezin, F. M., Barch, D. M., Raichle, M. E., Petersen, S. E., & Schlaggar, B. L. (2008). The maturing architecture of the brain's default network. *Proceedings of the National Academy of Sciences of the United States of America, 105*(10), 4028–4032.

Fowler, J. H., Settle, J. E., & Christakis, N. A. (2011). Correlated genotypes in friendship networks. *Proceedings of the National Academy of Sciences, 108*(5), 1993.

Fox, M. D., Snyder, A. Z., Vincent, J. L., Corbetta, M., Van Essen, D. C., & Raichle, M. E. (2005). The human brain is intrinsically organized into dynamic, anticorrelated functional networks. *Proceedings of the National Academy of Sciences of the United States of America, 102*(27), 9673–9678.

Fox, M. D., Corbetta, M., Snyder, A. Z., Vincent, J. L., & Raichle, M. E. (2006). Spontaneous neuronal activity distinguishes human dorsal and ventral attention systems. *Proceedings of the National Academy of Sciences of the United States of America, 103*(26), 10046–10051.

Franks, D. D. (2010). *Neurosociology: The nexus between neuroscience and social psychology*. New York: Springer.

Freese, J., Li, J. C. A., & Wade, L. D. (2003). The potential relevances of biology to social inquiry. *Annual Review of Sociology, 29*, 233–257.

Greicius, M. D., Supekar, K., Menon, V., & Dougherty, R. F. (2009). Resting-state functional connectivity reflects structural connectivity in the default mode network. *Cerebral Cortex, 19*(1), 72–78.

Grimm, S., Ernst, J., Boesiger, P., Schuepbach, D., Hell, D., Boeker, H., & Northoff, G. (2009). Increased self focus in major depressive disorder is related to neural abnormalities in subcortical cortical midline structures. *Human Brain Mapping, 30*(8), 2617–2627.

Guo, G., Cai, T., Guo, R., Wang, H., Harris, K. M., & Domschke, K. (2010). The dopamine transporter gene, a spectrum of most common risky behaviors, and the legal status of the behaviors. *PLoS One, 5*(2), 583–597.

Gusnard, D., Akbudak, E., Shulman, G., & Raichle, M. E. (2001). Role of medial prefrontal cortex in a default mode of brain function. *NeuroImage, 13*(6), S414–S414.

Gilbert, S. J., Dumontheil, I., Simons, J. S., Frith, C. D., & Burgess, P. W. (2007). Comment on "Wandering minds: the default network and stimulus-independent thought". *Science, 317*, 43b.

Hagmann, P., Cammoun, L., Gigandet, X., Meuli, R., Honey, C. J., Wedeen, V., & Sporns, O. (2008). Mapping the structural core of human cerebral cortex. *PLoS Biology, 6*(7), 1479–1493.

Hebb, D. O. (2002). *The organization of behavior: A neuropsychological theory*. Mahwah: L. Erlbaum Associates.

Iacoboni, M., Molnar-Szakacs, I., Gallese, V., Buccino, G., Mazziotta, J. C., & Rizzolatti, G. (2005). Grasping the intentions of others with one's own mirror neuron system. *PLoS Biology, 3*(3), e79.

Inder, M. L., Crowe, M. T., Moor, S., Luty, S. E., Carter, J. D., & Joyce, P. R. (2008). "I actually don't know who I am": The impact of bipolar disorder on the development of self. *Psychiatry: Interpersonal and Biological Processes, 71*(2), 123–133.

Jiao, Q., Lu, G. M., Zhang, Z. Q., Zhong, Y. A., Wang, Z. G., Guo, Y. X., Li, K., Ding, M. Z., & Liu, Y. J. (2011). Granger causal influence predicts BOLD activity levels in the default mode network. *Human Brain Mapping, 32*(1), 154–161.

Kawasaki, H., Adolphs, R., Kaufman, O., Damasio, H., Damasio, A. R., Granner, M., Bakken, H., Hori, T., & Howard, M. A. (2001). Single-neuron responses to emotional visual stimuli recorded in human ventral prefrontal cortex. *Nature Neuroscience, 4*(1), 15–16.

Kim, M. J., Gee, D. G., Loucks, R. A., Davis, F. C., & Whalen, P. J. (2011). Anxiety dissociates dorsal and ventral medial prefrontal cortex functional connectivity with the amygdala at rest. *Cerebral Cortex, 21*, 1667–1673.

Kompus, K., Olsson, C. J., Larsson, A., & Nyberg, L. (2009). Dynamic switching between semantic and episodic memory systems. *Neuropsychologia, 47*(11), 2252–2260.

Kwan, D., Carson, N., Addis, D. R., & Rosenbaum, R. S. (2010). Deficits in past remembering extend to future imagining in a case of developmental amnesia. *Neuropsychologia, 48*(11), 3179–3186.

Leech, R., Kamourieh, S., Beckmann, C. F., & Sharp, D. J. (2011). Fractionating the default mode network: Distinct contributions of the ventral and dorsal posterior cingulate cortex to cognitive control. *Journal of Neuroscience, 31*(9), 3217.

Lemogne, C., le Bastard, G., Mayberg, H., Volle, E., Bergouignan, L., Lehericy, S., Allilaire, J. F., & Fossati, P. (2009). In search of the depressive self: Extended medial prefrontal network during self-referential processing in major depression. *Social Cognitive and Affective Neuroscience, 4*(3), 305–312.

Levine, B., Turner, G. R., Tisserand, D., Hevenor, S. J., Graham, S. J., & McIntosh, A. R. (2004). The functional neuroanatomy of episodic and semantic autobiographical remembering: A prospective functional MRI study. *Journal of Cognitive Neuroscience, 16*(9), 1633–1646.

Leyhe, T., Muller, S., Milian, M., Eschweiler, G. W., & Saur, R. (2009). Impairment of episodic and semantic autobiographical memory in patients with mild cognitive impairment and early Alzheimer's disease. *Neuropsychologia, 47*(12), 2464–2469.

Machalek, R., & Martin, M. W. (2004). Sociology and the second Darwinian revolution: A metatheoretical analysis. *Sociological Theory, 22*(3), 455–476.

Mead, G. H. (1934). *Mind, self and society*. Chicago: University of Chicago Press.

Mor, N., & Winquist, J. (2002). Self-focused attention and negative affect: A meta-analysis. *Psychological Bulletin, 128*(4), 638.

Northoff, G. (2007). Psychopathology and pathophysiology of the self in depression – Neuropsychiatric hypothesis. *Journal of Affective Disorders, 104*(1–3), 1–14.

Northoff, G. (2011). The brain and its self: Concepts of self and the cortical midline structures. In S. Han & E. Pöppel (Ed.), *Culture and neural frames of cognition and communication* (pp. 41–63). Berlin/Heidelberg: Springer.

Northoff, G., & Bermpohl, F. (2004). Cortical midline structures and the self. *Trends in Cognitive Sciences, 8*(3), 102–107.

Ochsner, K. N., Beer, J. S., Robertson, E. R., Cooper, J. C., Gabrieli, J. D. E., Kihsltrom, J. F., & D'Esposito, M. (2005). The neural correlates of direct and reflected self-knowledge. *NeuroImage, 28*(4), 797–814.

Preminger, S., Harmelech, T., & Malach, R. (2011). Stimulus-free thoughts induce differential activation in the human default network. *NeuroImage, 54*(2), 1692–1702.

Prince, S. E., Tsukiura, T., & Cabeza, R. (2007). Distinguishing the neural correlates of episodic memory encoding and semantic memory retrieval. *Psychological Science, 18*(2), 144.

Raichle, M. E., & Gusnard, D. A. (2005). Intrinsic brain activity sets the stage for expression of motivated behavior. *The Journal of Comparative Neurology, 493*(1), 167–176.

Raichle, M. E., MacLeod, A. M., Snyder, A. Z., Powers, W. J., Gusnard, D. A., & Shulman, G. L. (2001). A default mode of brain function. *Proceedings of the National Academy of Sciences of the United States of America, 98*(2), 676–682.

Rameson, L. T., Satpute, A. B., & Lieberman, M. D. (2010). The neural correlates of implicit and explicit self-relevant processing. *NeuroImage, 50*(2), 701–708.

Schutz, A. (1967). *The phenomenology of the social world.* Evanston: Northwestern University Press.

Shanahan, M. J., Vaisey, S., Erickson, L. D., & Smolen, A. (2008). Environmental contingencies and genetic propensities: Social capital, educational continuation, and dopamine receptor gene "DRD2". *The American Journal of Sociology, 114*, 260–286.

Sheline, Y. I., Price, J. L., Yan, Z., & Mintun, M. A. (2010). Resting-state functional MRI in depression unmasks increased connectivity between networks via the dorsal nexus. *Proceedings of the National Academy of Sciences, 107*(24), 11020.

Smallwood, J., & Schooler, J. W. (2006). The restless mind. *Psychological Bulletin, 132*(6), 946.

Sporns, O. (2011). *Networks of the brain.* Cambridge: MIT Press.

Spreng, R. N., & Grady, C. L. (2010). Patterns of brain activity supporting autobiographical memory, prospection, and theory of mind, and their relationship to the default mode network. *Journal of Cognitive Neuroscience, 22*(6), 1112–1123.

Spreng, R. N., Mar, R. A., & Kim, A. S. N. (2009). The common neural basis of autobiographical memory, prospection, navigation, theory of mind, and the default mode: A quantitative meta-analysis. *Journal of Cognitive Neuroscience, 21*(3), 489–510.

Spreng, R. N., Stevens, W. D., Chamberlain, J. P., Gilmore, A. W., & Schacter, D. L. (2010). Default network activity, coupled with the frontoparietal control network, supports goal-directed cognition. *NeuroImage, 53*(1), 303–317.

Squire, L. R., & Kandel, E. R. (2009). *Memory: From mind to molecules.* Greenwood Village: Roberts & Co.

Stawarczyk, D., Majerus, S., Maquet, P., & D'Argembeau, A. (2011). Neural correlates of ongoing conscious experience: Both task-unrelatedness and stimulus-independence are related to default network activity. *PLoS One, 6*(2), 14.

Stets, J. E. (2010). The social psychology of the moral identity. In S. Hitlin & S. Vaisey (Eds.), *Handbook of the sociology of morality* (pp. 385–409). New York: Springer.

Stryker, S. (2002). *Symbolic interactionism: A social structural version.* Caldwell: Blackburn Press.

Turner, J. H., & Maryanski, A. (2005). *Incest: Origins of the taboo.* Boulder: Paradigm Publishers.

Udry, J. R. (1995). Sociology and biology: What biology do sociologists need to know? *Social Forces, 73*(4), 1267–1278.

van den Heuvel, M. P., Mandl, R. C. W., Kahn, R. S., & Pol, H. E. H. (2009). Functionally linked resting-state networks reflect the underlying structural connectivity architecture of the human brain. *Human Brain Mapping, 30*(10), 3127–3141.

Wagner, D. G. (1984). *The growth of sociological theories.* Thousand Oaks: Sage.

Wagner, D. G. (2007). The limits of theoretical integration. *Social Justice Research, 20*(3), 270–287.

Wagner, D. G., & Berger, J. (1985). Do sociological theories grow? *American Journal of Sociology, 90*, 697–728.

Wiggs, C. L., Weisberg, J., & Martin, A. (1998). Neural correlates of semantic and episodic memory retrieval. *Neuropsychologia, 37*(1), 103–118.

Chapter 13
The Emergent Self: How Distributed Neural Networks Support Self-Representation

Istvan Molnar-Szakacs and Lucina Q. Uddin

The self has been broadly conceptualized and divided into two main aspects—the mental/psychological and the physical/embodied self—with two different brain network, the default mode network (DMN) and the mirror neuron system (MNS) implicated as its neural foundations. As the self-face is the most identifiable marker of the physical aspect of the self, it has been the subject of extensive study at the behavioral and neural level. Recent functional magnetic resonance imaging (fMRI) studies of self-face recognition have consistently found right frontoparietal areas associated with identification of the self-face. These areas appear to broadly overlap the human MNS. Mirror neurons are active when someone performs an action and when they observe that same action being performed. This neural simulation of motor acts and related representations creates an agent-independent link between actor and observer. In fact accumulating evidence suggests that simulation processes are used in a multitude of cognitions that constitute the self including autobiographical memory and prospection, perspective taking, understanding other's actions and mental states, and embodied self-representation. Equally, components of the DMN - particularly the medial prefrontal cortex and the posterior cingulate cortex - have been implicated in representing aspects of the mental self, including autobiographical memory and self-knowledge. Thus interactions of the DMN and MNS may subserve the integration of self-relevant traits within the context of autobiographical memory as well as future action goals—positioning the self as a "center of gravity" of one's private and social behavior.

Scope and Limitations of Current Review

One major and useful distinction that has guided much research on the neural representation of the self is that between the physical and psychological aspects of the self (Gillihan and Farah 2005). Physical aspects of the self are typically examined in studies of self-face recognition, body recognition, agency, and perspective taking. Psychological aspects of the self tend to be operationalized

I. Molnar-Szakacs (✉)
Semel Institute for Neuroscience and Human Behavior, University of California, Los Angeles, CA 90095, USA

Tennenbaum Center for the Biology of Creativity, University of California, Los Angeles, CA, USA
e-mail: imolnar@ucla.edu

L.Q. Uddin
Psychiatry & Behavioral Science-Child and Adolescent Psychiatry, Stanford University School of Medicine, Stanford, CA 94305, USA

D.D. Franks and J.H. Turner (eds.), *Handbook of Neurosociology*, Handbooks of Sociology and Social Research, DOI 10.1007/978-94-007-4473-8_13,

with studies examining autobiographical memory and self-knowledge in the form of personality traits. This conceptual distinction bears out in neuroimaging work, which suggests that physical or embodied self-related processes and psychological or evaluative self-related processes rely on distinct yet interacting large-scale brain networks (Lieberman 2007; Molnar-Szakacs and Arzy 2009; Uddin et al. 2007).

In the current work, we review how recent advances in the study of large-scale human brain networks have contributed to our understanding of self-related cognition and representation. We will follow the useful distinction of physical versus psychological aspects of the self in this chapter by limiting its scope to illustrative studies on these two areas. We begin by discussing seminal theories on the self to provide a historical context. We give working definitions and descriptions of terminology relevant to our discussion of the self and the brain. Next, we describe the mirror neuron system (MNS) and the default mode network (DMN), two brain systems thought to subserve complementary aspects of self-representation. We then discuss paradigms that have been used to operationalize the study of the self, within the domains of self-face recognition and self-trait recognition—the most commonly examined physical and psychological aspects, respectively. We conclude with a discussion of autobiographical memory and propose a means by which an emergent self arises from complex interactions among seemingly disparate representations.

Historical and Contemporary Definitions of Self

A central feature of human experience is our sense of self that persists across space and time. Understanding the essence of this experience has challenged scholars for many centuries. What is the self? The term itself is difficult to define and has led some of the most influential thinkers in psychology to theorize about the self. William James wrote in *The Principles of Psychology* that the self is not a single primordial entity (James 1983). This early conceptualization of multiple aspects of the self set the stage for later work examining these different facets. Ulric Neisser, a social psychologist, suggests that people have access to five different kinds of self-knowledge, which may develop during different periods: (1) the ecological self, perceived with respect to the physical environment; (2) the interpersonal self, depending on emotional and other species-specific forms of communication; (3) the temporally extended self, based on memory and anticipation, implying a representation of self; (4) the private self, reflecting knowledge that our conscious experiences are exclusively our own; and (5) the conceptual self, based on sociocultural experience (Neisser 1995). His claim is that the self is not some special part of a person or brain, but rather a whole person considered from a particular point of view. For example, the ecological self is the individual considered as an agent in the environment, and the interpersonal self is that individual engaging in face-to-face contact with others. Key to this theory is that perception of oneself in these different ways is the first and most fundamental form of self-knowledge and self-awareness. This definition of self in terms of one's existence in the world shifts focus from an inward-looking view based on private experience to an outward-looking view of the self as ecologically and socially situated (Neisser 1993).

Dennett (1991) relates a language-based approach to the self, referring to the self as the center of narrative gravity. According to this view, humans, with our unique capacity for language, spin narratives that are the essence of ourselves: "Our fundamental tactic of self-protection, self-control, and self-definition is…telling stories, and more particularly concocting and controlling the story we tell others—and ourselves—about who we are." This center of narrative gravity posited as the self is analogous to a center of gravity in the physical sense—a simplified, single point of origin.

Expanding on the idea of a narrative self, Shaun Gallagher (2000) delineates a distinction called the "minimal" self versus the "narrative" self. Here, the "minimal" self is referred to as the self devoid of temporal extension, a consciousness of oneself as an immediate subject of experience, depending on brain processes and an ecologically embedded body. The "narrative" self, on the other hand, involves personal identity and continuity across time and is a self-image constituted with a past and future in stories that we and others tell about ourselves.

Marc Jeannerod, on the other hand, grounds the self in the body and, more specifically, the motor system. He holds the view that a key component of self-recognition in humans is recognizing oneself as the owner of a body and the agent of actions. These sensations of agency and ownership arise from congruence of proprioceptive feedback and sensory signals from body parts, and central signals that contribute to the generation of movements. He claims that the sense of agency provides a way for the self to build an identity independent of the external world (Jeannerod 2003).

An extreme view put forth by the philosopher Thomas Metzinger is that there are no such things as selves. Metzinger (2003) claims that nobody ever has or had a self and that all that exists are conscious self-models. He states, "the phenomenal self is not a thing, but a process—and the subjective experience of being someone emerges if a conscious information-processing system operates under a transparent self-model." This conscious self-model of human beings is a way of allowing an organism to conceive of itself as a whole and thus causally interact intelligently with its environment.

The last 20 years have seen enormous advances in our understanding of the human brain, and this has allowed cognitive neuroscientists and neuropsychologists to begin the study of linking the "self" to its neural substrates. Acquisition of much of this new knowledge has been facilitated by developments in brain imaging technology including methods that (1) measure neuronal firing patterns at the scalp (using electroencephalography, EEG); (2) measure the decay of an injected radioactive isotope, generally glucose which is taken up by active neurons in the brain, reflecting regional metabolic activity (using positron emission tomography, PET): (3) measure the changing levels of deoxygenated blood in response to neuronal firing patterns throughout the whole brain (using functional magnetic resonance imaging, fMRI); and (4) measure motor evoked potentials caused by electromagnetic induction from a rapidly changing magnetic field that leads to depolarization in the neurons of the brain (using transcranial magnetic stimulation, TMS). Researchers are using these methods to ask which brain regions and systems are critical to self-awareness and other forms of self-related processing, usually by focusing on one particular aspect of the self, such as visual self-recognition, to uncover the neural basis of that particular process.

Large-Scale Brain Networks and Methods in Cognitive Neuroscience

Recent years have witnessed a paradigm shift in cognitive neuroscience (Aminoff et al. 2009). Whereas early functional brain imaging work focused primarily on localization of function, revealing activation in specific brain regions during performance of cognitive tasks, interest has recently shifted toward developing a deeper understanding of intrinsic brain connectivity and the architecture of brain networks that influences cognitive and affective information processing. A network, generally, is any set of objects that interact or share some relationship with one another (Wig et al. 2011). We are familiar with networks in our daily lives in the form of the World Wide Web. A brain network consists of individual brain regions (or nodes) that interact via structural and/or functional connections. A brain network can be defined based on structural connectivity as measured in the human brain with diffusion tensor imaging (DTI), or functional connectivity as typically measured by fMRI (Bressler and Menon 2010). Conceptualizing the brain in terminology used to characterize large-scale networks has proven to have a great deal of explanatory value; thus many cognitive neuroscientists have adopted this way of thinking about brain function and cognitive processes.

Two networks that have been linked to social cognition and therefore deeply implicated in the study of the self in particular—the human mirror neuron system and the default mode network—are described in detail in the following sections.

The Neural Networks

The Mirror Neuron System in Humans

For the authors of this chapter, one of the most exciting recent developments to emerge from cognitive neuroscience, with the potential to impact significantly both our conceptualization and our understanding of the self, is the discovery of the so-called human mirror neuron system (MNS) (Rizzolatti and Sinigaglia 2010). In essence, the MNS allows us to understand and predict the behavior of others, by engaging the neural regions required to produce such behavior ourselves. That is, when we see *another's hand* grasping an object, we activate the regions of *our brain* that control grasping; when we hear sounds associated with *someone else's action*, we activate the appropriate movement regions of *our brain*; and by extension, when we observe the *emotional states of others, we can feel the same emotion* in empathy (Carr et al. 2003; Gazzola et al. 2006; Molnar-Szakacs et al. 2006). It has thus been suggested that "mirror neurons are a kind of 'neural wi-fi' that monitors what is happening in other people. This system tracks their emotions, what movements they're making, what they intend, and it activates, in our brains, precisely the same brain areas as are active in the other person. This puts us on the same wavelength and it does it automatically, instantaneously and unconsciously" (Goleman 2006).

Since the discovery of the MNS, the brain can no longer be considered as an independent input–output, perception–action machine—it is deeply, intrinsically connected with our bodies and, most intriguingly, with our understanding of the actions of other individuals (Cattaneo et al. 2011). Such a conceptual shift offers a helpful new framework for our understanding of the self as well, which of course is also hard to conceptualize as an independent entity, as we are each of us intrinsically connected with our physical environment as well as with those around us.

In the macaque monkey brain, neurons with mirror properties have been recorded using single-unit electrode recordings in both area F5 of the premotor cortex and in parietal area PF (Rizzolatti and Craighero 2004). These visuomotor neurons discharge both during the performance of an action and during the observation of another individual performing a similar action (di Pellegrino et al. 1992; Gallese et al. 1996; Rizzolatti et al. 1996a, b). Parietal mirror neurons have the special property of coding motor acts as belonging to an action sequence, predicting the intended goal of a complex action (Fogassi et al. 2005). In addition, it has been shown that a subset of mirror neurons are able to represent actions even when the final part of the action is unseen (Umilta et al. 2001) or from hearing sounds associated with particular actions (Kohler et al. 2002). Thus, area F5 of the ventral premotor cortex and area PF of the inferior parietal lobule in the macaque brain form a frontoparietal mirror neuron system critical to action understanding and intention attribution (Fogassi et al. 2005; Rizzolatti and Craighero 2004; Rizzolatti et al. 2001).

Soon after the discovery of mirror neurons in the monkey brain, neuroimaging using PET (Grafton et al. 1996; Rizzolatti et al. 1996a, b) and TMS studies (Fadiga et al. 1995) demonstrated a network with similar functional properties in the human brain. Fadiga and colleagues (1995) provided an elegant demonstration of the matching neural representation in humans for actions observed and those performed by the self. In their study, magnetic stimulation was delivered to the scalp above the motor cortex, the source of motor commands for action, (1) while participants were observing an experimenter perform various hand actions in front of them or (2) during control conditions including arm movement observation, object observation, and dimming detection, while motor evoked potentials (MEPs) were recorded from hand muscles. MEPs serve as a quantifiable index of activity within the motor cortex. Results showed that during hand action observation, but not in the other conditions, there was an increase in the amplitude of the MEPs in the *same* hand muscles that are used when the observed action is actually performed by the observer (Fadiga et al. 1995). This increase in MEPs resulting from observing a hand action reflects a facilitation or priming of the motor cortex—an increased potential for action—due to "inner imitation" of the hand action that was observed. Subsequent work then confirmed and extended these findings (Borroni and Baldissera 2008; Gangitano et al. 2001; Montagna et al. 2005; Strafella and Paus 2000), leading researchers to conclude that in humans there is a neural system, resembling the one described in the monkey, matching observed actions and executed actions.

During development and into adulthood, we acquire the basis of language and our culture, including a multitude of motor skills—like how to play sports and use everyday objects—through imitation, by reproducing observed behaviors (Greenfield 2006). It is very likely that most of these functions have, at their bases, a simulation mechanism (Iacoboni 2009; Pineda 2008; Rizzolatti and Craighero 2004). Simulation, or inner imitation, refers to the fact that the same neural resources are recruited while one observes *and* while one executes an action, allowing the brain to link the perceptions of an observer to the actions of an agent at the neural level. Furthermore, the fact that the same neural networks are activated within the brains of all individuals who participate in an interaction—the observer(s) *and* the agent(s)—allows for shared representations to be established among individuals, giving rise to a mutual understanding of the interaction.

Imitation is a ubiquitous human behavior composed of both an observation and an execution phase, making it an ideal paradigm for the study of neural activity associated with observation–execution matching in the human brain. fMRI was used to localize the brain regions matching observed action to executed action using an imitation paradigm. Basing their predictions on neural firing rates in the monkey (Gallese et al. 1996), Iacoboni and colleagues (1999) hypothesized that areas of the human brain that show mirror properties would show an increase of brain activity in the same area during action observation *and* during action execution, and action execution would lead to approximately double the increase in activity as action execution contains an overt motor response. In turn, imitation, which contains both an observation and execution component, would lead to the greatest increase in neural activity. This study found two cortical areas of the human brain that showed this predicted pattern of activity, the posterior inferior frontal gyrus (IFG) and the rostral sector of the inferior parietal lobule (IPL) (Iacoboni et al. 1999).

A wealth of additional studies have shown that mirror resonance mechanisms are involved in action perception and performance, meaning that the same neural network subserves the understanding of actions we observe, as in planning and executing our own movements (Aziz-Zadeh et al. 2006; Binkofski et al. 1999; Buccino et al. 2004; Chong et al. 2008; Fadiga et al. 1995; Grafton et al. 1996; Grezes et al. 2003; Iacoboni et al. 1999, 2005; Johnson-Frey et al. 2003; Koski et al. 2002; Molnar-Szakacs et al. 2005a; Nishitani and Hari 2000; Oberman et al. 2005; Rizzolatti et al. 1996a, b; Woodruff and Maaske 2010). The presence of an MNS in the human brain is also corroborated by the fact that its main neuroanatomical nodes—inferior frontal gyrus (IFG) and premotor cortex (PMC) anteriorly and inferior parietal lobule (IPL) posteriorly—are considered homologous to the areas forming the MNS in the monkey (Amunts et al. 1999; Mazziotta et al. 2001; Petrides and Pandya 1997; Rizzolatti and Matelli 2003; Tomaiuolo et al. 1999; VonBonin and Bailey 1947; VonEconomo and Koskinas 1925). Taken together, results of these studies imply that perception and action are inseparable, both in our daily behavior, as well as within the neural networks supporting them in our brain. In fact, it appears that the MNS may have further evolved to subserve more sophisticated functions in humans that are only rudimentarily present or even completely absent in monkeys, such as imitation, recognition of intransitive and symbolic gestures, language, intention understanding, and, of immediate interest to this chapter, self-representation. Based on the property of mirror neurons to internally simulate actions performed by others, it has been proposed that the

MNS may provide the link between the physical representation of the self as related to the physical image of others. Thus, the inner mirroring of other's actions and emotions, as supported by the human MNS, allows us to see the self, reflected in others.

The Default Mode Network

Another brain network that has been most consistently linked to self-related processing is the so-called default mode network (DMN). The repeated observation that the ventral medial prefrontal cortex (VMPFC), posterior cingulate cortex (PCC), lateral parietal cortices, and medial temporal lobes paradoxically exhibit high levels of activity during resting baseline and decreases in activity during externally oriented cognitive tasks led to the initial characterization of these regions as belonging to a "default mode network" (Esposito et al. 2006; Fransson 2006; Gusnard et al. 2001; McKiernan et al. 2003; Raichle et al. 2001). This network has also been referred to as the "task-negative network" (Fox et al. 2005), or the "cortical midline structures" (Northoff et al. 2006), and was originally proposed as a system for evaluating "information broadly arising in the external and internal milieu" (Raichle et al. 2001). The network has since been posited to underlie a variety of functions, many of them social cognitive in nature. The DMN has been linked to episodic memory (Greicius and Menon 2004) and memory consolidation (Miall and Robertson 2006) in some studies and social (Iacoboni et al. 2004; Uddin et al. 2005) or self-related processes (Buckner and Carroll 2007; Gusnard et al. 2001; Wicker et al. 2003) in others. Recent theories posit that this network is critical for self-projection, or thinking about the future (Buckner and Carroll 2007). It is also notable that the network resembles that which is activated during theory of mind or mentalizing tasks where participants are asked to consider the mental viewpoint of another, as well as tasks requiring moral social evaluations (Harrison et al. 2008). Still other studies associate default mode function with more general processes such as stimulus-independent (Mason et al. 2007) or task-unrelated thought (McKiernan et al. 2006).

It is difficult to envision one comprehensive theory explaining the DMN's ability to support such a diverse array of cognitive functions and be associated with such a wide range of psychiatric and neurological disturbances. One common thread that can be seen, however, is that the functions attributed to the DMN are all in some sense self-related and in particular involve the representation of the psychological aspects of the self and its relationship to the external world. At present, the authors take the view that the DMN may be involved in maintaining a self-representation in evaluative terms, which requires both self-referential processing and understanding of others' mental states. We further speculate that the DMN might support evaluative simulation in the same way that the MNS supports motor simulation (Uddin et al. 2007).

Aberrations in activity of the DMN have been linked to cognitive deficits in a number of clinical conditions that are related to disturbances of the self. To date, abnormalities in the DMN have been demonstrated in individuals with autism spectrum disorders (Kennedy et al. 2006), Alzheimer's disease (Greicius et al. 2004), and schizophrenia (Liang et al. 2006). These disorders all manifest as altered psychological self-related cognition in the realms of social, memory, and self-monitoring processes.

Interestingly, recent theories related to different aspects of self-representation, as well as to conditions which involve a disturbance of self-related processing, often invoke explanations that are based either in deficits of the DMN, the human MNS, or both. For example, theories of *how we understand other minds* have implicated both the DMN (Spreng et al. 2009) and the MNS (Gallese and Goldman 1998); theories about *moral thinking* have been linked to both the DMN (Harrison et al. 2008) and the MNS (Molnar-Szakacs 2011); and the DMN and the MNS have both been implicated in theories of *physical self-representation* (Molnar-Szakacs and Arzy 2009;

Uddin et al. 2007). In the realm of psychiatric or neurological disturbances, both the DMN (Cherkassky et al. 2006; Kennedy et al. 2006) and the MNS (Dapretto et al. 2006; Molnar-Szakacs et al. 2009) have been implicated in *autism spectrum disorders,* and aberrant DMN connectivity (Garrity et al. 2007) and MNS dysfunction have been suggested in *schizophrenia* (Iacoboni and Dapretto 2006). Taken together, this evidence from both the healthy and the atypical brain suggests that these two neural systems—the human MNS and the DMN—are functionally connected and are together profoundly implicated in self-related cognition.

Operationalizing the Self

Physical Self-recognition

A key component of the self is embodiment, and that part of the body that functions as the most unique identifier is the face. Thus, the self-face is a critical component of self-identity. Cole (1999) describes cases of individuals with various problems affecting visual face perception and the effects of these disorders on their sense of selfhood. He discusses a case of a congenitally blind patient who, though unable to extract visual information from faces, still understood the importance of face-to-face contact in social communication: "… in order to interact and talk with people you present your face to them. It's not just a place your voice comes out of, it allows contact with others." A patient who became completely blind later in life reflects on a loss of identity resulting from his blindness: "To what extent is the loss of the image of the face connected with loss of the image of the self?" A patient with Möbius syndrome, which renders the subject unable to move any of the muscles of facial expression, relates the feeling of "living entirely in [his] head" due to his inability to engage in social emotional interactions. Patients with Bell's palsy and autism are also discussed in light of their problems with emotional facial expression and subsequent issues with self-representation (Cole 1999). A particularly interesting case of aberrant self-recognition is that referred to as "mirror sign." Phillips and Howard (1996) describe a patient who exhibited some global cognitive impairment and was unable to recognize herself in a mirror. The patient had no insight into her condition. The authors classify the phenomenon as a delusion of self-misidentification on the basis of a normal CT scan (Phillips and Howard 1996). Others have reported on this strange phenomenon as well, some emphasizing right hemispheric dysfunction as being an underlying common thread (Breen et al. 2001). Thus, as these examples highlight, the self-face plays a seminal role in our physical, psychological, and social identity throughout the lifespan.

As early as 1889, Preyer used mirrors to assess the development of the self-concept, noting that "the behavior of the child toward his image in the glass shows unmistakably the gradual growth of the consciousness of self out of a condition in which objective and subjective changes are not yet distinguished from each other" (Preyer 1889). It has been demonstrated that infants around 2 years of age begin to show behavior indicative of self-recognition in front of a mirror (Amsterdam 1972). Early observations led Gallup to conclude that self-recognition is predicated on a sense of identity—that this capacity is indicative of an underlying self-concept (Gallup 1977). While these studies were among the first to systematically use mirrors to test hypotheses regarding self-awareness and self-concept, the use of mirrors to this end has also been reported by Charles Darwin (Darwin 1877). While the purported relationship between self-recognition and other forms of self-awareness has been discussed and evaluated for a long time, the ability to mirror-self-recognize has only been demonstrated in humans, chimpanzees (Gallup 1970; Povinelli and Gallup 1997), orangutans (Lethmate and Ducker 1973), elephants (Plotnik et al. 2006), the bottlenose dolphin (Reiss and Marino 2001), and for the first time in a nonmammalian species, the magpie (Prior et al. 2008).

We, as well as others, have shown that self-face recognition abilities rely on a frontoparietal network in the right hemisphere (Platek et al. 2006; Sugiura et al. 2005; Uddin et al. 2005). In particular, in our own work, we observed that the pattern of signal increases observed in the right inferior frontal gyrus (IFG) and right inferior parietal lobule (IPL) were related to the amount of self-face presented in morphed stimuli. In other words, the greater amount of "self" present in the stimulus, the greater the activation in right frontoparietal areas that overlap with the human MNS (Uddin et al. 2005). We proposed that mirror areas may be more active for stimuli containing more "self" because their role is to establish communication between individuals via a simulation mechanism that maps actions of others onto one's own motor repertoire, thereby making others "like me" (Meltzoff and Brooks 2001). Thus, when one sees one's own image, these mirror areas are more strongly activated because of the ease with which one can map oneself onto one's own motor system. This mapping produces the best match or correspondence, reflected in activity of the mirror neuron system, primarily in the right hemisphere (Uddin et al. 2005). Interestingly, we have also observed similar brain activation patterns distinguishing the self-voice from other voices, suggesting that the right hemisphere MNS may contribute to multimodal abstract self-representation (Kaplan et al. 2008). We have also used image-guided repetitive transcranial magnetic stimulation (rTMS) to create a "virtual lesion" over the parietal component of this self-recognition network to test whether the region is necessary for discriminating self-faces from other familiar faces. We showed that 1-Hz rTMS to the right IPL selectively disrupted performance on a self–other discrimination task, whereas applying 1-Hz rTMS to the left IPL had no effect. Thus it appears that activity in the right IPL is essential to the task, providing causal evidence for a relationship between the right IPL and self-face recognition (Uddin et al. 2006). Another recent rTMS study has also corroborated the finding of right hemisphere dominance for self-face recognition (Heinisch et al. 2011).

While recent neuroimaging reports have shown that several additional brain regions may contribute to self-face processing, a review of these studies highlights the common finding of right frontal and parietal activations accompanying self-face viewing, especially when compared to other familiar faces (Devue and Bredart 2011). Interestingly, a meta-analysis of studies of self-face recognition found that in addition to right frontoparietal regions which overlap the human MNS, the right precuneus is a region that is also associated with this task (Platek et al. 2008). This finding is particularly relevant to our hypothesis that the human MNS and the DMN give rise to an integrated self-representation—given the fact that the precuneus is often linked with the DMN. While useful for the purposes of study and discourse, the lines we have drawn between physical and psychological self-representation may not be as relevant in functional terms. Based on their recent review of the literature, Devue and Bredart conclude that it remains difficult to determine which specific cognitive operation is reflected by each recruited brain area and, thus, suggest that goals for future research should include understanding the precise cognitive operations induced by perception of the self-face in order to better determine the functional significance of brain activations in specific regions (Devue and Bredart 2011).

Self-trait Recognition

Unlike self-face recognition, which recruits autobiographical representations through specific visual processing invoking memory retrieval processes (Fink et al. 1996; Keenan et al. 2001), personality-trait words likely access a representation of the self predominantly through linguistic aspects of the self-schema (Faust et al. 2004; Molnar-Szakacs et al. 2005b; Moran et al. 2006). Self-schemata are cognitive representations of the self derived from past social interactions and experiences that promote elaboration and organization of stored information and may be used to guide behavior (Markus 1977). As traits are incorporated into the self-schema, subsequent memory for these trait words is increased (Rogers et al. 1977). For example, memory for previously presented trait adjectives (e.g., happy) was better if they had been processed with reference to the self (e.g., "does happy describe you?") than if they had been processed only for their general meaning (e.g., "does happy mean the same as optimistic?"), a phenomenon labeled the self-reference effect (Symons and Johnson 1997).

There have been two major competing explanations for the self-reference memory effect. The first view is that the self is a cognitive structure that possesses special mnemonic abilities, leading to the privileged status of material processed in relation to self. The contrasting view is that no distinct structure or neural process is dedicated to self-referential processing, and the memory enhancement that accompanies self-referential processing can be interpreted as a standard depth-of-processing effect. That is, because we know a lot of information about ourselves, we encode additional information about the self more deeply. In turn, this elaborative encoding enhances the memory for self-relevant information. Functional imaging studies have identified multiple regions that are responsive to various aspects of self-relevant processing. For example, within the category of self-related linguistic stimuli, regions of the left prefrontal cortex are involved in semantic encoding. But are there neural structures that are selective for self-relevant information?

Kelley and colleagues (2002) designed an fMRI study to look precisely at this question—whether knowledge about the self is unique in terms of its functional anatomic representation within the human brain. Participants were imaged while making judgments about trait adjectives under three experimental conditions—self-relevance, other-relevance, or case judgment (upper- versus lowercase letters). The authors found that while the semantic processing component found across all conditions activated left prefrontal regions, the self-trials were distinctive for their selective activity in areas of the medial prefrontal cortex (MPFC), suggesting that this region might be involved in processing self-referential linguistic information (Kelley et al. 2002).

The special role of the MPFC in processing self-related material has now been demonstrated in a variety of neuroimaging studies. For example, dorsomedial prefrontal cortex (DMPFC) activation was observed during evaluation of self-referential statements (e.g., "I like Leipzig: yes/no") compared with memory retrieval trials (Zysset et al. 2002). In an fMRI study using self-descriptive positive and negative words, it was found that processing related to the self recruited the right DMPFC and posterior cingulated cortex (PCC) specifically during self-referential evaluation irrespective of the valence of the presented words (Fossati et al. 2003). A subsequent study by the same group found that the correct recognition of self-related positive and negative words reactivated the right DMPFC. Activity in this region was driven by the self-negative words, indicating that self-characteristic negative stimuli may facilitate retrieval of the self-schema (Fossati et al. 2004). Indeed, in addition to the MPFC, neuroimaging studies suggest a role for the PCC and the adjacent precuneus in integrating self-referential stimuli. For example, activation in the PCC and precuneus was observed when subjects had to indicate whether a word or statement was self-descriptive or not (Fossati et al. 2003; Johnson et al. 2002; Kircher et al. 2000). Similarly, reflection on one's own personality traits was associated with activation in the precuneus and MPFC when compared with reflection on traits of the Danish queen (Kjaer et al. 2002).

A wealth of other functional brain imaging studies have revealed activations in this set of cortical midline structures—the VMPFC, the DMPFC, the posterior cingulate, and the precuneus—that form part of the DMN when people reflect on their psychological characteristics (Craik et al. 1999; D'Argembeau et al. 2005; Fossati et al. 2003; Johnson et al. 2002; Kelley et al. 2002; Kjaer et al. 2002; Lou et al. 2004; Mitchell et al. 2005; Moran et al. 2006; van Buuren et al. 2010; Whitfield-Gabrieli et al. 2010). In fact, the DMN structures are recruited when reflecting both on one's own characteristics as well as those of others (Amodio and Frith 2006; Jenkins et al. 2008). Jenkins and colleagues (2008) proposed that the reason for this was that individuals automatically refer to their own mental states when considering those of a similar other, and used the repetition suppression paradigm in fMRI to investigate this hypothesis. In support of their hypothesis, they found that ventral medial prefrontal cortex (VMPFC) response was suppressed when self-reflections followed either an initial reflection about self or a judgment of a similar, but not a dissimilar, other (Jenkins et al. 2008). Recently, Sugiura and colleagues (2008) used fMRI to investigate face-specific and domain-general neural responses to self, familiar and unfamiliar faces, and proper names. In addition to finding networks that respond to self-faces (as described above), they also found higher activation for the self and familiar other, compared to the unfamiliar other, in the medial cortical structures during face *and* name recognition (Sugiura et al. 2008). These data show that cortical midline structures respond to both self and familiar others' images *and* names, suggesting a domain-general role for the DMN in self-related cognitions. These results suggest that we use the self as a point of reference, even when thinking about the mind of another person; however, these brain structures do appear to be more engaged when referring to the self (see Northoff et al. (2006) for a meta-analysis). Taken together, these data suggest an important role for midline structures—including the VMPFC, DMPFC, PCC, and precuneus—that form part of the DMN in processing self-relevant information.

Autobiographical Memory

Self-reference and self-relevance—whether by visual self-face recognition or through the enhanced memory for trait adjectives that are part of the self-schema—invoke autobiographical memory processes (Molnar-Szakacs and Arzy 2009). Memory is vital to the survival of the "self" as we use our memory for past events to predict the future in a cogitation called "mental time travel" (for reviews, see Schacter et al. 2007, 2008). Accordingly, it has been found that patients with amnesia are markedly impaired relative to matched control subjects not only in retrieving past events but also at imagining new experiences. In fact, a particular aspect of this deficit in thinking about the future is an impairment in predicting events about one's *personal* future rather than public or world events (Hassabis et al. 2007).

Recent neuroimaging studies have started to investigate the neural networks subserving self-projection in time (Addis et al. 2007; Arzy et al. 2008; Buckner and Carroll 2007; Szpunar et al. 2007). For example, Arzy and colleagues used a paradigm that involved participants making self-projections to both past and future and found that self-location in time recruits a distributed neural network—including anterior temporal, occipitotemporal, and temporoparietal regions—that partly overlaps the DMN (Arzy et al. 2008). The authors also found an effect of "self" in the behavioral data whereby participants responded significantly faster to personal (self-relevant) events than to world (non-self-relevant) events. In terms of brain regions, the above results show an overlap with the regions recruited during other self-relevant tasks, such as visuospatial perspective taking and spatial self-location (Arzy et al. 2006; Blanke et al. 2005; Vogeley and Fink 2003).

The finding that DMN structures were recruited when reflecting on one's own image, traits, past, and future confirms the important role of these brain structures in processing self-relevant information and maintaining a sense of self that is continuous through time. D'Argembeau and colleagues also found that the degree of activity within this network varied significantly according to the target of reflection. More specifically, reflecting on the self in the present elicited greater activity in the ventral and dorsal MPFC and PCC compared to reflecting on the self in the past or reflecting on an intimate other (D'Argembeau et al. 2008). In fact, it has been proposed that not only does activity in MPFC track with self-referential processing but it also contributes to the encoding of self-relevant memories (Macrae et al. 2004). Thus, structures of the DMN may be important in indexing the degree to which a psychological trait corresponds to the self-schema or a physical image represents the self. The more strongly a stimulus is related to the self, the more activity it will elicit in DMN structures (Molnar-Szakacs and Arzy 2009; Moran et al. 2006; Northoff et al. 2006; Schmitz and Johnson 2007; Uddin et al. 2007).

The Emergent Self: Conclusions and Future Directions

The distinction drawn in this chapter between functions of the DMN and MNS serves merely as a practical division of labor for purposes of discourse between two networks that subserve related and interacting processes which are crucial to giving rise to a cohesive sense of self that is continuous through time (Molnar-Szakacs and Arzy 2009; Uddin et al. 2007). While the MNS provides the physical other-to-self mapping that is necessary for comprehending physical actions of intentional agents, the DMN maintains and supports processes that are related to understanding psychological states of others by reflecting on one's own attitudes (Jenkins et al. 2008; Mitchell et al. 2005). In a broad sense, these neural networks allow us to know about our own mind *and* others' minds, as well as to adopt a point of view that is different from our current experience of the world. Through the constant functional interaction of these networks, the "self" is able to make inferences about what is going on inside other people—their intentions, feelings, and thoughts—allowing us to thrive in our social world.

Questions for future work include understanding the precise conditions under which these two systems interact with each other and how this seamless interaction contributes to social cognition. Developments in the tools of research, such as diffusion tensor imaging (Jbabdi et al. 2007) which provides information about the structural connectivity of the human brain, and developments in computing, such as functional connectivity modeling (Friston et al. 2003) which provides estimates of information flow between structures, will continue to be an active area of research and integration. Just as the brain's networks integrate information within and among them, researchers must integrate information from many different approaches, techniques, and sources to be able to answer the eternal question of what is the "self" (Aminoff et al. 2009).

References

Addis, D. R., Wong, A. T., & Schacter, D. L. (2007). Remembering the past and imagining the future: Common and distinct neural substrates during event construction and elaboration. *Neuropsychologia, 45*(7), 1363–1377.

Aminoff, E. M., Balslev, D., Borroni, P., Bryan, R. E., Chua, E. F., Cloutier, J., et al. (2009). The landscape of cognitive neuroscience: Challenges, rewards, and new perspectives. In M. S. Gazzaniga (Ed.), *The cognitive neurosciences* (pp. 1255–1262). Boston: MIT Press.

Amodio, D. M., & Frith, C. D. (2006). Meeting of minds: The medial frontal cortex and social cognition. *Nature Reviews Neuroscience, 7*(4), 268–277.

Amsterdam, B. (1972). Mirror self-image reactions before age two. *Developmental Psychobiology, 5*(4), 297–305.

Amunts, K., Schleicher, A., Burgel, U., Mohlberg, H., Uylings, H. B., & Zilles, K. (1999). Broca's region revisited: Cytoarchitecture and intersubject variability. *The Journal of Comparative Neurology, 412*(2), 319–341.

Arzy, S., Thut, G., Mohr, C., Michel, C. M., & Blanke, O. (2006). Neural basis of embodiment: Distinct contributions of temporoparietal junction and extrastriate body area. *Journal of Neuroscience, 26*(31), 8074–8081.

Arzy, S., Molnar-Szakacs, I., & Blanke, O. (2008). Self in time: Imagined self-location influences neural activity related to mental time travel. *Journal of Neuroscience, 28*(25), 6502–6507.

Aziz-Zadeh, L., Koski, L., Zaidel, E., Mazziotta, J., & Iacoboni, M. (2006). Lateralization of the human mirror neuron system. *Journal of Neuroscience, 26*(11), 2964–2970.

Binkofski, F., Buccino, G., Stephan, K. M., Rizzolatti, G., Seitz, R. J., & Freund, H. J. (1999). A parieto-premotor network for object manipulation: Evidence from neuroimaging. *Experimental Brain Research, 128*(1–2), 210–213.

Blanke, O., Mohr, C., Michel, C. M., Pascual-Leone, A., Brugger, P., Seeck, M., et al. (2005). Linking out-of-body experience and self processing to mental own-body imagery at the temporoparietal junction. *Journal of Neuroscience, 25*(3), 550–557.

Borroni, P., & Baldissera, F. (2008). Activation of motor pathways during observation and execution of hand movements. *Social Neuroscience, 3*(3–4), 276–288.

Breen, N., Caine, D., & Coltheart, M. (2001). Mirrored-self misidentification: Two cases of focal onset dementia. *Neurocase, 7*(3), 239–254.

Bressler, S. L., & Menon, V. (2010). Large-scale brain networks in cognition: Emerging methods and principles. *Trends in Cognitive Science, 14*(6), 277–290.

Buccino, G., Lui, F., Canessa, N., Patteri, I., Lagravinese, G., Benuzzi, F., et al. (2004). Neural circuits involved in the recognition of actions performed by nonconspecifics: An FMRI study. *Journal of Cognitive Neuroscience, 16*(1), 114–126.

Buckner, R. L., & Carroll, D. C. (2007). Self-projection and the brain. *Trends in Cognitive Science, 11*(2), 49–57.

Carr, L., Iacoboni, M., Dubeau, M. C., Mazziotta, J. C., & Lenzi, G. L. (2003). Neural mechanisms of empathy in humans: A relay from neural systems for imitation to limbic areas. *Proceedings of the National Academy of Sciences of the United States of America, 100*(9), 5497–5502.

Cattaneo, L., Barchiesi, G., Tabarelli, D., Arfeller, C., Sato, M., & Glenberg, A. M. (2011). One's motor performance predictably modulates the understanding of others' actions through adaptation of premotor visuo-motor neurons. *Social Cognitive and Affective Neuroscience, 6*(3), 301–310.

Cherkassky, V. L., Kana, R. K., Keller, T. A., & Just, M. A. (2006). Functional connectivity in a baseline resting-state network in autism. *NeuroReport, 17*(16), 1687–1690.

Chong, T. T., Cunnington, R., Williams, M. A., Kanwisher, N., & Mattingley, J. B. (2008). fMRI adaptation reveals mirror neurons in human inferior parietal cortex. *Current Biology, 18*(20), 1576–1580.

Cole, J. (1999). On 'being faceless': Selfhood and facial embodiment. In S. Gallagher & J. Shear (Eds.), *Models of the self* (pp. 301–318). Thorverton: Imprint Academic.

Craik, F. I. M., Moroz, T. M., Moscovitch, M., Stuss, D. T., Winocur, G., Tulving, E., et al. (1999). In search of the self: A positron emission tomography study. *Psychological Science, 10*(1), 26–34.

D'Argembeau, A., Collette, F., Van der Linden, M., Laureys, S., Del Fiore, G., Degueldre, C., et al. (2005). Self-referential reflective activity and its relationship with rest: A PET study. *NeuroImage, 25*(2), 616–624.

D'Argembeau, A., Feyers, D., Majerus, S., Collette, F., Van der Linden, M., Maquet, P., et al. (2008). Self-reflection across time: Cortical midline structures differentiate between present and past selves. *Social Cognitive and Affective Neuroscience, 3*(3), 244–252.

Dapretto, M., Davies, M. S., Pfeifer, J. H., Scott, A. A., Sigman, M., Bookheimer, S. Y., et al. (2006). Understanding emotions in others: Mirror neuron dysfunction in children with autism spectrum disorders. *Nature Neuroscience, 9*(1), 28–30.

Darwin, C. (1877). A biographical sketch of an infant. *Mind, 2*, 285–294.

Dennett, D. C. (1991). *Consciousness explained.* Boston: Little, Brown and Company.

Devue, C., & Bredart, S. (2011). The neural correlates of visual self-recognition. *Consciousness and Cognition, 20*(1), 40–51.

di Pellegrino, G., Fadiga, L., Fogassi, L., Gallese, V., & Rizzolatti, G. (1992). Understanding motor events: A neurophysiological study. *Experimental Brain Research, 91*(1), 176–180.

Esposito, F., Bertolino, A., Scarabino, T., Latorre, V., Blasi, G., Popolizio, T., et al. (2006). Independent component model of the default-mode brain function: Assessing the impact of active thinking. *Brain Research Bulletin, 70*(4–6), 263–269.

Fadiga, L., Fogassi, L., Pavesi, G., & Rizzolatti, G. (1995). Motor facilitation during action observation: A magnetic stimulation study. *Journal of Neurophysiology, 73*(6), 2608–2611.

Faust, M., Kravetz, S., & Nativ-Safrai, O. (2004). The representation of aspects of the self in the two cerebral hemispheres. *Personality and Individual Differences, 37*, 607–619.

Fink, G. R., Markowitsch, H. J., Reinkemeier, M., Bruckbauer, T., Kessler, J., & Heiss, W. D. (1996). Cerebral representation of one's own past: Neural networks involved in autobiographical memory. *Journal of Neuroscience, 16*(13), 4275–4282.

Fogassi, L., Ferrari, P. F., Gesierich, B., Rozzi, S., Chersi, F., & Rizzolatti, G. (2005). Parietal lobe: From action organization to intention understanding. *Science, 308*, 662–667.

Fossati, P., Hevenor, S. J., Graham, S. J., Grady, C., Keightley, M. L., Craik, F., et al. (2003). In search of the emotional self: An FMRI study using positive and negative emotional words. *The American Journal of Psychiatry, 160*(11), 1938–1945.

Fossati, P., Hevenor, S. J., Lepage, M., Graham, S. J., Grady, C., Keightley, M. L., et al. (2004). Distributed self in episodic memory: Neural correlates of successful retrieval of self-encoded positive and negative personality traits. *NeuroImage, 22*(4), 1596–1604.

Fox, M. D., Snyder, A. Z., Vincent, J. L., Corbetta, M., Van Essen, D. C., & Raichle, M. E. (2005). The human brain is intrinsically organized into dynamic, anticorrelated functional networks. *Proceedings of the National Academy of Sciences of the United States of America, 102*(27), 9673–9678.

Fransson, P. (2006). How default is the default mode of brain function? Further evidence from intrinsic BOLD signal fluctuations. *Neuropsychologia, 44*(14), 2836–2845.

Friston, K. J., Harrison, L., & Penny, W. (2003). Dynamic causal modelling. *NeuroImage, 19*(4), 1273–1302.

Gallagher, S. (2000). Philosophical conceptions of the self: Implications for cognitive science. *Trends in Cognitive Sciences, 4*(1), 14–21.

Gallese, V., & Goldman, A. (1998). Mirror neurons and the simulation theory of mind-reading. *Trends in Cognitive Sciences, 2*(12), 493–501.

Gallese, V., Fadiga, L., Fogassi, L., & Rizzolatti, G. (1996). Action recognition in the premotor cortex. *Brain, 119* (Pt 2), 593–609.

Gallup, G. G. (1970). Chimpanzees: Self-recognition. *Science, 167*, 86–87.

Gallup, G. G., Jr. (1977). Self-recognition in primates: A comparative approach to the bidirectional properties of consciousness. *American Psychologist, 32*(5), 329–338.

Gangitano, M., Mottaghy, F. M., & Pascual-Leone, A. (2001). Phase-specific modulation of cortical motor output during movement observation. *NeuroReport, 12*(7), 1489–1492.

Garrity, A. G., Pearlson, G. D., McKiernan, K., Lloyd, D., Kiehl, K. A., & Calhoun, V. D. (2007). Aberrant "default mode" functional connectivity in schizophrenia. *The American Journal of Psychiatry, 164*(3), 450–457.

Gazzola, V., Aziz-Zadeh, L., & Keysers, C. (2006). Empathy and the somatotopic auditory mirror system in humans. *Current Biology, 16*(18), 1824–1829.

Gillihan, S. J., & Farah, M. J. (2005). Is self special? A critical review of evidence from experimental psychology and cognitive neuroscience. *Psychological Bulletin, 131*(1), 76–97.

Goleman, D. (2006). *Social intelligence: The new science of human relationships*. New York: Random House.

Grafton, S. T., Arbib, M. A., Fadiga, L., & Rizzolatti, G. (1996). Localization of grasp representations in humans by positron emission tomography. 2. Observation compared with imagination. *Experimental Brain Research, 112*(1), 103–111.

Greenfield, P. (2006). Implications of mirror neurons for the ontogeny and phylogeny of cultural processes: The examples of tools and language. In M. A. Arbib (Ed.), *Action to language via the mirror neuron system*. New York: Cambridge University Press.

Greicius, M. D., & Menon, V. (2004). Default-mode activity during a passive sensory task: Uncoupled from deactivation but impacting activation. *Journal of Cognitive Neuroscience, 16*(9), 1484–1492.

Greicius, M. D., Srivastava, G., Reiss, A. L., & Menon, V. (2004). Default-mode network activity distinguishes Alzheimer's disease from healthy aging: Evidence from functional MRI. *Proceedings of the National Academy of Sciences of the United States of America, 101*(13), 4637–4642.

Grezes, J., Armony, J. L., Rowe, J., & Passingham, R. E. (2003). Activations related to "mirror" and "canonical" neurones in the human brain: An fMRI study. *NeuroImage, 18*(4), 928–937.

Gusnard, D. A., Akbudak, E., Shulman, G. L., & Raichle, M. E. (2001). Medial prefrontal cortex and self-referential mental activity: Relation to a default mode of brain function. *Proceedings of the National Academy of Sciences of the United States of America, 98*(7), 4259–4264.

Harrison, B. J., Pujol, J., Lopez-Sola, M., Hernandez-Ribas, R., Deus, J., Ortiz, H., et al. (2008). Consistency and functional specialization in the default mode brain network. *Proceedings of the National Academy of Sciences of the United States of America, 105*(28), 9781–9786.

Hassabis, D., Kumaran, D., Vann, S. D., & Maguire, E. A. (2007). Patients with hippocampal amnesia cannot imagine new experiences. *Proceedings of the National Academy of Sciences of the United States of America, 104*(5), 1726–1731.

Heinisch, C., Dinse, H. R., Tegenthoff, M., Juckel, G., & Brune, M. (2011). An rTMS study into self-face recognition using video-morphing technique. *Social Cognitive and Affective Neuroscience, 6*(4), 442–449.

Iacoboni, M. (2009). Imitation, empathy, and mirror neurons. *Annual Review of Psychology, 60*, 653–670.

Iacoboni, M., & Dapretto, M. (2006). The mirror neuron system and the consequences of its dysfunction. *Nature Reviews Neuroscience, 7*(12), 942–951.

Iacoboni, M., Woods, R. P., Brass, M., Bekkering, H., Mazziotta, J. C., & Rizzolatti, G. (1999). Cortical mechanisms of human imitation. *Science, 286*(5449), 2526–2528.

Iacoboni, M., Lieberman, M. D., Knowlton, B. J., Molnar-Szakacs, I., Moritz, M., Throop, C. J., et al. (2004). Watching social interactions produces dorsomedial prefrontal and medial parietal BOLD fMRI signal increases compared to a resting baseline. *NeuroImage, 21*(3), 1167–1173.

Iacoboni, M., Molnar-Szakacs, I., Gallese, V., Buccino, G., Mazziotta, J. C., & Rizzolatti, G. (2005). Grasping the intentions of others with one's own mirror neuron system. *PLoS Biology, 3*(3), e79.

James, W. (1983). *The principles of psychology*. Cambridge: Harvard University Press.

Jbabdi, S., Woolrich, M. W., Andersson, J. L., & Behrens, T. E. (2007). A Bayesian framework for global tractography. *NeuroImage, 37*(1), 116–129.

Jeannerod, M. (2003). The mechanism of self-recognition in humans. *Behavioural Brain Research, 142*, 1–15.

Jenkins, A. C., Macrae, C. N., & Mitchell, J. P. (2008). Repetition suppression of ventromedial prefrontal activity during judgments of self and others. *Proceedings of the National Academy of Sciences of the United States of America, 105*(11), 4507–4512.

Johnson, S. C., Baxter, L. C., Wilder, L. S., Pipe, J. G., Heiserman, J. E., & Prigatano, G. P. (2002). Neural correlates of self-reflection. *Brain, 125*(Pt 8), 1808–1814.

Johnson-Frey, S. H., Maloof, F. R., Newman-Norlund, R., Farrer, C., Inati, S., & Grafton, S. T. (2003). Actions or hand-object interactions? Human inferior frontal cortex and action observation. *Neuron, 39*(6), 1053–1058.

Kaplan, J. T., Aziz-Zadeh, L., Uddin, L. Q., & Iacoboni, M. (2008). The self across the senses: An fMRI study of self-face and self-voice recognition. *Social Cognitive and Affective Neuroscience, 3*(3), 218–223.
Keenan, J. P., Nelson, A., O'Connor, M., & Pascual-Leone, A. (2001). Self-recognition and the right hemisphere. *Nature, 409*(6818), 305.
Kelley, W. M., Macrae, C. N., Wyland, C. L., Caglar, S., Inati, S., & Heatherton, T. F. (2002). Finding the self? An event-related fMRI study. *Journal of Cognitive Neuroscience, 14*(5), 785–794.
Kennedy, D. P., Redcay, E., & Courchesne, E. (2006). Failing to deactivate: Resting functional abnormalities in autism. *Proceedings of the National Academy of Sciences of the United States of America, 103*(21), 8275–8280.
Kircher, T. T., Senior, C., Phillips, M. L., Benson, P. J., Bullmore, E. T., Brammer, M., et al. (2000). Towards a functional neuroanatomy of self processing: Effects of faces and words. *Brain Research. Cognitive Brain Research, 10*(1–2), 133–144.
Kjaer, T. W., Nowak, M., & Lou, H. C. (2002). Reflective self-awareness and conscious states: PET evidence for a common midline parietofrontal core. *NeuroImage, 17*(2), 1080–1086.
Kohler, E., Keysers, C., Umilta, M. A., Fogassi, L., Gallese, V., & Rizzolatti, G. (2002). Hearing sounds, understanding actions: Action representation in mirror neurons. *Science, 297*(5582), 846–848.
Koski, L., Wohlschlager, A., Bekkering, H., Woods, R. P., Dubeau, M. C., Mazziotta, J. C., et al. (2002). Modulation of motor and premotor activity during imitation of target-directed actions. *Cerebral Cortex, 12*(8), 847–855.
Lethmate, J., & Ducker, G. (1973). Studies on self-recognition in a mirror in orangutans, chimpanzees, gibbons and various other monkey species. *Zeitschrift für Tierpsychologie, 33*(3), 248–269.
Liang, M., Zhou, Y., Jiang, T., Liu, Z., Tian, L., Liu, H., et al. (2006). Widespread functional disconnectivity in schizophrenia with resting-state functional magnetic resonance imaging. *NeuroReport, 17*(2), 209–213.
Lieberman, M. D. (2007). Social cognitive neuroscience: A review of core processes. *Annual Review of Psychology, 58*, 259–289.
Lou, H. C., Luber, B., Crupain, M., Keenan, J. P., Nowak, M., Kjaer, T. W., et al. (2004). Parietal cortex and representation of the mental self. *Proceedings of the National Academy of Sciences of the United States of America, 101*(17), 6827–6832.
Macrae, C. N., Moran, J. M., Heatherton, T. F., Banfield, J. F., & Kelley, W. M. (2004). Medial prefrontal activity predicts memory for self. *Cerebral Cortex, 14*(6), 647–654.
Markus, H. (1977). Self-schemata and processing information about the self. *Journal of Personality and Social Psychology, 35*(2), 63–78.
Mason, M. F., Norton, M. I., Van Horn, J. D., Wegner, D. M., Grafton, S. T., & Macrae, C. N. (2007). Wandering minds: The default network and stimulus-independent thought. *Science, 315*(5810), 393–395.
Mazziotta, J., Toga, A., Evans, A., Fox, P., Lancaster, J., Zilles, K., et al. (2001). A probabilistic atlas and reference system for the human brain: International Consortium for Brain Mapping (ICBM). *Philosophical Transactions of the Royal Society of London. Series B, Biological Sciences, 356*(1412), 1293–1322.
McKiernan, K. A., Kaufman, J. N., Kucera-Thompson, J., & Binder, J. R. (2003). A parametric manipulation of factors affecting task-induced deactivation in functional neuroimaging. *Journal of Cognitive Neuroscience, 15*(3), 394–408.
McKiernan, K. A., D'Angelo, B. R., Kaufman, J. N., & Binder, J. R. (2006). Interrupting the "stream of consciousness": An fMRI investigation. *NeuroImage, 29*(4), 1185–1191.
Meltzoff, A. N., & Brooks, R. (2001). "Like me" as a building block for understanding other minds: Bodily acts, attention, and intention. In B. F. Malle, L. J. Moses, & D. A. Baldwin (Eds.), *Intentions and intentionality: Foundations of social cognition* (pp. 171–191). Cambridge: MIT Press.
Metzinger, T. (2003). *Being no one: The self-model theory of subjectivity*. Cambridge: MIT Press.
Miall, R. C., & Robertson, E. M. (2006). Functional imaging: Is the resting brain resting? *Current Biology, 16*(23), R998–R1000.
Mitchell, J. P., Banaji, M. R., & Macrae, C. N. (2005). The link between social cognition and self-referential thought in the medial prefrontal cortex. *Journal of Cognitive Neuroscience, 17*(8), 1306–1315.
Molnar-Szakacs, I. (2011). From actions to empathy and morality – A neural perspective. *Journal of Economic Behavior and Organization, 77*, 76–85.
Molnar-Szakacs, I., & Arzy, S. (2009). Searching for an integrated self-representation. *Communicative and Integrative Biology, 2*(4), 365–367.
Molnar-Szakacs, I., Iacoboni, M., Koski, L., & Mazziotta, J. C. (2005a). Functional segregation within pars opercularis of the inferior frontal gyrus: Evidence from fMRI studies of imitation and action observation. *Cerebral Cortex, 15*, 986–994.
Molnar-Szakacs, I., Uddin, L. Q., & Iacoboni, M. (2005b). Right-hemisphere motor facilitation by self-descriptive personality-trait words. *European Journal of Neuroscience, 21*(7), 2000–2006.
Molnar-Szakacs, I., Kaplan, J., Greenfield, P. M., & Iacoboni, M. (2006). Observing complex action sequences: The role of the fronto-parietal mirror neuron system. *NeuroImage, 33*(3), 923–935.

Molnar-Szakacs, I., Wang, M. J., Laugeson, E. A., Overy, K., Wu, W. L., & Piggot, J. (2009). Autism, emotion recognition and the mirror neuron system: The case of music. *Mcgill Journal of Medicine, 12*(2), 87.

Montagna, M., Cerri, G., Borroni, P., & Baldissera, F. (2005). Excitability changes in human corticospinal projections to muscles moving hand and fingers while viewing a reaching and grasping action. *European Journal of Neuroscience, 22*(6), 1513–1520.

Moran, J. M., Macrae, C. N., Heatherton, T. F., Wyland, C. L., & Kelley, W. M. (2006). Neuroanatomical evidence for distinct cognitive and affective components of self. *Journal of Cognitive Neuroscience, 18*(9), 1586–1594.

Neisser, U. (1993). *The perceived self: Ecological and interpersonal sources of self-knowledge*. Cambridge: Cambridge University Press.

Neisser, U. (1995). Criterion for an ecological self. In P. Rochat (Ed.), *The self in infancy: Theory and research* (pp. 17–34). Amsterdam: Elsevier.

Nishitani, N., & Hari, R. (2000). Temporal dynamics of cortical representation for action. *Proceedings of the National Academy of Sciences of the United States of America, 97*(2), 913–918.

Northoff, G., Heinzel, A., de Greck, M., Bermpohl, F., Dobrowolny, H., & Panksepp, J. (2006). Self-referential processing in our brain–a meta-analysis of imaging studies on the self. *NeuroImage, 31*(1), 440–457.

Oberman, L. M., Hubbard, E. M., McCleery, J. P., Altschuler, E. L., Ramachandran, V. S., & Pineda, J. A. (2005). EEG evidence for mirror neuron dysfunction in autism spectrum disorders. *Brain Research. Cognitive Brain Research, 24*(2), 190–198.

Petrides, M., & Pandya, D. (1997). Comparative architectonic analysis of the human and the macaque frontal cortex. In F. Boller & J. Grafman (Eds.), *Handbook of neuropsychology* (Vol. IX, pp. 17–58). New York: Elsevier.

Phillips, M. L., & Howard, R. (1996). Mirror, mirror on the wall, who…?: Towards a model of visual self-recognition. *Cognitive Neuropsychiatry, 1*(2), 153–164.

Pineda, J. A. (2008). Sensorimotor cortex as a critical component of an 'extended' mirror neuron system: Does it solve the development, correspondence, and control problems in mirroring? *Behavioral and Brain Functions, 4*, 47.

Platek, S. M., Loughead, J. W., Gur, R. C., Busch, S., Ruparel, K., Phend, N., et al. (2006). Neural substrates for functionally discriminating self-face from personally familiar faces. *Human Brain Mapping, 27*(2), 91–98.

Platek, S. M., Wathne, K., Tierney, N. G., & Thomson, J. W. (2008). Neural correlates of self-face recognition: An effect-location meta-analysis. *Brain Research, 1232*, 173–184.

Plotnik, J. M., de Waal, F. B., & Reiss, D. (2006). Self-recognition in an Asian elephant. *Proceedings of the National Academy of Sciences of the United States of America, 103*(45), 17053–17057.

Povinelli, D. J., & Gallup, G. G., Jr. (1997). Chimpanzees recognize themselves in mirrors. *Animal Behavior, 53*, 1083–1088.

Preyer, W. (1889). *The mind of the child. Part II: The development of the intellect*. New York: Appleton.

Prior, H., Schwarz, A., & Güntürkün, O. (2008). Mirror-induced behavior in the magpie (*Pica pica*): Evidence of self-recognition. *PLoS Biology, 6*(8), e202. doi:10.1371/journal.pbio.0060202.

Raichle, M. E., MacLeod, A. M., Snyder, A. Z., Powers, W. J., Gusnard, D. A., & Shulman, G. L. (2001). A default mode of brain function. *Proceedings of the National Academy of Sciences of the United States of America, 98*(2), 676–682.

Reiss, D., & Marino, L. (2001). Mirror self-recognition in the bottlenose dolphin: A case of cognitive convergence. *Proceedings of the National Academy of Sciences of the United States of America, 98*(10), 5937–5942.

Rizzolatti, G., & Craighero, L. (2004). The mirror-neuron system. *Annual Review of Neuroscience, 27*, 169–192.

Rizzolatti, G., & Matelli, M. (2003). Two different streams form the dorsal visual system: Anatomy and functions. *Experimental Brain Research, 153*(2), 146–157.

Rizzolatti, G., & Sinigaglia, C. (2010). The functional role of the parieto-frontal mirror circuit: Interpretations and misinterpretations. *Nature Reviews Neuroscience, 11*(4), 264–274.

Rizzolatti, G., Fadiga, L., Gallese, V., & Fogassi, L. (1996a). Premotor cortex and the recognition of motor actions. *Cognitive Brain Research, 3*(2), 131–141.

Rizzolatti, G., Fadiga, L., Matelli, M., Bettinardi, V., Paulesu, E., Perani, D., et al. (1996b). Localization of grasp representations in humans by PET: 1. Observation versus execution. *Experimental Brain Research, 111*(2), 246–252.

Rizzolatti, G., Fogassi, L., & Gallese, V. (2001). Neurophysiological mechanisms underlying the understanding and imitation of action. *Nature Reviews Neuroscience, 2*(9), 661–670.

Rogers, T. B., Kuiper, N. A., & Kirker, W. S. (1977). Self-reference and the encoding of personal information. *Journal of Personality and Social Psychology, 35*(9), 677–688.

Schacter, D. L., Addis, D. R., & Buckner, R. L. (2007). Remembering the past to imagine the future: The prospective brain. *Nature Reviews Neuroscience, 8*(9), 657–661.

Schacter, D. L., Addis, D. R., & Buckner, R. L. (2008). Episodic simulation of future events: Concepts, data, and applications. *Annals of the New York Academy of Sciences, 1124*, 39–60.

Schmitz, T. W., & Johnson, S. C. (2007). Relevance to self: A brief review and framework of neural systems underlying appraisal. *Neuroscience and Biobehavioral Reviews, 31*(4), 585–596.

Spreng, R. N., Mar, R. A., & Kim, A. S. (2009). The common neural basis of autobiographical memory, prospection, navigation, theory of mind, and the default mode: A quantitative meta-analysis. *Journal of Cognitive Neuroscience, 21*(3), 489–510.

Strafella, A. P., & Paus, T. (2000). Modulation of cortical excitability during action observation: A transcranial magnetic stimulation study. *NeuroReport, 11*(10), 2289–2292.

Sugiura, M., Watanabe, J., Maeda, Y., Matsue, Y., Fukuda, H., & Kawashima, R. (2005). Cortical mechanisms of visual self-recognition. *NeuroImage, 24*(1), 143–149.

Sugiura, M., Sassa, Y., Jeong, H., Horie, K., Sato, S., & Kawashima, R. (2008). Face-specific and domain-general characteristics of cortical responses during self-recognition. *NeuroImage, 42*(1), 414–422.

Symons, C. S., & Johnson, B. T. (1997). The self-reference effect in memory: A meta-analysis. *Psychological Bulletin, 121*(3), 371–394.

Szpunar, K. K., Watson, J. M., & McDermott, K. B. (2007). Neural substrates of envisioning the future. *Proceedings of the National Academy of Sciences of the United States of America, 104*(2), 642–647.

Tomaiuolo, F., MacDonald, J. D., Caramanos, Z., Posner, G., Chiavaras, M., Evans, A. C., et al. (1999). Morphology, morphometry and probability mapping of the pars opercularis of the inferior frontal gyrus: An in vivo MRI analysis. *European Journal of Neuroscience, 11*(9), 3033–3046.

Uddin, L. Q., Kaplan, J. T., Molnar-Szakacs, I., Zaidel, E., & Iacoboni, M. (2005). Self-face recognition activates a frontoparietal "mirror" network in the right hemisphere: An event-related fMRI study. *NeuroImage, 25*(3), 926–935.

Uddin, L. Q., Molnar-Szakacs, I., Zaidel, E., & Iacoboni, M. (2006). rTMS to the right inferior parietal lobule disrupts self-other discrimination. *Social Cognitive and Affective Neuroscience, 1*(1), 65–71.

Uddin, L. Q., Iacoboni, M., Lange, C., & Keenan, J. P. (2007). The self and social cognition: The role of cortical midline structures and mirror neurons. *Trends in Cognitive Science, 11*(4), 153–157.

Umilta, M. A., Kohler, E., Gallese, V., Fogassi, L., Fadiga, L., Keysers, C., et al. (2001). I know what you are doing: A neurophysiological study. *Neuron, 31*(1), 155–165.

van Buuren, M., Gladwin, T. E., Zandbelt, B. B., Kahn, R. S., & Vink, M. (2010). Reduced functional coupling in the default-mode network during self-referential processing. *Human Brain Mapping, 31*(8), 1117–1127.

Vogeley, K., & Fink, G. R. (2003). Neural correlates of the first-person-perspective. *Trends in Cognitive Science, 7*(1), 38–42.

VonBonin, G., & Bailey, P. (1947). *The neocortex of Macaca mulatta*. Urbana: University of Illinois Press.

VonEconomo, C., & Koskinas, G. N. (1925). *Die Cytoarchitektonik der Hirnrinde des erwachsenen Menschen*. Berlin: Springer.

Whitfield-Gabrieli, S., Moran, J. M., Nieto-Castanon, A., Triantafyllou, C., Saxe, R., & Gabrieli, J. D. (2010). Associations and dissociations between default and self-reference networks in the human brain. *NeuroImage, 55*(1), 225–232.

Wicker, B., Ruby, P., Royet, J. P., & Fonlupt, P. (2003). A relation between rest and the self in the brain? *Brain Research. Brain Research Reviews, 43*(2), 224–230.

Wig, G. S., Schlaggar, B. L., & Petersen, S. E. (2011). Concepts and principles in the analysis of brain networks. *Annals of the New York Academy of Sciences, 1224*(1), 126–146.

Woodruff, C. C., & Maaske, S. (2010). Action execution engages human mirror neuron system more than action observation. *NeuroReport, 21*(6), 432–435.

Zysset, S., Huber, O., Ferstl, E., & von Cramon, D. Y. (2002). The anterior frontomedian cortex and evaluative judgment: An fMRI study. *NeuroImage, 15*(4), 983–991.

Chapter 14
The Human Mirror Neuron System, Social Control, and Language

Sook-Lei Liew and Lisa Aziz-Zadeh

> The self is not something ready-made but something in continuous formation through choice of action.
>
> -John Dewey *(American psychologist, philosopher, and educator, 1859–1952)*

How do we understand others' actions? Dewey's approach suggests that our own actions have a lot to do with how we perceive ourselves and those around us. However, what mechanisms in our brain allow us to smoothly navigate a complex social scene, ripe with dynamically changing intentions, actions, and contexts? How do our own actions affect factors influencing social control, such as how empathic we are and how we are able to understand others? While these questions beg an interdisciplinary approach from fields as diverse as sociological observations to molecular interactions, the field of cognitive neuroscience has provided new insights into understanding social interchanges. Though many networks are involved in social understanding, one prominent neural network is the mirror neuron system (MNS). The discovery of the MNS has propelled a widespread effort into understanding how our own actions play a role in understanding the actions of others. This line of research is in fact engulfed in a wider theory known as embodied cognition, in which higher cognition and our very thoughts are grounded in our bodily actions and experiences. In this chapter, we first review literature on what exactly mirror neurons are and how they were discovered. We will then delve into the research suggesting various ways in which mirror neurons might play a role in social cognition and social control and ways that mirror neurons may be linked to language, and conclude with an extended view of how shared networks in embodied cognition contribute to our overall social functioning.

The Mirror Neuron System

Mirror neurons were originally discovered in macaque monkeys through single-cell recordings in premotor area F5, an area that has neurons which are active when the monkey performs actions, like reaching for a piece of food. By chance, the experimenters continued to record from the neuron when picking up the food themselves and, to their surprise, found that the motor neuron also fired when the monkey simply observed the action being performed (di Pellegrino et al. 1992; Rizzolatti

S.-L. Liew (✉) • L. Aziz-Zadeh
The Brain and Creativity Institute, University of Southern California, Los Angeles, CA, USA
e-mail: sliew@usc.edu

D.D. Franks and J.H. Turner (eds.), *Handbook of Neurosociology*, Handbooks of Sociology and Social Research, DOI 10.1007/978-94-007-4473-8_14, © Springer Science+Business Media B.V. 2013

et al. 1996a; Gallese et al. 1996). These neurons, termed "mirror neurons" for their ability to respond both to performed actions and mirrored observations of the same actions, were hypothesized to match visual information about another's actions to one's own motor representations, thus supporting the observer's ability to understand the observed action (Rizzolatti and Craighero 2004; Fadiga et al. 2009).

Mirror neurons can be divided into strictly and broadly congruent categories, with approximately 1/3 of mirror neurons being strictly congruent (i.e., they respond to observations of a specific grasp that they also execute) and 2/3 being broadly congruent (i.e., they respond to observations of more diverse grasps, not just the ones that they motorically encode; Gallese et al. 1996; Rizzolatti and Craighero 2004). Similarly, recent findings demonstrate that some mirror neurons respond only to one view of an action (e.g., someone facing you at 180°), while others respond to several views (e.g., both someone facing you at 180° and someone facing to the side at 90°; Caggiano et al. 2011). This diversity within the mirror neurons in F5 suggests that some neurons encode a direct match between what is seen and what is performed, while others may encode a more generalized match—such as the same goal performed in different orientations or with different kinematics. The direct-matching neurons may provide an indication of how exactly the action is being performed, while the neurons that fire for the goal, regardless of the kinematics or view, may encode a more abstract understanding of the goal of the observed action.

In addition, mirror neurons may be active not only to visual and motor modalities but also to auditory stimuli. One study demonstrated that a proportion of mirror neurons may respond to auditory input as well, with some premotor neurons firing in response to the *sound* of actions, such as the cracking of a peanut. The same neuron might also respond to observing a peanut being cracked and the motor act of cracking a peanut, thus rendering it a highly multimodal neuron. The firing is specific, such that, for example, the latter neuron would respond only to cracking peanuts and not to tearing paper (Kohler et al. 2002). Altogether, these discoveries further emphasize that there may be different types of mirror neurons, some of which are highly multimodal, some that are more attuned to the abstract goals of actions, and some that are attuned to specific sensorimotor parameters of the movements themselves.

Mirror neurons are not only limited to premotor region F5 but are also found in a subset of parietal neurons in area 7b (or PF; a rostral portion of the inferior parietal lobule; Fogassi et al. 1998, 2005; Gallese et al. 2002; Rizzolatti and Craighero 2004. Neurons in this region typically respond to sensory stimuli, although some also fire during execution of motor actions. In addition, a subset of visual neurons here fired specifically for action observation, and a subset of these contained mirror properties (Gallese et al. 2002). Anatomically, visual input regarding human biological movement is processed in the superior temporal sulcus (STS) and then passed to the inferior parietal lobule (area 7b), where it is then relayed to premotor area F5 (Perrett et al. 1989, 1990; Jellema et al. 2000; Rizzolatti and Craighero 2004). This anatomical structure provides the pivotal connections needed between visual input and parietal and premotor mirror neurons.

While the discovery of motor neurons with visual properties is largely undisputed, it is important to point out a topic that has received a fair amount of debate: the origin and function of these mirror neurons. Some researchers propose that these neurons are an evolutionary adaptation that serve as a key way in which we understand one another (Gallese et al. 2004; Rizzolatti and Craighero 2004), while others suggest that mirror neurons are simply the product of learned associations from sensory and motor co-activations over time (Hickok 2009; Heyes 2010). In addition, while some schools of thought promote the idea that mirror neurons allow us to completely simulate another's experiences (Gallese et al. 2004), others suggest that mirror neurons alone are not adequate for understanding others' perspectives (Saxe 2005; Hauser and Wood 2010). We note that the views mentioned above are not mutually exclusive nor are all the researchers referenced here necessarily dogmatic to one viewpoint. Nevertheless, despite any debates, a wealth of recent research has demonstrated that both the mirror system and the regions involved in higher-level cognition, known as mentalizing regions, are active during the complex process of social cognition, with the mirror system playing an active role in action observation. In addition, many findings suggest that the activity of one network or another very likely depends on the particular task that the observer is engaged in at any given moment (de Lange et al. 2008; Hesse et al. 2009; Spunt et al. 2011; Liew et al. 2010), which may shed some light into why some studies find the mirror neuron system active and others do not. Here we focus on the role of the mirror neuron system in social cognition, keeping in mind that this particular network is one of several embedded in a far more complex arrangement of interconnected brain regions that support the complexity of human social cognition.

The Putative Human Mirror System

The term "mirror neuron system," as originally discovered, refers to motor neurons that also respond to the visual observation of motor actions. However, since it is not generally feasible to implant electrodes directly into single neurons in human participants' brains to record activity during the performance and observation of actions, research on the MNS in humans instead rely on indirect measures of neural activity, such as positron emission tomography (PET), functional magnetic resonance imaging (fMRI), and transcranial magnetic stimulation (TMS); see Fig. 14.1 for a depiction of the latter two methods. These methods in human participants have demonstrated overlapping activity patterns in motor-related brain regions both during the execution and observation of actions (Fadiga et al. 1995; Iacoboni et al. 1999; Baldissera et al. 2001; Buccino et al. 2001; Aziz-Zadeh et al. 2006a; Koski et al. 2002; Liew et al. 2010). Such activity is referred to generally as the putative human mirror neuron system (MNS) and is thought to be located in regions of the brain theorized to be homologues to the macaque mirror neuron regions, the inferior frontal gyrus (IFG) and inferior parietal lobule (IPL; Rizzolatti et al. 1996b; Buccino et al. 2001; Rizzolatti and Craighero 2004; see Fig. 14.2). Interestingly, while the core regions of the human mirror system may include the IFG and IPL, other regions of the brain may contain similar mirrorlike properties that contribute to understanding another's actions, such as regions of the somatosensory cortex and the insula (Wicker et al. 2003; Keysers et al. 2004, 2010).

One interesting difference between the macaque MNS and the human MNS is that while macaque mirror neurons have been shown to fire only for object-oriented transitive actions, such as reaching for a piece of food, in humans the MNS is active both for transitive, goal-directed actions and intransitive actions, such as a communicative gesture. Transitive actions, such as reaching for a piece of food, tend to be less abstract than intransitive actions, such as pantomiming an action, gesturing, or performing sign language which has definite symbolic and conceptual meaning. In monkeys, the exception to this trend is a subset of macaque mouth mirror neurons, which fire when the monkey observes communicative mouth actions (e.g., lip smacking; Ferrari et al. 2003) and which provide evidence of a suggestive link between goal-directed actions and communicative abilities (Rizzolatti and Craighero 2004). However, in general, monkey mirror neurons do not respond to pantomimes of actions (Fadiga et al. 1995; Umilta et al. 2001). In humans, however, mirror regions demonstrate activity whether reaching for an object, performing a meaningless intransitive action (Iacoboni et al. 1999), or making a symbolic hand gesture (Gentilucci and Dalla Volta 2008; Villarreal et al. 2008; Liew et al. 2010; Schippers et al. 2009; Skipper et al. 2009). The plausible evolution of this system from concrete actions to abstract gestures in humans has led some to propose that the development of the MNS and its interaction with other brain regions played a role in the formation and development of language (Rizzolatti and Arbib 1998; Gallese and Lakoff 2005; Arbib 2005, 2010; Fadiga et al. 2009. In addition, the encoding of even intransitive gestures suggests that the putative human MNS may be able to encode more abstract goals than that found in macaques.

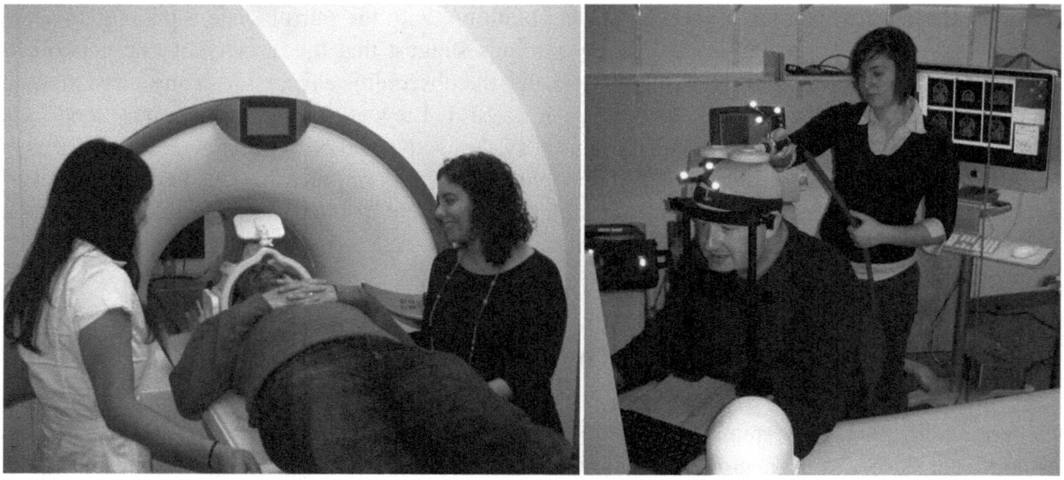

Fig. 14.1 An image of the typical setup for a functional magnetic resonance imaging (fMRI) study (*left*; photo courtesy of Kim Kelton) and an image of a typical setup for a transcranial magnetic resonance (TMS) study (*right*; photo courtesy of David Pitcher)

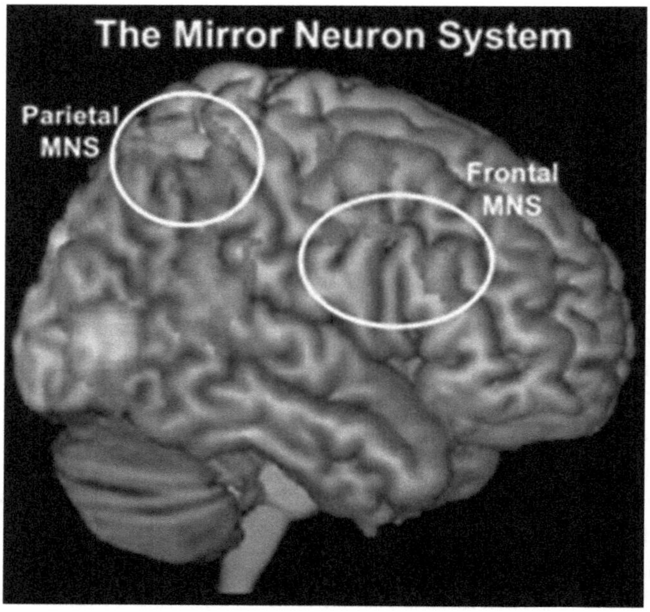

Fig. 14.2 A schematic of the mirror neuron system, consisting of a frontal MNS in the inferior frontal gyrus and a posterior MNS in the inferior parietal lobule

Furthermore, the MNS may be more sensitive to the goal of an action than to the way in which it is performed. Iacoboni et al. (2005) found a stronger mirror response in the posterior IFG and ventral PMC for actions embedded in a context (e.g., picking up a cup to clean it after having tea) than the same motor action performed outside of a context (e.g., simply picking up a cup), attributing a high level of specificity and intentionality to the premotor portion of the MNS (Iacoboni et al. 2005). This finding is extended to the posterior component of the MNS by another study suggesting that the anterior intraparietal sulcus also responds selectively to the goal of an action as opposed to the way in which the action is completed (Hamilton and Grafton 2006).

MNS and Experience

The activity of the MNS in response to both low-level motor kinematics and higher-level action goals and intentions suggests that these regions may play a very unique role in supporting social cognition, providing an automatic link between an observed action, and understanding the actor's more abstract goals. In addition, the ability to match observed or heard actions to one's own motor representations for producing the same actions has led researchers to speculate that a function of the human MNS might be to support motor simulations of other people's behaviors (Gallese et al. 2004). Simulating another person's behaviors allows us to generate an internal, first-person understanding of the observed action as though we ourselves performed the action and, based on that understanding, predict their next actions. In a sense, motor simulation is a vehicle through which one person can "get into the mind of another." If simulation is part of social cognition, then the MNS may be one of the principle regions involved in such processing. However, the ability to simulate an action may depend strongly upon our own existing motor repertoire, which is developed through experiences with prior actions and learning new actions. In the following section, we will explore ways that the MNS is modulated by our prior experiences, by the context of the task we are engaged in, during imitation and learning, and by activities that are impossible for us to do.

Prior Experiences

Our life experiences shape our perceptions of the world around us, how we take in new scenery, what stands out to us when reading a story, and the ways in which we interact with others. As experiences can consist of a variety of modalities—visual experience from having seen a desert sunrise, auditory experience from having heard the sounds of a symphony, motor experience from having thrown a baseball, or semantic experience from knowing the meaning of a specific gesture—all of these individual and social experiences affect the neural regions that underlie our ability to process the actions of others. In particular, the human MNS appears to be strongly modulated by one's prior experiences.

A wealth of literature has demonstrated that, during passive observation, actions that one has either seen before (visually familiar) or done before (motorically familiar) generate greater MNS activity than unfamiliar actions (Calvo-Merino et al. 2005, 2006; Cross et al. 2006, 2009), with both the IPL and ventral PMC being specifically modulated by one's experience with the actions (Cross et al. 2006). For example, in one study, it was found that when ballet dancers watched sequences of ballet moves as opposed to capoeira moves, there was greater MNS activity (Calvo-Merino et al. 2005). This experience-driven effect can be gained in a matter of days or even hours, as, in a different study, participants who learned simple dance patterns had greater MNS activity for their newly learned patterns than for unlearned patterns (Cross et al. 2009). Thus, there seems to be a clear modulatory role of experience on MNS activation.

In contrast, lack of experience, or difficult-to-understand actions, may recruit regions that are more strongly associated with effortful reasoning and intention understanding, such as those found in the medial prefrontal cortex (mPFC), posterior cingulate cortex (PCC), and bilateral temporoparietal junctions (TPJ; Brass et al. 2007; Liepelt et al. 2008). In some situations, the MNS alone may be insufficient to understand the observed actions. For instance, contextually appropriate actions, such as using one's knee to flip up a light switch when one's hands are full of books, may generate less activity in reasoning-related regions (e.g., the left pSTS and mPFC) than the same action in a contextually inappropriate situation, such as using one's knee to flip a light switch when one's hands are free (Brass et al. 2007; Kilner and Frith 2008). In addition, implausible, non-stereotypic finger movements were also shown to activate mentalizing regions in the TPJ and angular gyrus, as well as the pSTS (likely contributing to increased visual activity for the unusual movement), as opposed to MNS regions (Liepelt et al. 2008). However, these results are in direct contrast to prior literature suggesting that the IFG still represents observed actions, even when they are biologically implausible (Costantini et al. 2005). Further research is needed to better reconcile these disparate findings.

Task and Context

One's focus during a certain event also modulates neural activation. For instance, when participants are instructed to attend to how the *movement* is being performed, versus attending to the *goal* of the movement, different neural networks are active. There is greater MNS activity when attending to the movement but greater mentalizing activity when attending to the goal of the action (de Lange et al. 2008; Hesse et al. 2009). In accordance with this finding, new literature suggests that activity levels in the MNS are modulated by a number of complex factors, including experience and the nature of task. Normally more experience with a task, such as observing a familiar task, leads to more mirror activation, but when the task was trying to understand the *intention* of the action, participants watching familiar and unfamiliar hand gestures demonstrate greater mirror activation than when watching *unfamiliar* gestures (e.g., gestures like a thumbs up versus "The Netherlands" in sign language) (Liew et al. 2010; see Fig. 14.3). Taken together, these results suggest that there is a delicate interaction between familiarity and experience with an action and one's task when viewing the action. While usually more experience leads to more motor simulation, sometimes trying to understand an action that one has less experience with requires additional MNS activity. Further research is necessary to explore the neural activity underlying how we understand others' actions when they are embedded in a dynamic social context, as found in real-life social situations.

Imitation and Learning

The mirror system not only plays a role in understanding others' actions, but it has been hypothesized that it is involved with imitating others' actions. John Dewey, the American educator quoted at the beginning of this chapter, was a large proponent of learning through doing. In fact, the MNS is thought to be highly involved in imitation of new movements, with a wide body of literature demonstrating increased MNS activity when attempting to learn and imitate novel motor patterns (Buccino et al. 2004b; Iacoboni 2005; Vogt et al. 2007). Since imitation is a primary method of human motor learning (Meltzoff and Prinz 2002), the mapping of visual representations onto motor representations, as seen in the MNS, may be essential to this process (Iacoboni 2005). In one study, increased activity in mirror regions was found when participants observed specific guitar chord finger patterns with an intent to imitate the patterns shortly thereafter, as compared to passive observation with no intent to imitate (Buccino et al. 2004b). In contrast, when transcranial magnetic stimulation (TMS) was applied to the IFG using parameters that disrupt neural activity in that

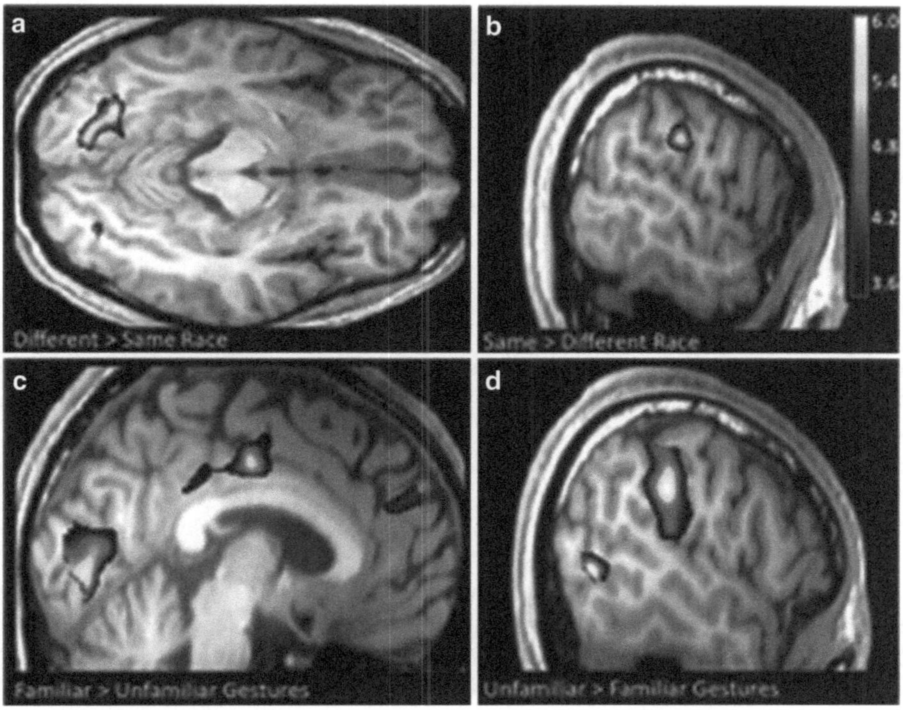

Fig. 14.3 Race and familiarity effects on the MNS. (**a**) Greater BOLD signal in the occipital regions when observing individuals from a different race make a gesture as compared to individuals of the same race; (**b**) greater BOLD signal in the posterior MNS when observing individuals from the same race as compared to a different race make a gesture; (**c**) greater BOLD signal in the medial prefrontal and posterior medial parietal regions (e.g., mentalizing system) for observing familiar versus unfamiliar gestures; and (**d**) greater BOLD signal in the posterior MNS and visual regions for observing unfamiliar versus familiar gestures (Adapted from Liew et al. 2010)

region, participants showed a decreased ability to imitate finger movements while their motor execution, in general, was unaffected (Heiser et al. 2003; Catmur et al. 2009). The latter results indicate that the IFG may be an essential component of imitation processing.

Iacoboni (2005) proposed that imitative learning involves the MNS as well as a region that is involved with working memory, the dorsolateral prefrontal cortex (DLPFC), along with other regions involved with motor preparation. Supporting this theory, Vogt et al. (2007) demonstrated that the MNS is most involved during the early stages of learning and that the DLPFC may direct attention and integrate information from MNS regions to other brain regions, thus helping to coordinate the learning process.

Impossible Actions

Given all this information on experience increasing the MNS response, how then does the MNS respond to observation of actions that are impossible for one to make due to the presence of a congenital disability? In a recent study, participants with congenital aplasia demonstrated an MNS response for hand actions despite being born without hands and thus never having performed hand actions (Gazzola et al. 2007a, b). This implies that this MNS activity is flexible and can be adapted to different bodies and different situations, depending on one's own personal experiences. Interestingly, these individuals mapped observed hand actions onto their own foot or mouth regions in the motor cortex, depending on which effector they used to achieve the goal of the movement (e.g., mapping the observation of someone using their hand to pick up a pen onto their own foot region, if they use their foot to pick up a pen; Gazzola et al. 2007a). Similar results were obtained in another study of action observation in a participant with congenital aplasia who was born without arms or legs (Aziz-Zadeh et al. 2012). However, while this participant showed activity in her MNS for actions whose goals were possible for her using another effector (e.g., flipping a book page with her mouth or upper stump instead of fingers), goals that were *impossible* for her even with another effector (e.g., cutting with scissors) not only activated the MNS but also recruited additional brain regions associated with higher-level reasoning abilities, such as the medial prefrontal cortex. Interestingly, when typically developed observers watched her perform actions using her upper stump as compared to observing a typically developed individual perform hand actions, these observers activated *more* MNS activity when watching stumps than hands—possibly due to an increase in attention and a stronger attempt to understand the kinematics of the novel body parts (Liew et al. 2011).

Do individuals with acquired disabilities also show activity in the MNS when observing an action they themselves can no longer make? One study showed that individuals who have suffered complete spinal cord injuries and lost the ability to voluntarily move their feet demonstrated activity within both the IPL and cerebellum when attempting and imagining foot actions, despite their loss of voluntary foot movements (Hotz-Boendermaker et al. 2008). This suggests mental imagery of foot actions still engages cortical motor representations, even when motor output from the cortex to the spinal cord is impeded. Similarly, we may expect observation of foot actions to also activate motor regions. Such findings have strong implications for rehabilitation after spinal and cortical injury (Garrison et al. 2010).

Similarly, fMRI and transcranial magnetic stimulation (TMS) studies have demonstrated that MNS activity in the IFG occurs during observation of actions that are both biomechanically possible and biomechanically impossible (Romani et al. 2005; Costantini et al. 2005). For example, the IFG is active during observation of finger movements whether they are within normal physical abilities or not, such as finger hyperextension (Costantini et al. 2005). TMS delivered to the left primary motor cortex when watching finger movements revealed that both possible and impossible finger movements increased corticospinal excitability in the specific muscles needed for the motion (Romani et al. 2005). This data suggests that regions of the mirror system may be invariant to the physical potential of the observer and may instead be active in response to observed movement patterns regardless of whether or not those patterns can be completed by those doing the observing. The researchers proposed that other regions, such as sensorimotor parietal regions, may encode the movement feasibility.

The findings on familiar, learned, and impossible actions suggest that mirror region activity may be greater for movements that match our individual motor repertoires than for movements we wish to add to our motor repertoires through imitation and learning. Given that action observation and imitation are essential to development and learning (Meltzoff and Prinz 2002), this may be an important function of the MNS. In addition, the MNS appears to sometimes assist when actions are novel and require additional effort to understand, thus engaging our own sensorimotor representations to understand new actions as well. Finally, the MNS also responds to the goals of observed actions even when the motor means to achieve the goal do not fall within our motor repertoire, whether we have never performed the action before, can no longer perform the action anymore, or when it is physically impossible to perform. In these instances, regions beyond the MNS may be additionally recruited to provide further processing for observation of an action outside our motor repertoire.

The MNS and the Social Domain

The mirror system has also been implicated in tasks beyond just watching and understanding actions. MNS activity has been associated with a multitude of diverse social processes, from simply observing motor actions (Buccino et al. 2001; Aziz-Zadeh et al. 2002, 2006a) to interpreting communicative gestures (Villarreal et al. 2008; Skipper et al. 2009; Liew et al. 2010), to processing sign language in both deaf and hearing sign language (Emmorey et al. 2003; Corina and Knapp 2006; Villarreal et al. 2008; Skipper et al. 2009; Liew et al. 2010), and to understanding the emotions in another's voice (Aziz-Zadeh et al. 2010). Activity in the MNS has further been correlated with behavioral measures of empathy (Gazzola et al. 2006; Shamay-Tsoory et al. 2009; Aziz-Zadeh et al. 2010), suggesting that MNS activity, or the ability to use one's own motor representations to understand another, plays a large role in empathy. This section will provide a review of findings suggesting that activity in the MNS is modulated by one's social group affiliation and correlated with behavioral measures of empathy.

Social Groups

Social group affiliations, such as one's racial group, cultural group, or social identity, may also modulate MNS activity. One facet of an early study by Buccino et al. (2004a) considered action observation of conspecifics compared to non-conspecifics (Buccino et al. 2004a). Human participants observed humans, monkeys, or dogs perform mouth actions and showed the greatest MNS activity in response to mouth actions performed by humans, followed by monkeys, followed by dogs. That is, MNS activity during the observation of mouth actions decreased as humans observed species that were less and less similar (Buccino et al. 2004a). This suggests that the MNS may match visual and kinesthetic features of actions such that the more physically similar one is to the actor, the more MNS activity that occurs.

In humans, race is a highly automatic and implicitly encoded social group affiliation (Phelps and Thomas 2003; Chiao et al. 2008). Experimental modulation of race between the actor and the observer has been shown to affect an array of neural responses depending on the task, including empathy for another's pain (Xu et al. 2009b), fear responses to others (Chiao et al. 2008), and social liking (Phelps and Thomas 2003). These data suggest that racial in-group/out-group associations can powerfully modulate neural responses to others in a variety of contexts. While little has been studied regarding race and the MNS, several studies suggest that there is a complex race-based modulation of mirror regions (Molnar-Szakacs et al. 2007; Liew et al. 2010). Using transcranial magnetic stimulation (TMS), Molnar-Szakacs et al. (2007) found increased corticospinal excitability when Euro-American participants observed actors of their own race compared to another race (e.g., Nicaraguan) perform the same actions, suggesting a racial in-group bias during action observation. Similarly, recent fMRI research in Chinese participants observing both Chinese and Caucasian actors demonstrated that action observation of members of one's own race compared to the other race led to increased activity in the IPL and the insula, associated with emotional processing (Liew et al. 2010; see Fig. 14.2). These findings indicate that one may more readily map observed actions onto their own motor representations if the actor is similar to them (Molnar-Szakacs et al. 2007; Liew et al. 2010), a conclusion that is consistent with results from the conspecific study discussed previously (Buccino et al. 2004a).

In addition to plausible modulations of the MNS related to race, however, the MNS also responds differently to observations of the self versus others. Researchers found that repetitive TMS (rTMS) applied to the right IPL significantly decreased the participants' abilities to distinguish their own face from the faces of others (Uddin et al. 2006). Notably, only the right hemisphere demonstrated this effect in decreased self-other distinction, leading researchers to surmise that a possible role for the MNS in self-processing may occur largely in the right hemisphere. Further support for this hypothesis was found in an fMRI study which found that the right hemisphere MNS was more active for the self even across modalities (Kaplan et al. 2008). Both observations of one's own face versus a friend's face and listening to one's own voice versus a friend's voice generated increased activity in the right IFG, suggesting that the MNS may play a role in distinguishing the self from others across both visual and auditory systems.

Empathy

Studies of humans and primates demonstrate that individuals tend to like those who imitate their mannerisms more than those who do not (Chartrand and Bargh 1999; Paukner et al. 2009). In a behavioral study, people preferred actors who copied their mannerisms, such as fidgeting with a pen or bouncing a foot, more than actors who did not (Chartrand and Bargh 1999). In another study, monkeys preferred humans who copied their movements, for example, humans who played with a ball in the same way as the monkey, more than those who did not (Paukner et al. 2009). People also tended to unconsciously adopt their partners' facial expressions, postures, and tics as they worked together, with the amount of mimicry increasing in more empathic participants (Chartrand and Bargh 1999). An interpretation of such findings is that increased implicit mimicry of another's actions and mannerisms, known as the chameleon effect, increases motor simulation between individuals, which may make it easier to understand, relate to, and empathize with others (Chartrand and Bargh 1999).

Given the mirror neuron system seems to play a prominent role in motor imitation and emulation (Heiser et al. 2003; Iacoboni 2005; Vogt et al. 2007), some researchers began to explore the role of the MNS in implicit imitation, social imitation, and empathy. One study found that the imitation of facial expressions showed increased activity in components of the MNS as well as the insula, a region associated with emotional processing (Carr et al. 2003). Similarly, another study demonstrated that observing emotional facial expressions engaged the MNS, particularly in the IFG (Schulte-Rüther et al. 2007). In the same study, activity in MNS correlated with increased scores on the Balanced Emotional Empathy Scale (BEES; Mehrabian 1996), a behavioral measure of empathy, suggesting that one's own empathic abilities may be correlated with activity in the MNS during emotion processing (Schulte-Rüther et al. 2007).

Several additional studies have found that increases in MNS activity during action perception—whether hearing action-related sounds, seeing emotional facial expressions, interpreting emotional stories, or watching others' motor actions—correlate with self-reported levels of empathy, suggesting that the MNS may be involved in empathic processing (Kaplan and Iacoboni 2006; Gazzola et al. 2006; Schulte-Rüther et al. 2007; Shamay-Tsoory et al. 2009; Aziz-Zadeh et al. 2010). For example, Gazzola et al. (2006) showed that increased MNS activity when listening to human action sounds compared to neutral environmental sounds correlated with increased scores on the Interpersonal Reactivity Index (IRI; Davis 1983), a self-report measure of empathy. This correlation suggests that there is a positive link between activity in regions that map others' actions onto one's own motor repertoire and empathic abilities. Kaplan and Iacoboni (2006) further demonstrated that activity in the right IFG during observation of actions in context—for example, a context that indicates that a cup is being grasped in order to wash it—is correlated with increased scores on the IRI. Similarly, increased empathic accuracy, which is the ability to accurately predict how another person is feeling as they are talking, was correlated to regions of the MNS (right IPL and bilateral dorsal premotor cortices) as well as the STS and mPFC (Zaki et al. 2009). A study by Shamay-Tsoory et al. (2009) evaluated patients with lesions in IFG and found that these lesions, compared to a non-MNS region, predictably demonstrated impairments in emotional aspects of empathy (Shamay-Tsoory et al. 2009). These results suggest that one's natural empathic tendencies are correlated with the amount of MNS activity one has when trying to understand others. While we do not suggest that the MNS is the only area involved in empathic processing, it does seem to be one network that is involved in empathizing with others.

In addition, understanding the role of the MNS in this diverse range of social functions is important because the MNS has also been hypothesized as being dysfunctional in a number of diseases associated with poor social processing, such as autism (Iacoboni and Dapretto 2006; Dapretto et al. 2006; Perkins et al. 2010; Fan et al. 2010), schizophrenia (Arbib and Mundhenk 2005; Greicius 2008; Lynall et al. 2010), and stroke (Foundas et al. 1995; Mukherjee et al. 2000; Heath et al. 2001; Buxbaum et al. 2008; Damoiseaux and Greicius 2009). However, many debates are still unsettled about the role of the MNS in each of these disorders. Notably, some researchers suggest that engaging MNS or mentalizing regions through other tasks may result in improvements in social functioning at large (Iacoboni and Dapretto 2006; Buccino et al. 2006; Iacoboni and Mazziotta 2007; Garrison et al. 2010), a line of research that warrants further development.

MNS, Language, and Embodied Cognition

The MNS not only may play a role in motor learning, action understanding, and social cognitive processing, but it has been further cited in theories about the development of language. We focus here on the literature suggesting the role of the MNS in language and gestures, which have been hypothesized by some to be the precursors to language, as well as the role of the MNS in embodied semantics, which demonstrates that words can activate parts of the brain that are associated with what the word describes (e.g., a motor part of the brain becomes active when reading about a motor action). Finally, we conclude by extending the discussion to embodied cognition beyond the MNS and noting other ways that our very thoughts may be grounded in our lived experiences.

The MNS, Language, and Embodied Semantics

> The origins of language are not to be found in the mouth alone but in the hand, and their mutual interaction.
> -Corballis (2002), as quoted in Franks (2010)

Several researchers have hypothesized that shared representations between individuals may have been one mechanism by which we evolved language. Recall that the IFG is located in Broca's area (BA 44), a primary language area. The IFG's dual roles of motor planning and expressive language abilities have been interpreted by some researchers to suggest an evolutionary link between action execution/observation and language (Rizzolatti and Arbib 1998; Gallese and Lakoff 2005; Arbib 2005; Fadiga et al. 2009). One interesting difference between the macaque MNS and the human MNS is that while macaque mirror neurons tend to fire only for object-oriented transitive actions, such as reaching for a piece of food, in humans the MNS is active both for transitive, goal-directed actions and intransitive actions, such as a communicative gesture. Transitive actions, such as reaching for a piece of food, tend to be less abstract than intransitive actions, such as a hand shape symbolizing a semantic or conceptual meaning. The exception to this trend is a subset of macaque mouth mirror neurons, which in fact fire when the monkey observes communicative mouth actions (e.g., lip smacking; Ferrari et al. 2003). This provides evidence of a suggestive link between goal-directed actions and communicative abilities (Rizzolatti and Craighero 2004). However, in humans, mirror regions demonstrate activity whether reaching for an object or making a symbolic hand gesture (Gentilucci and Dalla Volta 2008; Villarreal et al. 2008; Liew et al. 2010; Schippers et al. 2009; Skipper et al. 2009). The plausible evolution of this system from concrete actions to abstract gestures in humans has led some to propose that the evolution of MNS and its interaction with other brain regions played a role in the formation and development of language (Rizzolatti and Arbib 1998; Gallese and Lakoff 2005; Arbib 2005, 2010; Fadiga et al. 2009). Several of these hypotheses emphasize the development of gestures (first concrete, then abstract) as leading the way for abstract language formation, a phenomenon which is supported by observations in child development with infants developing gestural abilities prior to language (for a review, see Franks 2010).

Further support for the role of the MNS in language may be found in the encoding of even *intransitive* (symbolic) gestures, like a thumbs up, which links an abstract meaning (e.g., good) with a hand shape. The ability of the MNS to encode both concrete and abstract gestures suggests that it may be able to encode more *abstract goals* as well. Work on the goal-specific nature of the MNS supports this, with the finding that there is a stronger MNS response in the posterior IFG and ventral PMC for actions embedded in a context (e.g., picking up a cup to clean it after having tea) than the same motor action performed outside of a context (e.g., simply picking up a cup), attributing a high level of specificity and intentionality to the premotor portion of the MNS (Iacoboni et al. 2005). In addition, regions in the IFG and IPL also demonstrate different responses to meaningful and meaningless object-directed actions, again promoting the idea of regions of the MNS being modulated by high-level action goals (Newman-Norlund et al. 2010). In general, these studies support the idea that our manual hand actions may have, over time, given rise to the meaning associated with our verbal utterances, thus forming a basis for language.

Gestures: Linking Action and Language

Thus, special attention should be given to the study of gesture, which can be considered as much a part of language as spoken or written words and grammar and can be thought of in development with gestures leading the way and speech following (Iacoboni 2008; Franks 2010). There are many types of gestures, including co-speech gestures (i.e., gestures that naturally accompany speech without holding inherent meaning in and of themselves), pantomimes (i.e., gestures that mimic actual object or tool use), intransitive, communicative gestures (i.e., gestures that connotate abstract meanings, which are also known as emblems), and sign language (i.e., gestures that hold linguistic meaning; McNeill 1992, 2005). Communicative gestures have strong ties to language, and many studies have examined the neural activity evoked by communicative gestures, showing activation of mirror, mentalizing, and language-related regions based on different task and stimulus conditions (Gallagher and Frith 2004; Villarreal et al. 2008; Straube et al. 2009; Skipper et al. 2009; Flaisch et al. 2009; Schippers et al. 2010). For a review of literature linking action and speech over several types of gestures, see Willems and Hagoort (2007). Particularly, communicative gestures have been shown to share an overlapping network with their spoken description (e.g., a thumbs up and the phrase "it's good") in the left IFG and bilateral pMTG extending into the pSTS, with greater left-sided activation (Xu et al. 2009a). Uniquely, however, gestures versus speech activated the fusiform gyrus and inferior temporal cortex bilaterally, while speech versus gestures additionally activated the anterior STS, MTG, and bilateral STG, closer to the auditory cortex. In addition, functional connectivity between these regions demonstrated a similar pattern, with gestures activating a network between the left IFG and left ventral temporal regions, and speech activating a network between the left IFG and pMTG and STS (Xu et al. 2009a). Xu et al. (2009a) suggested that such results support the idea of a modality-independent communication system that is not specifically tied to language processing but is more general for communication of many types. This is in line with suggestions by several researchers that language-related regions, such as Broca's area in the IFG, may be functionally involved in a variety of networks that allow the one region to partake in many flexible processes as opposed to being only speech-specific or even language-specific (Corina and Knapp 2006; Willems and Hagoort 2007).

There is also a wealth of literature that supports these hypotheses regarding the relationship between action and gesture. One well-studied class of gestures consists of manual systems designed specifically for language, as found in sign language used by native signers (American Sign Language, British Sign Language, etc.). Observing sign language in congenitally deaf native signers generally produced activations along the perisylvian cortex (including Broca's and Wernicke's areas) and the pSTS, regions that are also associated with processing spoken language in hearing individuals (Corina et al. 1992a, b; MacSweeney et al. 2002a, b, 2004; Corina and Knapp 2006; Willems and Hagoort 2007), with activity generally lateralized to the left hemisphere (Corina et al. 1992b).

Interestingly, in addition to speech/language-related regions, several other regions have been associated with sign language comprehension, namely, the left IPL. Observations of British Sign Language (BSL) have been shown to activate the IFG and middle and superior temporal cortex, as well as the IPL (supramarginal gyrus), more strongly than watching another manualized communication system that did not have linguistic meaning, which indicated that these regions are not simply active for gestural observations of any sort (MacSweeney et al. 2004). Left parietal activity is commonly found in a number of sign language studies (Corina et al. 1999; MacSweeney et al. 2002a, b, 2004; Emmorey et al. 2004; Emmorey et al. 2005; Corina and Knapp 2006), with researchers proposing that the IPL and SPL may play an important role in extracting hand configurations as well as hand/arm spatial positions (MacSweeney et al. 2002a). In addition, lesions in the supramarginal gyrus (SMG) in deaf American Sign Language users impair sign comprehension, suggesting that it plays a unique role in representing semantic meaning attached to signs (Chiarello et al. 1982; Corina et al. 1992a).

In addition to the left IPL, the left inferior frontal cortex is also active during sign language observation (MacSweeney et al. 2002a, b, 2004), as well as during observation of communicative gestures and even co-speech gestures (Willems and Hagoort 2007; Willems et al. 2007). Lesions in the IFG produce deficits in sign production, similar to Broca's aphasia in hearing individuals, but not deficits in sign comprehension (Poizner et al. 1987; Corina and Knapp 2006), suggesting that the frontal component is not necessarily needed for sign language comprehension. These results were also demonstrated using cortical stimulation mapping in an individual undergoing treatment for a seizure disorder, demonstrating impairments in motor execution of signs with stimulation to Broca's area and sign comprehension deficits and difficulties with semantic-phonological decisions with stimulation of the SMG, suggesting a role for the parietal component in binding linguistic features (Corina et al. 1999). Neuroimaging results also support the role of a generalized frontoparietal network in sign language production and comprehension which is stronger on the left hemisphere but, under certain conditions, it is represented bilaterally (MacSweeney et al. 2002a, b, 2004; Corina and Knapp 2006; Villarreal et al. 2008; Straube et al. 2009; Skipper et al. 2009). Regardless, it appears that manualized language may involve both regions typically involved in language processing in hearing individuals, as well as unique contributions from the supramarginal gyrus among other regions, which may contribute to the binding of linguistic and motor activity.

One might expect a similar finding, then, for intransitive, communicative gestures observed by typically developed individuals, and indeed, this is generally supported by the literature (Molnar-Szakacs et al. 2007; Villarreal et al. 2008; Schippers et al. 2009; Straube et al. 2009, 2010). Most studies find activity in IFG, pSTS, and IPL for gestural observation, with one study finding greater BOLD activity in the left IFG when watching intransitive communicative gestures as compared to pantomimes of transitive gestures, suggesting a possible modulation of this region by abstract meaning (Villarreal et al. 2008). In fact, evidence of a shared single communication system that integrates gesture and speech comes from observations that producing an emblem and saying the word that describes the emblem have cross-modal effects that are not seen when either producing the emblem or saying the word separately (Bernardis and Gentilucci 2006). That is to say, when participants pronounced words and performed emblems at the same time, the acoustic range of the word increased while the duration of the gesture's movement patterns decreased, suggesting that the two are not contained in isolated systems but interact within the brain. Following up on this finding, the same research group then used TMS to disrupt the left IFG, which modulated the acoustic production of the word pronounced in response to a symbolic gesture, an effect which was not seen when TMS was delivered to the right IFG or not delivered at all (Gentilucci et al. 2006). These findings suggest a role for the left IFG in multimodal semantic processing.

In addition to increased activity in the IFG, communicative gestures also showed greater hippocampal activity when observing metaphoric, compared to free or unrelated, gestures that accompanied speech, also with a strong left lateralization suggesting that the left hemisphere may be involved in semantic integration of gesture with speech (Straube et al. 2009). However, even co-speech gestures, which accompany speech and may not contain meaning in themselves, demonstrate increased processing in Broca's area when verbal and gestural information do not match, potentially indicating integration of action and language processing at this region (Willems et al. 2007). In an EEG experiment, when subjects were presented with either a word or a gesture that didn't fit the context of a sentence, an N400 effect (associated with difficulty processing a word based on the surrounding context) was found (Ozyurek et al. 2007). Importantly, the timing of this effect did not differ between speech and gesture conditions, suggesting that the "timecourse" of integration for a gesture is similar to that of integration for a spoken word (Ozyurek et al. 2007). Supporting this is recent evidence that incongruent speech and co-speech gestures elicit the N400 effect only within a certain time window (e.g., when speech and gesture are presented simultaneously or within 160 ms, but not at 360 ms). This suggests that the two are processed in an integrative manner in order to resolve each (Habets et al. 2010). In addition, these effects were found to increase activation in the left IFG for both speech and gesture, suggesting a common neural basis for integrating semantic information, regardless of the modality (Willems et al. 2007). Such co-speech gestures were also found to increase accuracy of memory retrieval for stories and have been shown to increase functional connectivity between MNS regions in both the IFG and IPL and anterior regions of the superior temporal sulcus. These have been associated with semantic aspects of language comprehension (Skipper et al. 2009).

Furthermore, task instructions or contextual components of the gesture may also increase mentalizing activity to gestural observations, as found in several studies (Gallagher and Frith 2004; Schippers et al. 2009; Straube et al. 2010; Liew et al. 2010). The task of inferring others' intentions from charades evoked activity in both MNS (premotor, parietal) and mentalizing (TPJ) regions (Schippers et al. 2009). Granger causality between the charades actor and the charades guesser also demonstrated that MNS activity in the actor's brain was Granger causally related to both MNS and mentalizing activity in the guesser's brain (Schippers et al. 2010). In addition, Liew et al. (2010) demonstrated greater mentalizing activity when participants inferred the meaning of familiar gestures, but greater MNS activity when participants inferred the meaning of unfamiliar gestures, suggesting that the task demands, as well as prior experience with the gestures, modulated the activity of MNS and mentalizing systems in understanding the gestures. Greater mentalizing activity was also found for gestures that are expressive (e.g., "I am angry") compared to motor-related (e.g., "Come here"), particularly in the anterior paracingulate cortex, bilateral temporal poles, right pSTS, and amygdala (Gallagher and Frith 2004). In contrast to, and in line with the reviewed literature, motor-related gestures compared to expressive gestures more strongly activated a left-lateralized frontoparietal system associated with language and motor imitation. Finally, it appears that social cues, such as face/body orientation, may influence the neural regions related to processing gestures, due to top-down modulations by the mentalizing system (Straube et al. 2010). Thus, it appears clear that while MNS regions are commonly activated in response to observations of many types of gestures, mentalizing regions may also be activated based on the social context, task, and content of the gestures themselves.

Notably, these findings are reinforced by studies of individuals with apraxia, in which brain damage to specific portions of the brain can cause difficulty performing and/or comprehending actions, including actions involving conceptual knowledge of tool use, imitation of movements, pantomimed actions with tool use, and/or symbolic gestures, among many other types (Heilman et al. 1975, 1982; Geschwind 1975; Rothi and Heilman 1984; Rothi et al. 1985). Some patients with apraxia were found with specific deficits in the comprehension of intransitive gestures (see Heath et al. 2001). Interestingly, deficits in gestural comprehension were linked to limb apraxia due to deficits of the left IFG and not apparent when individuals had lesions in the left IPL or the right IFG/IPL, a finding which is controversial with the existing literature but explained by differences between this novel gestural recognition task and prior studies (Pazzaglia et al. 2008b).

The MNS and Embodied Semantics

In addition, a role for the MNS in conceptual representations and semantic processing has been suggested. This is largely based on the possibility that having multimodal representation (visual, auditory, motor) in the premotor cortex may lead to the possibility of abstract, conceptual representations. It is furthermore linked to a theory in linguistics known as embodied semantics, which proposes that concepts are represented in the same neural sensory-motor circuits that support the enactment of that concept. For example, embodied semantics predicts that phrases related to foot actions activate cortical motor representations of the foot. This theory extends beyond the motor system, for example, predicting that hearing a phrase about color activates cortical areas responsible for color processing (Damasio 1989; Damasio and Tranel 1993; Barsalou 1999; Lakoff and Johnson 1999; Glenberg and Kaschak 2002; Feldman and Narayanan 2004; Gallese and Lakoff 2005; Pulvermuller et al. 2005a, b; Pulvermuller 2005; Pulvermuller and Hauk 2006; Aziz-Zadeh and Damasio 2008).

There is some evidence that phrases related to actions activate the MNS in an effector-specific manner, consistent with the theory of embodied semantics for actions. In one study, Aziz-Zadeh et al. (2006b) showed that reading phrases that focused on foot, hand, or mouth actions also activated premotor regions that were most strongly active for observation of foot, hand, or mouth actions, respectively. Furthermore, this activation occurred in the left hemisphere, where language is largely supported (Aziz-Zadeh et al. 2006b). Other studies have also supported this effect with similar paradigms (Tettamanti et al. 2005; Pulvermuller 2005; Pulvermuller et al. 2005a, b).

Such *embodied semantics* not only show somatotopic properties, mapping onto the specific regions associated with the actions, but also are body-specific, such as right-handed individuals will activate left premotor cortex when hearing actions, while left-handed individuals activate right premotor cortex for the same actions (Willems et al. 2010). In addition, individuals with apraxia are also impaired in their ability to match action sounds with the appropriate action photo, and this effect is body-part-specific such that individuals with limb apraxia are impaired in matching limb action sounds/pictures, while individuals with buccofacial apraxia are impaired in matching mouth action sounds/pictures (Pazzaglia et al. 2008a). This body-specific representation is also true during motor imagery (Willems et al. 2009), suggesting that words can evoke one's own motor representations during comprehension. There is a wealth of literature on embodied semantics that is not discussed here for the purposes of brevity (for a review see Aziz-Zadeh and Damasio 2008), but overall these findings suggest that there are multimodal representations of conceptual information that may involve sensorimotor information (Fig. 14.4).

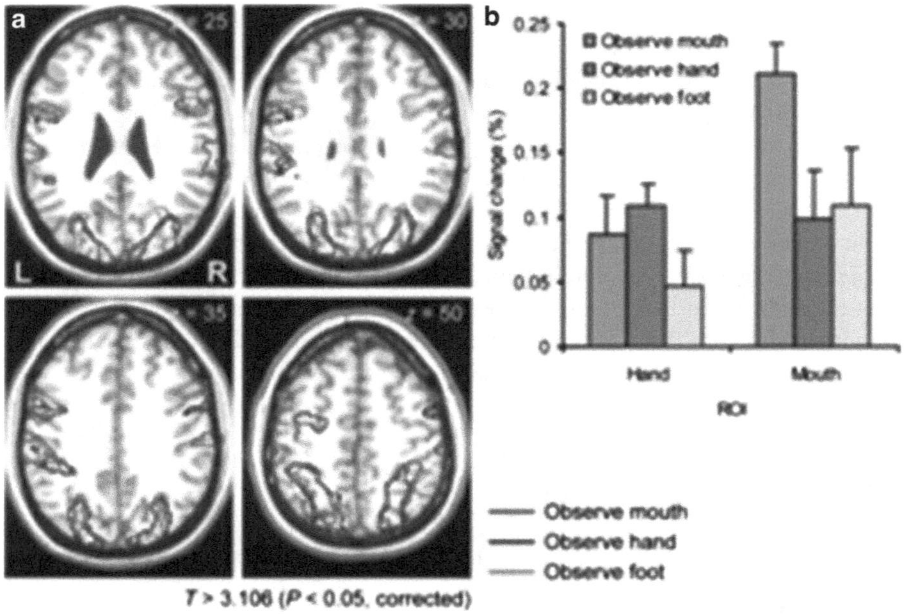

Fig. 14.4 Areas activated by observation of mouth, hand, and foot actions. (**a**) Brain regions activated by observing mouth, hand, and foot actions compared to a resting baseline. (**b**) Amount of neural activation, measured as % signal change, in each of the defined regions of interest when observing each effector (Adapted from Aziz-Zadeh et al. 2006b)

Embodied Cognition and Shared Representations Beyond the MNS

Since the discovery of the MNS, there have also been an increasing number of studies investigating the possibility that other brain regions and neural systems outside the motor system also respond to both one's own sensations and observations of another's sensations. Overall, these studies suggest that there are shared representations between the self and other for other perceptions and states, such as emotions, somatosensation, and pain processing. It has been proposed that such shared representations are important for simulation of other people's emotional and mental states, and that these shared representations may be modulated by our experiences and by our social groups (e.g., race, political affiliation), in a manner similar to the modulation of the MNS discussed previously (Singer et al. 2004, 2006; Singer and Frith 2005; Cheng et al. 2007; Serino et al. 2009; Xu et al. 2009b). While it is yet unknown how many regions have shared representations, it is possible that these "as-if body loops," initially proposed by Damasio (1994), are common in many sensory and motor regions (Damasio 1994; Damasio and Damasio 2006). Much of this work is beyond the scope of this chapter; however, as it is of interest in understanding social interactions, we briefly introduce some of this research here.

One novel fMRI study explored shared representations for disgust (Wicker et al. 2003). When participants observed videos of actors with disgusted facial expressions after sniffing presumably noxious odors, they showed increased activation in the same brain regions that were active when they themselves smelt noxious odors (e.g., the anterior insula; Wicker et al. 2003). This suggests that participants may have simulated the feeling of disgust when they simply observed the disgusted facial expressions of another.

Similarly, brain regions that are active when one experiences pain may also become active when observing another person in pain. Numerous studies have demonstrated that observing others undergo a painful situation activates regions that are active when we ourselves experience pain, namely, the "pain matrix," which consists of the anterior insula, anterior cingulate, and somatosensory cortex (Singer et al. 2004; Singer and Frith 2005; Bufalari et al. 2007; Cheng et al. 2007; Xu et al. 2009b). Neural activity in regions that may support shared representations for pain empathy is correlated with behavioral measures of empathy such as the IRI and BEES (Singer et al. 2004) and can be modulated by experience (Cheng et al. 2007), race (Xu et al. 2009b) and opinions of another (Singer et al. 2006). For example, Cheng et al. (2007) found that experience modulates pain empathy such that understanding the benefits of the pain or knowing that a painful situation is actually not painful (e.g., needles inserted into the skin during acupuncture) decreases the amount of activity within the pain matrix when watching an individual receive painful stimulation. Similarly, there is increased activity in the pain matrix when observing someone of the same race receiving painful stimulation than when observing someone of a different race (Xu et al. 2009b).

Observations of tactile sensations may also be processed by one's own somatosensory cortices. Participants demonstrated activity in the somatosensory cortex both when they were touched as well as when observing others being touched (Keysers et al. 2004; Blakemore et al. 2005). Interestingly, some individuals have what is termed "mirrored touch" synesthesia and demonstrate significantly greater activity in the somatosensory cortex compared to typical individuals when they observe others being touched (Blakemore et al. 2005). In a task where they received tactile sensation on their faces and simultaneously observed someone else receive tactile sensation to his or her face, participants with mirrored touch synesthesia made more errors when discerning which side of their own face was touched (Banissy and Ward 2007). Individuals with mirrored touch synesthesia also scored higher on behavioral measures of empathy than typical individuals, again suggesting that the ability to simulate another's sensations may be linked to increased empathy for others. In addition, typical individuals may be biased to more strongly represent observed touch when the observed individual is a political leader from one's own political affiliation than when the individual is a political leader from a different political affiliation, again demonstrating modulation of shared representation systems based on social group membership (Serino et al. 2009). In sum, it appears that how we perceive the world around us depends largely on how we ourselves enact the sounds, sensations, and images that we see. As Damasio wrote (1994: 225), "Perception is more involved in action than we think. Perceiving is as much about acting on the environment as it is about receiving signals from it" (quoted in Franks 2010).

Conclusion

While there is much left to be discovered about the neural processing involved in social understanding, the results provided here strongly suggest that we understand other people—their actions, sensations, emotions, and experiences—through evoking aspects of our own experiences that resonate with theirs. The mirror neuron system may be one way in which we do this as it is active both during our own experiences of actions and during our observations of others' actions. As discussed in this chapter, one interesting aspect of these regions is that they not only become active when observing, imitating, or learning actions but are also modulated by a myriad of social factors and, additionally, are correlated with individual differences in empathy. Furthermore, shared representations that allow us to evoke our own experiences to understand others may extend beyond the motor system, such as in brain regions underlying somatosensation, pain, and disgust. Altogether, these findings suggest that the actions that we take play a strong role in allowing us to understand and interact with those around us.

References

Arbib, M. A. (2005). From monkey-like action recognition to human language: An evolutionary framework for neurolinguistics. *Behavioral Brain Sciences, 28*(2), 105–124; discussion 125–167.

Arbib, M. A. (2010). Mirror system activity for action and language is embedded in the integration of dorsal and ventral pathways. *Brain and Language, 112*(1), 12–24.

Arbib, M. A., & Mundhenk, T. N. (2005). Schizophrenia and the mirror system: An essay. *Neuropsychologia, 43*(2), 268–280.

Aziz-Zadeh, L., & Damasio, A. (2008). Embodied semantics for actions: Findings from functional brain imaging. *Journal of Physiology, Paris, 102*(1–3), 35–39.

Aziz-Zadeh, L., Maeda, F., Zaidel, E., Mazziotta, J., & Iacoboni, M. (2002). Lateralization in motor facilitation during action observation: A TMS study. *Experimental Brain Research, 144*(1), 127–131.

Aziz-Zadeh, L., Koski, L., Zaidel, E., Mazziotta, J., & Iacoboni, M. (2006a). Lateralization of the human mirror neuron system. *Journal of Neuroscience, 26*(11), 2964–2970.

Aziz-Zadeh, L., Wilson, S. M., Rizzolatti, G., & Iacoboni, M. (2006b). Congruent embodied representations for visually presented actions and linguistic phrases describing actions. *Current Biology, 16*(18), 1818–1823.

Aziz-Zadeh, L., Sheng, T., & Gheytanchi, A. (2010). Common premotor regions for the perception and production of prosody and correlations with empathy and prosodic ability. *PLoS One, 5*(1), e8759.

Aziz-Zadeh, L., Sheng, T., Liew, S.-L., & Damasio, H. (2012). Understanding otherness: The neural basis of action understanding and empathy in a congenital amputee. *Cerebral Cortex, 22*(4), 811–819.

Baldissera, F., Cavallari, P., Craighero, L., & Fadiga, L. (2001). Modulation of spinal excitability during observation of hand actions in humans. *European Journal of Neuroscience, 13*(1), 190–194.

Banissy, M. J., & Ward, J. (2007). Mirror-touch synesthesia is linked with empathy. *Nature Neuroscience, 10*(7), 815–816.

Barsalou, L. W. (1999). Perceptions of perceptual symbols. *The Behavioral and Brain Sciences, 22*(04), 637–660.

Bernardis, P., & Gentilucci, M. (2006). Speech and gesture share the same communication system. *Neuropsychologia, 44*(2), 178–190.

Blakemore, S. J., Bristow, D., Bird, G., Frith, C., & Ward, J. (2005). Somatosensory activations during the observation of touch and a case of vision-touch synaesthesia. *Brain, 128*(7), 1571.

Brass, M., Schmitt, R. M., Spengler, S., & Gergely, G. (2007). Investigating action understanding: Inferential processes versus action simulation. *Current Biology, 17*(24), 2117–2121.

Buccino, G., Binkofski, F., Fink, G. R., Fadiga, L., Fogassi, L., Gallese, V., et al. (2001). Action observation activates premotor and parietal areas in a somatotopic manner: An fMRI study. *European Journal of Neuroscience, 13*(2), 400–404.

Buccino, G., Lui, F., Canessa, N., Patteri, I., Lagravinese, G., Benuzzi, F., et al. (2004a). Neural circuits involved in the recognition of actions performed by nonconspecifics: An FMRI study. *Journal of Cognitive Neuroscience, 16*(1), 114–126.

Buccino, G., Vogt, S., Ritzl, A., Fink, G. R., Zilles, K., Freund, H. J., et al. (2004b). Neural circuits underlying imitation learning of hand actions: An event-related fMRI study. *Neuron, 42*(2), 323–334.

Buccino, G., Solodkin, A., & Small, S. L. (2006). Functions of the mirror neuron system: Implications for neurorehabilitation. *Cognitive and Behavioral Neurology, 19*(1), 55–63.

Bufalari, I., Aprile, T., Avenanti, A., Di Russo, F., & Aglioti, S. M. (2007). Empathy for pain and touch in the human somatosensory cortex. *Cerebral Cortex, 17*(11), 2553–2561.

Buxbaum, L. J., Haaland, K. Y., Hallett, M., Wheaton, L., Heilman, K. M., Rodriguez, A., et al. (2008). Treatment of limb apraxia: Moving forward to improved action. *American Journal of Physical Medicine & Rehabilitation, 87*(2), 149–161.

Caggiano, V., Fogassi, L., Rizzolatti, G., Pomper, J. K., Thier, P., Giese, M. A., et al. (2011). View-based encoding of actions in mirror neurons of area f5 in macaque premotor cortex. *Current Biology, 21*(2), 144–148.

Calvo-Merino, B., Glaser, D. E., Grezes, J., Passingham, R. E., & Haggard, P. (2005). Action observation and acquired motor skills: An FMRI study with expert dancers. *Cerebral Cortex, 15*(8), 1243–1249.

Calvo-Merino, B., Grezes, J., Glaser, D. E., Passingham, R. E., & Haggard, P. (2006). Seeing or doing? Influence of visual and motor familiarity in action observation. *Current Biology, 16*(19), 1905–1910.

Carr, L., Iacoboni, M., Dubeau, M. C., Mazziotta, J. C., & Lenzi, G. L. (2003). Neural mechanisms of empathy in humans: A relay from neural systems for imitation to limbic areas. *Proceedings of the National Academy of Sciences of the United States of America, 100*(9), 5497.

Catmur, C., Walsh, V., & Heyes, C. (2009). Associative sequence learning: The role of experience in the development of imitation and the mirror system. *Philosophical Transactions of the Royal Society of London. Series B, Biological Sciences, 364*(1528), 2369–2380.

Chartrand, T. L., & Bargh, J. A. (1999). The chameleon effect: The perception-behavior link and social interaction. *Journal of Personality and Social Psychology, 76*, 893–910.

Cheng, Y., Lin, C. P., Liu, H. L., Hsu, Y. Y., Lim, K. E., Hung, D., et al. (2007). Expertise modulates the perception of pain in others. *Current Biology, 17*(19), 1708–1713.

Chiao, J. Y., Iidaka, T., Gordon, H. L., Nogawa, J., Bar, M., Aminoff, E., et al. (2008). Cultural specificity in amygdala response to fear faces. *Journal of Cognitive Neuroscience, 20*(12), 2167–2174.

Chiarello, C., Knight, R., & Mandel, M. (1982). Aphasia in a prelingually deaf woman. *Brain, 105*(Pt 1), 29–51.

Corballis, M. C. (2002). *From hand to mouth: The origins of language*. Princeton: Princeton University Press (See Aziz- Zadeh, L., & Ivr R. B. (2008) The humnan mirror neuron system and embodied representations. In D. Sternad (Ed.), *Progress in motor control*. New York: Springer).

Corina, D. P., & Knapp, H. (2006). Sign language processing and the mirror neuron system. *Cortex, 42*, 529–539.

Corina, D. P., Poizner, H., Bellugi, U., Feinberg, T., Dowd, D., & O'Grady-Batch, L. (1992a). Dissociation between linguistic and nonlinguistic gestural systems: A case for compositionality. *Brain and Language, 43*(3), 414–447.

Corina, D. P., Vaid, J., & Bellugi, U. (1992b). The linguistic basis of left hemisphere specialization. *Science, 255*(5049), 1258–1260.

Corina, D. P., McBurney, S. L., Dodrill, C., Hinshaw, K., Brinkley, J., & Ojemann, G. (1999). Functional roles of Broca's area and SMG: Evidence from cortical stimulation mapping in a deaf signer. *NeuroImage, 10*(5), 570–581.

Costantini, M., Galati, G., Ferretti, A., Caulo, M., Tartaro, A., Romani, G. L., et al. (2005). Neural systems underlying observation of humanly impossible movements: An FMRI study. *Cerebral Cortex, 15*(11), 1761–1767.

Cross, E. S., Hamilton, A. F., & Grafton, S. T. (2006). Building a motor simulation de novo: Observation of dance by dancers. *NeuroImage, 31*(3), 1257–1267.

Cross, E. S., Kraemer, D. J. M., Hamilton, A. F. C., Kelley, W. M., & Grafton, S. T. (2009). Sensitivity of the action observation network to physical and observational learning. *Cerebral Cortex, 19*(2), 315.

Damasio, A. R. (1989). Time-locked multiregional retroactivation: A systems-level proposal for the neural substrates of recall and recognition. *Cognition, 33*(1–2), 25–62.

Damasio, A. R. (1994). *Descartes' error*. New York: Penguin.

Damasio, A., & Damasio, H. (2006). Minding the body. *Daedalus, 135*(3), 15–22.

Damasio, A. R., & Tranel, D. (1993). Nouns and verbs are retrieved with differently distributed neural systems. *Proceedings of the National Academy of Sciences, 90*(11), 4957.

Damoiseaux, J. S., & Greicius, M. D. (2009). Greater than the sum of its parts: A review of studies combining structural connectivity and resting-state functional connectivity. *Brain Structure & Function, 213*(6), 525–533.

Dapretto, M., Davies, M. S., Pfeifer, J. H., Scott, A. A., Sigman, M., Bookheimer, S. Y., et al. (2006). Understanding emotions in others: Mirror neuron dysfunction in children with autism spectrum disorders. *Nature Neuroscience, 9*(1), 28–30.

Davis, M. H. (1983). Measuring individual differences in empathy: Evidence for a multidimensional approach. *Journal of Personality and Social Psychology, 44*(1), 113–126.

de Lange, F. P., Spronk, M., Willems, R. M., Toni, I., & Bekkering, H. (2008). Complementary systems for understanding action intentions. *Current Biology, 18*(6), 454–457.

Dewey, J. (1976–1988). In J. A. Boydston (Ed.), *The middle works: 1899–1924* (Vol. 9, p. 361). Carbondale: Southern Illinois University Press.

di Pellegrino, G., Fadiga, L., Fogassi, L., Gallese, V., & Rizzolatti, G. (1992). Understanding motor events: A neurophysiological study. *Experimental Brain Research, 91*(1), 176–180.

Emmorey, K., Grabowski, T., McCullough, S., Damasio, H., Ponto, L. L., Hichwa, R. D., et al. (2003). Neural systems underlying lexical retrieval for sign language. *Neuropsychologia, 41*(1), 85–95.

Emmorey, K., Grabowski, T., McCullough, S., Damasio, H., Ponto, L., Hichwa, R., et al. (2004). Motor-iconicity of sign language does not alter the neural systems underlying tool and action naming. *Brain and Language, 89*(1), 27–37.

Emmorey, K., Grabowski, T., McCullough, S., Ponto, L. L., Hichwa, R. D., & Damasio, H. (2005). The neural correlates of spatial language in English and American Sign Language: A PET study with hearing bilinguals. *NeuroImage, 24*(3), 832–840.

Fadiga, L., Fogassi, L., Pavesi, G., & Rizzolatti, G. (1995). Motor facilitation during action observation: A magnetic stimulation study. *Journal of Neurophysiology, 73*(6), 2608–2611.

Fadiga, L., Craighero, L., & D'Ausilio, A. (2009). Broca's area in language, action, and music. *Annals of the New York Academy of Sciences, 1169*, 448–458.

Fan, Y. T., Decety, J., Yang, C. Y., Liu, J. L., & Cheng, Y. (2010). Unbroken mirror neurons in autism spectrum disorders. *Journal of Child Psychology and Psychiatry, 51*(9), 981–988.

Feldman, J., & Narayanan, S. (2004). Embodied meaning in a neural theory of language. *Brain and Language, 89*(2), 385–392.

Ferrari, P. F., Gallese, V., Rizzolatti, G., & Fogassi, L. (2003). Mirror neurons responding to the observation of ingestive and communicative mouth actions in the monkey ventral premotor cortex. *European Journal of Neuroscience, 17*(8), 1703–1714.

Flaisch, T., Schupp, H. T., Renner, B., & Junghofer, M. (2009). Neural systems of visual attention responding to emotional gestures. *NeuroImage, 45*(4), 1339–1346.
Fogassi, L., Gallese, V., Fadiga, L., & Rizzolatti, G. (1998). Neurons responding to the sight of goal-directed hand/arm actions in the parietal area PF (7b) of the macaque monkey. *Social Neuroscience Abstract, 24*, 257.
Fogassi, L., Ferrari, P. F., Gesierich, B., Rozzi, S., Chersi, F., & Rizzolatti, G. (2005). Parietal lobe: From action organization to intention understanding. *Science, 308*(5722), 662–667.
Foundas, A. L., Macauley, B. L., Raymer, A. M., Maher, L. M., Heilman, K. M., & Rothi, L. J. (1995). Gesture laterality in aphasic and apraxic stroke patients. *Brain and Cognition, 29*(2), 204–213.
Franks, D. (2010). *Neurosociology: The nexus between neuroscience and social psychology*. New York: Springer Press New York. I heard recently it received the ASA subjection award for the best faculty Book.
Gallagher, H. L., & Frith, C. D. (2004). Dissociable neural pathways for the perception and recognition of expressive and instrumental gestures. *Neuropsychologia, 42*(13), 1725–1736.
Gallese, V., & Lakoff, G. (2005). The brain's concepts: The role of the sensory-motor system in conceptual knowledge. *The Multiple Functions of Sensory-Motor Representations, 22*(3/4), 455.
Gallese, V., Fadiga, L., Fogassi, L., & Rizzolatti, G. (1996). Action recognition in the premotor cortex. *Brain, 119*(Pt 2), 593–609.
Gallese, V., Fogassi, L., Fadiga, L., & Rizzolatti, G. (2002). Action representation and the inferior parietal lobule. In W. Prinz & B. Hommel (Eds.), *Attention & performance XIX. Common mechanisms in perception and action* (pp. 247–266). Oxford: Oxford University Press.
Gallese, V., Keysers, C., & Rizzolatti, G. (2004). A unifying view of the basis of social cognition. *Trends in Cognitive Sciences, 8*(9), 396–403.
Garrison, K. A., Winstein, C. J., & Aziz-Zadeh, L. (2010). The mirror neuron system: A neural substrate for methods in stroke rehabilitation. *Neurorehabilitation and Neural Repair, 24*(5), 404–412.
Gazzola, V., Aziz-Zadeh, L., & Keysers, C. (2006). Empathy and the somatotopic auditory mirror system in humans. *Current Biology, 16*(18), 1824–1829.
Gazzola, V., Rizzolatti, G., Wicker, B., & Keysers, C. (2007a). The anthropomorphic brain: The mirror neuron system responds to human and robotic actions. *NeuroImage, 35*(4), 1674–1684.
Gazzola, V., van der Worp, H., Mulder, T., Wicker, B., Rizzolatti, G., & Keysers, C. (2007b). Aplasics born without hands mirror the goal of hand actions with their feet. *Current Biology, 17*(14), 1235–1240.
Gentilucci, M., & Dalla Volta, R. (2008). Spoken language and arm gestures are controlled by the same motor control system. *Quarterly Journal of Experimental Psychology, 61*(6), 944–957.
Gentilucci, M., Bernardis, P., Crisi, G., & Dalla Volta, R. (2006). Repetitive transcranial magnetic stimulation of Broca's area affects verbal responses to gesture observation. *Journal of Cognitive Neuroscience, 18*(7), 1059–1074.
Geschwind, N. (1975). The apraxias: Neural mechanisms of disorders of learned movement. *American Scientist, 63*(2), 188–195.
Glenberg, A. M., & Kaschak, M. P. (2002). Grounding language in action. *Psychonomic Bulletin and Review, 9*(3), 558–565.
Greicius, M. (2008). Resting-state functional connectivity in neuropsychiatric disorders. *Current Opinion in Neurology, 21*(4), 424–430.
Habets, B., Kita, S., Shao, Z., Ozyurek, A., & Hagoort, P. (2010). The role of synchrony and ambiguity in speech-gesture integration during comprehension. *Journal of Cognitive Neuroscience, 23*(8), 1845–1854.
Hamilton, A. F., & Grafton, S. T. (2006). Goal representation in human anterior intraparietal sulcus. *Journal of Neuroscience, 26*(4), 1133–1137.
Hauser, M., & Wood, J. (2010). Evolving the capacity to understand actions, intentions, and goals. *Annual Review of Psychology, 61*(303–24), C1.
Heath, M., Roy, E. A., Black, S. E., & Westwood, D. A. (2001). Intransitive limb gestures and apraxia following unilateral stroke. *Journal of Clinical and Experimental Neuropsychology, 23*(5), 628–642.
Heilman, K. M., Schwartz, H. D., & Geschwind, N. (1975). Defective motor learning in ideomotor apraxia. *Neurology, 25*(11), 1018–1020.
Heilman, K. M., Rothi, L. J., & Valenstein, E. (1982). Two forms of ideomotor apraxia. *Neurology, 32*(4), 342–346.
Heiser, M., Iacoboni, M., Maeda, F., Marcus, J., & Mazziotta, J. C. (2003). The essential role of Broca's area in imitation. *European Journal of Neuroscience, 17*(5), 1123–1128.
Hesse, M. D., Sparing, R., & Fink, G. R. (2009). End or means–the "what" and "how" of observed intentional actions. *Journal of Cognitive Neuroscience, 21*(4), 776–790.
Heyes, C. (2010). Where do mirror neurons come from? *Neuroscience and Biobehavioral Reviews, 34*(4), 575–583.
Hickok, G. (2009). Eight problems for the mirror neuron theory of action understanding in monkeys and humans. *Journal of Cognitive Neuroscience, 21*(7), 1229–1243.
Hotz-Boendermaker, S., Funk, M., Summers, P., Brugger, P., Hepp-Reymond, M. C., Curt, A., et al. (2008). Preservation of motor programs in paraplegics as demonstrated by attempted and imagined foot movements. *NeuroImage, 39*(1), 383–394.
Iacoboni, M. (2005). Neural mechanisms of imitation. *Current Opinion in Neurobiology, 15*(6), 632–637.

Iacoboni, M. (2008). *Mirroring People: The new science of how we connect with others*. New York: Farrar, Struaa and Giroux.
Iacoboni, M., & Dapretto, M. (2006). The mirror neuron system and the consequences of its dysfunction. *Nature Reviews Neuroscience, 7*(12), 942–951.
Iacoboni, M., & Mazziotta, J. C. (2007). Mirror neuron system: Basic findings and clinical applications. *Annals of Neurology, 62*(3), 213–218.
Iacoboni, M., Woods, R. P., Brass, M., Bekkering, H., Mazziotta, J. C., & Rizzolatti, G. (1999). Cortical mechanisms of human imitation. *Science, 286*(5449), 2526–2528.
Iacoboni, M., Molnar-Szakacs, I., Gallese, V., Buccino, G., Mazziotta, J. C., & Rizzolatti, G. (2005). Grasping the intentions of others with one's own mirror neuron system. *PLoS Biology, 3*(3), e79.
Jellema, T., Baker, C. I., Wicker, B., & Perrett, D. I. (2000). Neural representation for the perception of the intentionality of actions. *Brain and Cognition, 44*(2), 280–302.
Kaplan, J. T., & Iacoboni, M. (2006). Getting a grip on other minds: Mirror neurons, intention understanding, and cognitive empathy. *Social Neuroscience, 1*(3–4), 175–183.
Kaplan, J. T., Aziz-Zadeh, L., Uddin, L. Q., & Iacoboni, M. (2008). The self across the senses: An fMRI study of self-face and self-voice recognition. *Social Cognitive and Affective Neuroscience, 3*(3), 218–223.
Keysers, C., Wicker, B., Gazzola, V., Anton, J. L., Fogassi, L., & Gallese, V. (2004). A touching sight: SII/PV activation during the observation and experience of touch. *Neuron, 42*(2), 335–346.
Keysers, C., Kaas, J. H., & Gazzola, V. (2010). Somatosensation in social perception. *Nature Reviews Neuroscience, 11*(6), 417–428.
Kilner, J. M., & Frith, C. D. (2008). Action observation: Inferring intentions without mirror neurons. *Current Biology, 18*(1), R32–R33.
Kohler, E., Keysers, C., Umilta, M. A., Fogassi, L., Gallese, V., & Rizzolatti, G. (2002). Hearing sounds, understanding actions: Action representation in mirror neurons. *Science, 297*(5582), 846–848.
Koski, L., Wohlschlager, A., Bekkering, H., Woods, R. P., Dubeau, M. C., Mazziotta, J. C., et al. (2002). Modulation of motor and premotor activity during imitation of target-directed actions. *Cerebral Cortex, 12*(8), 847–855.
Lakoff, G., & Johnson, M. (1999). *Philosophy in the flesh: The embodied mind and its challenge to Western thought*. New York: Basic Books.
Liepelt, R., Von Cramon, D. Y., & Brass, M. (2008). How do we infer others' goals from non-stereotypic actions? The outcome of context-sensitive inferential processing in right inferior parietal and posterior temporal cortex. *NeuroImage, 43*(4), 784–792.
Liew, S.-L., Han, S., & Aziz-Zadeh, L. (2010). Familiarity modulates mirror neuron and mentalizing regions during intention understanding. *Human Brain Mapping, 32*, 1986–1997.
Liew, S. -L., Seckin, M., & Aziz-Zadeh, L. (2011). The effects of experience on the observation of novel effectors. In *Proceedings of the Cognitive Neuroscience Society annual meeting*, San Francisco.
Lynall, M. E., Bassett, D. S., Kerwin, R., McKenna, P. J., Kitzbichler, M., Muller, U., et al. (2010). Functional connectivity and brain networks in schizophrenia. *Journal of Neuroscience, 30*(28), 9477–9487.
MacSweeney, M., Woll, B., Campbell, R., Calvert, G. A., McGuire, P. K., David, A. S., et al. (2002a). Neural correlates of British sign language comprehension: Spatial processing demands of topographic language. *Journal of Cognitive Neuroscience, 14*(7), 1064–1075.
MacSweeney, M., Woll, B., Campbell, R., McGuire, P. K., David, A. S., Williams, S. C., et al. (2002b). Neural systems underlying British Sign Language and audio-visual English processing in native users. *Brain, 125*(Pt 7), 1583–1593.
MacSweeney, M., Campbell, R., Woll, B., Giampietro, V., David, A. S., McGuire, P. K., et al. (2004). Dissociating linguistic and nonlinguistic gestural communication in the brain. *NeuroImage, 22*(4), 1605–1618.
McNeill, D. (1992). *Hand and mind: What gestures reveal about thought*. Chicago: University of Chicago Press.
McNeill, D. (2005). *Gesture and thought*. Chicago: University of Chicago Press.
Mehrabian, A. (1996). *Manual for the balanced emotional empathy scale (BEES)*. Available from Albert Mehrabian, 1130, Monterey, USA.
Meltzoff, A. N., & Prinz, W. (2002). *The imitative mind: Development, evolution, and brain bases*. Cambridge: Cambridge University Press.
Molnar-Szakacs, I., Wu, A. D., Robles, F. J., & Iacoboni, M. (2007). Do you see what I mean? Corticospinal excitability during observation of culture-specific gestures. *PLoS One, 2*(7), e626.
Mukherjee, P., Bahn, M. M., McKinstry, R. C., Shimony, J. S., Cull, T. S., Akbudak, E., et al. (2000). Differences between gray matter and white matter water diffusion in stroke: Diffusion-tensor MR imaging in 12 patients. *Radiology, 215*(1), 211–220.
Newman-Norlund, R., van Schie, H. T., van Hoek, M. E., Cuijpers, R. H., & Bekkering, H. (2010). The role of inferior frontal and parietal areas in differentiating meaningful and meaningless object-directed actions. *Brain Research, 1315*, 63–74.
Ozyurek, A., Willems, R. M., Kita, S., & Hagoort, P. (2007). On-line integration of semantic information from speech and gesture: Insights from event-related brain potentials. *Journal of Cognitive Neuroscience, 19*(4), 605–616.

Paukner, A., Suomi, S. J., Visalberghi, E., & Ferrari, P. F. (2009). Capuchin monkeys display affiliation toward humans who imitate them. *Science, 325*(5942), 880.

Pazzaglia, M., Pizzamiglio, L., Pes, E., & Aglioti, S. M. (2008a). The sound of actions in apraxia. *Current Biology, 18*(22), 1766–1772.

Pazzaglia, M., Smania, N., Corato, E., & Aglioti, S. M. (2008b). Neural underpinnings of gesture discrimination in patients with limb apraxia. *Journal of Neuroscience, 28*(12), 3030–3041.

Perkins, T., Stokes, M., McGillivray, J., & Bittar, R. (2010). Mirror neuron dysfunction in autism spectrum disorders. *Journal of Clinical Neuroscience, 17*(10), 1239–1243.

Perrett, D. I., Harries, M. H., Bevan, R., Thomas, S., Benson, P. J., Mistlin, A. J., et al. (1989). Frameworks of analysis for the neural representation of animate objects and actions. *Journal of Experimental Biology, 146*, 87–113.

Perrett, D. I., Mistlin, A. J., Harries, M. H., & Chitty, A. J. (1990). Understanding the visual appearance and consequence of hand actions. In M. A. Goodale (Ed.), *Vision and action: The control of grasping* (pp. 163–342). Norwood: Ablex.

Phelps, E. A., & Thomas, L. A. (2003). Race, behavior, and the brain: The role of neuroimaging in understanding complex social behaviors. *Political Psychology, 24*(4), 747–758.

Poizner, H., Klima, E. S., & Bellugi, U. (1987). *What the hands reveal about the brain*. Cambridge, MA: MIT Press.

Pulvermuller, F. (2005). Brain mechanisms linking language and action. *Nature Reviews Neuroscience, 6*(7), 576–582.

Pulvermuller, F., & Hauk, O. (2006). Category-specific conceptual processing of color and form in left fronto-temporal cortex. *Cerebral Cortex, 16*(8), 1193.

Pulvermuller, F., Hauk, O., Nikulin, V. V., & Ilmoniemi, R. J. (2005a). Functional links between motor and language systems. *European Journal of Neuroscience, 21*(3), 793–797.

Pulvermuller, F., Shtyrov, Y., & Ilmoniemi, R. (2005b). Brain signatures of meaning access in action word recognition. *Journal of Cognitive Neuroscience, 17*(6), 884–892.

Rizzolatti, G., & Arbib, M. A. (1998). Language within our grasp. *Trends in Neurosciences, 21*(5), 188–194.

Rizzolatti, G., & Craighero, L. (2004). The mirror-neuron system. *Annual Review of Neuroscience, 27*, 169–192.

Rizzolatti, G., Fadiga, L., Gallese, V., & Fogassi, L. (1996a). Premotor cortex and the recognition of motor actions. *Cognitive Brain Research, 3*(2), 131–141.

Rizzolatti, G., Fadiga, L., Matelli, M., Bettinardi, V., Paulesu, E., Perani, D., et al. (1996b). Localization of grasp representations in humans by PET: 1. Observation versus execution. *Experimental Brain Research, 111*(2), 246–252.

Romani, M., Cesari, P., Urgesi, C., Facchini, S., & Aglioti, S. M. (2005). Motor facilitation of the human cortico-spinal system during observation of bio-mechanically impossible movements. *NeuroImage, 26*(3), 755–763.

Rothi, L. J., & Heilman, K. M. (1984). Acquisition and retention of gestures by apraxic patients. *Brain and Cognition, 3*(4), 426–437.

Rothi, L. J., Heilman, K. M., & Watson, R. T. (1985). Pantomime comprehension and ideomotor apraxia. *Journal of Neurology, Neurosurgery & Psychiatry, 48*(3), 207–210.

Saxe, R. (2005). Against simulation: The argument from error. *Trends in Cognitive Science, 9*(4), 174–179.

Schippers, M. B., Gazzola, V., Goebel, R., & Keysers, C. (2009). Playing charades in the fMRI: Are mirror and/or mentalizing areas involved in gestural communication? *PLoS One, 4*(8), e6801.

Schippers, M. B., Roebroeck, A., Renken, R., Nanetti, L., & Keysers, C. (2010). Mapping the information flow from one brain to another during gestural communication. *Proceedings of the National Academy of Sciences of the United States of America, 107*(20), 9388–9393.

Schulte-Rüther, M., Markowitsch, H. J., Fink, G. R., & Piefke, M. (2007). Mirror neuron and theory of mind mechanisms involved in face-to-face interactions: A functional magnetic resonance imaging approach to empathy. *Journal of Cognitive Neuroscience, 19*(8), 1354–1372.

Serino, A., Giovagnoli, G., & Ladavas, E. (2009). I feel what you feel if you are similar to me. *PLoS One, 4*(3), e4930.

Shamay-Tsoory, S. G., Aharon-Peretz, J., & Perry, D. (2009). Two systems for empathy: A double dissociation between emotional and cognitive empathy in inferior frontal gyrus versus ventromedial prefrontal lesions. *Brain, 132*(3), 417–427.

Singer, T., & Frith, C. (2005). The painful side of empathy. *Nature Neuroscience, 8*(7), 845–846.

Singer, T., Seymour, B., O'Doherty, J., Kaube, H., Dolan, R. J., & Frith, C. D. (2004). Empathy for pain involves the affective but not sensory components of pain. *Science, 303*(5661), 1157–1162.

Singer, T., Seymour, B., O'Doherty, J. P., Stephan, K. E., Dolan, R. J., & Frith, C. D. (2006). Empathic neural responses are modulated by the perceived fairness of others. *Nature, 439*(7075), 466.

Skipper, J. I., Goldin-Meadow, S., Nusbaum, H. C., & Small, S. L. (2009). Gestures orchestrate brain networks for language understanding. *Current Biology, 19*(8), 661–667.

Spunt, R. P., Satpute, A. B., & Lieberman, M. D. (2011). Identifying the what, why, and how of an observed action: An fMRI study of mentalizing and mechanizing during action observation. *Journal of Cognitive Neuroscience, 23*(1), 63–74.

Straube, B., Green, A., Weis, S., Chatterjee, A., & Kircher, T. (2009). Memory effects of speech and gesture binding: Cortical and hippocampal activation in relation to subsequent memory performance. *Journal of Cognitive Neuroscience, 21*(4), 821–836.

Straube, B., Green, A., Jansen, A., Chatterjee, A., & Kircher, T. (2010). Social cues, mentalizing and the neural processing of speech accompanied by gestures. *Neuropsychologia, 48*(2), 382–393.

Tettamanti, M., Buccino, G., Saccuman, M. C., Gallese, V., Danna, M., Scifo, P., et al. (2005). Listening to action-related sentences activates fronto-parietal motor circuits. *Journal of Cognitive Neuroscience, 17*(2), 273–281.

Uddin, L. Q., Molnar-Szakacs, I., Zaidel, E., & Iacoboni, M. (2006). rTMS to the right inferior parietal lobule disrupts self-other discrimination. *Social Cognitive and Affective Neuroscience, 1*(1), 65.

Umilta, M. A., Kohler, E., Gallese, V., Fogassi, L., Fadiga, L., Keysers, C., et al. (2001). I know what you are doing: A neurophysiological study. *Neuron, 31*(1), 155–165.

Villarreal, M., Fridman, E. A., Amengual, A., Falasco, G., Gerscovich, E. R., Ulloa, E. R., et al. (2008). The neural substrate of gesture recognition. *Neuropsychologia, 46*(9), 2371–2382.

Vogt, S., Buccino, G., Wohlschläger, A. M., Canessa, N., Shah, N. J., Zilles, K., et al. (2007). Prefrontal involvement in imitation learning of hand actions: Effects of practice and expertise. *NeuroImage, 37*(4), 1371–1383.

Wicker, B., Keysers, C., Plailly, J., Royet, J. P., Gallese, V., & Rizzolatti, G. (2003). Both of us disgusted in My insula: The common neural basis of seeing and feeling disgust. *Neuron, 40*(3), 655–664.

Willems, R. M., & Hagoort, P. (2007). Neural evidence for the interplay between language, gesture, and action: A review. *Brain and Language, 101*(3), 278–289.

Willems, R. M., Ozyurek, A., & Hagoort, P. (2007). When language meets action: The neural integration of gesture and speech. *Cerebral Cortex, 17*(10), 2322–2333.

Willems, R. M., Toni, I., Hagoort, P., & Casasanto, D. (2009). Body-specific motor imagery of hand actions: Neural evidence from right- and left-handers. *Frontiers in Human Neuroscience, 3*, 39.

Willems, R. M., Hagoort, P., & Casasanto, D. (2010). Body-specific representations of action verbs: Neural evidence from right- and left-handers. *Psychological Science, 21*(1), 67–74.

Xu, J., Gannon, P. J., Emmorey, K., Smith, J. F., & Braun, A. R. (2009a). Symbolic gestures and spoken language are processed by a common neural system. *Proceedings of the National Academy of Sciences of the United States of America, 106*(49), 20664–20669.

Xu, X., Zuo, X., Wang, X., & Han, S. (2009b). Do you feel my pain? Racial group membership modulates empathic neural responses. *Journal of Neuroscience, 29*(26), 8525–8529.

Zaki, J., Weber, J., Bolger, N., & Ochsner, K. (2009). The neural bases of empathic accuracy. *Proceedings of the National Academy of Sciences, 106*, 11382–11387.

Chapter 15
A Neurosociological Model of Weberian, Instrumental Rationality: Its Cognitive, Conative, and Neurobiological Foundations

Warren D. TenHouten

Introduction

This chapter presents a neurosociological model of Weberian, instrumental rationality that identifies and explains its accompanying social relations and its predominant underlying neurocognitive processes. It adopts the meaning and usage of the term "instrumental rationality" provided by twentieth-century German sociologist, Max Weber, the preeminent scholar of rationality. Weber considered instrumental rationality to principally involve calculation and planning. We shall see that these cognitive activities can be generalized to the modal neuroscientific concepts of logical analysis and central-executive functioning and their interactions. Logical analysis includes propositional logic, linear and numerical thinking, linear time consciousness, and the quantitative assessment of value. Central-executive functioning, the locus of planning, involves motivation and conation. This chapter additionally discusses Weber's fourfold typology of the main forms of rationality, briefly considers their sociorelational and neurocognitive underpinnings, and explains both their difference from and commonality with instrumentally rational social action. This chapter's principal aims are to advance understanding of the neurobiological substrate, or underlying brain processes, of rational action, and to tie Weber's classic insights into rationality to recent empirical brain research concerning rational decision-making, evaluative processes, authority relations, and goal selection.

Rationality is a dominant theme in Weber's (1904–1905, 1905, 1918, 1921) economic, political, and religious writings, where he provided many different, and occasionally conflicting, definitions (Lukes 1967: 207; Eisen 1978; Kronman 1983; Brubaker 1984: 14–15). Weber envisioned a great historical struggle to establish a rational, yet just, social order, as premodern societies shed their irrational foundations. For Weber, this process of rationalization involved two main phases. First, through gradual sociohistorical development, traditional societies' belief systems shifted emphasis from supernatural and religious percepts to science and precise calculation. Second, newly dominant bureaucratic forms of organization emerged, involving impersonal organizational-level decision-making, universally applicable rules, technocratic skills, and the use of means–ends rationality.

Weber's (1904–1905) preoccupation with rationality and rationalization led him to conduct a protracted study of rationality and irrationality in the world's religions and to articulate his famous

W.D. TenHouten (✉)
Department of Socialogy, University of California, Los Angeles,
CA 90095, USA
e-mail: tenhouten@soc.ucla.edu

D.D. Franks and J.H. Turner (eds.), *Handbook of Neurosociology*, Handbooks of Sociology and Social Research, DOI 10.1007/978-94-007-4473-8_15, © Springer Science+Business Media B.V. 2013

insights into what he considered the special relationship between the "Protestant ethic" and the development of modern capitalism. Weber saw modern capitalism as emerging synchronistically with Calvinist Protestantism—not in causal interdependence but rather in an "elective affinity,"[1] wherein these two contemporaneous historical processes developed alongside, and synergistically reinforced, one another. For Weber, instrumental rationality is a key component of modern capitalism, and the closely related formal rationality a crucial feature of modernity. In particular, Weber characterized bureaucratic administration as an institutionalization of formal, rule-bound rationality, which he considered indispensable to the efficient functioning of modern society. At the same time, Weber considered bureaucratization potentially dehumanizing, even malevolent, in its impersonality and possible elevation of economic efficiency and profit maximization over human values and social justice. With good reason, Weber was genuinely alarmed by the prospect that an institutionalized, morally vacuous or pernicious instrumental, and especially, formal, rationality might be used to justify morally questionable ends. Such rationality could potentially strip individuals of their freedom and dignity, placing them, metaphorically, in an "iron cage" (ein stahlhartes Gehäuse) of despair.

For seventeenth-century natural philosophers, the term "instrumental rationality" narrowly denoted the use of mathematical modeling together with scientific instruments to measure variables in scientific inquiry. Weber considerably transformed this notion and applied it to the social sphere. Weber (1921/1978: 24) described instrumentally rational (zweckrational) social action as "…determined by expectations as to the behavior of objects in the environment, including other human beings." Weber further noted that "these expectations are used as 'conditions' or 'means' for the attainment of the actor's own rationally pursued and calculated ends," and that "the end, the means, and the secondary results are all rationally taken into account and weighed" (ibid., p. 26). Weber equated instrumental rationality with "deliberate planning" (ibid., 63) and stated that instrumentally rational action involves the evaluation and selection of the most expedient and efficient means to achieve an anticipated and planned-for-future objective (ibid., 86). Weber emphasized that the methodical attainment of practical ends is accomplished by means of precise calculation of adequate means that are consistent with one's chosen objective, and that "the orientation of action [is] wholly to the rational achievement of ends without relation to fundamental values" (ibid., 26). Weber's emphasis on identifying, evaluating, or choosing the means to some end continues to inform our understandings of instrumental action and instrumental rationality. In contemporary social theory, instrumental rationality denotes the individual's selection of the most efficient and expedient means to achieve a desired, planned end, without regard for affectual or value considerations, and with maximum certitude that the chosen actions are consistent with the desired objective.

Weber viewed all forms of rationality as efforts to order the world into meaningful regularities that can be translated into patterns of social action. In addition to his discussions of instrumental rationality and instrumentally rational social action, he identified four main kinds of rationality—practical, theoretical, substantive, and formal (see Kalberg 1980: 1145). Practical rationality closely aligns with instrumentally rational social action, as it refers to means–ends, calculative, goal-directed behavior in everyday life. Practical ends are attained through the careful weighting and increasingly precise calculation of available, adequate means (Weber 1918). The individual views, judges, and plans worldly activity in relation to purely egoistic interests and pragmatically accepts the realities of existing

[1] Rather than use a vocabulary of causality, Weber invoked Goethe's notion of "elective affinity" to explain the perceived relationship between the this-worldly asceticism and precepts of Calvinist Protestantism, and the ethos, business practices, and behaviors associated with and propitious to the emergence of modern industrial capitalism in nineteenth century Germany. "Elective affinity" refers to an hypothesized non-deterministic and synergistic interaction between two independent yet convergent historical processes. Weber elaborated on the key facets of this beneficent and mutually reinforcing dynamic interaction, through which, for Calvinists, hard work and the resultant capital accumulation represented evidence of one's salvation and membership in the Elect. Delacroix and Nielsen (2001) note the lack of empirical evidence for Weber's "common interpretation" that Protestantism's presence positively correlated with a country's early development of industrial capitalism.

conditions. This realistic orientation is unconcerned with impractical and transcendent values of "the beyond," or with the abstract explanations of science (Kalberg 1980: 1151). Besides its emphasis on calculation, practically rational behavior can require insight into the competitive or cooperative beliefs, intentions, and feelings of others. As with other forms of rationality that we will discuss, it can depend upon social cognition in order to discover others' intentions. This has motivated social neuroscientists' quest to find the neural underpinnings of social exchange and mutual cooperation (Montague et al. 2002; Fehr and Gächter 2000; Fehr 2009).

In contrast to practical rationality's egoistic emphasis, mundane orientation, pragmatism, and concreteness, theoretical rationality endeavors to uncover, comprehend, and even master the underlying laws of nature and the structure and meaning of reality. It involves the use of abstract concepts and theoretical frameworks. In early civilizations, theoretical rationality encompassed the efforts of sorcerers and priests, whose rituals, including astrology and numerology, aspired to tame nature and the putatively supernatural realm. At later points in history, these endeavors were motivated by the metaphysical needs and irrepressible quests of systematizers. Theoretical rationality yields insights into reality and can enable individuals to transcend the quotidian routine by supplying everyday life with an overall coherence and meaning. In modernity, theoretical rationality primarily denotes scientific theorizing, through which communities of individuals endeavor to discover knowledge and truth, often independently of any potential practical application. Because theoretical rationality emphasizes calculation and planning, it aligns with instrumental rationality and depends upon some of the same underlying neurocognitive processes. But particularly as it pertains to modern scientific inquiry, it differs from instrumental rationality, where means to ends are chosen according to principles of efficiency and expediency. In scientific inquiry, in contrast, research norms govern and constrain the choice of means, which are cooperatively established research methodologies and protocols. Theoretical rationality also partially aligns with substantive rationality (which, as we shall see below, endorses ultimate values or ultimate ends) because its goals effectively concern ultimate ends, namely, the uncovering, discovery, or creation of knowledge and truth. Indeed, an argument can be made that the research process itself, through which truth and knowledge are uncovered, constitutes an ultimate end.

"Substantive," or value, rationality, according to Weber (1921/1978: 85–6), involves decision-making and actions that acquire meaning and validity within the context of one's belief system, rather than on the basis of formal rules, expediency, or common-sense-based premises. Weber defined substantive rationality as uniquely involving "ultimate" values, or ultimate ends, which may or may not be economic in nature. Value rationality is thus not "goal oriented rational calculation with adequate technical means" but rather the endeavor to ensure that the actions, the choices, and the means/ends calculations one makes in order to attain a goal uphold one's core beliefs. Substantive rationality's principal emphasis, then, is on chosen ends, which also govern and constrain one's choice of means. In this sense, substantive rationality radically differs from instrumental rationality, where means represent the principal focus and where they are chosen according to how efficiently or expediently they lead to successful attainment of a desired objective.

Formal rationality, for Weber, is the process of using efficient, carefully calculated, and scientifically and technically valid means, or methods, to achieve chosen objectives. Weber believed that formal rationality became prominent with modern capitalism and its economic, legal, and scientific spheres, and particularly characterizes bureaucratic organizations. Within formal, especially bureaucratic, organizations, formal rationality denotes the process of selecting and implementing efficient, calculable, and scientifically valid means. This involves impartial, impersonal, and universally applicable action based on explicitly codified rules, regulations, and laws that govern the organization's decision-making and behavior, and necessitates the technical expertise critical to ensuring the maximum technical efficiency and calculability of selected means. Formal rationality thus principally focuses on the process through which means to ends are chosen and implemented, and closely aligns with instrumental rationality in its emphasis on precise calculation, efficiency and planning, and the endeavor to choose scientifically valid, efficient means. Weber fully understood that formal rationality and value rationality are in practice not mutually exclusive, insofar as the ends pursued by a "rational-legal" organization might include "vague justificatory ideas (culture, education)" (Rueschemeyer 1986: 63), or conflicting goals that must be adjudicated (Luhmann 1982: 26–7). Thus, while formal rationality aims to maximize the calculability of actions, such actions can be oriented to a range of possible substantive ends—from capitalistic enterprise, to technology and science, to religious pursuits. Weber thus acknowledged an ever-present tension between the formal rationality of the modern capitalist order and the value-rational approach of alternative orientations. Because of this relational tension, Weber saw irreconcilable, often conflicting, judgments of rationality and irrationality, and, consequently, limits to rationality as an organizing principle of modern capitalist society (Schlucter 1981: 131).

Both formal and instrumental rationality emphasize calculability and impersonality, are focused on technologies designed to increase efficiency, are oriented toward "success," and are, in principle, but sometimes not in practice, unconcerned with substantive considerations. Formal rationality can be considered tantamount to instrumental rationality when this is implemented within the constraints of universally applicable formal rules, and institutionalized within a formal organization, rather than as practiced by the individual, acting alone.[2]

The Cognitive Foundations of Instrumental Rationality: Calculation and Planning

Two forms of information processing, working together in productive interaction, namely, logical analysis and central-executive functioning, are the most modal cognitive foundations of instrumental rationality. It can be indirectly inferred that Weber invoked these two modes of cognition in his model of rationality. More specifically, the cognitive processes intrinsic to instrumentally rational action are "calculation" and "deliberate planning" (Weber 1904–1905, 1905, 1921). On the economic level, planning pertains to the identification, ranking, and selection of objectives; budgetary management; the evaluation of the means of achieving a productive purpose in terms of present and expected market situations; the expected and actual results of various courses of economic action; a comparison of the goods and assets controlled by an economic unit at a given time; and an ex ante estimate and ex post verification of receipts and expenditures (Weber 1921/1978: 86–7). According to Weber, calculation helps identify effective means (to achieve ends) using quantifiable, numerical measures of efficiency and productivity (ibid., 107). Where economic issues were important, instrumental rationality also involved arithmetic, accounting practices and market-based social relations. Weber considered a market economy and the use of money to be "the most rational means of orienting economic activity" (ibid., 86). He defined "economic action" (*Wirtschaften*) as "any peaceful exercise of an actor's control over resources which is in its main impulse oriented toward economic ends" (ibid., 63). "Rational economic action," he added, "requires *instrumental rationality* in this orientation, that is, *deliberate planning*" (ibid., italics added).

Weber saw modern capitalism as an economic system whose development requires a cognitive *integration of calculation and control*. This is attainable in three ways. First, the production process is calculable; so also is the legal and administrative environment, as this affects economic behavior. The calculability of the production process rests on the capitalist's legally assured control over the workplace, tools, sources of power, and other aspects of the means of production. Exact calculation and maximum efficiency also depend on the centralization of control over the processes of production,

[2] Weber (1921) saw the individual actor as the fundamental "atom" in all social and civilizational processes. Social collective entities such as states, businesses, neighborhoods, and families are not themselves capable of social action but exist simple as a result of the actions of individual persons.

distribution, and marketing. Second, calculability of the production process requires control of highly refined technical knowledge, based on the exact sciences and their precise rational foundations (Weber 1919/1970: 139), including mathematical and statistical modeling. And third, calculability of the production process depends on the uniquely Western system of "free" labor and on the "disciplined control" of workers by capitalists. Slave labor does not allow this calculability; only labor that is formally free yet economically compelled to sell its services on the market can be subjected to the discipline and control of "scientific management" (Weber 1921/1978: 338). Instrumental rationality—the hallmark of modern capitalist society for Weber—thus requires logical–analytic and central-executive reasoning, and a third level of cognitive functioning, namely, a productive interaction between these two kinds of reasoning.

In addition to these three ways in which the logical analysis of meaning is linked to planning, purpose, and intention, especially with respect to rationally organized formal–legal systems, there is a fourth requirement linking the logical analysis of meaning to planning, purpose, and intention. According to Weber's epistemology of values, values that do not inhere in facts and an individual cannot acquire his values by knowledge alone. Instead, values must be legislated into existence by imposition of human will on a morally neutral world. Thus, legal rationality requires that the logical analysis of meaning places human beings (as opposed to gods, oracles, and magical injunctions) at the center of things, "by assuming that facts have legal significance only insofar as they are related to purposive human attitudes" (Kronman 1983: 87). Only when logically interpreted meanings are linked to human plans, purposes, and intentions, and controlled by human intellect, can they be arranged in a fully systematic manner. In this connection, calculability and logical analysis, while not identical, are closely related and are both essential, for example, for the establishment of a rational–legal order meeting the ideals of comprehensiveness and organizational quality, which makes calculable the consequences of every social action. Such rationality, for example, in a legal system, Weber argued, can only emerge when legal thinking is based upon the logical analysis of meaning. As Kronman (ibid., 90) concludes, "There is...a connection between calculability and the logical analysis of meaning: the latter is the only type of legal thinking that leads, even potentially, to the systematic organization of the law and it is only through its systematization that the legal order can achieve a maximum degree of calculability." Weber only indirectly explained how a legal system can attain such comprehensiveness through the logical analysis of meaning, but he essentially argued that a collection of concrete, rules, decisions, and principles having a high degree of generality must be developed and deployed so that all situations can be handled in a comprehensive, gapless manner. Individual cases come to be interpreted according to one or more "principles," that is, "legal propositions" (Weber 1921/1978: 655).

The Sociorelational and Neurobiological Foundations of Instrumental Rationality

Market-Oriented and Authority-Ranked Social Relations and Rationality

In modern capitalist society, instrumental rationality primarily concerns formal social organization and typically involves the world of money and power, economic and political action, and market-priced and authority-ranked social relations (Fiske 1991). It is helpful to review the interrelationships between these elementary forms of sociality and general cognitive structures (TenHouten 1999a, b, 2005). Three propositions are required here: (1) participation in positively valenced market-priced social relations (MP+) involves a logical–analytic cognitive style; (2) positively valenced authority-ranked (AR+) social relations involve episodic, central-executive thought; and (3) simultaneous participation in MP+ and AR+ involves rational cognition, particularly the integration of logical–analytic and central-executive functioning. We can note, further, that ordinary–linear, episodic–futural, and rational forms of time consciousness were shown to be aspects of three larger cognitive structures: logical analysis, central-executive or episodic cognition, and the integration of the two (as shown empirically in TenHouten 2005: 205–6, 219). A description of the biological infrastructure of these three modal cognitive styles will now be provided, in order to provide criterion validation for this chapter's explanation of the neurosociological foundation of instrumental rationality. Specifically, instrumental rationality is hypothesized to depend on these three large-scale cognitive structures, namely, logical analysis, central-executive cognition, and their interaction.

Logical–Analytic Information-Processing, Utility, and the Quantitative Assessment of Value

Logical–analytic cognition has been identified as a general mode of cognition specialization dominating the brain's left cerebral hemisphere. It is based on the clear use of language and is essential to rational calculation. As conceptualized by neurosurgeon Joseph Bogen (1985), dual-brain theory centers on two main generalizations: hemispheric independence/integration and hemispheric specialization. Hemispheric independence/integration has far-ranging implications for the unity or duality of attention and for an understanding of the dynamics of hemispheric interaction (Bogen ibid.; Sperry 1984). Our focus here, however, is rather on hemispheric specialization. Inference about hemispheric specialization has been a two-step process: (1) two inventories of higher cognitive functioning have been found to be closely associated with the two sides of the brain, and (2) these left-hemisphere- and right-hemisphere-dependent functions have been interpreted as implying two distinct cognitive styles or modes of thought. The processing of the two hemispheres has been contrasted as "propositional" vs. "appositional" (Bogen 1969a, b), "logical–analytic" vs. "gestalt–synthetic," (Levi-Agresti and Sperry 1968), and "analytic" vs. "holistic" (Bradshaw and Nettleton 1983). These terminological conveniences and preferences are unimportant because they hardly capture the underlying complex realities.

Experimental validation of hemispheric specialization emerged through study of "split-brain" (technically, cerebral commissurotomy or corpus callosotomy) patients, whose cerebral hemispheres had been surgically divided as a treatment of last resort for severe, drug-refractory epileptic seizures. Bogen's (1969a) dramatic finding of dysgraphia and dyscopia in the left and right hands, respectively, of his and Dr. Vogel's callosotomy patients (illustrated in Fig. 15.1) provided crucial experimental evidence for this theory's claim of hemispheric specialization.

We now examine indicators of logical–analytic reasoning that are proposed to be essential to instrumental rationality. These include propositional logic, deductive reasoning, linear thinking, numerical cognition, linear time consciousness, and the quantitative assessment of value.

Propositional Logic

Weber made clear that in order to generate a legal system, a particular type of legal analysis is required. Legal analysis, a basis of formal rationality, was described as a two-step process. First, those aspects of events or states of affairs that are potentially relevant must be distinguished from irrelevant aspects of events or affairs. Second, the potentially relevant aspects of the facts must be assigned a specific, juristic meaning, in Kronman's (1983: 74) terms, "through the construction of legal rules or…general legal propositions." Weber made clear that such propositional reasoning was both logical and analytic. He wrote that legal analysis of meaning can only be identified as "the logical analysis of [their] meaning" (Weber 1921/1978: 657).

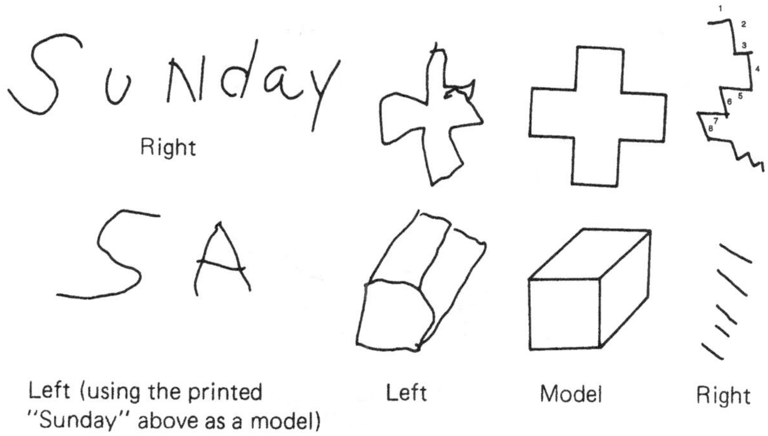

Fig. 15.1 Split-brain patients, shortly after surgery, were unable to write (dysgraphia) with the left hand, but not with the right hand, suggesting that writing is controlled by the left hemisphere. These patients also experienced a reduced capacity to copy figures (dyscopia) with the right hand, but not with the left, suggesting that this ability is controlled by the left hemisphere. To measure dysgraphia, a written model of the word "SUNDAY" was presented. The patient could only manage a crude "S A" with his left hand. His dyscopia is illustrated by his effort to copy a solid cube and a Greek cross. For the cube, it is as if he had merely used his left-hemisphere recognition of a number of connected lines and then made a visual gesture of stacking some of them up. For the cube, he showed no configurational ability to recognize the figure as a whole or gestalt. He copied it adequately with his left hand but failed with his right hand (His *right* hand showed no configurational ability. If we can assume that he started at the top (*line segment 1*) and then proceeded clockwise, he drew the first seven lines correctly; but at the end of line 7 (*the bottom line*), he made a wrong turn, leftward, instead of upward and to the right. After that, he drew the rest of the lines correctly. It is as if the left side of the cross fell off its axis between lines 7 and 8. It appears as if he grasps the figure as a sequence of line segments that turn either to the left or right, but made just one wrong turn. Thus, this would appear to be an effort at linear direction finding, but it most certainly was not an exercise in gestalt completion) (Source: Bogen 1969a, Fig. 5b, page 83)

Impressive evidence shows that the brain's left hemisphere is specialized for logical–analytic information processing and generates higher forms of cognitive activity pertaining to the production and comprehension of language. Its specializations include analysis of speech sounds, analysis of lexical and grammatical materials, arithmetic operations and other mathematical–analytic tasks, and analysis of causal relations between moving objects. To perform logical deductions, conduct cause-and-effect reasoning, and frame propositions, language is used in a logically consistent manner, free of contradiction.[3] The left hemisphere dominates phonetic processing of incoming auditory and visual language-related stimuli and language's syntactic and semantic processing (but not pragmatic) aspects. It also governs numerical cognition, calculation, quantitative reasoning, and logical inference. Logical reasoning is a learned, not innate, ability and is prone to error even among highly educated individuals (Johnson-Laird 1983; Evans 1989, 1998; Houdé et al. 2000).

[3] The law of noncontradiction states: It cannot be the case that something is both the case and not the case (Fogelin 2003, Chap. 1). To abandon this principle is to abandon reason itself, which is the very feature that sets human beings apart from all other animals. It is irrational to accept a proposition that is inconsistent within itself or with other propositions in an axiomatic system. Yet this assertion need not be followed at all cost. As a well-known example, it was not irrational for Gottlob Frege to accept the axioms of naïve set theory, even though they harbor Russell's paradox. For informal purposes, naïve set theory is easier to use than later, more complex formulations devised to avoid this paradox. Such local irrationality can be justified if it facilitates a computationally efficient or globally rational strategy, and can avoid "cognitive paralysis" of the kind that would result from taking Descartes' skeptical model of universal doubt serious (Cherniak 1986: 100–4). Descartes' rationalism was opposed to sense-based empiricism, as he rather advocated a rationalism in which truths could be discovered about the world by means of pure intellect, independent of the senses.

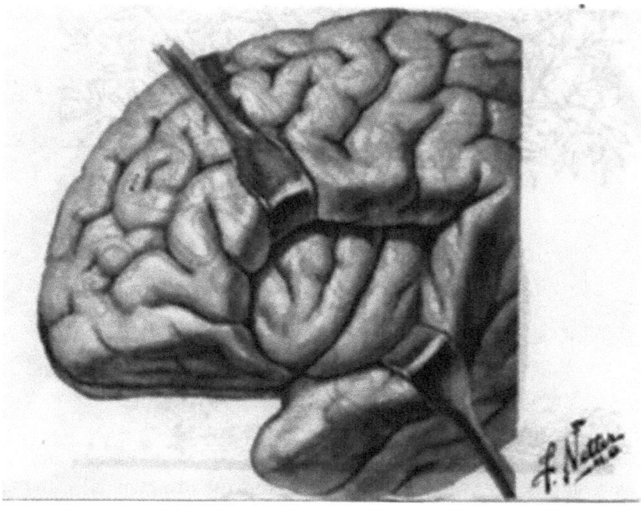

Fig. 15.2 The insular cortex (abbreviated as the *insula*), which lies deep in the brain's lateral surface. Parts of the frontal, temporal, and parietal lobes cover this structure and are known as opercula ("lids")

Even when these and other rules of logic are mastered, the attainment of rationality requires a capacity to evaluate the truth-value of propositions, a process includes an affective component. Spinoza (1677) conjectured that merely comprehending a proposition entails tacit acceptance of its being true, whereas disbelief requires the same tacit acceptance, followed up by a mental process of rejection. Spinoza's conjecture has been validated by a recent neuroimaging study by Harris et al. (2008). Fourteen adults were presented with factual statements from various content domains, which were true, false, or uncertain. Examples were: "Most people have 10 fingers and 10 toes," "Eagles are common pets," and "The Dow Jones Industrial Average rose 1.2% last Tuesday," respectively. The main finding was the contrasting levels of functional magnetic resonance imaging (fMRI) brain activation in the ventromedial prefrontal cortex (VMPFC) for propositions judged true and false.[4] This signal was greater during belief trials than during disbelief trials. The inferred involvement of the VMPFC in belief processes suggests there is an anatomical linkage between the purely cognitive aspects of belief and emotion/reward. Even judging the truth-value of emotionally neutral propositions activates brain regions that are strongly connected to the limbic system. These results make sense because the VMPFC has strong reciprocal connections with the limbic system, the basal ganglia, and parietal association cortex. This area of the frontal lobes is instrumental in linking factual knowledge with relevant emotional associations, in changing behavior in response to changing prospects for rewards, and in choosing between goal-directed courses of actions. It is also activated by reasoning tasks with high emotional salience (Northoff et al. 2004). In the Harris et al. study, both uncertain and disbelieved statements, compared to believed statements, were characterized by bilateral activation of the anterior insular, an area below the frontal lobes that is responsible

[4] Neuroscientific evidence shows that right ventromedial prefrontal (VMPFC) damage impairs reasoning, decision-making, emotion, and feeling (Damasio 1994). This right-frontal area helps put the mind on a "logical track." The VMPFC is involved in emotional processes that involve moral judgments (Koenigs et al. 2007). It also integrates emotional feelings concerning costs and benefits, whether one's choices involve economic goods or "noneconomic" social preferences such as showing concern for the welfare of those worse off, acting altruistically and with fairness, a desire for social equality, and kindness, or engaging in reciprocally cooperative behavior (see Fehr 2009: 216–7). Thus, the right side of the prefrontal cortex can be expected to become involved in the (substantive) rationality of decision-making and action where values, beliefs, and ethical standards are involved.

for the sensation of taste and pain and is involved in empathy for the pain of others (see Singer 2009). The insular regions, shown in Fig. 15.2, together with the left frontal operculum (also activated in the "disbelief–belief" contrast), appear to mediate negatively valenced feelings such as disgust. The opposite primary emotions acceptance and disgust are both closely related to oral stimuli, and disgust is associated with unpleasant odors. In a sense, to adjudge a statement false is to assess the statement as unpleasant and unfit for incorporation into one's belief system. Thus, just as Spinoza conjectured, evaluating a statement as not true involves a process of rejection, which is both cognitive and emotional. Statements seen as "true" are evaluated more quickly than statements judged "false" or "uncertain." To merely comprehend a statement is to tacitly accept it as true, but disbelief requires additional processing as the statement is adjudged false and then rejected. Thus, while "propositional" reasoning was seen by Bogen (1969a, 1977) as a descriptive term for the mode of information processing of the left hemisphere, even this putatively "pure" process involves the two opposite primary emotions, acceptance and disgust (Plutchik 1962; TenHouten 2007, in press). Those statements which are first tacitly accepted, then rejected, are processed by the brain in much the same way as an unpleasant odor. Harris et al. (2008: 141) conclude: "[T]ruth may be beauty, and beauty truth, in more than a metaphorical sense, and false propositions may actually disgust us."

Spatial and Language-Based Modes of Deductive Reasoning

Intuitively, deductive reasoning appears to require the left hemisphere's logical–analytic capabilities. However, evidence suggests that problems solvable through logical analysis can also be solved through the use of spatial reasoning, which takes place in the parietal lobes of both the left and right hemispheres. Two distinct theories addressing this issue of hemispheric specialization have dominated the cognitive science literature. The mental logic theory assumes this hemispheric specialization, but the mental models theory does not. They differ in the knowledge they draw upon, the brain mechanisms they invoke, the mental representations they postulate, and the neuroanatomical predictions they make (Goel et al. 2000: 504).

The *mental models* theory (Johnson-Laird 1983; Byrne and Johnson-Laird 1991) hypothesizes that the mapping of the world to model involves spatial encoding of the structural properties that sentences address, and that this activates (especially the right hemisphere's) parietal lobe's spatial reasoning abilities. According to this model, logical inferences involve knowledge of logical terms (e.g., "some," "none," "and"), which are used to construct and search alternative scenarios. These scenarios are based on internal representations of the structural properties of the deductive argument, through manipulation of spatial relations (possibly represented as Venn diagrams), using a brain network involving the right parietal lobe (according to early proponents of this theory).

The alternative, *mental logic,* or *syntactic* theory (Braine 1978; Rips 1994), holds that language processes underlie human reasoning. According to this theory, the left hemisphere's rule-governed syntactic processes internally represent linguistic strings, in which premises are stated and conclusions commensurate with rules of logic, inferred. Here, the subject understands the meaning of language's logical terms and uses this knowledge to deduce conclusions from premises. This theory predicts that left-lateralized neuroanatomical mechanisms of semantics and syntax undergird deductive reasoning.

These spatial and linguistic models of the cognitive processes underlying deductive cognition have been widely portrayed as competing. Yet deductive cognition possibly occurs through either spatial or linguistic processing. Recent neuroimaging studies have transcended this debate. Goel et al. (1998) appeared to have determined whether deductive reasoning is inherently spatial or sentential, through a positron-emission tomography (PET) study of 12 subjects working on deductive, logical reasoning tasks. Subjects were presented with three types of deductive argument forms: categorical syllogisms, three-term spatial relations items, and three-term nonspatial relational arguments. In these deductive reason conditions, subjects determined whether the third sentence was entailed by the first two sentences. Regional cerebral blood flow patterns were analyzed using [$^{15}O\ H_2O$] PET imaging. These three kinds of stimuli each activated a distributed computational network confined to the left hemisphere, which were similar to each other and showed a distributional network of activation reported in earlier studies. The activated areas included the left inferior frontal gyrus, a region of the left lateral inferior temporal gyrus and superior temporal gyrus, and a portion of the left cingulate gyrus. There was not significant right-hemisphere or parietal activation, suggesting that spatial regions were not involved in reasoning about linguistically presented spatial relations. Despite the small sample size, the theoretical implication of this study seemed clear. The spatial model appeared to have failed. Deductive reasoning, whether or not it was spatially evocative, had been carried out by a propositional, language-based, left-hemisphere-dependent capability of the human mind.

In a subsequent fMRI study, Goel and Dolan (2004) scanned the brains of 16 subjects as they performed inductive and deductive reasoning tasks. They found that engagement in both types of logical reasoning activated a neural network comprising left lateral prefrontal cortex, but also bilaterally activated dorsal frontal, parietal, and occipital cortices (a complex network that had been identified in many other studies: see, e.g., Goel et al. 2000; Goel and Dolan 2003; Knauff et al. 2002; Kroger et al. 2002). Both deductive and inductive reasoning thus shared a wide computational network, but there was a significant difference between the two categories of thought: for the deductive, but not inductive, tasks, the left inferior frontal gyrus (Broca's area) was involved; for inductive, but not deductive, tasks, the left dorsolateral prefrontal gyrus was activated (along with right superior occipital gyrus). Goel and Dolan (2004: B119–20) explain why Broca's area, the locus for the production of inner speech and speech, is more involved in deduction than induction: Broca's area is part of the phonological loop of working memory, and deductive reason requires more working memory than does inductive reasoning (Gilhooly et al. 1993). Broca's area is also involved in syntax and the logical forms encoded within syntactic structures. Thus, the heavy involvement of Broca's area in deductive reasoning might well be a function of its requirement for syntactic processing and working memory. Inductive reasoning, in contrast, draws on background knowledge rather than logical form, for it is well known that the *dorsolateral* prefrontal cortex is involved in the generation and evaluation of the kind of hypotheses dealt with in everyday life, which are largely of an inductive nature (Grafman 2002), whereas the *lateral* prefrontal cortex is more involved in logical, deductive reasoning (Goel et al. 1998, 2000; Goel and Dolan 2003, 2004: B120).

Other studies suggest the presence of both spatial- and language-based models of deductive reasoning, which are utilized by subjects under different circumstances. Goel et al. (2000) demonstrated this possibility in an fMRI study of 11 right-handed normal subjects. Here, the mental logic model's predicted language system (activating Broca's and Wernicke's areas of the left frontal and left temporal lobes) was found for 30 syllogisms with semantic content (e.g., "All dogs are pets; all pets are furry; therefore, all dogs are furry."). A comparison set of 30 logically equivalent content-free syllogisms (e.g., "All A are B; all B are C; therefore, All A are C.") instead activated a network incorporating *bilateral* occipital, left parietal, bilateral dorsal frontal, and other bilateral frontal areas. Additional studies have uncovered a second pattern of activation involved in the internal representation and manipulation of spatial information (see Laeng 1994), similar to the activation pattern of certain kinds of mathematical reasoning involving approximated numerical quantities (Dehaene et al. 1999). For example, Noveck et al. (2004) carried out a PET-imaging study of 16 subjects given conditional propositional reasoning tasks with arbitrary, nonmeaningful content (with emphasis on modus ponens ("If *P*, then *Q*; *P*, therefore *Q*.") and modus tollens ("If *P*, then *Q*; not *Q*, therefore not *P*.")). Replicating the results of the Goel and Dolan (2003) study, they found that a left-lateralized parietal–frontal network was activated by two conditional forms—modus ponens and modus tollens, with the highest level of activation occurring during the more difficult tasks. In adults, neuroimaging studies have shown that the inferior parietal cortex is active during most number processing and the left basal ganglia are active during multiplication (Kunzig 1997). For two kinds of deductive reasoning, relational and conditional, Knauff et al. (2002) found surprising results consistent with the spatial model theory, which further complicates the picture of brain mechanisms underlying deductive reasoning. Knauff et al. found evidence of a bilateral occipitoparietal–frontal network (along with activation of parts of the prefrontal cortex, the cingulate gyrus, the precuneus, and the visual association cortex). They interpreted this as corroborative evidence for the mental model theory of reason. Their results suggested that subjects spatially represented sentences (especially if they are abstract and devoid of content) possibly by using internal representations of Venn diagrams or Euler circles or, more generally, by using spatial models, as earlier predicted by the Johnson-Laird (1983) spatial model.

There thus appear to be two dissociable networks for reasoning, sharing common circuits in basal ganglia nuclei and in the cerebellum[5] and which involve the lingual gyri (more involved in processing letter length for longer content words) and the left prefrontal cortical regions. These results show that the presence of content engages the left hemisphere's language system in reasoning, while the absence of content (for identical tasks) engages the spatial system.

The functional neuroanatomy of reasoning is still poorly understood, and brain imaging studies have indicated that there are both spatial and linguistic models of deductive reasoning (Parsons and Osherson 2001: 954). Problems of logical inference can be solved either by a language-based semantic and syntactical method or through spatial reasoning. The two models need not compete, because these two methods of problem-solving are fundamentally complementary. A larger complementarity exists between the logical–analytic and gestalt–synthetic models of information processing partially lateralized to the left and right sides of the brain.

Linear and Numerical Thinking

Weber (1921) identified computational reasoning as a key component of instrumental rationality. Computational reasoning is linear, numerical, and sequential. It typically involves the brain's left hemisphere's more general mode of information processing, which has been variously described as logical–analytic, propositional, and, occasionally, "linear" (e.g., Rotenberg 1995; Schore 1997). Linear thinking implies a spatial conceptualization, a line existing in a space, and a single dimension that can be equated with an infinite set of numbers. Galton (1880) hypothesized the existence of a literal representation of numbers in the brain, a stable linear space, which expresses the very essence of the meaning of number. Since Galton, the intuition that the mental representation of numbers contains a series of visuospatial properties has found significant empirical support (reviewed by de Hevia et al. 2008). Even monkeys possess a parietal–frontal network for visual numerical information (Dehaene et al. 1998). Göbel et al. (2006) report that left parietal repetitive transcranial magnetic stimulation (rTMS) distorts the mental number line and temporarily interferes with cognitive functioning. This suggests that the number line is a capability of the stimulated left parietal lobe, while spatial representations, akin to a mental number line, are involved in basic numerical processing tasks.

The studies reviewed above point to a left-parietal–frontal network for logical tasks (Goel and Dolan 2004; Noveck et al. 2004). Dehaene (1997) has hypothesized that the internal representation of the number line might have a literal spatial representation that is "hard-wired" into the brain, such that each number corresponds to a dedicated cluster of neurons, arranged in the same orders as the numbers themselves. Gallistel and Gelman (2000; see also Dehaene 2003) present evidence that this

[5] A circuit involving dorsolateral frontal cortex, caudate nucleus, and thalamus has been linked to working memory- and rule-based learning (Cummings 1993). Both the basal ganglia, a "reptilian" structure, and the cerebellum play a critical role in high-level cognition activities, including timekeeping with respect to sequences of bodily movements (Rao et al. 1997).

"number line" is logarithmically compressed. Through reverse inference from neuroimaging data, the existence of this "number line" has been linked to complex processes involving the left inferior parietal cortex. Dehaene et al. (2003) have addressed this complexity by identifying three distinct parietal circuits of number processing.[6] Inferior left parietal lesions disorganize the number line and also contribute to confusion about left and right and to acalculia (Dehaene et al. 1998). Thus, while numerical cognition is essential to rational calculation, it appears to draw on both logical analysis and on the spatial representation of a number line.

Linear Time Consciousness

> [T]he most important distinction between the left and right hemisphere modes is the extent to which a linear *concept* of time participates in the ordering of thought.
>
> —Joseph E. Bogen (1977: 141)

Time is created in the interaction of people and things; it is the crucial dimension of all human social actions (Gosden 1994: 122). Numerical cognitive processing is necessary for utilizing the two key metrics of a cash economy—time and money, which are calculated as a ratio-level, linear dimension. The left hemisphere surpasses the right in duration perception (in reaction time and accuracy) of musical and speech sounds (Brancucci et al. 2008), the detection of fine temporal events (e.g., detecting a temporal gap in a light exposure) (Nicholls and Whelan 1998), and the estimation of elapsed duration (Contreras et al. 1985). Fine temporal resolution for auditory stimuli (e.g., detecting the temporal order of two clicks) is predominantly associated with posterior regions of the left hemisphere (Steinbüchel et al. 1999). More specifically, the fMRI signals increased in the rostral regions of the gyrus rectus and the orbitomedial gyrus, predominantly in the left hemisphere.

In economic activity, linear time becomes a quantity that can be used, allocated, and exchanged. The quantitative time of a modern economic system is in one way similar to currency, whose value partly varies with supply and demand. Capitalist work discipline abstracts both work and time from their sociorelational contexts and associated meanings and imposes on workers a straight-jacket of linear clock time. Nowotny (1975) sees this abstraction from meaningful context as achieved only through the quantification of time, expressed in numerical values. E. P. Thompson (1967) saw that lived time, the substance of being and social life, has become extended by a second notion of time, time as an abstract quantity of pure duration. This quantitative time is freely exchangeable with all other times and serves as an abstract medium of exchange. Linear time, despite its commodification, can be conceptualized as "a spatial quantity by which time is measured as distinct" and which exists as "resource, as money, and as clock-based rhythm" (Adam 1995: 75).

The quantified timing of work (and pay) has led to the timing of other spheres of everyday life. In my comparative study of time consciousness among Aboriginal- and Euro-Australians (TenHouten 2005), a seven-part model of linear time was constructed and validated by finding that Euro-Australians, compared to Aborigines, were specialized for all seven aspects of linear time. Linear time (1) is a single dimension; (2) separates past, present, and future; (3) is regular, continuous, and homogeneous; (4) is measured by clocks and calendars; (5) diachronically orders events as exhibiting posteriority, simultaneity, or priority; (6) is quantitative, with an invariant anchor point; and (7) is

[6] In addition to its internal "number line," a left angular gyrus area, in conjunction with other left-hemisphere perisylvian areas, supports the manipulation of numbers in verbal form. Also, a bilateral posterior superior parietal system supports attentional orientation of the mental number line, just like any other spatial dimension.

conceptualized as fleeting and flying by, as a succession of moments (ibid., Chap. 5). It was also found that the positive experience of market-pricing social relations was predictive of the use of word categories under all seven indicators of linear time consciousness and for an overall index.[7] Thus, the emergence of linear, quantitative, clock- and calendar-based time has also contributed to the development of the rationalization of work and life in the modern world.

The Quantitative Assessment of Value**

An assessment of the anticipated economic value of various possible decisions or courses of action is a component of rational decision-making. Is there a neural network in the human brain that performs this estimation? Critics of rational-choice and subjective-utility theorizing doubt that humans come close to accurately estimating the value of possible courses of action. But such a dismissive judgment might be mistaken, both in terms of economic decision-making and associated brain functioning. In highly competitive situations, and especially when economic decision-makers' preferences are strongly economic, rather than social (Fehr 2009), exotic (Loewenstein 2007), or political (Green and Shapiro 1994), expected subjective-utility models do quite well.

The concept of utility, as used in modern economics, represents the basis for choice between various adaptive behaviors. If an animal, or a person, chooses option "x" in a given set of circumstances, then that option is inferred to have had a higher utility than the competing alternatives at the time the decision was made. Ability to compute utilities of behavioral choices contributes to fitness, survival, and reproductive opportunities (Shizgal 1997: 198).[8] Shizgal and Murray (1989) have searched for traces of subjective-utility estimation in the animal economy. Experimental studies of choice behavior in monkeys show that, in order to decide which of two or more incommensurable goals to pursue, the factors favoring one goal over another appear to be reduced to a common valuation on a single scale of subjective utility. Deaner et al. (2005) showed this in studies using a "pay per view" task, in which thirsty male rhesus macaques' gazing behavior was observed in order to measure their preference for one of two visual targets. Orienting to one yielded fruit juice; orienting to the other, fruit juice and a picture of a familiar monkey. By systematically changing the juice amounts and the pictures, it was possible to estimate the value of different types of reproductive and social stimuli in a liquid currency. The monkeys would forego larger juice rewards in order to view pictures of females communicating sexual signals or faces of high-ranking males, but required smaller rewards to choose to view faces of low-ranking males and females. Hayden et al. (2007) have extended these findings to humans, by demonstrating that human males will pay more to view pictures of attractive members of the opposite sex than to view pictures of unattractive ones even when the reward cues are left implicit. These and related results indicate that decisions based on value operate on a common currency that is independent of both the modality of the goods under consideration and of the actions they motivate.

[7] It was also found that the positive (but not the negative) experiences of equality matching, communal sharing, and authority ranking, as hypothesized, were predictive, on the basis of lexical-level indicators of immediate–participatory, patterned–cyclical, and episodic–futural kinds of time consciousness, respectively (TenHouten 2005: 58, Ch. 3). Patterned–cyclical time consciousness was conceptualized as the opposite of ordinary–linear time, as it (1) is dualistic; (2) fuses past, present, and, to a lesser extent, future; (3) is irregular, discontinuous, and heterogeneous; (4) is event-oriented; (5) is synchronic in the ordering of events, cyclical, patterned, and oscillatory; (6) is qualitative (now the anchor point); and (7) emphasizes the experience of long duration.

[8] British philosopher Jeremy Bentham (1776, 1825) had a broader definition of utility, as he included a bipolar, hedonic dimension of pleasure and pain as the basis for decision. Cabanac (1992) is one of a few who still holds to this definition of utility.

Shizgal and his coworkers suggest that, in computing the subjective utility of positive behavioral outcomes, brains compute an intensity of reinforcement signals. These signals mediate the rewarding effects of focal electrical stimulation of the brains' medial forebrain bundle, stimulating a system that functions to compute the utility of outcomes. They are carried by myelinated axons in the medial forebrain, which runs through the lateral hypothalamus and connects the ventral midbrain to the frontal lobes. It is no accident that, in humans, dopamine neurons carrying timed reward anticipations run through the medial forebrain bundle. The human mind must be able to aggregate nonlogical aspects of mental activity on a linear scale of value.

In addition to the demonstrated role of the medial forebrain, the orbital frontal cortex (OFC) is also highly involved in the process of value estimation. The OFC takes in signals from many emotion-linked brain structures and, in Zimmer's (2004: 280) terms, "...crunches them like a hedge fund manager, making calculations about the relative value of things. It puts value on abstract things such as money by associating with them all the things they signify." Patients with OFC damage, such as frontotemporal dementia, assign the "wrong value" to appetitive stimuli, display abnormal behavior in gambling tasks (suggesting inability to assess risks and rewards), and generally make poor choices, both in game-theory-based experimental settings and in everyday life (Damasio 1994). The brain, it appears, transforms information about disparate options into a common currency of value in which behavioral options are compared and evaluated (Platt and Padoa-Schioppa 2009: 450).

The linear assessment of value is highly adaptive and contributes to actions possessing high behavioral utility. These assessment processes occur in the medial- and orbitofrontal lobes, but also involve the parietal cortex and other structures. The left parietal cortex incorporates the value of each possible behavioral option and links its associated sensory signals with motor commands and behavioral choices. In humans and primates alike, visual cues are important for interpreting others' social displays and assessing their values. For example, signals of social rank are interpreted in ways that facilitate the establishment of preferential relationships with more dominant individuals. In discussing the neural basis for these valuation processes, Platt and Padoa-Schioppa (ibid., 450) also implicate the posterior cingulate cortex (PCC); neurons in this area encode ongoing estimates of the subjective values and consequent preferences for extra-personal events and actions, and the resulting signals update value-related information in several brain areas, importantly including the parietal cortex. The PCC is strongly interconnected to other brain areas involved in learning and motivation, and its signals serve to update value-related information in the parietal cortex and other areas sensitive to reinforcement (ibid., 452–3). The PCC is especially oriented to the motivational and emotional significance of visual events. It reacts to the presentation of both positive and negative emotion-laden words, to errors in reward anticipation during risky decision-making processes, and to spatial cues in peripheral attention tasks in a way that takes spatial location into account. PCC neurons thus play a role in signaling motivationally significant events and actions and are perhaps involved in the consideration of their subjective value in guiding future behavior by informing lateral inferior parietal neurons that contribute to behavioral choices.

Another brain structure necessary for value assessment is the striatum, where reward values are likely first encoded. This is the primary target of dopaminergic signals of reward values and motivation, which regulate corticostriatal synaptic transmission (Knutson et al. 2009; Doya and Kimura 2009). The striatum might well be the locus of the representation of subjective value, and its activation should contribute to subsequent choice (both economic and social). Direct infusion of dopamine-releasing agents into the ventral (but not dorsal) striatum of rats contributes to motivation to approach stimuli that had previously predicted reward, and unconditionally evokes appetitive behaviors such as sniffing and forward movement. The ventral striatum assesses expected gains, while the dorsal striatum uses these value estimates to inform future cognitions and conative actions (ibid.). If stimulated during outcome processing (but not during reward anticipation), the ventral striatum's caudate improves monkeys' ability to learn the next appropriate response for gaining a reward (Nakamura and Hikosaka 2006).

This section has discussed key indicators of what dual-brain theorists have called propositional and logical–analytic cognition, which Weber saw as an essential component of means–ends rationality. It is proposed that this general mode of cognition is highly involved in deductive and inductive reasoning, linear and numerical cognition, linear time measurement, and the quantitative assessment of value. However, our brief review of neuroscientific studies of these cognitive activities reveals that these mental activities do not derive from a specific center or from a single mode of thought or area of the brain. Rather, they are all complex modalities of cognition; they involve spatial as well as verbal reasoning and draw on widely distributed neural networks.

Modern capitalistic societies are based above all on three modal categories: (1) *money*, the abstract medium of exchange value; (2) *ordinary–linear time*, based on clocks and calendars; and (3) *number*, the basis of all quantitative reasoning. The common neural substrate of these categories of reason, it is hypothesized here, resides in the ability to calculate the *value* of various behavioral alternatives. Rationality, as this means choosing between cognized alternative courses of behavior with differing anticipated reward values, crucially depends upon linear conceptualizations of money, time, and number.

Executive Functions of the Frontal Lobes

> The prefrontal cortex plays the central role in forming goals and objectives and then in devising plans of action required to attain these goals. It selects the cognitive skills required to implement the plans, coordinates these skills, and applies them in a correct order. Finally, the prefrontal cortex is responsible for evaluating our actions as success or failure relative to our intentions.
>
> —Elkhonen Goldberg (2001: 24)

Goldberg (ibid., 2) considers the frontal lobes "the organ of civilization." He notes that motivation, foresight, drive, and vision are central to success in any walk of life. In the development of the individual, the frontal cortex is the last area of the cortex to myelinate, is phylogenetically the most recently evolved cortical area, and, through evolution, has increased in size more than any other brain area (Passingham 1973). The brain regions most involved in intentional functions are the dorsolateral and orbital prefrontal cortex (Fuster 1997). The principal and lateral dorsal limbic nuclei have an absolutely, and a relatively, greater number of nerve cells in modern humans than in species of great apes (*pongids*) and lesser apes (*hylobatids*). These structures' larger size helps modulate the integration of emotion and cognition. If the prefrontal lobes are disconnected from subcortical limbic structures involved in emotion, the individual has difficulty making reasoned decisions of any kind and is rendered incapable of acting with sustained intentionality. But frontal activation does not guarantee an individual is thinking rationally. Even an individual with an intact brain can make fundamentally irrational choices, using irrelevant information from the past, which the frontal lobes can retroactively justify and rationalize (Bailey 2007: 132).

The frontal lobes constitute the meeting place par excellence for information from two functional realms of the brain: (1) the posterior regions involved in the processing of sensory information (excluding olfaction) and (2) the limbic systems, where motivational and emotional functions are housed and internal affective states are generated. With this connectivity, the frontal lobes constitute "the realm where neural networks representing the individual's inner milieu – personal feelings, motivations, and subjective knowledge – converge with the systems representing the external milieu – the sights, sounds, tastes, and mores of the world" (Gardner 1983: 262). Through their connections and location, the frontal lobes are thus the brain's central integration station and hold a special status for the individual's sense of personhood.[9] They exert command and control over our impulses, passions, and emotions and lead us in our projects and plans. As the seat of the brain's leadership, its veritable CEO, they provide the individual novelty, innovation, achievement, and, hopefully, success in the adventures and endeavors of life.

Crucial to the frontal lobes' capacity for executive functioning, developing intentions, forming plans and programs of action, and evaluating and modifying performances directed to these plans and ends, are their ongoing comparisons and evaluations of the perceived outcomes and efficacy of actions. Executive functioning requires mental flexibility, the ability to filter out irrelevant information, and the capacity to anticipate the consequences of one's actions (Ardila et al. 2007). In order for the frontal lobes' higher-order representations to guide behavior, their architecture must be multimodal, broadly integrative, and highly plastic (Harris et al. 2008: 141). The executive functioning of the prefrontal cortex is the latest, and highest, development in the evolution of the human brain. Humans' extraordinary level of cognitive control stems from the active maintenance of patterns of activity in frontal cortex that represent goals and the means to achieve them. The frontal cortex signals other brain structures whose overall effect is to guide activity along neural pathways that establish the necessary mappings between sensory inputs, internal states, and outputs, in order to perform tasks and attain goals. Finding reasoned means to attain goals is definitive of rationality (Miller and Cohen 2001).

But what *is* central-executive functioning? Decades earlier, Jerry Fodor (1983) declared it impossible to provide a satisfactory functional account of the central-executive system, partly because measures of executive functioning have low test–retest reliability and uncertain validity. Neuroscientists have found it difficult to distinguish between executive and nonexecutive functions (Rabbitt 1997: 1). Consequently, it has been difficult to relate behaviors considered indicative of executive function to specific neurophysiological systems and neuroanatomical brain areas. Research on executive function is further hindered by conceptual terminology that remains vague and laden with historical constraints. Rabbitt (ibid.) points out that descriptions of executive impairment associated with prefrontal lesions are strikingly similar to the definitions of "willed," "voluntary," and "purposive" behavior that have preoccupied Western theologians and philosophers for more than two millennia.[10] Rabbitt (ibid., 3) argues that the key difference between executive and nonexecutive brainwork is determined not by the task's level of complexity but by the ability to choose among *alternative* courses of action.[11] Executive processing possesses an adaptive flexibility

---

[9] We have seen that beliefs are important to rationality, and they are also important to personhood, which is in turn central to choosing and pursuing our ends and goals. The human brain, Harris et al. (2008: 146) note, "is a prolific generator of beliefs. Indeed, personhood is largely the result of the capacity of a brain to evaluate new statements of propositional truth in light of other that it already accepts." Moreover, by evaluating truth, falsity, uncertainty, logical necessity, and contradiction, it becomes possible to cobble together coherent views and models of the world and one's place in the world as a social being.

[10] Rabbitt's (1997: 2) statement on this matter merits citation:

> Contemporary catalogues of the functions of the hypothetical "central executive" arestrikingly similar to the formal criteria for commission of mortal sin given by RomanCatholic theologians…. The minimal functional processes involved in the commission of amortal sin are awareness of the self as the intending perpetrator of the act; recognition ofunpleasant implications of the act for others by possession of a theory of mind;recognition of its moral repulsiveness by possession of a theory of the mind of God; anability simultaneously to represent alternative acts and their possible outcomes inworking memory in order to efficiently to choose between them; conscious formulationof a well-articulated plan to perform the act successfully; self-initiative and execution ofsequences of appropriate actions to consummate this plan during which recognition ofpersonal culpability is maintained by continuous monitoring; recognition of attainmentof the vile goal state as an intention to use what has been learned in its pursuit to performit again if opportunity occurs.

[11] Executive functioning, as an essential component of instrumental rationality, is not universal in human society but is rather a concomitant of modern, industrial society. For tribal-living Australian Aborigines, decision-making typically does not involve choosing between multiple courses of action but rather focuses on a single action, which is taken if and only if a full, and congenial, consensus is reached (see TenHouten 2005: 55–6).

to rescue plans when the environment, including other people, does not respond as anticipated. This can involve changing one's model of the world or adding alternative courses of action. Executive functioning requires highly developed, yet flexible, models of the world that can enable command and control over episodes of life involving interpersonal competition over outcomes and efforts to attain goals.

A second important hallmark of central-executive functioning is its retrieval of structured information from memory (Burgess 1997). Knowledge from the past enables individuals to attain foresight and to control and manage transactions with the external world. The internal environment of long-term memory restructures information from the past necessary to exert control over the future and provides useful models of the world which can be modified as present circumstances dictate. In referring to executive functioning as "episodic," Pribram (1981) emphasized the importance of the remembered past and anticipated future. Planning pertains to episodic reasoning and the manner in which elements of plans are implemented in a step-by-step manner, in which the temporal ordering of the steps is rearranged as needed. Patients with extensive frontal lesions have little concern for either the past or the future; they are "stuck" in the present and suffer impaired temporal ordering, disorganizing the sequencing of both remote and recent events (McAndrews and Milner 1991). Persons with limited executive function, in contrast, generate categories of words by simple association between successively occurring events or by inscrutably random processes, both of which reveal a lack of effective goal-directed memory search strategies. Lesions in prefrontal brain areas are often associated with dysexecutive functioning and impaired learning, suggesting impaired strategic control of mental processes related to recall and recognition (ibid., 4).

A third feature of executive functioning is its ability to interrupt or otherwise inhibit automatic, involuntary, and habitual sequences of responses and to initiate new sequences of task-appropriate behavior. Executive functioning can "switch" attention to new sources of information, particularly in complex tasks in which a variety of demands must be immediately met (Lowe and Rabbitt 1997). Fourth, evidence suggests that executive functioning affects the choice and execution of socially appropriate behavior. Clinical observations show that frontal-lobe lesions result in socially inappropriate behaviors. Social lapses have logical analogues in frontal lesion patients' bizarre responses to cognitive tests[12] (Shallice and Evans 1978) and in the propensities of older, executively challenged persons to score false-positives in memory recognition experiments, suggesting a loss of strategic control of memory (Parkin 1997). Persons with limited executive capacity respond without considering available information or alternatives and fail to monitor their insights' plausibility in the social contexts within which decisions must be made (Rabbitt 1997: 5).

Fifth, executive functioning is always conscious, whereas nonexecutive functioning might or might not be. While executive functioning presupposed consciousness, it is not the case that consciousness is merely a function of the frontal lobes (see Bogen 1995a, b).

And sixth, executive functioning enables sustained attention to long-range personal goals. This attentiveness facilitates the prediction of outcomes for long-term, slowly developing sequences of events. Diligence and persistence, for example, are necessary to work toward an academic degree over several years. Executively challenged persons, in contrast, exhibit "goal neglect," both in experimental situations and in the real world (Duncan 2005).

While Rabbitt provides a global view of frontal-lobe functioning, special attention must be directed to two frontal areas that are deeply involved in rationality. The dorsolateral prefrontal cortex (DLPFC) is important for situations in which one's model of the world is maladaptive and where it

[12] It is not measure of intelligence, the "intelligence quotient" (IQ), but personhood that depends on the frontal lobes. That frontal-lobe-damaged patients behave in inappropriate and bizarre ways in testing situations, which contribute to their poor test performances, which in turn, unfortunately, contribute to the mistaken view that IQ centrally depends on the frontal lobes. It is not that the patients cannot solve problems in which analytic reasoning is required, but they are not seriously motivated to do so.

is necessary to reappraise situations, to edit, and to detect inconsistencies between models and data. The DLPFC might well have the last word in decision-making (see D'Esposito and Postle 2002). The orbital frontal cortex, especially in the left hemisphere, is also particularly important to rationality because it is involved in the encoding of economic value and the calculation of utilities (Padoa-Schioppa and Assad 2006), the assessment of risk–reward contingences (Schultz et al. 1998), the social salience of linguistic communications, and the regulation of the limbic system's affects linked to cultural belief systems (Bailey 2007).

Conation and Motivation

Rabbitt's sixth criterion of executive functioning can be elaborated by the concept he does not use, that of "conation." Reitan and Wolfson (2000: 443) argue that conation, along with cognition and emotion, is a central feature of the human mind. They (ibid.) define conation as "the ability to apply oneself diligently and productively to the completion of a task over time." Conation is the ability to persistently and protractedly focus one's intellectual energy in order to successfully confront complex situations requiring a considerable time and mental energy, to solve a problem, to complete a task, or to attain a goal. There is, in conation, a purposeful striving, or the exercise of focused, highly motivated willpower, directed toward a desired future state of affairs. Over the last half century, the concept of conation has been largely ignored, yet adding conation to the present model of executive functions—as a component of instrumental rationality—is important. Without conation, there is only a weak correlation between cognitive ability and practical performance. Conation might well be a missing link in neuropsychological assessment, and also might contribute to explaining why rational-choice models so badly misfit data in the rare studies attempting to test these prescriptive models (see Green and Shapiro 1994). Individuals with frontal-lobe damage deplete their available intellectual energy and intellectual endurance more rapidly than normal controls. While component actions of complex behavior are sometimes spared, purposeful integration suffers (Bechara et al. 1994). Focus is easily lost, concentration wanes, and the individual's mental energies rapidly dissipate. Such an individual is apt to engage in inappropriate and impulsive behavior. Damage to the frontal lobes can alter reciprocal relationships between cortical functioning and the reticular activating system, so that the individual with injury to the frontal cortex might not attain a sufficient level of arousal for complex, goal-directed behavior requiring sustained attention.

Integration of Analytic and Episodic Information Processing

We have now established the biological infrastructures for logical analysis and central-executive functioning. A third task is to explore interactions of these two models of cognition on the level of brain function, in order to validate the claim that instrumentally rational cognition requires integration of executive functioning (including conation) and left-hemisphere-dominant verbal processing. The frontal lobes—which give the human species its great conceptual powers—are largely dependent upon language. The frontal lobes are responsible for motivating speech production and writing, guiding and controlling search activities associated with purposive reasoning, and formulating prelinguistic ideas. While powerful, they are also fragile and easily damaged, and patients with severe frontal-lobe damage can utter grammatically correct sentences but often lack motivation to do so, and also experience difficulty sustaining interest in projects and problems whose solution requires language production.

A fundamental interdependence exists between the episodic information processing of the frontal lobes and the processing of the left hemisphere's language areas. Neural activity passes through the frontal lobes to the primary motor cortex, where impulses are transmitted into motor routines and speech patterns. It is in this premotor area that neural activity organizes movements (e.g., clasping the hands, forming words and sentences), before passing through structures of the primary motor cortex to the periphery, where action is initiated. Jason Brown (1982) found that cortical areas posterior to the prefrontal cortex can dominate mental dialogue and that levels of EEG waveform desynchronization generated by activation of motor cortex were positively correlated with subjects' feeling of intentional control over their thought processes. While speech production has long been considered a function of the "verbal" left hemisphere, it is particularly dependent upon prefrontal functioning, especially upon Broca's area. Allen Schore (1994: 237) notes that recent brain imaging and blood flow studies clearly indicate that verbal fluency is linked to increases in metabolic activity of the left dorsolateral prefrontal cortex (Frith et al. 1991), an area operative in analyzing sequences of speech sounds (phonemes) (Alexander et al. 1989). Consistent with the model presented here, Bailey (2007: 127) refers to the "integration" of executive working memory and language areas of the brain in attaining choice–outcome determinations.

In order to act with intentionality—to carry out plans and realize goals—one needs to mentally represent the future. This requires that the brain use the generative power of language to retrieve from memory, and then reconfigure, elements of past experience. As Goldberg (2001: 25) explains,

> The ability to manipulate and recombine internal representations critically depends on the prefrontal cortex and the emergence of this ability parallels the evolution of the frontal lobes.... Therefore, the roughly contemporaneous development of executive function and language was adaptively highly fortuitous. Language provided the means for building models and the executive functions the means of manipulating them and conducting operations of the models.

Thus, the development of the frontal lobes was necessary to implement the generative capacity of language. Goldberg (ibid.) concludes, "For the believers in drastic discontinuity in language as a major factor in evolution, the confluence between the development of language and executive functions may have been the definitive force behind the quantum leap that was the advent of man." The left hemisphere's premotor and motor areas are important not only for speech production but also for reasoning. These cortical areas are located along the top of the left frontal lobe, then merge posteriorly into the motor or precentral cortex, which is considered anatomically intermediate between the frontal and posterior cortices. These motor areas enable us to articulate individual sounds and also provide the "rhythmic structures" (Brown 1982) necessary for arranging words in a grammatically proper, linear order. Thus, while ideas are nonlinear, the translating functions of the frontal cortex's posterior portion are able to linearize them and represent them with sequentially organized utterances consistent with grammatical rules. Patients with damage to this brain region have difficulty organizing their ideas in a meaningful temporal order. Without this melodic/rhythmic structure, words are apt to be deleted or used in the wrong order, confusing both the listener and the speaker. It is therefore not the case that speech is formed separately and then transmitted to the motor areas for expression; instead, formulation of the motor flow of the intended utterances is part and parcel of the circuit necessary for an idea to be verbally articulated (Boyle 1985: 68–9). Thus, both frontal and posterior areas of the brain participate in mental dialogue. The translation of ideas into inner speech leads to a rhythmically articulated, grammatically elaborated structure, which can be represented to the frontal cortex as higher-order perception. Language provides vocabularies for encoding meanings and logical structures for elaborating and sequentially ordering them. Language's great importance for intentionality, and for rationality, means that such conative mental processes cannot be strictly "localized" in the prefrontal cortical areas of the brain.

Conclusions

This chapter has presented a multidimensional model of the cognitive and neurobiological foundations of Weber's classic concept of instrumental rationality, whose principal features are calculation and planning. It has been possible to generalize these concepts, to show that calculation is a prominent aspect of the more general, logical–analytic mode of information processing and that planning is a key aspect of central-executive, episodic processing. According to dual-brain theory, calculation is largely lateralized to the human brain's left cerebral hemisphere,[13] while episodic processing largely depends upon the frontal lobes. Just as Weber demonstrated the important "interaction" between calculation and planning, we have shown the crucial interaction between logical–analytic processing and central-executive functioning. This interaction suggests that instrumental rationality involves three, not two, modal levels of cognition. By identifying and explaining the cognitive and neuroscientific underpinnings of instrumental rationality, we have elevated a sociological model of instrumental rationality—using an elaborated version of Weber's classic concept—to a contemporary neurosociology of rationality (TenHouten 1997, 1999a, b). This conceptualization is extended to a consideration of affect and its importance to rationality elsewhere (TenHouten 2006, 2007, in press, Chaps. 7–8).

References

Adam, B. (1995). *Timewatch: The social analysis of time*. Cambridge: Polity Press.

Alexander, M. P., Benson, D. F., & Stuss, D. T. (1989). Frontal lobes and language. *Brain and Language, 37*, 656–691.

Ardila, A., Surloff, C., & Mark, V. W. (2007). *Dysexecutive syndromes*. San Diego: Medlink Neurology.

Bailey, C. E. (2007). Cognitive accuracy and intelligent executive function in the brain and in business. In C. Senior & M. J. R. Butler (Eds.), *The social cognitive neuroscience of organizations* (pp. 122–141). Boston: Blackwell.

Bechara, A., Damasio, A. R., Damasio, H., & Anderson, S. W. (1994). Insensitivity to future consequences following damage to human prefrontal cortex. *Cognition, 50*, 7–15.

Bentham, J. ([1776] 1969). In M. P. Mack (ed.). *A Bentham reader*. New York: Pegasus.

Bentham, J. (1825). *The rationale of reward*. London: John and H. L. Hunt.

Bogen, J. E. (1969a). The other side of the brain. I. Dysgraphia and dyscopia following cerebral commissurotomy. *Bulletin of the Los Angeles Neurological Society, 34*, 73–105.

Bogen, J. E. (1969b). The other side of the brain. II. An appositional mind. *Bulletin of the Los Angeles Neurological Society, 34*, 135–162.

Bogen, J. E. (1977). Some educational implications of hemispheric specialization. In M. C. Wittrock (Ed.), *The human brain* (pp. 135–152). Englewood Cliffs: Prentice-Hall, Inc.

Bogen, J. E. (1985). The dual brain: Some historical and methodological aspects. In D. F. Benson & E. Zaidel (Eds.), *The dual brain: Hemispheric specialization in humans* (pp. 133–152). New York: Guilford Press.

Bogen, J. E. (1995a). On the neurophysiology of consciousness: I. An overview. *Consciousness and Cognition, 4*, 52–62.

Bogen, J. E. (1995b). On the neurophysiology of consciousness: II. Constraining the semantic problem. *Consciousness and Cognition, 4*, 137–158.

Boyle, R. P. (1985). The dark side of mead: Neuropsychological foundations for immediate experience and mystical consciousness. In N. K. Denzin (Ed.), *Studies in symbolic interaction: A research annual* (Vol. 6, pp. 59–78). Greenwich: JAI Press.

Bradshaw, J. L., & Nettleton, N. C. (1983). *Human cerebral asymmetry*. Englewood Cliffs: Prentice-Hall.

[13] This theory has been subjected to some rather harsh, even extreme, criticism, especially by Efron (1990) and Corballis (1991). I (1992) have presented an extensive critique of their and others' criticisms of this theory, which was developed out of several long conversations with neurosurgeon and theorist Joe Bogen.

Braine, M. D. S. (1978). On the relation between the natural logic of reasoning and standard logic. *Psychological Review, 85*, 1–21.
Brancucci, A., D'Anselmo, A., Martello, F., & Tommasi, L. (2008). Left hemisphere specialization for duration discrimination of musical and speech sounds. *Neuropsychologia, 46*, 2013–2019.
Brown, J. W. (1982). Hierarchy and evolution in neurolinguistics. In M. A. Arbib, D. Caplan, & J. C. Marshall (Eds.), *Neural models of language processes* (pp. 447–467). New York: Academic.
Brubaker, R. (1984). *The limits of rationality: An essay on the social and moral thought of Max Weber*. London: Allen & Unwin.
Burgess, P. W. (1997). Theory and methodology in executive function research. In P. Rabbitt (Ed.), *Methodology of frontal and executive function* (pp. 81–116). Hove, East Sussex: Psychology Press.
Byrne, R. M. J., & Johnson-Laird, P. N. (1991). *Deduction*. Hillsdale: Lawrence Erlbaum.
Cabanac, M. (1992). Pleasure: The common currency. *Journal of Theoretical Biology, 155*, 173–200.
Cherniak, C. (1986). *Minimal rationality*. Cambridge: MIT Press.
Contreras, C. M., Mayagoitia, L., & Mexicano, G. (1985). Interhemispheric changes in alpha rhythm related to time perception. *Physiology & Behavior, 34*, 525–529.
Corballis, M. C. (1991). *The lopsided ape: Evolution of the generative mind*. Oxford: Oxford University Press.
Cummings, J. L. (1993). Frontal-subcortical circuits and human behavior. *Archives of Neurology, 50*, 873–880.
D'Esposito, M., & Postle, B. R. (2002). The organization of working memory function in lateral prefrontal cortex: Evidence from event-related functional MRI. In D. T. Stuss & R. T. Knight (Eds.), *Principles of frontal lobe function* (pp. 168–187). New York: Oxford University Press.
Damasio, A. R. (1994). *Descartes' error: Emotion, reason, and the human brain*. New York: Putnam.
de Hevia, M., Dolores, G. V., & Girelli, L. (2008). Visualizing numbers in the mind's eye: The role of visuo-spatial processes in numerical abilities. *Neuroscience and Biobehavioral Reviews, 32*, 1361–1372.
de Spinoza, B. ([1677] 1982). In S. Feldman, & S. Shirley (Eds.). *The ethics and selected letters*. Indianapolis: Hackett.
Deaner, R. O., Khera, A. V., & Platt, M. L. (2005). Monkey pay per view: Adaptive valuation of social images by rhesus macaques. *Current Biology, 15*, 543–548.
Dehaene, S. (1997). *The number sense: How the mind creates mathematics*. New York: Oxford University Press.
Dehaene, S. (2003). The neural basis of the Weber-Fechner law: A logarithmic mental number line. *Trends in Cognitive Sciences, 7*, 145–147.
Dehaene, S., Dehaene-Lambertz, G., & Cohen, L. (1998). Abstract representations of numbers in the animal and human brain. *Trends in Neuroscience, 21*, 355–361.
Dehaene, S., Spelke, E., Pinel, P., Stanescu, R., & Tsivkin, S. (1999). Sources of mathematical thinking: Behavioral and brain-imaging evidence. *Science, 284*, 970–974.
Dehaene, S., Piazza, M., Pinel, P., & Cohen, L. (2003). Three parietal circuits for number processing. *Cognitive Neuropsychology, 20*, 487–506.
Delacroix, J., & Nielsen, F. (2001). The beloved myth: Protestantism and the rise of industrial capitalism in nineteenth-century Europe. *Social Forces, 80*, 509–553.
Doya, K., & Kimura, M. (2009). The basal ganglia and the encoding of value. In P. W. Glimcher, C. F. Camerer, E. Fehr, & R. A. Poldrack (Eds.), *Neuroeconomics: Decision making and the brain* (pp. 407–416). Amsterdam: Elsevier.
Duncan, J. D. (2005). Frontal lobe function and general intelligence: Why it matters. *Cortex, 41*, 215–217.
Efron, R. (1990). *The decline and fall of hemispheric specialization*. Hillsdale: Lawrence Erlbaum.
Eisen, A. (1978). Meanings and confusions of weberian rationality. *The British Journal of Sociology, 29*, 57–70.
Evans, J. St. B. T. (1989). *Bias in human reasoning: Causes and consequences*. Hillsdale: Lawrence Erlbaum.
Evans, J. St. B. T. (1998). Matching bias in conditional reasoning: Do we understand it after 25 years? *Thinking and Reasoning, 4*, 45–82.
Fehr, E. (2009). Social preferences and the brain. In P. W. Glimcher, C. F. Camerer, E. Fehr, & R. A. Poldrack (Eds.), *Neuroeconomics: Decision making and the brain* (pp. 215–232). Amsterdam: Elsevier.
Fehr, E., & Gächter, S. (2000). Fairness and retaliation: The economics of reciprocity. *Journal of Economic Perspectives, 14*, 159–181.
Fiske, A. P. (1991). *Structures of social life: The four elementary forms of human relations: Communal sharing, authority ranking, equality matching, market pricing*. New York: The Free Press.
Fodor, J. (1983). *The modularity of mind: An essay on faculty psychology*. Cambridge, MA: MIT Press.
Fogelin, R. (2003). *Walking the tightrope of reason: The precarious life of a rational animal*. New York: Oxford University Press.
Frith, C. D., Friston, K. J., Liddle, P. F., & Frackowiak, R. S. (1991). A PET study of word finding. *Neuropsychologia, 29*, 1137–1148.
Fuster, J. M. (1997). *The prefrontal cortex: Anatomy, physiology, and neuropsychology of the frontal lobe* (3rd ed.). Philadelphia: Lippincott-Raven.

Gallistel, C. R., & Gelman, R. (2000). Non-verbal numerical cognition: From reals to integers. *Trends in Cognitive Sciences, 4*, 59–65.
Galton, F. (1880). Visualized numerals. *Nature, 21*, 59–65.
Gardner, H. (1983). *Frames of mind: The theory of multiple intelligences*. New York: Basic Books.
Gilhooly, K. J., Logie, R. H., Wetherick, N. E., & Wynn, V. (1993). Working memory and strategies in syllogistic-reasoning tasks. *Memory and Cognition, 21*, 115–124.
Göbel, S. M., Calabria, M., Farnè, A., & Rossetti, Y. (2006). Parietal rTMS distorts the mental number line: Simulating 'Spatial' neglect in healthy subjects. *Neuropsychologia, 44*, 860–868.
Goel, V., & Dolan, R. J. (2003). Explaining modulation of reasoning by belief. *Cognition, 87*, B11–B22.
Goel, V., & Dolan, R. J. (2004). Differential involvement of left prefrontal cortex in inductive and deductive reasoning. *Cognition, 93*, B109–B121.
Goel, V., Gold, B., Kapur, S., & Houle, S. (1998). Neuroanatomical correlates of human reasoning. *Journal of Cognitive Neuroscience, 10*, 293–302.
Goel, V., Buchel, C., Frith, C., & Dolan, R. J. (2000). Dissociation of mechanisms underlying syllogistic reasoning. *NeuroImage, 12*, 504–514.
Goldberg, E. (2001). *The executive brain: Frontal lobes and the civilized mind*. Oxford: Oxford University Press.
Gosden, C. (1994). *Social being and time*. Cambridge: Blackwell.
Grafman, J. (2002). The structured event complex and the human prefrontal cortex. In D. T. Stuss & R. T. Knight (Eds.), *The frontal lobes* (pp. 292–310). New York: Oxford University Press.
Green, D. P., & Shapiro, I. (1994). *Pathologies of rational choice theory: A critique of applications in political science*. New Haven: Yale University Press.
Harris, S., Sheth, S. A., & Cohen, M. S. (2008). Functional neuroimaging of belief, disbelief, and uncertainty. *Annals of Neurology, 63*, 141–147.
Hayden, B. Y., Parikh, P. C., Deaner, R. O., & Platt, M. L. (2007). Economic principles motivating social attention in humans. *Proceedings of the Royal Society, Biological Sciences, Series B, 274*(1619), 1751–1756.
Houdé, O., Zago, L., Mellet, E., Moutier, S., Pineau, A., Mazoyer, B., & Tzourio-Mazoyer, N. (2000). Shifting from the perceptual brain to the logical brain: The neural impact of cognitive inhibition training. *Journal of Cognitive Neuroscience, 12*, 721–728.
Johnson-Laird, P. N. (1983). *Mental models: Toward a cognitive science of language, inference, and consciousness*. New York: Harvard University Press.
Kalberg, S. (1980). Max Weber's types of rationality: Cornerstones for the analysis of rationalization processes in history. *The American Journal of Sociology, 85*, 1145–1179.
Knauff, M., Mulack, T., Kassubek, J., Salih, H. R., & Greenlee, M. W. (2002). Spatial imagery in deductive reasoning: A functional MRI study. *Cognitive Brain Research, 13*, 203–212.
Knutson, B., Delgado, M. R., & Phillips, P. E. M. (2009). Representation of subjective value in the striatum. In P. W. Glimcher, C. F. Camerer, E. Fehr, & R. A. Poldrack (Eds.), *Neuroeconomics: Decision making and the brain* (pp. 389–406). Amsterdam: Elsevier.
Koenigs, M., Young, L., Adolphs, R., Tranel, D., Cushman, F., Hauser, M., & Damasio, A. (2007). Damage to the prefrontal cortex increases utilitarian moral judgments. *Nature, 446*, 908–911.
Kroger, J. K., Sabb, F. W., Fales, C. L., Bookheimer, S. Y., Cohen, M. S., & Holyoak, K. J. (2002). Recruitment of anterior dorsolateral prefrontal cortex in human reasoning: A parametric study of relational complexity. *Cerebral Cortex, 12*, 477–485.
Kronman, A. T. (1983). *Max Weber*. London: Edward Arnold.
Kunzig, R. (1997). A head for numbers. *Discover Magazine, 18*, 108–115.
Laeng, B. (1994). Lateralization of categories and coordinate spatial functions. A study of unilateral stroke patients. *Journal of Cognitive Neuroscience, 6*, 189–203.
Levi-Agresti, J., & Sperry, R. W. (1968). Differential perceptual capacities in major and minor hemispheres. *Proceedings of the National Academy of Sciences USA, 61*, 1151.
Loewenstein, G. (2007). *Exotic preferences: Behavioral economics and human motivation*. Oxford: Oxford University Press.
Lowe, C., & Rabbitt, P. (1997). Cognitive models of ageing and frontal lobe deficits. In P. Rabbitt (Ed.), *Methodology of frontal and executive function* (pp. 39–59). Hove, East Sussex: Psychology Press.
Luhmann, N. (1982). *The differentiation of society* (S. Holmes & C. Larmore, Trans.). New York: Columbia University Press.
Lukes, S. (1967). Some problems about rationality. *Archives Européennes de Sociologie, VIII*, 247–264.
McAndrews, M. P., & Milner, B. (1991). The frontal cortex and memory for temporal order. *Neuropsychologia, 29*, 849–859.
Miller, E. K., & Cohen, J. D. (2001). An integrative theory of prefrontal cortex function. *Annual Review of Neuroscience, 24*, 167–202.

Montague, P. R., Berns, G. S., Cohen, J. D., et al. (2002). Hyperscanning: Simultaneous fMRI during linked social interactions. *NeuroImage, 16*, 1159–1164.

Nakamura, K., & Hikosaka, O. (2006). Facilitation of saccadic eye movements by postsaccadic electrical stimulation in the primate caudate. *Journal of Neuroscience, 26*, 12885–12895.

Nicholls, M. E. R., & Whelan, R. R. E. (1998). Hemispheric asymmetry for the temporal resolution of brief tactile stimuli. *Journal of Clinical and Experimental Neuropsychology, 4*, 445–456.

Northoff, G., Heinzel, A., Bermpohl, F., Niese, R., Pfennig, A., Pascual-Leone, A., & Schlaug, G. (2004). Reciprocal modulation and attenuation in the prefrontal cortex: An fMRI study on emotional-cognitive interaction. *Human Brain Mapping, 21*, 202–212.

Noveck, I. A., Goel, V., & Smith, K. W. (2004). The neural basis of conditional reasoning with arbitrary content. *Cortex, 40*, 613–622.

Nowotny, H. (1975). Time structuring and time measurement: On the interrelations between timekeeping and social time. In J. T. Fraser & N. M. Lawrence (Eds.), *The study of time* (Vol. 2, pp. 325–361). New York: Springer.

Padoa-Schioppa, C., & Assad, J. A. (2006). Neurons in the orbitofrontal cortex encode economic value. *Nature, 441*, 223–226.

Parkin, A. J. (1997). Normal age-related memory loss and its relation to frontal lobe dysfunction. In P. Rabbitt (Ed.), *Methodology of frontal and executive function* (pp. 177–190). Hove, East Sussex: Psychology Press.

Parsons, L. M., & Osherson, D. (2001). New evidence for distinct right and left brain system for deductive versus probabilistic reasoning. *Cerebral Cortex, 11*, 954–965.

Passingham, R. E. (1973). Anatomical differences between the neocortex of man and other primates. *Brain, Behavior and Evolution, 7*, 337–359.

Platt, M., & Padoa-Schioppa, C. (2009). Neuronal representation of value. In P. W. Glimcher, C. F. Camerer, E. Fehr, & R. A. Poldrack (Eds.), *Neuroeconomics: Decision making and the brain* (pp. 44–62). Amsterdam: Elsevier.

Plutchik, R. ([1962] 1991). *The emotions* (Rev. ed.). Lanham: University Press of America.

Pribram, K. H. (1981). Emotions. In S. B. Filskov & T. J. Boll (Eds.), *Handbook of clinical neuropsychology* (Vol. 1, pp. 102–134). New York: Wiley.

Rabbitt, P. (1997). Introduction: Methodologies and models in the study of executive function. In P. Rabbitt (Ed.), *Methodology of frontal and executive function* (pp. 1–38). Hove, East Sussex: Psychology Press.

Rao, S. M., Bobholz, J. A., Hammeke, T. A., Rosen, A. C., Woodley, S. J., Cunningham, J. M., Cox, R. W., Stein, E. A., & Binder, J. R. (1997). Functional MRI evidence for subcortical participation in conceptual reasoning skills. *Neuroreport, 8*, 1987–1993.

Reitan, R. M., & Wolfson, D. (2000). Conation: A neglected aspect of neuropsychological functioning. *Archives of Clinical Neuropsychology, 15*, 443–453.

Rips, L. (1994). *The psychology of proof: Deductive reasoning in human thinking*. Cambridge, MA: MIT Press.

Rotenberg, V. S. (1995). Right hemisphere insufficiency and illness in the context of search activity. *Dynamic Psychiatry, 150*, 54–63.

Rueschemeyer, D. (1986). *Power and the division of labour*. Stanford: Stanford University Press.

Schlucter, W. (1981). *The rise of western rationalism: Max Weber's developmental history*. Berkeley: University of California Press.

Schore, A. N. (1994). *Affect regulation and the origin of the self: The neurobiology of emotional development*. Hillsdale: Lawrence Erlbaum.

Schore, A. N. (1997). Early organization of the nonlinear right brain and development of a predisposition to psychiatric disorders. *Developmental Psychopathology, 9*, 595–631.

Schultz, W., Tremblay, L., & Hollerman, J. L. (1998). Reward prediction in primate basal ganglia and frontal cortex. *Neuropharmacology, 37*, 421–429.

Shallice, T., & Evans, M. E. (1978). The involvement of the frontal lobes in cognitive estimation. *Cortex, 14*, 294–303.

Shizgal, P. (1997). Neural basis of utility estimation. *Current Opinion in Neurobiology, 7*, 198–208.

Shizgal, P., & Murray, B. (1989). Neuronal basis of intracranial self-stimulation. In J. M. Liebman & S. J. Cooper (Eds.), *The neuropharmacological basis of reward* (pp. 106–163). Oxford: Clarendon Press.

Singer, T. (2009). Understanding others: Brain mechanisms of theory of mind and empathy. In P. W. Glimcher, C. F. Camerer, E. Fehr, & R1. A. Poldrack (Eds.), *Neuroeconomics: Decision making and the brain* (pp. 251–268). Amsterdam: Elsevier.

Sperry, R. W. (1984). Consciousness, personal identity and the divided brain. *Neuropsychologia, 22*, 661–673.

TenHouten, W. D. (1992). "Cerebral lateralization: A scientific paradigm in crisis?" Critique of Efron and Corballis. *Journal of Social and Evolutionary Systems, 15*, 319–326.

TenHouten, W. D. (1997). Neurosociology. *Journal of Social and Evolutionary Systems, 20*, 7–37.

TenHouten, W. D. (1999a). Explorations in neurosociological theory: From the spectrum of affect to time-consciousness. In D. D. Franks & T. S. Smith (Eds.), *Mind, brain, and society: Toward a neurosociology of emotion* (pp. 41–80). Stamford: JAI Press.

TenHouten, W. D. (1999b). The four elementary forms of sociality, their biological bases, and their implications for affect and cognition. In L. Freese (Ed.), *Advances in human ecology* (Vol. 8, pp. 253–284). Stamford: JAI Press.
TenHouten, W. D. (2005). *Time and society*. Albany: State University of New York Press.
TenHouten, W. D. (2006). From alexithymia, borne of trauma and oppression, to symbolic elaboration, the creative expression of emotion, and rationality. In J. C. Kaufman & J. Baer (Eds.), *Creativity and reason in cognitive development* (pp. 177–201). New York: Cambridge University Press.
TenHouten, W. D. (2007). *A general theory of emotions and social life*. London: Routledge.
TenHouten, W. D. (in press). *Emotion & reason: Mind, brain, and the social domains of work and love*. London: Routledge.
Thompson, E. P. (1967). Time, work-discipline, and industrial capitalism. *Past & Present, 38*, 56–97.
von Steinbüchel, N., Wittman, M., Strasburger, H., & Szelag, E. (1999). Auditory temporal-order judgment is impaired in patients with cortical lesions in posterior regions of the left hemisphere. *Neuroscience Letters, 264*, 168–171.
Weber, M. ([1904–1905] 1958). *The protestant ethic and the spirit of capitalism* (T. Parsons, Trans.). New York: Charles Scribner's Sons.
Weber, M. ([1905] 1975). Knies and the problem of irrationality. In G. Oakes (Ed.), *Roscher and knies: The logical problems of historical economics* (pp. 97–205). New York: Free Press.
Weber, M. ([1918] 1970). Politics as a vocation. In H. H. Gerth & C. W. Mills (Ed. & Trans.). *From max weber: Essays in sociology* (pp. 77–128). New York: Oxford University Press.
Weber, M. ([1919] 1970). Science as a vocation. In H. H. Gertz & C. W. Mills (Eds. and Trans.) *From max Weber: Essays in sociology* (pp. 129–179). New York: Oxford University Press.
Weber, M. ([1921] 1978). In G. Roth & C. Wittich (Ed. & Trans.). *Economy and society: An outline of interpretive sociology* (2 vols.). Berkeley: University of California Press.
Zimmer, C. (2004). *Soul made flesh: The discovery of the brain — And how it changed the world*. New York: Free Press.

Chapter 16
Neurosociology and Theory of Mind (ToM)

Rosemary L. Hopcroft

Theory of mind (henceforth ToM) is the term used to describe a person's ability to understand another person's mental states. It is somewhat cryptically called "ToM" because it shares some features with the explanations scientists call theories. That is, ToM refers to the ability to use the observed behavior of another person to develop an explanation for the behavior involving that person's mental processes, make predictions about the person's future behavior, and use those predictions to explain new observations (Malle 2005: 225). For example, "She chose only the chocolate candy because she likes chocolate. In the future, if she has a choice of candy, she will choose chocolate candy again." "She took the strawberry candy because there was no chocolate candy."

ToM forms an important bridge between those sciences focusing on the individual (e.g., neuroscience and psychology) and those sciences focusing on the group (e.g., sociology). For neuroscientists and cognitive psychologists, development of ToM allows the individual to be fully social and function normally in the social world. Individuals who have trouble understanding other peoples' mental states have social difficulties, while severe lack of understanding of others' mental states becomes pathological. For sociologists and other social scientists, ToM may be regarded as the psychological underpinning of the social nature of individuals. Microsociology is grounded in the ability of individuals to think about and anticipate the behaviors of others, and hence in ToM.

In this chapter, I first discuss the standard test used to gauge a child's ToM, the false belief test. Then I trace the typical development of ToM in young children as revealed by the work of neuroscientists and cognitive psychologists. I also discuss theories of ToM development, its association with language learning, the brain bases of ToM, and the lack of development of ToM in autistic children. Last, I discuss the relevance of the development of ToM in children for sociological work on the emergence of the self, as well as other sociological perspectives.

I discuss the work of social neuroscientists and cognitive psychologists first because it deals with a lower level of analysis (the individual) while sociology deals with a higher level of analysis (the group). Given this, sociology both subsumes and transcends the more micro-oriented disciplines, and thus logically is discussed last. An important subtheme of this chapter, as that of this whole volume, is that sociological work builds on, but does not contradict, the findings of neuroscience, psychology, and the other life sciences. In this view, ToM is a cornerstone of the sociological edifice.

R.L. Hopcroft (✉)
Department of Sociology, University of North Carolina, Charlotte, NC, USA
e-mail: rlhopcro@uncc.edu

D.D. Franks and J.H. Turner (eds.), *Handbook of Neurosociology*, Handbooks of Sociology and Social Research, DOI 10.1007/978-94-007-4473-8_16, © Springer Science+Business Media B.V. 2013

The False Belief Task

The standard task used to assess a child's ToM is the "false belief task." There are many versions of the task, but all seek to determine whether the child can understand that another person may be motivated by a false belief. For example, the task may be as follows: The child is told that Max puts his candy in a green cabinet; then he goes outside to play. While Max is outside, his mother moves the candy into a blue cabinet. The child is asked, "When Max comes back in, where will he look for his candy?" A child who answers correctly that Max will look for his candy in the wrong, i.e., green, cabinet is said to have passed the false belief test. Although the age of success differs somewhat depending on the particular version of the task used, generally children can pass this test by the time they are 4 years old (Saxe 2006). Ability to pass the ToM test occurs at about the same age in developed and undeveloped countries (Lillard and Skibbe 2005, p. 281). Three-year-olds generally do not pass the test and have trouble understanding that Max will not know everything that they themselves know (i.e., the true location of the candy). To them, all others see the world like they do, and there is no notion of other people having another perspective.

Children younger than 4 years old can pass an implicit version of this task, however. Garnham and Perner (2001) use an ingenious method of finding out whether children can understand another's false belief, even if they cannot express it to someone else. In this version of the false belief test, children are confronted with a model containing two slides reached by one stair. On the side of each slide is a box. The child is told the following story, which is also acted out by two small figurines and a toy ball: Alan plays with a ball and places it in one of the boxes by one of the slides, then goes away to climb the stair to the slides. While he is doing this, Rebecca comes and switches the ball to the box by the other slide, but Alan does not see this. Meanwhile, Alan decides he wants to play with the ball again. Some of the children were asked, "Which slide will Alan come down?" Others were told, "Quick! Move the mat to catch Alan as he comes down the slide!" Children were much more likely to move the mat to the correct slide than to answer the question correctly and explicitly, showing an implicit understanding of Alan's false belief. The fact that children can pass an implicit version of the test before they can verbally give the correct answer suggests that the child's "ToM" is acquired slowly over early childhood. A great deal of research in neuroscience and psychology tracks this development in young children.

Development of ToM in Normal Children

Before a child can understand the content of another person's mental states, they have to understand that others have mental states. Babies are very interested in other people and quickly learn to differentiate them from objects (Legerstee 1992). Even new born babies imitate the actions of other people, such as sticking out a tongue (Meltzoff and Moore 1977). Babies imitate a person sticking out a tongue, but not similar behaviors produced by an object, suggesting that they know the difference (Legerstee 1991). Babies also understand that some behaviors that people perform have motivations and some behaviors do not (i.e., they are accidental). Children can differentiate between intentional and accidental behavior by the age of 9 months (Woodward 1999).

Many authors argue that gaze following is an important precursor to ToM as this suggests that children know that the gazer is intentionally paying attention to something. It suggests some understanding of the intentional states of others (Baron-Cohen 1995). In humans, gaze-following behavior emerges between 3 and 18 months of age (Scaife and Bruner 1975; Butterworth and Cochran 1980; Corkum and Moore 1995). By 14–18 months of age, children can follow adult eye gaze by following the eyes only (Moore and Corkum 1998; Brooks and Meltzoff 2002). By 18 months, children start to call adults' attention to objects by pointing to them or holding them up (Golinkoff 1986).

At approximately the same age (18–24 months), infants develop the capacity to recognize themselves in mirrors (Gallup et al. 2002; Brüne and Brüne-Cohrs 2006; Reddy et al. 2010). This is demonstrated by the mirror test, where a researcher applies a red mark to the child's nose and then puts the child in front of a mirror. Children show self-recognition when they investigate the mark on their nose in the mirror, and most children can do this by the age of two. Before the age of two, children react socially to the reflection as if it were another person, smiling, making noises, etc. (Gallup et al. 2002). The ability to pass the mirror test is not limited to humans, since all of the great apes, as well as dolphins and elephants, can pass this test and recognize themselves in a mirror. It is very likely that whales can do so as well (de Waal 2009).

Self-recognition is the beginning of the development of a sense of self. Embarrassment and the use of personal pronouns also occur at about this age, suggesting that children are developing self-awareness and some understanding of how others see them. Being aware of oneself is an important step in being aware of your own thinking. Being aware of your own thoughts also means you are more likely to be able to infer what others are likely to think in similar situations (Gallup et al. 2002). This is also the time that pretend play begins (Lewis and Ramsay 2004; Stone 2006). Pretend play often involves pretense about the desires, beliefs, and intentions of others and so is an important step in understanding the minds of others. In pretend play, however, children at this age often make mistakes about what others can see and think (Flavell et al. 1981).

Pretend play often involves role playing, where children practice being another person in a different role with a different repertoire of behaviors. This not only allows children to practice seeing things from another person's point of view (the beginning of role taking), but it allows children to practice the socially acceptable behaviors of people in their social group. The sociologist George Herbert Mead thought this role playing was vital to the full development of sense of self, as discussed below.

Between two and three, children develop their understandings of the desires and emotions of others (Saxe et al. 2004). They now understand that some people like certain things and others do not, and these preferences influence what those people subsequently do (Flavell 1999, p. 34). This is an important step in acquiring a full understanding of the mental and emotional states of others. By 3, they can pass the implicit version of the false belief task, and by 4 normal children can pass the standard version of the false belief test.

The development of ToM appears to be similar across cultures, although some argue that there are important differences between cultures in adult's ToM (Flavell 1999). For example, Americans are more likely to give trait reasons for why a person behaved a certain way, while people from other cultures are more likely to give situational reasons (Lillard and Skibbe 2005, p. 285). Liu et al. (2008) found similar patterns of development of ToM as measured by passing the false belief test in China, Hong Kong, and the United States, although the timing did differ across regions. The timing of development for children in mainland China and children in the United States was very similar, while Canadian children developed earlier and Hong Kong Chinese children developed later. The factors responsible for the differences in timing were not clear, however.

ToM also seems to improve with age. As children age, they are more likely to explain actions with reference to the actor's goals. In particular, between middle childhood and adulthood, the recognition of psychological goals of others increases greatly (Lovett and Pillow 2010). Part of the brain called the "default network" that is implicated with ToM and empathy become significantly more integrated as children get older, suggesting that the ability to understand others continues to improve with age (Fair et al. 2008).

Theories of Development of ToM

There are currently three theories of the development of ToM from psychology and neuroscience (Flavell 1999). The first is "theory theory" (Gopnick and Wellman 1994; Wellman and Gelman 1998). This is the idea that ToM is the development of an informal theory of other minds and is developed much as a scientist develops a theory through prediction and testing. According to this view, children revise and improve their ToM in response to experience.

The second theory of development of ToM is the modular theory (Fodor 1992; Leslie 1994; Baron-Cohen 1995). Modular theory posits that ToM is supported by a distinct, separate, brain module that matures along with other brain modules. The development of ToM is thus overwhelmingly a process of physical maturation. Experience may trigger the ToM mechanism but does not determine the mechanism itself (Brüne and Brüne-Cohrs 2006). Sometimes, modular theory is treated as a subset of Theory theory. In this view, the developing ToM in a child is seen as the maturation of a domain-specific module rather than the refining of a general-purpose algorithm (Gallese and Goldman 1998).

The last theory is "simulation theory" (Harris 1992). This is the theory that children use their own mental states to simulate the mental states of others. Thus, the ability to pass the false belief test is an extension of self-understanding. Experience is crucial in this development. It is especially through role taking in pretend play that children learn to simulate the experiences of others.

It is likely that all three theories are correct to a certain extent. The existence of specific regions of the brain associated with ToM tasks supports the modular theory, while evidence of increasingly developed understandings of others' beliefs supports the theories that stress the role of experiential learning (theory theory and simulation theory). The fact that thinking about the self uses the same brain regions as thinking about others, particularly the mPFC, supports the simulation theory of ToM (Mitchell et al. 2006). Others have argued that the role of mirror neurons in ToM reasoning also supports the simulation theory (Gallese and Goldman 1998).

ToM: Association with Language

Development of the ability to pass the false belief test is also associated with word learning (Malle 2005, p. 229). After all, the primary difference between the explicit and implicit versions of the false belief task is that the explicit version requires children to verbalize what they know. Some have even argued that developing the ability to pass the false belief is entirely about learning the ability to represent one's thoughts verbally.

However, a study by Andrew S. Gordon (2006) plotted the frequency of usage of words related to knowledge and beliefs by age of children. He assumed that if the ability to represent others' thoughts in words was necessary for the child to understand the mind states of others, then before a child was able to pass the false belief test, there would be a large increase in knowledge and use of words related to knowledge and beliefs. He found that children do increase the number of words related to knowledge and beliefs from the age of two, but it is a steady increase indicative of developing linguistic competency. There is no qualitative shift right before children can pass the false belief test. Nor does the type of concepts expressed by children change dramatically at any time. This suggests that the ability to pass a false belief test is not because of a large improvement in the ability to express ones' understanding of the mental states of others.

Others have argued the reverse: that it is children's developing ToM that aids in their developing linguistic competency. Bloom (2000, p. 61) writes, "children use their naïve psychology or ToM to figure out what people are referring to when they use words." This means that children do not simply take literal interpretations of statements such as "Would you mind telling me what time it is?" and respond with "I don't mind at all." Instead, they understand the intent of the speaker, and so respond appropriately with the time. A variety of research shows that children use the eye gaze, body language, and facial expressions of the speaker to understand the meaning and intentions of the speaker, and this helps them learn new words (Tomasello and Barton 1994).

Continued improvement of language abilities fuels the interactional process involved in role playing, role taking, socialization, and the emergence of the self, as language is the most effective symbol system available to humans. The full development of language capabilities is also necessary for adult role performance and presentation of self as described by Goffman and others (see below).

Brain Bases of ToM

In typically developing adults, ToM is supported by a consistent group of four brain regions: the bilateral temporoparietal junction (TPJ), both left and right; medial prefrontal cortex (mPFC); posterior cingulate (PC); and anterior temporal sulci (aSTS) (Saxe 2006). These regions are sometimes collectively called the ToM (or mentalizing) system (Bedny et al. 2009; Vistoli 2011).

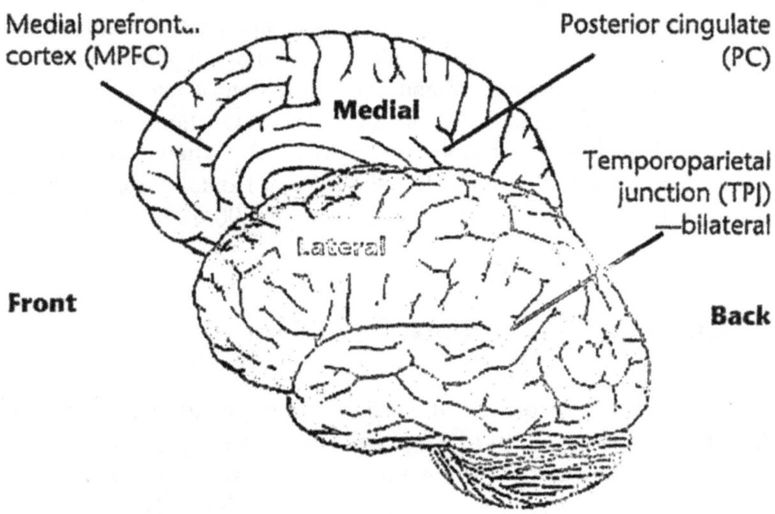

Further, a collection of brain regions referred to as the "default network" is also implicated with introspective mental activity, including ToM. Brain regions involved include the medial prefrontal cortex (mPFC), also posterior cingulate, and lateral parietal cortex (Fair et al. 2008). Empathy has also been shown to involve the mPFC (Schulte-Rüther et al. 2007).

Given the association between language and development of ToM, it may be assumed that similar brain regions are used in both language and ToM tasks. Although similar regions appear to be involved in both tasks, they do not appear to be exactly the same regions. For example, it has been found that stroke victims with impaired language capabilities can still pass a nonverbal false belief task (Varley and Siegal 2000).

Mirror Neurons and ToM (Make Regular Font)

It has been argued that mirror neurons play an important role in ToM. Mirror neurons are neurons in the brain that fire when a person performs an action and also fire when a person sees another person performing the same action (see Chap. 12 of this volume and Rizzolatti and Craighero 2004;

Montgomery et al. 2007). Mirror neurons are found concentrated in the inferior parietal lobule, the lower part of the precentral gyrus, and the posterior part of the inferior frontal gyrus (Rizzolatti and Craighero 2004; Vistoli 2011). Rizzolatti et al. (2001) suggest that by mirroring observed actions, we can understand another's actions from our own point of view, and we can also better predict the mental state of the person performing the action. Thus mirror neurons may be directly involved in ToM (Oberman et al. 2005; Iacoboni et al. 2005; Oberman and Ramachandran 2007; Rizzolatti et al. 2001; Pineda and Hecht 2009).

Mirror neurons have been shown to be involved with empathy (Schulte-Rüther et al. 2007). There is evidence that ToM does use some of the regions belonging to the mirror neuron system in addition to the mentalizing system (Vistoli et al. 2011). This suggests that the two systems are complementary, although not exactly the same.

Mirror neurons may also be involved in language learning (Le Bel et al. 2009; Oberman and Ramachandran 2007; Ramachandran 2011, p. 144). These authors suggest that mirror neurons help an infant repeat a word or words he or she has just heard, as mirror neurons in the baby's brain echo the neurons of the speaker. As evidence of the role of mirror neurons in language learning, Stephens et al. (2010) found that there is similar neural activity in the same regions of both the speaker's and listener's brains. Although on average this mirroring was delayed, there was also a predictive, anticipatory response in the listener's brain regions. The more this anticipatory response occurred, the more the communication was successful. The regions where the neural coupling was observed were those that have been implicated in the production and comprehension of language: early auditory areas (A1+), superior temporal gyrus, angular gyrus, temporoparietal junction (these areas are also known as Wernicke's area), parietal lobule, inferior frontal gyrus (also known as Broca's area), and the insula. Two of these areas, the parietal lobule and the inferior frontal gyrus, have also been associated with the mirror neuron system (Rizzolatti and Craighero 2004).

Lack of Development of ToM in Autistic Children

Autism has been described as a problem of the lack of development of ToM (Baron-Cohen 1995; Schroeder et al. 2010; Moran et al. 2011). Autistic children have great difficulty in understanding the mental states of others and, with the exception of those with high-functioning autism, cannot pass the false belief test. Although they can recognize themselves in a mirror at about the same age as normal children, they have a variety of problems regarding sense of self and tend to treat the self as a thing rather than a person (Reddy et al. 2010). They find the most basic social interactions difficult and stressful. Unlike normal children, they do not engage in pretend play. This deficit in autistic individuals seems to be independent of other cognitive abilities and general intelligence.

Poor ToM skills in autistic children are also associated with impaired language use. About 30% of autistic individuals have no language at all (Bloom 2000, p. 78). Individuals diagnosed with milder forms of autism, such as Asperger's syndrome, often talk like "little professors." Their language is highly literal. So they will answer a question put by a telephone caller "Is your mother available?" by saying "yes" and hanging up. Some autistic individuals are almost normal in their language, and these individuals tend to have better skills at understanding the thoughts of others (Frith et al. 1994).

A variety of evidence shows that autism involves deficits in the mirror neuron system (Williams et. al. 2001; Cattaneo et al. 2007; Le Bel et al. 2009; Schroeder et al. 2010). The neuroscientist V. S. Ramachandran and his colleagues showed this as follows (Ramachandran 2011): As the brain works, it emits various waves, including a mu wave. Whenever a person engages in a voluntary action, the mu wave is suppressed. It is also suppressed when a person watches another person engage in a voluntary action – so it appears to reflect mirror neuron activity. Ramachandran used EEG (electroencephalography) to measure the mu waves. A medium-functioning autistic child who was willing to have electrodes placed on his scalp was the subject. As in normal individuals, the child's mu wave was suppressed when he engaged in any simple voluntary action. Yet when he watched someone else perform the same action, the mu wave was not suppressed, suggesting that his mirror neurons were not firing. Deficits in the mirror neuron system may also help account to the difficulties autistic children have with language (Oberman and Ramachandran 2007).

ToM and the Emergence of the Sociological Self

Development of a ToM is key to developing a sense of self. Here, work on ToM links directly with sociological understandings of the emergence of the mind and self, particularly the work of the early sociologists Charles Horton Cooley (1864–1929) and George Herbert (1863–1931). Understanding the symbolic minds of others is central to Charles Horton Cooley's and George Herbert Mead's sociological theories of the self. Through understanding other people's mental states, we make self-appraisals, for example, we see ourselves as others see us, if we think what they think is important. This shapes both how we understand ourselves and what we choose to do. Cooley (1902) referred to this as the "looking glass self." Understanding the minds of others also helps us learn the rules and culture of our society and become fully socialized selves.

Cooley (1902) thought that members of our primary group were particularly important in constructing his looking glass self. That is, it is our understanding of their views of us that does most to influence our self-understanding. George Herbert Mead (1934) incorporated much of Cooley's thinking into his own understanding of the emergence of the mind and self. By understanding the minds of others and their views of us, we develop a sense of ourselves as an individual with a mind of our own. Mead stressed that this was a maturational, interactional, and adaptational process. An individual acts, sees the response of others, and then adjusts future behaviors to elicit desired responses (Turner et al. 1989: 440–441). With time, a stable repertoire of behavior and sense of self emerges. Both Cooley and Mead saw this feedback process as symbolic and linguistic. The discovery of the role of mirror neurons in this process and their role in helping individuals understand the minds of others provides an additional biological basis for the process (Franks 2010, p. 89).

For Mead and Cooley, our understanding of other people's minds is also an important source of social control, as we usually, but not always, want others' views of us to be favorable. Initially, the primary group is the most important group of others, but as an individual grows older, the relevant groups become larger and more diverse until an individual understands the larger "community of attitudes" among the wider social group. Thus, an individual can take the role of the other, but also the larger community – what Mead called "the generalized other." At this stage, the self is fully formed and relatively stable. This self, in turn, becomes an important determinant of what a person chooses to do.

For Mead, role playing was central to this process of the emergence of the self (Turner et al. 1989: 448). Understanding the minds of others helps us play roles because we know what others expect us to do. This is first manifested in pretend play, which occurs as noted above about 18–24 months of age. Thus, children literally practice the behaviors of people in their group in various roles. So a child at this age may pretend to be a mother or father, baby or child, teacher or waiter. It is by practicing roles that one learns them and also learns to put oneself in the place of another in different roles and thus can more readily anticipate the behaviors of others in other roles.

This helps us adjust our own behaviors in the various roles we play and helps us anticipate the behaviors of others. Mead thought this is what ultimately made a large society possible.

Both Cooley and, to a lesser extent, Mead avoid an oversocialized view of the self in which the self is entirely a construction of the social environment. A largely forgotten qualification of his "looking glass self" made by Cooley himself is that a self constructed totally from the looking glass self is weak and incomplete (Franks and Gecas 1992). The initial sense of self for Cooley is an inborn sense of aggression and assertiveness associated with the more aggressive words like "mine" and "my" instead of reflective appraisals. Mead's concept of role taking as critical to self-development is more completely social, placing more emphasis on a prior sharing of what he called a "universe of discourse" or significant symbols. Thus, for Mead in contrast to Cooley, role taking was the process of responding to oneself as the actor thinks others would and using this anticipated response to control ones' future behavior. This fosters flexible behavior necessary in a social animal. Without sharing significant symbols, this would be impossible (see Chap. 4 above). However, Mead did allow for some autonomous agency in his work separating the "I" from the "me" (Turner et al. 1989: 444–445). The "me" is our understanding of how others see us; the "I" is the actor capable of spontaneous, individual behavior. It is through the behavior of the actor "I" that individual differences and individual drives are manifest. Mead thought that high-status, better-integrated members of a social group have a stronger sense of "me," while low-status, more poorly integrated members of a social group have a stronger sense of (Turner et al. 1989, p. 445).

ToM and Other Sociological Perspectives

Understanding the minds of others is crucially involved in a variety of other sociological perspectives. Most of these are microsociological. First of all is the symbolic interactionist school closely associated with Cooley and Mead. This perspective examines the use of symbols in interaction: how actors use symbols to obtain desired results and how they read and react to others' use of symbols. Symbols include words, but also nonverbal signs such as physical behavior, facial expressions, styles of dress and self-presentations, and so on. Erving Goffman (1959) particularly studied role performance and how individuals use a variety of symbols and props in their role performances.

The exchange theory tradition within sociology sees all social behaviors as exchange of valued goods. For example, a friendship may be seen as the exchange of liking and support between individuals, or a marriage may be seen as an exchange of love and fidelity between husband and wife. Such exchanges depend on the anticipation of the behavior of the exchange partner – a friend offers support because she wants to be supported in the future, and she knows that if she does not offer her support, this may not happen. This involves an understanding of the other's mind and that person's reaction if support is withheld. Similarly, perspectives such as game theory and rational choice theory all involve understandings of a person's future course of behavior and how that behavior is likely to be influenced by the behavior of another person.

Expectation states theory likewise is grounded in our assumptions about others' mental processes and behavior given their social characteristics (Berger et al. 1977). High-status individuals, for example, are expected to think and behave in a competent manner, and lower-status individuals are expected to think and behave less competently. There is also research within the expectation tradition that examines how our understanding of other peoples' expectations of us, given our various characteristics, shapes our subsequent behavior (Webster and Whitmeyer 2002). ToM is also implicated in a variety of other important sociological processes, for example, the formation of self-concept (e.g., Burke 1991), judgments about morality (Guglielmo, Monroe and Malle (2009)), normative processes, and many others.

Conclusion

A sea change happens in the abilities of children between 3 and 4, in that they learn that other people have different beliefs compared to their own and that these beliefs may be false. The development of ToM is the end product of the development of prosocial skills by infants and toddlers that all normal children undergo. It is associated with the development of language abilities, but it is not the same thing. Just as the child's language abilities improve with age, evidence suggests that children become progressively better at understanding the mental states of others as they grow older.

Neuroimaging studies show that certain regions of the brain are involved in ToM tasks. These overlap with the mirror neuron system, the system of neurons in the brain that mirror the neurons of others performing actions, and the functioning of the mirror neuron system seems to support ToM tasks. However, the mirror neuron system complements, but does not replace, the ToM system in the brain.

Problems with ToM abilities are notable in individuals with autism, who also show deficits in brain-related regions and mirror neuron systems. Autistic children also have problems with language. As a result, severely autistic children cannot function normally in the social world.

ToM is an important cognitive development that underlies many of the social processes theorized about and researched by sociologists, including emergence of the mind and self as discussed by Cooley and Mead. Any sociological or other social scientific perspective that involves actors' anticipations of the behavior and response of others is of necessity grounded in this cognitive development. ToM thus links sociology and other social sciences to the life sciences. At the same time, cognitive psychology and neuroscience can learn much about group dynamics from sociologists.

References

Baron-Cohen, S. (1995). *Mindblindness: An essay on autism and ToM*. Cambridge, MA: MIT Press.

Bedny, M., Pascual-Leone, A., & Saxe, R. (2009, July 7). Growing up blind does not change the neural bases of ToM. *Proceedings of the National Academy of Sciences, 106*(27), 11312–11317.

Berger, J., Hamit Fisek, M., Norman, R. Z., & Zelditch, M., Jr. (1977). *Status characteristics and social interaction: An expectations states approach*. New York: Elsevier.

Bloom, P. (2000). *How children learn the meaning of words*. Cambridge, MA: MIT Press.

Brooks, R., & Meltzoff, A. N. (2002). The importance of eyes: How infants interpret adult looking behavior. *Developmental Psychology, 38*(6), 958–966.

Brüne, M., & Brüne-Cohrs, U. (2006). Theory of mind-evolution, ontogeny, brain mechanisms and psychopathology. *Neuroscience and Biobehavioral Reviews, 30*(4), 437–455.

Burke, P. J. (1991). Identity processes and social stress. *American Sociological Review, 56*(6), 836–849.

Butterworth, G., & Cochran, E. (1980). Towards a mechanism of joint visual attention in human infancy. *International Journal of Behavioral Development, 3*(3), 253–272.

Cattaneo, L., Fabbri-Destro, M., Boria, S., Pieraccini, C., Monti, A., Cossu, G., & Rizzolatti, G. (2007). Impairment of actions chains in autism and its possible role in intention understanding. *Proceedings of the National Academy of Sciences, 104*(45), 17825–17830.

Cooley, C. H. (1902). *Human nature and the social order*. New York: Charles Scribner's Sons.

Corkum, V., & Moore, C. (1995). Development of joint visual attention in infants. In C. Moore & P. J. Dunham (Eds.), *Joint attention: Its origins and role in development* (pp. 61–83). Hillsdale: Lawrence Erlbaum Associates, Inc.

de Waal, F. (2009). *The age of empathy*. New York: Crown Publishing.

Fair, D. A., Cohen, A. L., Dosenbach, N. U. F., Church, J. A., Miezin, F. M., Barch, D. M., Raichle, M. E., Petersen, S. E., & Schlaggar, B. L. (2008, March 11). The maturing architecture of the brain's default network. *Proceedings of the National Academy of Sciences, 105*(10), 4028–4032.

Flavell, J. H. (1999). Cognitive development: Children's knowledge about the mind. *Annual Review of Psychology, 50*, 21–45.

Flavell, J. H., Everett, B. A., Croft, K., & Flavell, E. (1981). Young children's knowledge about visual perception-Further evidence for the level 1-level 2 distinction. *Developmental Psychology, 17*, 99–103.

Fodor, J. A. (1992). A theory of the child's ToM. *Cognition, 44*, 283–296.
Franks, D. D. (2010). *Neurosociology*. New York: Springer.
Franks, D. D., & Gecas, V. (1992). Autonomy and conformity in Cooley's self theory: The looking glass self and beyond. *Symbolic Interaction, 15*, 49–68.
Frith, U., Happé, F., & Siddons, F. (1994). Autism and theory of mind in everyday life. *Social Development, 3*(2), 108–124.
Gallese, V., & Goldman, A. (1998). Mirror neurons and the simulation ToM-reading. *Trends in Cognitive Sciences, 2*(12), 493–501.
Gallup, G. G., Jr., Anderson, J. R., & Shillito, D. J. (2002). The mirror test. In M. Bekoff, C. Allen, & G. M. Burghardt (Eds.), *The cognitive animal* (pp. 325–334). Cambridge: MIT Press.
Garnham, W. A., & Perner, J. (2001). Actions really do speak louder than words—But only implicitly: Young children's understanding of false belief in action. *British Journal of Developmental Psychology, 19*, 413–432.
Giacomo, R., & Craighero, L. L. (2004). The mirror-neuron system. *Annual Review of Neuroscience, 27*, 169–92.
Goffman, E. (1959). *The presentation of self in everyday life*. New York: Double Day.
Golinkoff, R. M. (1986). I beg your pardon?: The preverbal negotiation of failed messages. *Journal of Child Language, 13*(3), 455–476.
Gopnick, A., & Wellman, H. M. (1994). The 'theory' theory. In L. A. Hirschfield & S. A. Gelman (Eds.), *Mapping the mind: Domain specificity in cognition and culture* (pp. 257–93). Cambridge: Cambridge University Press.
Gordon, A. S. (2006). Language evidence for changes in a theory of mind. In M. Arbib (Ed.), *Action to language via the mirror neuron system*. Cambridge: Cambridge University Press.
Guglielmo, S., Monroe, A., & Malle, B. F. (2009). At the heart of morality lies folk psychology. *Inquiry, 52*(5), 449–466.
Harris, P. L. (1992). From simulation to folk psychology: The case for development. *Mind Language, 7*, 120–144.
Iacoboni, M., Istvan, I. M.-S., Gallese, V., Buccino, G., Mazziotta, J. C., & Rizzolatti, G. (2005). Grasping the intentions of others with one's own mirror neuron system. *PLoS Biology, 3*(3), e79.
Le Bel, R. M., Pineda, J. A., & Sharma, A. (2009). Motor–auditory–visual integration: The role of the human mirror neuron system in communication and communication disorders. *Journal of Communication Disorders, 42*, 299–304.
Legerstee, M. (1991). The role of person and object in eliciting early imitation. *Journal of Experimental Child Psychology, 51*, 423–433.
Legerstee, M. (1992). A review of the animate-inanimate distinction in infancy: implications for models of social and cognitive knowing. *Early Development Parenting, 1*, 59–67.
Leslie, A. M. (1994). ToMM, ToBy and Agency: Core architecture and domain specificity. In L. A. Hirschfield & S. A. Gelman (Eds.), *Mapping the mind: Domain specificity in cognition and culture* (pp. 119–148). Cambridge: Cambridge University Press.
Lewis, M., & Ramsay, D. (2004). Development of self-recognition, personal pronoun use, and pretend play during the 2nd year. *Child Development, 75*, 1821–1831.
Lillard, A. S., & Skibbe, L. (2005). ToM: Conscious attribution and spontaneous trait inference. In R. R. Hassin, J. S. Uleman, & J. A. Bargh (Eds.), *The new unconscious* (pp. 277–305). New York: Oxford University Press.
Liu, D., Wellman, H. M., Tardif, T., & Sabbagh, M. A. (2008). ToM development in Chinese children: A meta-analysis of false-belief understanding across cultures and languages. *Developmental Psychology, 44*(2), 523–531.
Lovett, S. B., Pillow, B. H. & Bradford H. (2010). Age-related changes in children's and adults' explanations of interpersonal actions. *The Journal of Genetic Psychology: Research and Theory on Human Development, 171*(2), 139–167.
Malle, B. F. (2005). Folk ToM: Conceptual foundations of human social cognition. In R. R. Hassin, J. S. Uleman, & J. A. Bargh (Eds.), *The new unconscious* (pp. 225–255). New York: Oxford University Press.
Mead, G. H. (1934). *Mind, self and society*. Chicago: University of Chicago Press.
Meltzoff, A. N., & Keith Moore, M. (1977). Facial imitation in infants. *Science, 198*, 75–78.
Mitchell, J. P., Malia Mason, C., Macrae, N., & Banaji, M. R. (2006). Thinking about others: The neural substrates of social cognition. In J. T. Cacioppo, P. S. Visser, & C. L. Pickett (Eds.), *Social neuroscience* (pp. 63–82). Cambridge, MA: MIT Press.
Montgomery, K. J., Isenberg, N., & Haxby, J. V. (2007). Communicative hand gestures and object-directed hand movements activated the mirror neuron system. *Social Cognitive and Affective Neuroscience, 2*(2), 114–122.
Moore, C., & Corkum, V. (1998). Infant gaze following based on eye direction. *British Journal of Developmental Psychology, 16*(4), 495–503.
Moran, J. M., Young, L. L., Saxe, R., Lee, S. M., O'Young, D., Mavros, P. L., & Gabrieli, J. D. (2011, January). Impaired ToM for moral judgment in high-functioning autism. *Proceedings of the National Academy of Sciences*, Early Edition.
Oberman, L. M., & Ramachandran, V. S. (2007). The simulating social mind: The role of the mirror neuron system and simulation in the social and communicative deficits of autism spectrum disorders. *Psychological Bulletin, 133*(2), 310–327.

Oberman, L. M., Hubbard, E. M., McCleery, J. P., Altschulera, E. L., Ramachandrana, V. S., & Pineda, J. A. (2005). EEG evidence for mirror neuron dysfunction in autism spectrum disorders. *Cognitive Brain Research, 24*, 190–198.

Pineda, J. A., & Hecht, E. (2009). Mirroring and mu rhythm involvement in social cognition: Are there dissociable subcomponents of ToM? *Biological Psychology, 80*(3), 306–314.

Ramachandran, V. S. (2011). *The tell-tale brain*. New York: Norton Company.

Reddy, V., Williams, E., Costantini, C., & Lang, B. (2010). Engaging with the self: Mirror behaviour in autism, Down syndrome and typical development. *Autism, 14*(5), 531–546.

Rizzolatti Giacomo, L., Fogassi, L., & Gallese, V. (2001). Neurophysiological mechanisms underlying the understanding and imitation of action. *Nature Review of Neuroscience, 2*, 661–670.

Saxe, R. (2006). Four brain regions for one ToM? In J. T. Cacioppo, P. S. Visser, & C. L. Pickett (Eds.), *Social neuroscience* (pp. 83–102). Cambridge, MA: MIT Press.

Saxe, R., Carey, S., & Kanwisher, N. (2004). Effect of subjective perspective taking during simulation of action: A PET investigation of agency. *Nature Neuroscience, 4*(5), 546–550.

Scaife, M., & Bruner, J. S. (1975). The capacity for joint visual attention in the infant. *Nature (London), 253*(5489), 265–266.

Schroeder, J. H., Desrocher, M., Bebko, J. M., & Catherine Cappadocia, M. (2010). The neurobiology of autism: Theoretical applications. *Research in Autism Spectrum Disorders, 4*, 555–564.

Schulte-Rüther, M., Markowitsch, H. J., Fink, G. R., & Piefke, M. (2007). Mirror neuron and ToM mechanisms involved in face-to-face interactions: A functional magnetic resonance imaging approach to empathy. *Journal of Cognitive Neuroscience, 19*(8), 1354–1372.

Stephens, G. J., Silbert, L. J., & Hasson, U. (2010). Speaker–listener neural coupling underlies successful communication. *Proceedings of the National Academy of Sciences, 107*(32), 14425–14430.

Stone, V. E. (2006). ToM and the evolution of social intelligence. In J. T. Cacioppo, P. S. Visser, & C. L. Pickett (Eds.), *Social neuroscience: People thinking about thinking people* (pp. 103–129). Cambridge: MIT Press.

Tomasello, M., & Barton, M. (1994). Learning words in nonostensive contexts. *Developmental Psychology, 30*(5), 639–650.

Turner, J. H., Beeghley, L., & Powers, C. H. (1989). *The emergence of sociological theory*. Belmont: Wadsworth.

Varley, R., & Siegal, M. (2000). Evidence for cognition without grammar from causal reasoning and "ToM" in an agrammatic aphasic patient. *Current Biology, 10*, 723–726.

Vistoli, D., Brunet-Gouet, E., Baup-Bobin, E., Hardy-Bayle, M.-C., & Passerieux, C. (2011). Anatomical and temporal architecture of ToM: A MEG insight into the early stages. *NeuroImage, 54*, 1406–1414.

Webster, M., & Whitmeyer, J. M. (2002). Modeling second-order expectations. *Sociological Theory, 3*, 306–327.

Wellman, H. M., & Gelman, S. A. (1998). Knowledge acquisition in foundation domains. In D. Kuhn & R. S. Siegler (Eds.), *Handbook of child psychology, Vol. 2, Cognition, perception and language* (pp. 523–573). New York: Wiley.

Williams, J. H. G., Whiten, A., Suddendorf, T., & Perrett, D. I. (2001). Imitation, mirror neurons and autism. *Neuroscience and Biobehaviour Review, 25*, 287–295.

Woodward, A. L. (1999). Infants' ability to distinguish between purposeful and non-purposeful behaviors. *Infant Behavior & Development, 22*(2), 145–160.

Chapter 17
Attachment, Interaction, and Synchronization: How Innate Mechanisms in Attachment Give Rise to Emergent Structure in Networks and Communities

Thomas S. Smith

The Eclipse of Individualism

James Coleman (1964) once argued that theoretical coherence in sociology depended on adopting a research agenda that would allow the discipline to make advances akin to those made in nineteenth-century thermodynamics. On his mind was the micro-macro problem—building theoretical models that could take us from an analysis of the behavior of individuals to an understanding of larger systems and structures. That is what happened in nineteenth-century thermodynamics. More specifically, research in statistical mechanics had shown how the properties of whole systems of gases—the pressure, temperature, and volume of Boyle's law—could be synthesized from an analysis of the interactions of individual gas molecules. What Coleman wanted was a theory of the individual person and of individual interaction from which the properties of groups and collectivities could be synthesized.

Coleman's project eventually failed because he lacked the computational power to simulate large numbers of interacting persons. But that was in the late 1960s. Already by then, he had turned to economics and rational choice theory as an alternative—a normative way to explore sociological questions. By the beginning of the 1970s, however, breakthroughs in the biological sciences had begun to foreshadow developments that would make it possible to reconsider Coleman's ambitious project from entirely different foundations. The first of these discoveries came in 1972 when work in Solomon Snyder's laboratories at Johns Hopkins isolated receptor sites for a new class of naturally occurring molecules that resembled morphine—endogenous morphine, as it was then called, later shortened to endorphins. These morphine-like chemicals yield soothing and calming effects, reduce anxiety, and enhance comfort. Most interestingly, they can be elicited by many kinds of social behavior, including, most famously, physical exercise. But, from the perspective of social theory, chief among the kinds of behavior that elevate activity in the endogenous opioid system is attachment. By the end of the 1970s, the basis for a new field—social neuroscience—was being established. Though this field has always been open to research on many links between neural and social behavior, the link between attachment and opioids led to increasingly sophisticated understandings of the dynamics of the dyad of newborn infant and caregiver.

Not individual persons, then, but infant-caregiver dyads were the units of analysis in this early research. The dyad more generally conceived has never been missing from sociology, though it has

T.S. Smith (✉)
Department of Anthropology, University of Rochester, Rochester, NY 14627, USA
e-mail: thomas.smith@rochester.edu

D.D. Franks and J.H. Turner (eds.), *Handbook of Neurosociology*, Handbooks of Sociology and Social Research, DOI 10.1007/978-94-007-4473-8_17, © Springer Science+Business Media B.V. 2013

been weakly theorized and left under the control of nonbiological explanations. But what happens when we shift the focus of social theory away from individualist assumptions and toward the core dynamics of dyads? The infant-caregiver dyad then becomes the prototype for understanding attachment and interaction in general. Embracing human physiology and innate mechanisms in attachment does not necessarily invalidate other approaches to their understanding, but it does suggest that these alternatives are vulnerable to specification error when it comes to causal argument. We must first see how far studying the innate mechanisms can take us, and only then will we know where social psychology and other considerations should weigh in. And we shall have set ourselves on a road that enables us to solve the micro-macro problem in a way that sharply differs from Coleman's approach.

This chapter provides an introduction to the study of opioid-based innate mechanisms in attachment behavior. We shall begin with a formal model of the infant-caregiver dyad and then proceed to derive this model's implications for understanding other core dynamics in social life. These include the architecture of small networks, the dynamics of power-dependency relationships, conformity pressures in interaction, and the social integration of communities.

Hyperstructures

In the 1960s, speculation about the existence of "endogenous morphine" fueled a race among laboratories to find these previously unidentified molecules. When they were discovered and then, a few years later, when their molecular structure was identified, research began on many fronts about how they functioned in the human body. Many experiments showed that opioid antagonists and agonists, for example, produced effects in laboratory animals that mirrored those we see in heroin addicts. For example, mothers injected with opioid receptor agonists become unresponsive to their infants' distress calls. This finding mirrored the unresponsiveness of human mothers who are addicted to heroin. By contrast, opioid antagonists can bring on distress calls that are a weaker equivalent of withdrawal symptoms seen in heroin addicts. Much of this research is reviewed in previous publications (e.g., Smith 1994; Smith and Gregory 1996, 1999; Reite and Field 1985).

The shift from person-focused to dyad-focused experimentation was ripe for formal modeling. When I first began reading this literature in the 1980s, I struggled to find a way to represent the obvious feedback loops present in attachment behavior. The basic observation I had to work with was that a newborn's distress calls communicated themselves to their mothers, who were subsequently moved to attach themselves to their babies—to pick them up, cuddle them, and perhaps feed them. There seemed no mystery in this. But what was missing from the commonsense understanding of infant-caregiver attachment was the part played by endogenous opioids. A mother's attachment had the effect of causing an increase in the activity of the baby's opioid system, diminishing the baby's distress and elevating its comfort.

But it was not just the infant who benefitted from maternal attachment. Attachment behavior increased opioid activity in mothers as well. The mechanism thus involved a double-sided comfort incentive. The distress felt by the mother upon hearing her baby's cries was diminished not only in the infant but in herself as well.

The trigger to this mechanism was the distress call. Opioid levels fluctuate in sinusoidal patterns in all of us. Their equilibrium levels are zones of activity the brain has become habituated to. Dip far below these levels, and the infant will begin to show the classic signs of opioid withdrawal—not, of course, in the same degree observed in heroin addicts, but nonetheless cumulating from mild discomfort to agitation accompanied by distress calls.

The other interesting discovery about the opioid system was that it was inversely coupled, functionally speaking, to activity in the arousal system—more familiarly the HPA axis. A marker for activity in the arousal system is fluctuations in norepinephrine levels. In the infant-caregiver dyad, the arousal system is stimulated by a mother's separation from her baby, a behavior that is typically caused by the appearance of symptoms of withdrawal in the arousal system. These are not the same as those observed in opioid withdrawal but start out as signals that the baby is bored, symptoms that grow into ever more agitated states of fussiness and crying.

Since a mother cannot be simultaneously both attached and separated from her infant, the opioid system and the arousal system function reciprocally. When one is active, the other is shut down. The effects of either system are lagged into the period marking the onset of activity in the other, and these lag periods, when arousal and soothing are mixed, are actually the optimal states of comfort for the infant and the mother.

In Fig. 17.1a and b, we show schematics of the opioid cycle and the arousal cycle, respectively. For the special case of infant-caregiver interaction, these two cycles can be understood as competing with one another; each cycle depends on one of the two mutually exclusive forms of maternal behavior, attachment or separation. The implications of these cycles can be grasped by reconsidering them in relation to the concept of a depot—a reservoir of neurochemicals (opioids or, for the arousal system, norepinephrine) that can be depleted or can be replenished depending on the phase of cycle.[1] The symbols O and N in these graphics refer to opioids and norepinephrine, respectively; A and S refer to attachment and separation, and the subscripts i and c refer to infant and caregiver. The opioid depot cycle, in Fig. 17.1a, shows that opioid depletion, O_i^-, gives rise to withdrawal pressure in the infant, $W(O)_i$. This indicates a need for depot recovery and stimulates maternal attachment, A_c, commencing at a depot threshold point we have called O_i^*. When this threshold is passed, maternal attachment accelerates depot recovery up to the depot maximum, O_i^+, where subsequent use of the depot again depletes it.

A comparable logic appears in the arousal depot cycle. Like an empty reading on a fuel gauge, N_i^-, the depot minimum, is associated with the gradual onset of pressures, $W(N)_i$, for depot recovery. When the N_i^* depot threshold is passed, maternal separation S_c begins to stimulate depot recovery up to N_i^-, where use of the depot again gradually depletes it.

Hyperstructures couple these two neurosocial oscillators, and each then works in oscillatory anti-phase to the other. That is, when the depot associated with either system is depleted, the depot associated with the other becomes inactive; when the second depot in turn is depleted, the first again becomes inactive. The coupling functions to synchronize the infant-caregiver dyad, optimizing comfort. The complete coupled hyperstructure appears in Fig. 17.2.

This graphic is labeled so as to draw attention to various forms of feedback and coupling at work in this mechanism. For example, we speak here of entropy pressures behind the metabolism of resources in each depot, indicating that the mechanism is driven by basic thermodynamic forces

[1] What is a depot? In abstract terms, a depot is simply a reservoir of resources. In the context of studying complex systems, however, the concept of a depot is important and useful. With depots, or reservoirs of resources, systems have the ability to accomplish work, such as maintaining equilibrium, responding to various demands, or damping and distributing distress. Logically, depots also require associated depot structures that generate and store depot resources for later use and can detect when more resources are required. By identifying depot mechanisms and depot structures and by analyzing their related macroscopic functions, it becomes much easier to understand and model the behavior of complex systems. For example, the human immune system consists of a depot of multiple, specialized cell types (monocytes, lymphocytes, basophils, neutrophils, eosinophils). These cells arise from our primary blood-building depot structure, the bone marrow, a depot of stem cell-generating tissue. There are also associated central and peripheral immune structures, such as the thymus, spleen, and lymphatic system. Our immune systems respond to, and are recruited by, physiological distress. Also, since immune cells have a finite life, they must be constantly renewed, or the result would be eventual immune depot decay.

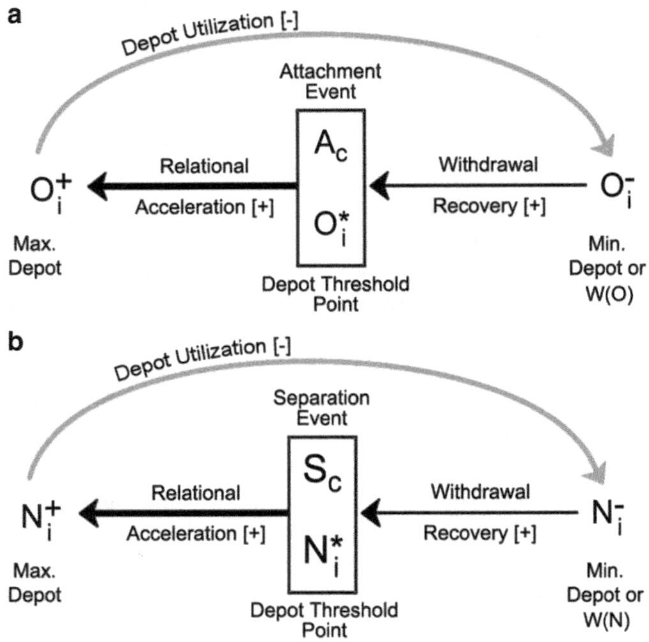

Fig. 17.1 (**a**) Opioid depot cycle. (**b**) Arousal depot cycle

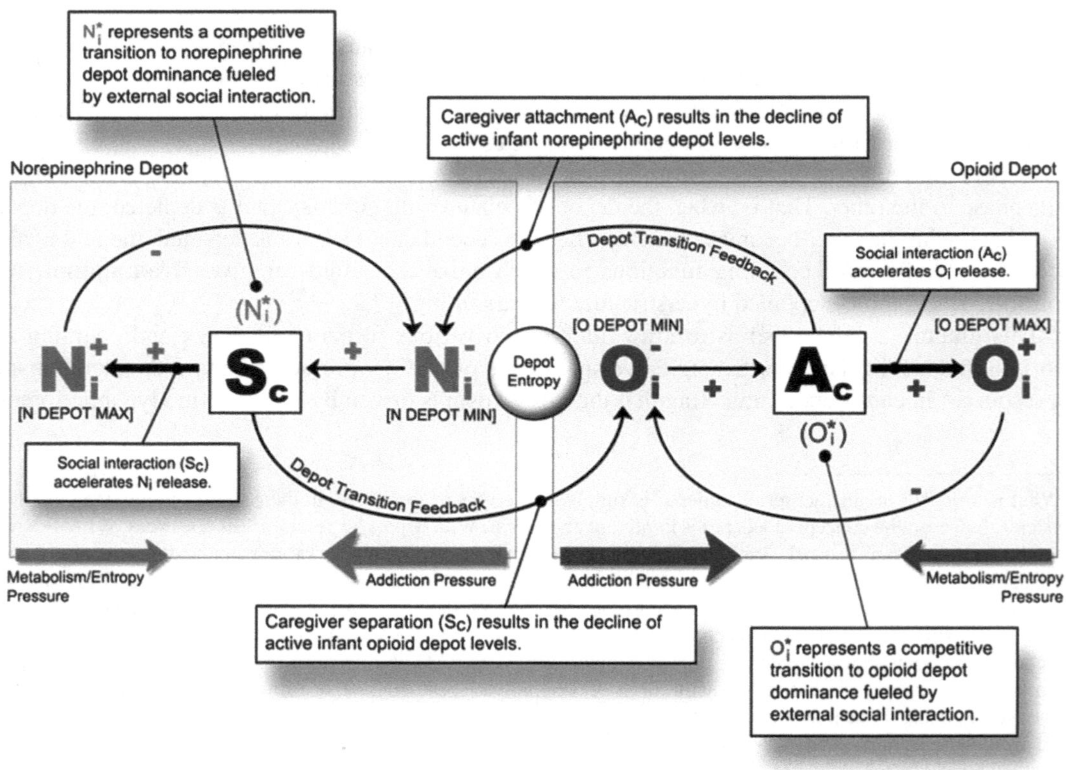

Fig. 17.2 Coupled oscillators produce the dyadic hyperstructure

that appear both interpersonally and physiologically. This is the significance of the entropy sphere at the center of the mechanism—an acknowledgment of the pressures behind metabolism and replenishment of the depots. For example, "addiction pressure" appears in relation to the replenishment of both depots and refers to the pressures behind both attachment and separation (or their various substitutes). In the case of attachment, this is an obvious extrapolation from the morphine-like structure of endogenous opioids. There are equivalent addictive liabilities associated with dependence on chemicals of the HPA axis or their surrogates.

In recent years, evidence from social neuroscience has also established the central part played in bonding and attachment by oxytocin (see especially Cacioppo et al. 2002). The oxytocinergic system controls contractions of smooth muscle tissue and is involved not only in uterine contraction and parturition but also in milk letdown during nursing. As oxytocin passes into the infant with mother's milk, it enhances infant comfort and augments secure bonding. In this way, it can be thought to work along with endogenous opioids in producing dynamic patterns in infant-caregiver attachment and might be modeled as part of the coupled dyadic hyperstructure.

Caregiver substitutes. The tight coupling of infant and mother weakens as the child develops capacities for modulating distress apart from direct attachment. At first, of course, babies are not very good at taking care of themselves. But even in the first months of life, they can be seen to develop weak cognitive and emotional strengths that allow them to decrease dependence for their own comfort on the presence of their mothers. These strengths appear initially in behaviors that babies use to self-soothe or self-stimulate themselves—thumb-sucking and rocking, for example—but they come ultimately under the control of thinking. As cognitive activity acquires increasing coherence over the first 2 years of life, memory and fantasies begin to become "places"—in effect, attractor states—a baby can turn toward to comfort itself. At first, these cognitive structures supplement in only small and insignificant ways the major direct effect of a mother's attachment. But with increases in the child's capacity to think, the hold of the attachment hyperstructure is loosened. Eventually, babies make use of all manner of caregiver substitutes, and they begin to regulate their own internal states with increasing efficiency.

Emergent Structure: Networks as Nonequilibrium Systems

Distress regularly spills beyond the boundaries of dyads. When the demands of infant care overwhelm parents, for example, nearby family members, older children, and even neighbors can be recruited to help manage an infant's needs. When this happens, preexisting links in a social network are being activated. In effect, the family's network is a depot of potential attachment resources, brought to life by signals of distress, pain, suffering, or anxiety.[2] The propagation of such distress into this depot serves to energize and renew the network, refreshing the links. Without the spreading of distress, the network would eventually collapse, its links unused. 'Out of synch, lose the link', as with the Hebbian rule describing neural networks.

Behind attachment, therefore, stands a mechanism that is a structure generator. A link is forged each time distress is communicated to an observer. Not just any observer, of course, but one with the emotional strengths to act responsively to a person in distress. Depending on the interpersonal paths taken as distress spreads, we can observe the growth of networks with differing structures. When we see networks created or activated by the propagation of distress, we have a good example

[2] We refer to this sharing of the burdens of infant care "pooling." Pooling of emotional resources under conditions of stress enables networks to carry higher stress burdens and makes them less vulnerable to breakdowns. For computational evidence that supports this argument, see Smith and Gregory (2002).

of how hyperstructures bridge levels of analysis—one example, that is, of how one solution to the micro-macro problem can be derived from studying the hyperstructure.

Networks generated by coupled oscillators, such as those produced by hyperstructures, are examples of nonequilibrium systems. In the physical world, the most studied example of such a system is water that is slowly heated. At a certain threshold temperature, the water will suddenly exhibit self-organization. A toroidal convection current will emerge (see Prigogine 1978, 1980; Prigogine and Stengers 1984). This self-organized pattern of movement arises because of the circulation of water molecules—first upward and away from the heat source and then, as the molecules cool, downward and back toward it. Beneath the temperature threshold where the convection current first appears, the molecules display only Brownian motion; just above it, we find self-organization; heated still more, the water's temperature will eventually pass another threshold, and the water will boil. Self-organization is then displaced by turbulence and chaos.

Social networks are also nonequilibrium systems. When we study their dynamics, we find similar phenomena of self-organization. The fuel for these dynamics is anxiety and distress—the social equivalents of the heat that produces self-organized features in water. The equivalence is straightforward: no anxiety, no attachment, no interaction. By fueling the hyperstructure mechanism, anxiety produces the patterns seen in infant-caregiver attachment, and these are interaction's basic self-organized features—together and apart, attach and separate. Experimental evidence in support of this claim can be found in the literature of social psychology from 50 years ago. In Schachter's (1959) famous experiments with norepinephrine, for example, it was shown that, even in a group of strangers, increased arousal led to increased interaction. Just as in the case of anxious infants seeking parental attachment, Schachter's subjects looked to one another as their anxiety increased. Dyads are formed and then larger systems of interaction. Through the links in this emergent network spread the anxiety that the experiment had induced. In general, this spreading of anxiety interpersonally is at the core of all interaction's self-organized patterns. These regularly appear so long as anxiety levels surpass the stress-buffering capacity of individuals considered alone. New networks will be generated and preexisting networks reactivated.[3]

As anxiety spreading connects persons, it can have dramatic structural effects on networks. Increasing numbers of persons can come into touch, a step or two, sometimes six steps, away from one another until all are linked into the same world. This is the network structure of the social world—as it turns out, a structure marked by small worlds, or many such small worlds, at least as measured by the average path distance between any two persons. Along the bridges and links connecting these worlds pass information and signals of all kinds.

The social world's network structure thus has a few core architectural features (Smith and Gregory 1999). Most importantly, it is marked by modular organization. Each of its small worlds is a network cluster—persons linked to one another more closely than they are linked to outsiders. When there are bridges between these clusters, these worlds potentially come within reach of each other. Since intercluster bridges are typically weak ties, as Granovetter (1973) famously argued in the 1970s, weak ties can be seen to strengthen social integration, chiefly by increasing the connectivity of a society and enabling the diffusion of information.

Yet there is also another face to weak links. Network scientists working in other fields have shown that if we continue to introduce stress into a network, it will eventually collapse. And, perhaps not surprisingly, a network's weakest links are where the collapse will begin. In social systems, there is long-standing evidence that shows this happening. When we raise the stress passing through a

---

[3] This energy dependence of networks is a common network property throughout the natural world. Network theorists call this "housekeeping energy"—the level of energy required to prevent a network from collapsing. Fluctuation in energy below the housekeeping threshold will cause networks to collapse. See, for example, Peter Csermely's discussion (2006).

community, for example, persons tend to retreat into local clusters—withdrawing from the larger community and retreating into smaller networks of family and kin (Fellin and Litwak 1963).

What is important about this? The tendency of weak links to collapse under increasing stress localizes or confines the dissipation of stress, an outcome that adds to network stability. If weak links never collapsed under heavy stress burdens, networks would be subject to cascading events like panics, even topological phases transitions—the quakes, slides, fractures, and population collapses commonly found in other complex systems. There are many illustrations of this kind of effect in social life. As downturns in the economy gain speed, for example, Stark and Vedres (2002) have shown that business consortiums undergo topological phase transitions. Their networks, once fully connected, suddenly reorganize into star-shaped structures, with one leading firm at the center. The star pattern has reduced the number of links in the network, with each business then connected only to one central firm. This same star shape emerges in other situations where stress increases. Where resources grow short, for example, a star network takes shape in communities that were formerly more densely connected. In the military and other "emergency systems" such as police agencies and fire departments—systems in which stress is chronic—a star network is institutionalized in the hierarchy of rank. One α-leader is at the center of each team of responders—each platoon, each engine company, each station house, each barracks. Increases in stress, thus, are associated with the collapse of weak links and a general reduction of the number of links in a network, with hierarchy entering into the picture. But is this the full story? When it comes to social networks, the story has a familiar twist.

Synchronization as a Network Emergent

Key to this story are regular temporal shifts in network structure. Some of these are calendrical, with weekends and holidays serving as times for relaxation, for de-stressing. Modest increases in the stress burden of a network, confined to the work week, are followed by days when stress is relieved, as on weekends. Family, religion, and community life—not work—are at the center of these retreats. As in the case of an infant's distress calls, stress creates pressures to attach, to strengthen ties—or, as we have come to understand this in view of the hyperstructure, to service the opioid depot. Once distress is modulated, then the arousal depot will again require servicing, and pressure subsequently supports separation. Oscillation between attachment and separation is the result, equating to the synchronization between infant and caregiver. Yet in other terms altogether, we have known something like this for a century or more.

At least since Durkheim's work on religion (1995 [1912]), sociologists have understood the importance to social integration of regular oscillations between communal and individual life. Times of intensified social or communal activity, in Durkheim's famous analysis, reinforced social integration, chiefly through ritual practices that synchronized a community—dance, chant, worship, religious theater, reconstructions of communal memory, and reenactments of communal myth. Again, synchronization is what is at issue here. If we synchronize even weakly connected persons, we strengthen their ties. Synchrony of the sort Durkheim addressed in religious or communal settings, therefore, is far more ubiquitous in social life than he might have imagined. Any performance ensemble—chamber orchestra, dance troupe, choir, theater company, marching band—yields examples. So does the unification of minds engaged in thinking together, the synchrony of a crew in a regatta, the reconstruction of autobiographical memory at a family event, and the emotional unification of an audience at a football game–all yield synchrony that can strengthen interpersonal links and deepen personal and social integration. Yet when these moments of synchronization end, persons drift apart and their merger into their shared world slowly attenuates, like a memory that fades. They separate, and the strength of their links weakens.

The mechanics of synchronization always require the coupling of multiple oscillators. Networks, therefore, are where synchronization sets up—no network, no synchronization. Significantly, in settings where synchronization is orchestrated by a community, exogenous oscillators such as dance or music piggyback onto the attachment mechanism. The result is that some of what a mother and an infant feel when synchronized is also felt where others are synchronized—comfort, security, attachment, and membership. Examples such as drumming, chant, or communal prayer highlight how this can happen. The beats of a drum or the recitals of a chant are clocks, and they slave persons to themselves. When persons are yoked to these exogenous clocks, the strength of their ties to one another increases. Notably, they also find themselves merged into a synchronized system of interaction, whether it is an ensemble making music together, a group of actors performing a play, or a whole community celebrating its history.

This pattern of merger-into-a-whole exhibits features of dissociation, not too far removed from what Mead (1934) discussed as "taking the role of the other." The person, usually adrift in secular pursuits, is displaced, a figure who is drawn into a story or a collectivity, captivated by the drama and enlivened by the strong feelings—the effervescence, as Durkheim called it—that are driven by synchrony with others. Identity fades, supplanted by membership in a wider world that has history, narratives, roles, and statuses that are usually acknowledged only at a distance, familiar and yet lost to sight when quotidian involvements turn the eye away from the "center" of society (to borrow Edward Shils's [1978] expression).

Perhaps because there is a tendency for weak links to collapse under extreme conditions, communities regularly orchestrate these occasions for stress to be dissipated and for links to be strengthened by synchronization. The oscillation is between a phase of social dispersion—Durkheim's profane time, marked by weaker links to the collective world—and a phase of intensified social and communal activity—Durkheim's sacred time, marked by increasing tie strength.[4] Durkheim saw that this phase of intense synchronization drew persons into contact with the sacred symbols of their social world and hence augmented moral and cultural integration. But there is still more that synchronization can teach us about moral life, though discovering it requires us to make a conceptual leap—to understand that attachment behavior is actually an extension of the human immune response. How can that be?

Synchronization, the Immune Response, and the Sixth Sense

J. E. Blalock (1984, 2005), one of the leading investigators of human immune responses, has described the immune system itself as a "sixth sense," meaning that it can plausibly be likened to organs or structures of the body that perceive threatening stimuli, detecting, in this case, invaders such as bacteria or viruses. Having registered one of these invaders, the immune system deploys its own cellular agents—complement proteins, cytokines, macrophages, neutrophils, and natural killer cells, in combination with other chemical and biological players—as part of a concerted biological response to engulf and rid the body of them. Such pathogens are stressors, signaled at

[4] These correspond to two different phases of moral culture. On the one hand, the secular phase is consistent with the view of culture as a "toolkit" from which, once persons act or make decisions, they can retrofit their decision with explanations or justifications (Swidler 1986). This is a phase of diluted social integration and fragmented morality. By contrast, the phase of intensified synchrony and strengthening ties corresponds to those occasions when exogenous oscillators like dance or worship synchronize whole communities and leverage social and moral integration from interaction. An outcome of this phase is that persons use moral reasoning to guide their activities and reach decisions.

the site of their invasion by a growing inflammatory response that is orchestrated by the immune system—increased blood flow, recruitment of macrophages from the blood in nearby tissue, the release of cytokines and other chemical messengers of the immune system, and other effects as well. An important result of inflammation is the release and subsequent recruitment of endogenous opioids—chiefly, β-endorphins—whose effect is to reduce pain at the site of the invasion. Once the body (and the person) has been alerted to the assault, opioids then soothe and reduce the inflammation.

Distress as inflammation. Attachment can also be seen as an immune response. The argument in support of this hypothesis is straightforward and indeed generalizes Blalock's concept of a sixth sense (Smith and Silon 2005, 2007). A summary will serve our purposes here. The same way the sting of a bee or an abrasion to the elbow will recruit analgesic substances to inflamed tissue, distress in social life also recruits analgesic resources to sites of social "inflammation" (Bildack 2000). A baby's distress call is the interpersonal equivalent of an inflammation—a signal that first recruits maternal attachment and that in turn stimulates the release of endogenous opioids. Attachment is thus analgesic—or, more properly put, it stimulates the release of analgesic substances. Alleviation of distress, in this way, is equivalent to the reduction of inflammation. The implication is then obvious: Attachment functions as a social extension of an innate immune response. Whether the cries of a baby, the sting of bee, or the suffering of a friend, attachment makes things better.

When we translate this argument into a network perspective, we begin to uncover some remarkable phenomena. Synchronization extrapolates the effects of attachment across networks, as when an anthem, taken up by an audience, induces a feeling of membership, affiliation, or citizenship—each an extract of attachment itself. If a crew in a regatta rows with perfectly timed strokes, its members have trained to the point where they can feel any deviation from synchrony. When monks sing together in chapel, any monk whose song stands out from the rest has failed his duties by establishing himself apart from the assembly. He is the deviant figure in an otherwise anonymous collectivity. The same is true of musical ensembles—their members are trained to make music together, not as individual performers. If an individual performer surfaces, synchrony is lost. Notably, this is exactly the opposite of how persons behave when they are away from these assemblies, ensembles, or teams. In those other settings, they strive for distinctiveness, achievement, and visibility. But in a synchronized setting, individualism and personal distinctiveness are disturbing. They are the equivalent of attachment loss or abandonment, the worst fear of a child.

Strong integrative pressures therefore exist in synchronized assemblies. Each member is linked to others by synchrony itself, a condition that recapitulates the perfect unity of infant and caregiver. There is family feeling in such assemblies. On the larger boats that compete in regattas, for example, there is always a coxswain who establishes a stroke, and the stroke brings every member into synchrony. The coxswain is the exogenous oscillator and the crew her slave. In these boats and in smaller boats without coxswains—boats where there are only two, four, or six rowers—there is also another form of communication that is not auditory or verbal at all. It occurs through detecting deviations in the balance of the boat. These indicate loss of perfect synchrony, a condition in which the boat loses power and speed. Such deviations are sensed as changes in the boat's balance—feelings to which a crew's training has sensitized them. In maintaining synch in larger teams or performance ensembles of all kinds, exogenous clocks become especially important. As the size of symphony orchestras increased in the second half of the eighteenth century, for example, orchestras stayed in sync first by relying on cues from concert masters (usually first chair violinists) and then, when orchestra size increased beyond the point where all players could be in visual and auditory contact, finally on the ultimate α-leader, the conductor (see Raynor 1978a, b; Hart 1973). Orchestras thus shifted away from being more densely connected networks to acquiring the familiar star-shaped structure we have discussed above, where stress levels cause weak links to collapse and a central core figure appears.

Coxswains and conductors are metronomic. They are clocks, akin to those we associate with the oscillations in every community between togetherness-and-apartness, attachment and separation. But there is another way to understand how they function, and this arises from the analgesic benefits to collective life network synchronization enables. Networks of persons synchronized to one another extend the comforting immune function of attachment. With synchronization coincides a network-based enhancement to the modulation of distress, akin to the "sixth sense" that Blalock has described. A monk out of sync with his brothers is a disturbance to the whole community. The member of a crew off stroke stresses other persons in the boat. As with these stressors and irritants, other disturbances to communal integration—invasion, illness, deaths, attacks, assaults, accidents, crimes, and threats of all kinds—will also spill distress into the links of synched networks from the point at which they are detected. When this happens, the network is activated. An inflammatory signal like a distress call transforms it into an attachment depot. And as in the body, inflammation then recruits analgesic resources from nearby links, sometimes from the whole depot. This is a network phenomenon and can only occur where networks are periodically strengthened by synchronization.

Lost Dogs, Missing Children, Strangers, and Thieves

Such shifts and oscillations are important because they reinvigorate local network organization and functioning. In the examples we have reviewed here, synchronization is obvious. But recall, now, the underlying oscillator. The integration of social life does not require ongoing and continuous synchronization. Instead, periodic synch is enough to maintain a network-based social immune response. The rest of the time, networks are latent, waiting to be activated. Their immune response does not even require that all persons in a community be part of a network. But those in it must have their links renewed from time to time.

Socially integrated neighborhoods show their network properties all the time. The best illustrations of this come from studies of crime. Jane Jacobs (1992 [1962]) reported that there were some neighborhoods in Milwaukee where the mere appearance of a stranger on the streets was enough to trigger phone calls to the police. By contrast is the indifference to crime that appears in fragmented neighborhoods, where persons have been observed watching from their windows as crimes take place below them. In general, social control and law enforcement in any society are network outcomes. They can never rest on the shoulders of policemen alone. Civic vigilance and social networks are always involved—reports to the police, witnesses to crimes, concerns for safety and order, and, from time to time, communal action itself. In the frontier settings of the American West, law enforcement was sometimes missing altogether. But taking its place were the posses that would form when news of a crime passed through a community. In European countries, the enforcement of law remained a rotating civic responsibility until the appearance of the first urban police systems at the end of the eighteenth century, or it similarly depended on the *posse comitatus* (Banton 1965; Buisson 1958).

So networks strengthened by synchronization are responsive to stressors of many kinds. Lost dogs and missing children are enough to mobilize whole neighborhoods into search parties. Strangers are always sources of apprehension. Telephone calls pass from house to house if someone unknown appears on a neighbor's property. If illness afflicts a neighbor, he will awake in the morning to find soup on his doorstep; a death will bring everyone in the network to a wake or a funeral; a house fire will inspire donations of clothes and food. Commonplace though these examples may seem, each illustrates a network response to distress—the recruitment of analgesic resources to sites of social inflammation.

Even in the largest cities, therefore, there are always functioning neighborhoods. People identify themselves as residents of a place with a name—Hyde Park or Greenwich Village or the Latin Quarter or the Trastevere. Such knowledge restores them to home and to the links for which kinship and family are the prototypes. In some of these settings, networks are so synchronized that they function as a kind of social sensory system, akin to the workings of the immune system. They are the sixth sense on which social life depends.

Conclusions

Synchronization in social networks underlies social integration. I have argued in this chapter that the prototype for this is infant-caregiver attachment. In the functioning of healthy families, innate forces always synchronize newborns and their mothers. From their synchrony emerge the remarkable analgesic effects of endogenous opioids. If we extrapolate the functioning of the innate mechanism behind attachment to social networks, we find similar outcomes. Synchronized networks are those Durkheim first described in his great studies of religion, but such networks are ubiquitous in social life.

Synchrony appears wherever networks are activated from time to time by stressors. Powerful illustrations of this are found in ensembles, monasteries, teams, theaters, conversation, and the reconstruction of memories. A few of these illustrations have been developed in detail above. But what they point toward are network phenomena that are far more pervasive in social life. We cannot really speak of neighborhood or community without recognizing how social life leverages social integration from attachment and interaction. No *gemeinschaft*, no village, no tribe, no congregation, and no society can really be fathomed without first understanding how its core properties derive from coupled oscillators—from synchronization that builds and strengthens social networks.

References

Banton, M. (1965). *The policeman in the community*. New York: Basic Books.

Bildack, J. (2000). Detection and function of opioid receptors on cells from the immune system. *Clinical and Diagnostic Laboratory Immunology, 7*, 719–723.

Blalock, J. E. (1984). The immune system as a sensory organ. *Journal of Immunology, 132*, 1067–1070.

Blalock, J. E. (2005). The immune system as a sixth sense. *Journal of Internal Medicine, 257*, 126–138.

Buisson, H. (1958). *La Police, son histoire*. Paris: Nouvelles Éditiones Latine.

Cacioppo, J., & Berentson, G. (2002). Social neuroscience. In J. Cacioppo, G. Berentson, et al. (Eds.), *Foundations in social neuroscience* (pp. 3–10). Cambridge: MIT Press.

Cacioppo, J., Berntson, G., Adolphs, R., Sue Carter, C., Davidson, R., McClintock, M., McEwen, B., Meany, M., Schachter, D., Sternberg, E., Suomi, S., & Taylor, S. (Eds.). (2002). *Foundations in social neuroscience*. Cambridge: MIT Press.

Coleman, J. S. (1964). *Introduction to mathematical sociology*. New York: The Free Press of Glencoe.

Csermely, P. (2006). *Weak links: Stabilizers of complex systems from proteins to social networks*. Berlin: Springer.

Durkheim, E. ([1912] 1995). *The elementary forms of religious life* (Trans. K. Fields). New York: The Free Press.

Fellin, P., & Litwak, E. (1963). Neighborhood cohesion under conditions of mobility. *American Sociological Review, 28*, 364–376.

Field, T. (1985). Attachment as psychobiological attunement: Being on the same wavelength. In M. Reite & T. Field (Eds.), *The psychobiology of attachment separation* (pp. 415–454). New York: Academic.Granovetter, M. (1973). The strength of weak ties. *The American Journal of Sociology, 78*, 1360–1380.

Hart, P. (1973). *Orpheus in the new world: The symphony orchestra as an American cultural institution*. New York: W. W. Norton.

Jacobs, J. ([1962] 1992). *The death and life of great American cities*. New York: Vintage.

Mead, G. H. (1934). *Mind, self, and society* (C. Morris, Ed.). Chicago: University of Chicago Press.
Prigogine, I. (1978). Time, structure, fluctuation. *Science, 201*, 777–785.
Prigogine, I. (1980). *From being to becoming: Time and complexity in the physical sciences*. New York: W. H. Freeman and Company.
Prigogine, I., & Stengers, I. (1984). *Order out of chaos*. New York: Bantam Books.
Raynor, H. (1978). *A social history of music* and *music in society*, 2 vols. New York: Taplinger.
Raynor, H. (1978b). *The orchestra: A history*. New York: Charles Scribner's Sons.
Reite, N., & Field, T. (1985). *The psychobiology of attachment and separation*. Orlando: Academic.
Schachter, S. (1959). *The psychology of affiliation*. Stanford: Stanford University Press.
Shils, E. (1978). *Center and periphery: Essays in macrosociology*. Chicago: University of Chicago Press.
Smith, T. S. (1994). Catastrophes in interaction. *Social Psychology Quarterly, 57*, 274–282.
Smith, T. S., & Franks, David. (1999). Emergence, reduction, and levels of analysis. In D. Franks & T. Smith (Eds.), *Mind, brain, and society* (Volume 5 of Social perspectives on the emotions, pp. 3–18). Stamford: JAI Press.
Smith, T., & Silon, D. (2005, July 12–14) *Neurosociological foundations of social networks: Distress-dependent synaptic architecture in social systems*. Paper discussed at Plenary Session, International Sociological Association, Interim Meetings, University of Rome.
Smith, T., & Silon, D. (2007). *Biological oscillators, circadian clocks, and sacred time*. Farfa Abbey: Plenary address made at meetings of the International Sociological Association.
Smith, T. S., & Gregory, T. S. (1996, July). Emergence, self-organization, and social interaction: Arousal-dependent structure in social systems. *Sociological Theory, 14*, 131–153. [Also: Santa Fe Institute Publication Series, Paper No. 94-08-046.]
Smith, T. S., & Gregory, T. S. (1999). The architecture of small networks: Strong interaction and dynamic organization in small social systems. *American Sociological Review, 64*, 403–420.
Smith, T. S., & Gregory, T. S. (2002). Hyperstructures and the biology of interpersonal dependence: Rethinking reciprocity and altruism. *Sociological Theory, 20*, 106–130.
Smith, T. S., Stevens, G. T., Caldwell, S. C., Franks, D., & Smith, T. (1999). The familiar and the strange: Hopfield network models for prototype entrained attachment-mediated neurophysiology. In *Mind, brain, and society* (Volume 5 of Social perspectives on the emotions, pp. 213–245). Stamford: JAI Press.
Stark, D., & Vedres, B. (2002). *Pathways of property transformation* (Santa Fe Institute Working Paper No. 20112081).
Swidler, A. (1986). Culture in action. *American Sociological Review, 51*, 273–286.

Part III
Evolution of the Brain

Chapter 18
The Secret of the Hominin Mind: An Evolutionary Story

Alexandra Maryanski

Nothing comes from nothing

Emile Durkheim 1895

In the playing field of evolution, organisms are favored or "selected out" depending on environmental demands. Natural selection acting on variant phenotypes is what drives change in a species' gene pool, or what Darwin called "descent with modification" *between* generations.[1] And no better example of its potent influence exists than the hominin (i.e., human) brain and mind.[2] Why do humans have such big and complex brains? The brain's infrastructure is difficult to see, and so knowledge about its intricate mechanics is limited, but one portal into its inner workings is to trace its evolutionary foundations.

Humans have encephalized brains—1,350 cubic centimeters (cm^3 or cc) on average—and stuffed with 100 billion neurons which are way beyond the expected relationship between brain and body size (Kass 2008).[3] As the command center for human activity, our brain directs all body and cognitive functions as well as states of emotional arousal. It also houses a rare "self-consciousness" among animals, which lends itself to a personal identity and to taking the role of others. The social mind also engages in two distinctive types of sociality: (1) an ability to form tight-knit kinship bonds or strong ties, a trait shared with social mammals, and (2) an ability to form loose-knit friendship bonds or weak ties, a rare trait but essential for the creation of large-scale societies with millions of individuals.

As our neocortex makes up 80% of our brain volume, this chapter provides an evolutionary account of the selection processes that created this large cortex. As our brain is metabolically (energetically) costly, its encephalized size and complexity must reflect seldom seen challenges in

---

[1] Selection is the architect of evolution, of course, but the variation it needs to act is generated by the other forces of evolution—mutation, genetic drift, and gene flow.

[2] A *hominin* is the new taxonomic term now used when referring to living humans and past humans as well as all extinct relatives that walked upright (e.g., *Australopithecus)*). It replaces the term *hominid* which is now more inclusive and includes our closest living and extinct ape relatives.

[3] The relationship between brain size and body size for most species is closely correlated, so it would be expected that larger animals would have bigger brains than smaller ones. The encephalization quotient (EQ) is the ratio of observed brain size to the predicated size given the overall body size of a species. The encephalization quotient in humans is tremendous.

A. Maryanski (✉)
Department of Sociology, University of California, Riverside 92521, CA, USA
e-mail: alexandra.maryanski@ucr.edu

D.D. Franks and J.H. Turner (eds.), *Handbook of Neurosociology*, Handbooks of Sociology and Social Research, DOI 10.1007/978-94-007-4473-8_18, © Springer Science+Business Media B.V. 2013

past habitats, and if so, clues should still exist in the context of the problems it evolved to solve. Consequently, to get answers on how our mind was put together as well as what was put together, we need to trace its earliest forms starting with the evolution of the *primate neocortex*. For surprisingly, most facets of our neocortex so often portrayed as special to humans alone are actually *the* hallmarks of primate evolution in general and according to Le Gros Clark (1962: 227) "undoubtedly the most distinctive trait of the primates." Hence, notions that the human cerebral cortex is a "gift of the gods," a result of large random mutations, or mostly assembled during the Pleistocene are misguided. The human brain was configured in a variegated mosaic pattern by a series of far-reaching evolutionary novelties—none of which evolved as adaptations to the modern world.

Instead, during the course of primate evolution, the evidence points to three adaptive landscapes that ultimately shaped the cognitive software that lies beneath its convoluted surface:

- The sensory legacy of early anthropoids and the shift to visual dominance
- The great hominoid die-off and the shift to forelimb dominant locomotion
- The hominin shift to bipedalism and an open-country niche

As natural selection is linked to environmental settings and what happens to populations over time, we can use the fossil record to trace the physical development of the hominin brain. Brain tissue is rarely preserved, but endocranial casts of fossilized cranial cavities are available that accurately reproduce intriguing bulges, fissures, and depressions to determine the brain's shape and size. Skulls can also be chronometrically dated in calendar years, and primate paleoenvironments can be recreated using fossilized seeds, pollens, dust, plants, and the diversities in habitat sediments drawn from the geological and archaeological records (for discussions, see Jerison 2007; Halloway 1978; Falk 2007; Rakic and Kornack 2001, 2007; Preuss 2007; Sousa and Wood 2007; Potts 2004).

While paleoneurological materials can chart identifiable brain changes, these data are limited when making inferences about internal organization and functioning. Here, we need data on the neuroanatomy, social behaviors, and the molecular record on living Old World primates. As higher primates (i.e., monkeys and apes) share most of our evolutionary history, they can provide a yardstick and clues into how the hominin brain is organized and what is truly distinct about human neuroanatomy. However, before venturing into deep time and a "stranger than fiction" tale that will call up images of *The Rise of the Planet of the Apes,* let me start with some necessary background on living primates.

The Primates

The primate order includes well over 200 living species who for our purposes can be placed into two major groups. The first group is prosimians (*Prosimii*), and they make up about 15% of all primate species, ranging from the lemurs of Madagascar (*Lemuroidea*) through the lorises and galagos of Asia and Africa (*Lorisoidea*) to the tarsiers (*Tarsioidea*) of Asia. Prosimians (which means before monkeys and apes) are the living descendants of a diverse radiation of early primates who are easily distinguished from higher primates by their night-active habits (except for lemurs), stiff upper lips and bonny facial expressions, distinctive dentition, "deer in the headlight" wide staring eyes, a moist patch on their noses (except for tarsiers), and by their amusing way of moving about by leaping and hopping.

The second group is anthropoids (*Anthropoidea*). They are day active (except for one night monkey) and divided up into three distinctive subgroups: (1) New World monkeys (or *Ceboidea*) of Mexico and South America; (2) Old World monkeys (*Cercopithecoidea*) of Asia, Africa and Europe; and (3) ape-human hominoids (*Hominoidea*) of Asia, Africa, and worldwide for humans.

One other necessary classification is that New World monkeys are known as *Platyrrhini* (which means flat shaped noses) while Old World monkeys and ape-humans are called *Catarrhini* (or primates with down-facing noses). Together, subgroups (1) and (2) of monkeys are a great evolutionary success because they make up about 80% of all primate species. New and Old World monkeys are distinctive in many ways, but both subgroups are characterized by a quadrupedal anatomy, a tail (some are prehensile in New World monkeys), a narrow rib cage, immobile shoulder joints, a short collarbone, and four limbs of near equal length. Monkeys, thus, reveal a morphology designed for "above branch" walking or ground walking with palms and soles flat on the surface.

The third subgroup of anthropoids is apes and humans (or the *hominoids*) who comprise just 5% of primate species. Hominoids are a tailless family that includes little gibbons (*Hylobates*) and four large-bodied hominoids—orangutans (*Pongo*), gorillas (*Gorilla*), chimpanzees (*Pan*), and humans (*Homo*). Apes and humans are characterized by a unique upper anatomy that includes a short, deep trunk and spine, an elongated collarbone, and specialized hands and shoulders. This anatomy is originally designed for vertical climbing and "below-branch" suspension in arboreal habitats, with an emphasis on the hands for support. Chimps, gorillas, orangutans, and humans are also very large-bodied primates with much bigger brains than monkeys—both in absolute terms and relative to their respective body weights.

Primates are socially inclined, and most live in year-round stable societies. For mammals that are highly intelligent, mature slowly, and live a long time, such permanent living arrangements require the integration of not just the adult males and females but also infants, juveniles, and adolescents. Primates rely primarily on socialized learned behaviors, but species usually keep to a modal group size, a typical sex and age ratio, and a sex-biased dispersal pattern. A male dispersal pattern is the norm for most primate societies (and for most social mammals), while females stay home with their blood relatives and provide the bedrock for cohesion and continuity over intergenerational time. In contrast, hominoid societies (ape and human) deviate from this pattern by dispersing both sexes (or only females) at puberty. As one would expect, the dispersal of both males and females after puberty results in a very different relational structure than those found in monkey or prosimian societies, with implications for the evolution of the hominin mind (for discussions, see Stanford et al. 2013; Fleagle 1999; Falk 2000; Turner and Maryanski 2008; Relethford 2010). Let us now turn to a brief account of the origin of the mammalian neocortex and sense modalities before we examine the three major evolutionary events that essentially created the hominin mind.

The Sensory Legacy of Early Anthropoids and the Shift to Visual Dominance

The Sensory System of Mammals

During most of biological evolution, organisms had little sensory equipment, but this changed about 250 million years ago when a six-layered neural structure—a "neocortex" (or new bark)—appeared on top of an ancestral mammalian forebrain.[4] Four paired specialized lobes (that are parts of the cerebrum) became distinguished—frontal, occipital, temporal, and parietal—with vast tracts of cortical tissue given over for sensory perception. The visual system in the occipital lobe (and partly in temporal lobe) has receptors in the retina responsive to different wavelengths of light energy; the auditory system in the temporal lobe has receptors in the ears sensitive to vibrations activated by

[4] Early vertebrate brains had a structure of three divisions—a forebrain, a midbrain, and a hindbrain.

sound waves; and the generalized somatic sense in the parietal lobe has receptors on the skin sensitive to vibrations and to touch. The tissue of simple mammals was composed mostly of motor and sensory zones, but with advanced mammals, their respective neocortices were enhanced with specialized regions or "association tissue" for complex, higher-order functioning.[5]

In turn, the mammalian forebrain (or limbic system) consists of a complex ring of subcortical tissue deep in the brain that houses not only the centers for emotions but two ancient chemical modalities: (1) the olfactory system with receptors in the nose sensitive to airborne chemical substances and (2) the gustatory system with taste receptors sensitive to substances in the mouth. While the limbic fields are engaged more in visceral tasks associated with emotional and motivational functions and the internal well-being of an organism, the neocortical fields are more involved in the extraction of general information about the outside environment. All mammals have the same sensory fields, but mammalian lines differ dramatically in how much tissue is devoted to each sensory domain compared to other sensory domains (Krubitzer and Kaas 2005; Molnár et al. 2007; Hoover 2010; Rakic and Kornack 2007).

Still, all sensory inputs into the real world remain profoundly restricted for any mammal because it is physically impossible—no matter how sophisticated the sensory tools—to detect all possible environmental stimuli.[6] Sensory receptors then *must be selective* in their representational inputs, with the quality and type of sensory equipment a function of environmental demands. The sensory organ that takes the leading role, especially for object recognition, is the one essential for survival and reproductive success. With these limitations in mind, let us start with our primate sensory legacy. Since our sensory equipment is nearly the same as Old World monkeys and apes, we can trace its phylogenetic progression for insights into why and how it evolved, the relations among modalities, and the essential role played by our dominant visual system in the creation of a "language-ready" hominin mind. And, by all accounts, this process was triggered when a small, ground-living rodent first took up life in the trees.[7]

The Origins of Primates

About 65 million years ago *primate-like* mammals appear in the Paleocene fossil record.[8] Called *Plesiadapiformes*,[9] these taxa lived in tropical habitats in Europe and North America (especially Wyoming and Montana) during a warm period of the earth's history. The *Plesiadapiformes* were

---

[5] Fossilized endocasts and comparative data on living mammalian groups indicate that early mammals were tiny creatures with small brains that contained only a tiny neocortex with only about 20–25 cortical areas and no hemispheric specializations. Modern humans are estimated to have at least 200 cortical areas (see Kass 2008).

[6] For example, most mammals cannot detect ultraviolet light, although many insects, fish, birds, and snakes are aware of its presence. One exception are Arctic reindeer (Rangifer) who live in a relatively rich UV light environment with rods and cones sensitive to ultraviolet light. With selection for an enhanced visual range that included short-wavelength vision, Arctic reindeer were able to adapt to the extreme seasonal differences in light levels (Hogg et al. 2011).

[7] Although evolution is a population concept that refers to change, and not progress or complexity, the fossil record reveals a very slow but increasing complexity of the cortical senses that virtually exploded with the rise of higher mammals, especially primates.

[8] Each epoch of the Cenozoic Era is associated with the radiation of a particular grouping of primates, although overlap exists in time period lines.

[9] The Plesiadapiformes refer to a large grouping of extinct mammalian families who appear to be the stem population for the origin of true primates in the late Paleocene and early Eocene. Not everyone agrees with this assessment, although recently discovered fossils and a reinterpretation of older fossils lend weight to this assessment (see Ross and Martin 2007: 74).What nobody doubts is that the first true primates relied upon a tree-living niche for survival and reproduction.

small, with bushy tails, short hind limbs and forelimbs, a quadrupedal gait, and dentition that reflect a diet of fruits and vegetables. One family member, *Carpolestes* (who is perhaps a relative of true primates), had a grasping foot with a nail on its big toe for a tighter grip (Ross and Martin 2007). Cranial materials show that *Carpolestes* was small brained, with eyes to the side, a face that protruded well in front of its braincase, and an impressive, elongated snout with a large and dominant olfactory bulb. Yet the *Plesiadapiformes* reveal a small decrease in olfaction and a small increase in cortical tissue over other Paleocene mammals, while its grasping foot hints at more precise muscular movement (for discussions, see Silcox et al. 2010; Bloch and Boyer 2007; Silcox 2007; Ross and Martin 2007).

About 55 million years ago, true primates appear in the Eocene when the earth was covered with wet tropical forests that invited a huge radiation of diverse primate species in Europe, Asia, and North America. Two superfamilies are well represented—the "lemur-like" *Adapoidea* that resembles living prosimians in body and brain size and the *Omomyoidea* that may be ancestral to monkeys, apes, and humans. By the Eocene, both superfamilies had evolved enhanced neocortical features that included bigger brains compared to *Plesiadapiformes*, with notable expansions in both the parietal (i.e., general somatic and motor) and in the occipital (visual) regions. In a three-dimensional habitat, a sense of balance coupled with a more finely tuned motor cortex relying on visual cues would be pivotal for grasping and hand-to-mouth feeding. By this time, nails had replaced claws on most primate digits to accompany the soft underlying finger pads used for touch discrimination. Primate eyes had also evolved a postorbital bar (a bony rim) for greater protection. The orbits also shifted to a forward position, and this overlapping of the eyes enhanced depth perception, signaling a greater reliance on the visual cortex.[10] Thus, the origin of primates involved some enhancement of neocortical motor and visual zones for movement in a precarious world of length, breadth, and depth. It also involved a locomotion pattern of vertical leaping and hopping, a form of movement that would necessitate an emphasis on vision acuity and coordinated body movement (for discussions, see Kass 2007; Whishaw 2003; Ross and Martin 2007; Lemelin and Jungers 2007; Seiffert et al. 2009).

The Evolution of Basal Anthropoids

By the late Eocene, global temperatures had dropped sharply, and as the vast northern rain forests began to shrink, prosimian species declined. But in the warmer, southerly regions, the winds of change ushered in a new kind of primate—a basal *anthropoid* (the suborder of living monkeys, apes, and humans). The richest fossil beds for the recovery of early *anthropoids* is the Fayum Depression in Egypt, which is the Sahara desert today, but during the Oligocene epoch (33–23 Mya), it was a "garden of paradise." This depression was a heavily forested landscape, with meandering streams and a lush variety of flora and fauna. Prosimians are found in the Fayum, but most beds are filled with anthropoids, with two families of special interest: the *Parapithecidae* and the *Propliopithecidae*. *Parapithecidae* are the oldest family and typified by the genus *Apidium* whose dentition resembles both prosimians and New World monkeys—including a dental formula of 2.1.3.3 or 36 teeth.[11] This genus marks a crucial timeline because it lived before the evolutionary

[10] While a few other mammalian species have postorbital bars, this is a major distinction from early proto-primates (see Ross and Martin 2007).

[11] Mammals have four kinds of teeth—incisors, canines, premolars, and molars. A dental formula is an abbreviated way of denoting the number and types of teeth in each quadrant of the mouth. A primate with a 2.1.3.3 dental formula means it has 2 incisors, 1 canine, 3 premolars, and 3 molars in *each* quadrant or 36 teeth. As teeth are completely under genetic control and get preserved more than any other body part, primate dentition provides information on habitat, body size, diet, and even "life history characteristics" (to be discussed).

split between New World monkeys (infraorder *Platyrrhini*) and Old World monkeys, apes, and humans (infraorder *Catarrhini*).

The *Propliopithecidae* family, however, is important for hominin evolution because it contains the genus *Aegyptopithecus*, a catarrhine anthropoid who may be ancestral to Old World monkeys, apes, and humans. *Aegyptopithecus* was day active and lived about 30 million years ago in an arboreal niche. It was heavily muscled, weighting in at about 15 pounds, with a tail and short forelimbs and hind limbs—a gait similar to Old World monkeys today. But in looks, *Aegyptopithecus* is an evolutionary jumble with some prosimian features and *monkey-ape*-like features that include a monkey-like skull with an expanded visual cortex. In addition, it possesses what would become the signature dentition of Old World monkeys, apes, and humans—a dental formula of 2.1.2.3 or 32 teeth (see note 11). Yet, it has jaws that resemble those of later apes. Was *Aegyptopithecus* a prosimian, a monkey, or an ape? It was a transitional form, a hybrid concoction—a basal anthropoid.

The traits of *Aegyptopithecus* are innovative as they show a marked decrease in the olfactory bulb and a definitive increase in the visual cortex. This change cannot be overstated for it represents a true qualitative shift in primates toward cortical-centered behavior. Recall that olfactory receptors project to subcortical brain regions with information received in these zones used more to regulate the internal emotional state of an animal. In contrast, information received from visual receptors is projected directly to the neocortex, a structure geared toward understanding a variable and changing external environment. So in *Aegyptopithecus*, there is a visible heightening of the cortical senses, a trend that was to culminate in the capacity of *Homo sapiens* to represent the world symbolically (for discussions, see Williams et al. 2010; Falk 2000; Fleagle 1999; Maclatchy 2004; Wolpoff 1999; Simons 1987; Stein and Rowe 2011).

The Sensory System of the Stem Anthropoid of Old World Monkeys and Apes/Humans: A Cladistic Analysis

By the late Oligocene, many ancestral prosimians had gone extinct, migrated to Madagascar (where few mammals reside), or shifted to a nocturnal lifestyle. In their vacated niches were newcomers—the *anthropoids*. How did *anthropoids* oust prosimians that had reigned supreme in the trees for millions of years? Only a few cranial materials have been recovered from this time period, but it is reasonable to infer that even a modest uptick in sensory enhancements would give basal anthropoids like *Aegyptopithecus* a competitive edge over ancestral prosimians. And if these traits increased fitness, they would subsequently be passed on to later anthropoid generations, including the stem ancestor of contemporary Old World higher primates *before* monkeys and apes/humans split into two separate lineages.

The molecular clock indicates that the stem ancestor or last common ancestor (LCA) of Old World anthropoids (i.e., monkeys, apes, and humans) lived during the late Oligocene or Early Miocene (circa 28–24 million years ago). What sensory equipment did this mother population possess and subsequently pass on to all its descendants? Despite the lack of preserved cranial materials (as this stem ancestor has not yet been identified), we can still retrieve this information by using a methodology that is the standard tool in such fields as linguistics, textual criticism, and in biology, where it is called cladistic analysis. This procedure involves the following steps: First, a group of entities known to be the end points of a developmental or evolutionary process are identified. Second, an analysis is done to isolate out what are called "evolutionary novelties" or the *derived* characteristics held in common by this group of related entities. For example, in historical linguistics, a language family is identified (e.g., Indo-European languages). Then, the jointly held linguistic characters or "evolutionary novelties" of these daughter languages are isolated out and used to reconstruct the ancestral mother tongue. In biology, a cladistic analysis—which classifies organisms on an evolutionary tree according to their order of branching—is undertaken to identify a group of living species (or a clade) with known phylogenetic relationships.[12] Then, what these daughters hold in common or the *derived* characters that they share are used to determine what they inherited from their mother lineage. To assure that these derived characters have a clear-cut evolutionary or developmental connection, this method has two built-in assumptions: (1) a *relatedness hypothesis* that assumes that any similarities found in a class of objects are not due to chance but are the outcome of descent from a common ancestor and (2) a *regularity hypothesis* that assumes that modifications from the ancestral form to the descendant forms are not randomly acquired but follow a clear systemic bias that links descendants to each other and to their last common ancestor (see Jeffers and Lehiste 1979; Norman and Cameron 1977). A final step is to include a closely related "out-group population" as a further basis for comparison. As our objective is to identify the inherited sensory traits of *both* Old World monkeys and apes/humans *before* the split, our reference out-group is living prosimians (and where appropriate New World monkeys) as they are a close lineage to Old World anthropoids.

A related point is that Old World monkeys, apes, and humans will naturally share *ancestral* (or primitive) traits with prosimians—as all anthropoids sprung from a remote prosimian lineage. A cladistic analysis is designed to cordon off *only* derived features or "evolutionary novelties" originally held by a stem ancestor population *before* its descendants branched away. We could start by entertaining the null hypothesis that the suite of sensory modalities held in common by Old World higher primates was acquired independently by each living monkey, ape, and human species *after* the split, but this is an implausible prospect. As Kaas (2008: 385) plainly put it, "if a feature or trait is present in all members of a group of phylogenetically related mammals, then it is parsimonious to infer that the common ancestor of the group also had that trait, and that it was retained in all the preserved lines of evolution from that ancestor." Thus, using the cladistic methodology, we can be quite confident that the suite of sensory traits to be discussed was inherited from the LCA of Old World anthropoids because the sensory software of this rootstock anthropoid is still intact in all Old World monkeys, apes, and humans. As it is well established that sensory modalities effect the brain's "functional architecture" (Kirk 2006: 76; Kirk 2004 and see Purvis et al. 1996) and are obviously the gateway for human awareness and, hence, inputs to the brain, we need to examine the functional properties of each one, and in particular the dominant role played by vision in the synchronizing of information (for discussions on cladistic analysis, see Kass 2008; Preuss 2007; Maryanski and Turner 1992; Platnick and Cameron 1977).

The Chemical Sense Modalities: Olfaction and Taste

The Olfactory Sense Modality

Olfaction is the dominant sense for most mammals, and although early primate endocasts indicate a decrease in its importance, the olfactory bulb is still relatively large in living prosimians who scent mark their habitat (Hoover 2010).[13] But by the Oligocene, basal anthropoids like *Aegyptopithecus* were relying less on smelling their environment, thereby initiating an evolutionary trend since

[12] A clade refers to a group of species that are descended from a single common ancestor. Clades are considered to be monophyletic—that is, they contain one last common ancestor and all its descendants. The LCA can be inferred (until discovered) or a known species.

[13] With the rise of mammals, the olfactory bulb and olfactory cortex underwent a foremost and elaborate expansion (Rowe et al. 2011).

the olfactory bulbs of extant monkeys, apes, and humans are significantly reduced compared to prosimians.[14] A related point is that the vomeronasal organ (VNO)—a chemoreceptor accessory to the olfactory system and used to detect pheromones and other chemosensory functions—is highly developed in prosimians (and to a lesser degree in New World monkeys) but is merely a vestige organ in Old World monkeys, apes, and humans. In timing, the loss of the VNO "...(is) consistent with the hypothesis that the catarrhine VNO became non-functional (in pheromonal communication) prior to the divergence of the cercopithecoid and hominoid lineages" (Bhatnagar and Smith (2007: 147)). These authors also surmise that the "cercopithecoid-hominoid common ancestor possessed a VNO vestige similar to that seen in extant chimpanzees and humans" (for discussions, see Zhang and Webb 2003; Smith et al. 2002; Rossie 2005). Of course, monkeys, apes, and humans still use olfactory cues for information involving sexual functions (e.g., hormonal states), for avoiding toxic chemicals, and for judging the tastefulness and contents of foods (and for humans to activate emotionally charged memories). Yet higher primates rarely use olfaction for social communication, for detecting predators, or for gathering any extensive information about their environment (for discussions, see Preuschoft and Preuschoft 1994; Stein and Rowe 2011; Maryanski 1997; Rouquier and Giorgi 2007).

The Taste Sense Modality

Taste (or gustation) is also a subcortical modality, often working with olfaction to recognize different qualities in foods and other chemical substances. Gustation is also essential for controlling and maintaining food intake, triggering taste-induced sensations (of pleasure, annoyance, or displeasure), and for monitoring the nutritional energy required for an organism. As olfaction and taste work in close correspondence, what possible effect might a reduction in olfactory receptors have on gustatory receptors?

Comparative molecular research on living primates has helped to answer this question by isolating out some derived gustatory modifications that can be linked to the last common ancestor (LCA) of Old world monkeys, apes, and hominins. A key finding is an uptick in sweetness receptors. What researchers found is that compounds that tasted sweet to all or most primates were activated by ancestral receptors, while more complex classes of sweet compounds went *undetected* by both prosimians and New World monkeys. To detect these refined sweet-tasting compounds required a more sophisticated receptor, and this "innovative sweetness receptor" is found only in the gustatory systems of Old World anthropoids (Nofre et al. 1996: 761). The researchers conclude that Old World monkeys, apes, and humans comprise a clade or an evolutionary "compact group," and this novelty was inherited during the Oligocene from the "ancestral stock of the Old World simians" or early *anthropoids* (Glaser 2007: 128; Nofre et al. 1996: 761; Glaser et al. 1995). Yet why favor Old World anthropoids with a more high-powered sweet-taste receptor? One possibility is that a diminished smell potency was compensated for by a heightened sweet potency for the detection of nutritious, energy-rich foods. In a related note, Nofre et al. (1996: 761) found that the molecular features of this sweet-taste receptor are "remarkably adapted to the specific detection of fructose and sucrose...." And they argue that this sweet-taste receptor was instrumental in catarrhine (i.e., Old World monkey, ape, and human) evolution because it improved "food search efficiency and dietary choice... for highly energetic nutriments, especially fruit, which could have favoured...[anthropoid]... mental development, and,

[14] As a subcortical sense, the olfactory bulb is located far from cortical regions—such as Broca's area in the left inferior frontal lobe where the programs for speech sounds are located. This probably accounts for why humans often have difficulty classifying and naming customary smells (Falk 2000:68). It is also why the aromas of familiar smells will evoke at times strong emotional sensations.

later, the emergence of humans." However, as we will see, this heightened sweet potency which made soft, ripe fruits especially attractive to Old World anthropoids would later come to have unintended consequences for the survival and reproductive success of Miocene apes.[15]

The Auditory Sense Modality

The auditory system is sensitive to traveling sound waves created by the air that surrounds us. Three properties of these acoustic sounds are *wavelength* (the distance between any two waves), *frequency* (the number of cycles completed passing a point per second), and *amplitude* (the amount of acoustic energy or height in a sound wave) (Hackett 2003, 2008; King and Nelken 2009). Wave frequency affects pitch and tone quality. A higher frequency has a higher pitch than a lower frequency, and pitch aids in distinguishing between low and high sounds and sound quality (e.g., a pleasing or unpleasing tone).

In pre-mammal vertebrate evolution, wave sensitivity was in the low-frequency range, but this changed with the rise of mammals who were sensitive to high-frequency sound waves. As high-frequency waves are strongly directional and easily reflected, this gave early mammals a competitive edge for the localization of sounds. But this competitive edge came at the loss of detecting low-frequency sounds. Today, most mammals hear above 10 kHz (or cycles per second), a distinctive mammalian trait (Heffner and Heffner 2008; Heffner 2004). Some higher mammals, however, have specialized threshold levels—from bats who rely on high-frequency sounds for locating objects by echolocation to elephants who communicate with conspecifics using very-low-frequency sounds.

Primate hearing is not as specialized as bats or elephants, but the fossilized ears of early primates reveal steady selection pressures for a *decrease* in high-frequency sensitivity and a corresponding *increase* in low-frequency sensitivity. Today, all primates hear low-frequency sounds, but their auditory thresholds differ along the following phylogenetic lines:

- Prosimians have much better high-frequency sensitivity with less low-frequency sensitivity than monkeys, apes, and humans.
- New World monkeys (platyrrhines) have a low-frequency sensitivity that falls middle range between prosimians, on the one side, and Old World monkeys, apes, and humans, on the other, which represents a significant variation not attributed to body size differences.
- Old World monkeys, apes, and humans (catarrhines) have similar acoustic profiles for both high- and low-frequency hearing (for discussions, see Coleman 2009; Coleman and Colbert 2010; Heffner 2004; Coleman et al. 2010; Heidegard et al. 2009).

Given the small differences in auditory sensitivity among Old World anthropoids, it follows that this acoustic design was in place before the split between Old World monkeys and apes/humans, with only very slight modifications (especially for humans) over time (Coleman 2009: 88; Heffner 2004).

Why a shift to lower-frequency hearing? The odds are on the nature of the primate landscape with its overlapping canopy, filtering sunlight, thick humidity, heavy rainfall, and dense year-round foliage. In this setting, higher frequencies are *less* valuable because they are strongly directional and fade quickly. Instead, lower frequencies in a cluttered multi-canopied forest travel better *around* objects

[15] Today, we can thank (or blame) our Oligocene stem ancestor for our remarkable sweet-taste receptor which may explain our fondness for rich, yummy desserts and other "junk foods" where sucrose or other refined sugars are the usual ingredient.

and can also be heard from longer distances—like the slithering noise made by a snake. In fact, as most sounds in the natural world are made by animals, and perhaps as compensation for the loss of olfactory sensitivity, primates are exceedingly sensitive to any abrupt and sudden sounds, like the bangs, clunks, thuds, hisses, and crackles that signal animated movement and possible danger (Masterton and Diamond 1973: 419; Heffner 2004). Seen in this adaptive light, the primary function of the anthropoid auditory system is to serve as a wide-ranging early warning system (see Masterton and Diamond 1973: 431; Glendenning and Masterton 1998).[16]

A related point is that the auditory modality is also an efficient receptor for close range and long distance social communication. Both Old World monkeys and apes have a wide repertoire of vocalizations that vary in content and acoustic properties—from warning conspecifics of impeding danger through localizing neighboring groups to general contact calls within groups. While primate *sound input* and *perception* are both under cortical control and adapted for higher cognitive processing, the auditory channel for primate vocal production is under limbic control for emotionally laden vocalizations. Humans also emit vocalizations with emotional affect, but their vocal-auditory channel is under cortical control for volitional, flexible responses (for discussion, see Snowdon 1990; Cheney and Seyfarth 1980), and thus, during human evolution, the vocal-auditory channel was subject to selection for a cortically controlled vocalization system. But it is important to emphasize that human speech rests on two critical early anthropoid precursors or preadaptations: (1) a sensitivity to low-frequency sounds and (2) the early placement of sound *perception* (but not sound production) under neocortical control.

The Haptic Sense Modality

Tactile perception is the "reality sense," a massive network of general body receptors that are responsive to continuous pressure, to electrical and chemical stimuli, and to temperature changes, with vibration powers so acute that for deaf individuals (like Helen Keller) it can serve as a surrogate for many hearing functions (Taylor et al. 1973; Keller 1904). Haptic refers to "active" touch or the manipulation of a stationary object using prehensile hands (and feet) for impressions about objects in space. Haptic also has an inherent and pronounced adeptness to recognize objects in space by processing temporal sequential patterns. In fact, individuals blind from birth can rely on their prehensile hands to guide their movements through space by stringing together a succession of independent chucks of information to perceive a pattern or a structure of the whole by connecting individual parts to each other in a linear time frame rather than a three-dimensional spatial frame (Von Senden 1960; and see Keller 1904: 135 who wrote of this phenomenon). However, this remarkable capacity is rarely used unless vision is deficient (Freides 1973: 302 and see Taylor et al. 1973; Preuss 2007; Kass and Pons 1988; Hamrick 2001).

The primate order originated with committed arborealists, and so the evolution of "exceptional sensorimotor abilities" would seem to be a crucial adaptation (Kass 2007 and see Bloch and Boyer 2002; Wise 2007; Dominy et al. 2004). Endocasts can chronicle these changes, but we need living primates to interpret these evolutionary trends. Extant prosimians have good grasping and climbing abilities, and they also use their hands for putting food into their mouth. But the evolution of anthropoids marked an elaborate expansion of the posterior parietal cortex for far more sophisticated somatosensory equipment. Anthropoids have superior prehensile hands with sensitive fingertip receptors for detecting what lies under the surface of objects, exploring their environment, and for establishing and maintaining bonds by social grooming. Along with active touch, primate digits also evolved raised ridges or unique "fingerprints" on their soft pad fingertips to

[16] According to Hackett (2008: 775), "there is both direct and indirect evidence that the major features of nonhuman primate auditory cortex organization are conserved in humans."

reduce slippage, with anthropoid dermal ridges clearly augmented for improved traction (Preuss 2007: 18–19 and see Taylor et al. 1973; Lederman and Klatzky 2009; Gibson 1962; Holloway 1968; Radinsky 1970 for discussions).

New World monkeys (i.e., the descendants of early platyrrhines) share with Old World monkeys, apes, and humans (i.e., the descendants of early catarrhines) greatly enhanced grasping skills over the basic prosimian adaptation, but New World monkeys have limited finger flexibility. In contrast, Old World anthropoids have finely tuned prehensile hands with independent control of their digits for very delicate motions and for some species (especially humans) a well-developed opposable thumb (Falk 2000). Haptic perception is under cortical control for learned, voluntary, and intentional responses to environmental stimuli (Preuschoft and Preuschoft 1994).

The Visual Sense Modality

All primates have excellent vision when compared to most mammals. Early in primate evolution, selection worked to protect primate eyes with a postorbital bar (or bony rim) and anthropoid orbits with a postorbital closure (a complete bony cup) (Martin 1990). Primate eyes also moved to a forward facing position to enhance binocular and depth perception. Anthropoids also underwent a suite of neurological enhancements that included full stereoscopic or three-dimensional vision, and the expansion of visual association areas for retaining information on the stability and distance of objects in space over time. Anthropoids also evolved a finely tuned fovea, a depression in the macula of the retina that imparts an exceptional clarity of vision to "see" the fine details of objects (Ross and Martin 2007).

Another key component of the primate eye is color vision. Once the primate retina is activated by light waves, the form, shape, and kind of image are determined by the number of rod photoreceptors for black and white vision and the number of cone photoreceptors for day-adapted vision. While most prosimians and New World monkeys have a two-cone system (dichromatic), all Old World monkeys, apes, and humans possess a three-cone system (i.e., known as routinely trichromatic) with spectral sensitivities that signal very high daylight acuity (Jacobs 2004). This rich number of photoreceptors enhances sensitivity both to dim light and to contrast in brightness, giving Old World monkeys, apes, and humans the ability to detect slight color variations and the fine particulars of objects and scenes (Campbell and Maffer 1976 and see Preuss 2007; Jacobs 1993, 2004, 2007; Jacob and Williams 2001; Lucas et al. 2003).

As light travels quickly, a three-cone system is also useful for other functions that include the detection of predators and prey at longer distances as light travels quickly; the identification of foods by shape, texture, *and* hues such as high-energy fruits which are very colorful; and the ability to perceive a "continuum of hues" for distinguishing, for example, young leaves with optimal nutrition from mature leaves (Dominy et al. 2004; Vorobyev 2004: 235; Jacobs and Deegan 1999; Lucas et al. 2003). All Old World anthropoids share these remarkable visual properties, and despite the attention given to the human auditory system because of its language function, visual perception *is* the premier sense for *all* monkeys, apes, and humans because it takes the lead in object recognition. And while the anthropoid modalities are integrated to prevent sensory disharmony, serving different functions depending on perceptual demands, vision is overwhelmingly dominant when it comes to spatial information[17] (Conroy 1990; Martin 1990; Radinsky 1974; Lyon 2007; Kirk 2004).

[17] Although each modality is distinct in its own right, all primates utilize more than one modality simultaneously, which increases the likelihood that a message will be properly received. Thus, cross-modal and intermodal activity is the norm. However, vision is the major integrator of most environmental stimuli, and it is so powerful that it even influences the information pickup for the auditory and tactile sense; for in cases of intersensory conflict, the visual system will usually determine what is perceived.

The visual modality is also dominant for primate communication through "body language" or sequences of gestures made by the hand or face such as the "play face" (a precursor to human laughter) and, of course, by direct eye-to-eye contact. As vision is under cortical control and specialized to input information about the outside world, the dominance shift from smelling the world to seeing the world buttressed Old World anthropoid cognition by promoting voluntary and purposeful patterns of behavior for a greater understanding of a variable and changing environment.

Summary

The evolution of the primate senses and their operation are linked to a three-dimensional habitat where a misstep is often fatal. This process was initiated with Eocene prosimians that began to rely less on their olfactory and more on their visual sense modality. By the Oligocene, the LCA of Old World monkeys, apes, and humans had evolved an amazing suite of sophisticated sensory receptors that included a taste for complex, sweet-tasting compounds, especially high-energy soft, ripe fruits; a hearing range sensitive to high- and low-frequency sounds; remarkable prehensile hands and feet; and a superior system of eyesight that included stereoscopic vision, a fovea for up-close viewing, and three prong color receptors. These sweeping transformations in the LCA of Old World anthropoids set the stage for continued expansion of association cortices within the neocortex, increased reliance upon stored individual learning experiences, and new capacities for more purposeful, flexible, and voluntary behavioral responses.

But who was this elusive LCA? No one knows, but a recently unearthed anthropoid that lived in Afro-Arabia between 28 and 30 million years ago is a fossil of interest. The preserved remains of *Saadanius hijazensis* (an Arabic collective term for monkey and ape) include substantial craniodental materials that are intermediate between *Aegyptopithecus* and Early Miocene apes (to be discussed). As *Saadanius* is within the range of the predicted traits of the LCA, it offers new evidence that the split between Old World monkeys and apes/humans took place between 28 and 24 million years ago, a date in line with the molecular clock (Zalmout et al. 2010 and see Glazko and Nei 2003; Casagrande and Khaytin 2007; Janečka et al. 2007).

Yet in a matter of time, the LCA went extinct despite its "avant-garde" modalities—but fortunately for us, *after* it had spun off two daughter lineages: the Old World monkeys (Cercopithecoidea) and the ape/hominins (Hominoidea). What happened next is a classic "who-done-it" mystery story with the hominoids starting out as evolutionary blockbusters and ending as evolutionary failures, a turnabout that, oddly enough, would pave the way for the evolution of the hominin mind. Let us turn now to the second revolutionary event—the great ape die-off and the shift to forelimb dominant locomotion.

The Cognitive Legacy of the Miocene Apes

The Origins of Old World Monkeys, Apes, and Hominins

The Early Miocene (24–15 million years ago) opened with rising global temperatures and some new players on the evolutionary field. The first group of players is Old World monkeys represented by *Victoriapithecidae*, a small monkey that weighted about 10 pounds (Miller et al. 2009). *Victoriapithecus* had a longest facial profile, a cranial capacity of 54 cm^3, and a 2.1.2.3 dental formula (shared by all living Old World monkeys, apes, and humans), with a *bilophodont molar cusp pattern*—a signature character of living Old World monkeys (this dentition may have originated with *Victoriapithecus*).[18] *Victoriapithecus* was a generalized quadruped who foraged on the ground or in trees with dentition adapted for ripe fruits, but its teeth also denote that it forged on the ground for harder and more compacted foods (Benefit 1999; Benefit and McCrossin 1997; Dean and Leakey 2004: 406). One fossil bed that preserved an entire *Victoriapithecus* population (due to some catastrophic event) indicates a social structure common to Old World monkeys today: adult males, females and assorted juveniles, adolescents, and infants, with speculation that males migrated out of their natal group at puberty as adolescent monkeys still do today (Benefit 1999; Dean and Leakey 2004). The second set of players is early hominoids (e.g., apes and humans today), represented by *Proconsul*, an ape weighing in at well over 100 pounds (although some species were smaller) that lived in Africa beginning about 18–20 million years ago. *Proconsul* was arboreal and adapted to multi-canopied evergreen forest with a dietary dependence on soft, ripe fruits. It was tailless with a prognathous face, no brow ridges, and an estimated cranial capacity of over 160 cm^3 with enhancements in the visual cortex and frontal lobe. *Proconsul's* dental formula was 2.1.2.3 or 32 teeth (shared by all Old World anthropoids), but it had a unique Y-5 cusp pattern on its lower molars, a signature trait of all living apes and humans[19] (Potts 2004; Walker et al. 1983; Harrison and Andrews 2009).

The Land of the Apes

The Miocene was a "Golden Age" for apes—the opposite of today where monkeys overwhelmingly dominate the forests. The fossil record tells us that *Victoriapithecus* and *all* monkey species are very scarce during both the Early and Middle Miocene while ape species like *Proconsul* are *extremely* abundant. But there are other unexpected doings. Miocene fossil beds show that early apes had "monkey-like" body plans with a pronograde posture (i.e., monkey-like lumbar vertebrae) and long torsos, with forelimbs and hind limbs of near equal length. In contrast, the body plan of *living* apes and humans is characterized by highly mobile shoulder joints, an elongated collarbone (essential for stabilizing the shoulder joints for overhead locomotion using the hands), a short deep trunk, and limbs of *unequal length* (living apes have arms longer than their legs and humans have legs longer than arms). As primate limb bones reflect their mode of locomotion, early apes like *Proconsul* were built to run along the tops of tree branches like monkeys do today. So while early ape dentition and cranial features say *hominoid*, they barely resembled living apes (Gebo et al. 1997; Kagaya et al. 2010; Filler 2007). However, living alongside *Proconsul* types but on the sidelines was a rare ape with a different body plan. Called *Morotopithecus bishopi* and dated to about 20 million years, it was also a large-bodied frugivore, but unlike early apes, this hominoid had a suspensory orthograde posture with flexibility in the forelimbs for climbing and suspending from tree branches (Deane 2009; Maclatchy et al. 2000, 2004).

Apes today are a handful of primates compared to monkeys. So things were very different in the Miocene. How did apes start out with such a competition edge? The answer, of course, lies with what happened after monkeys and apes split from the LCA. All indications are that monkeys were forced to adapt to a marginal niche where they had to subsist on less nutritious foods and this is reflected in

[18] As discussed earlier, *all* Old World monkeys, apes, and humans have a 2.1.2.3 dental formula or 32 teeth (see note 11). How then would you separate monkey teeth from ape/human teeth? The answer is—look at lower molar construction. All Old World monkeys will have four cusps arranged in two pairs, each linked by a loph so they have a *bilophodont*, four-cusp pattern on their molars. In contrast, ape and human lower molar construction will have an additional cusp that is shaped like a Y and called a *Y-5* cusp pattern.

[19] See note 18.

their dentition, whereas *Proconsul* apes were feasting and living high on energy-rich soft fruits. *Victoriapithecus's* anatomy also indicates greater mobility, with evidence that this taxon was comfortable both in the trees and on the ground. In fact, according to Benefit (1999: 168), after monkeys and apes separated, "It was the origin of Old World monkeys, not apes, that was linked to a shift in locomotor adaptation" (and see Andrews and Kelley 2007; Filler 2007; MacLatchy 2004; Ward 1993; Deane 2009; Temerin and Cant 1983).

Early monkeys were also very little, and apes were mostly very large, and this differential is still evident today. What advantage is there to a larger body size? For openers, larger animals are much less likely to be victimized by predators, and especially ones living in the treetops. And with lots of wholesome fruits and little competition, an animal can grow much larger and *reach maturity much later.* Today, monkeys and apes (and humans) vary strikingly in their reproduction and maturation rates. For example, if we compare the *life history traits* of an Old World monkey and a great ape, we get these average differences: A 50-pound adult male baboon (a very large monkey) has a gestation of 175 days and is breast-fed for 420 days, with an infant phase of 1.6 years, a juvenile phase of 4.4 years, and an adult phase of 23 years, with intervals of 1.7 years between births for a single offspring. In contrast, a 115-pound adult chimpanzee has a gestation of 228 days and is breast-fed for 1,460 days, with an infant phase of 3 years, a juvenile phase of 7.0 years, and an adult phase of 34 years, with 5.6 years between births. And this variance is not purely a body size distinction. The little 15-pound gibbon ape (who is smaller than many monkeys) has a gestation of 205 days and is breast-fed 730 days, with an infant phase of 2 years, a juvenile phase of 6.5 years, and an adult phase of 23 years, with 2.7 years between births (all demographic profiles are from Wolpoff 1999; Falk 2000).

This wide disparity in *life history traits* obviously occurred after monkeys and apes split from their anthropoid stem ancestor. But how far back? To answer this question, researchers analyzed the preserved teeth of young Miocene apes who died when their first molar cusps were emerging. After determining the age of each specimen, their molar morphology (solely under genetic control) was then used as a yardstick for other traits because life history characters are systemic such that a long gestation period is bundled with prolonged nursing, an extended infant and juvenile phase, longer maturation, and a longer life span (see Kelley and Smith 2003; Kelley 2002 for details). After applying this technique to a variety of early hominoid species who lived from the Early to Late Miocene (circa 17.5 million years to about 10 million years ago), the researchers established that young Miocene ape molars erupted *outside* the range of living Old World monkeys and well *within* (or above) the mean of living chimpanzees. This discovery that Early Miocene ape eruptions fall within the chimpanzee range is significant because, according to Kelley and Smith (2003: 326), "(t)his… (finding)…is compatible with the hypothesis that there was a shift to the prolonged life histories that characterize extant apes early in the evolution of the *Hominoidea*." Hence, the prolonged *life history characteristics of* living hominoid species (including humans) seemingly reflect a long-standing phylogenetic trend *built into the hominoid line,* and probably as a consequence of a stable rain forest environment with little predation (Maclatchy 2004). These extended life history characters probably account for the increase in early hominoid brain volume—and size matters. A large-bodied, visually dominant ape with expanded cortical tissue can process more sensory information for greater perceptual awareness than a little bodied primate. In fact, a number of scholars have argued that increased input from visual receptors has been a prime factor in the evolution of the primate brain (Kirk 2006; Jerison 1973; Barton 1998). Slower maturation rates would also entail a longer parental investment and a longer socialization period. And once in place, these characters could be subjected to directional selection and following Potts (2004: 211) serve as "a precondition for selection leading to brain enlargement and enhanced cognition." Thus, early in ape/proto-hominin evolution, several preadaptations or precursors for further encephalization of the neocortex were already in place: larger-sized bodies which correlate with brain size, slower gestation, longer nursing, and longer infant and juvenile developmental phases that would, if selected on, help prepare the way for the evolution of the hominin mind.

The Downfall of the Apes

During the Middle Miocene, the paradisial world of hominoids took a negative turn as Africa became drier and cooler, causing many ape populations to decline. But ape fortunes improved when a land bridge opened up between Africa and Eurasia, giving apes the means to leave Africa and migrate to the warmer, southern parts of Europe and Asia (Andrews and Kelley 2007). This new colonization into European rain forests led to another population explosion, with some apes even taking up a semiterrestrial lifestyle by moving into woodlands and swamps. And with this diversification, some species grew enormous such as the mysterious *Gigantopithecus* of Asia who was surely the "King Kong" of apes with its huge teeth and massive jaws with speculation that it weighted over 600 pounds and was 9 feet tall, whereas other apes ventured out into forestlands with patches of grasslands like the large-bodied *Kenyapithecus* of Africa who lived about 14 million years ago (Andrews 2007; Andrews and Kelley 2007; Potts 2004; Meldrum 2006).[20] And so the forests of the Old World became " the lands of the apes." Yet all too soon as the earth's climate grew steadily cooler, ape populations began to plummet as the forests receded, limiting the food supply for frugivores. For despite some dietary diversity (evidenced by heavy jaws, shearing crests, and thick molar enamel), ape dentition reflected a diet still dependent on soft ripe fruits, which is confirmed by the fact that ape continued to live primarily in rain forest or wooded localities with lots of fruit-bearing trees (Potts 2004; Pilbeam and Young 2004).

By the Late Miocene (11–7 million years ago), hominoid populations experienced an ominous decline. Some Eurasian apes competed for the remaining pockets of forests, while others returned to Africa, but the death toll only mounted with a catastrophic decline of apes in numbers and species. Part of this great ape die-off was caused by climate change as other forest animals went extinct as well, but was the heavy loss of apes tied solely to climate change? (Fortelius and Hokkanen 2001) To be sure, the forests contracted, but they did not disappear; they simply became more concentrated near to the equator (Keller and Barron 1987: 147). Another aberration is that Old World monkeys, who were rare during most of the Miocene, began to proliferate both in Eurasia and in Africa and, in fact, moved into the vacated niches of extinct apes. Today, Old World monkeys still sit in the former ape niches, while the hominoid family tree has only ape leftovers who can survive only in limited and restricted forest niches, *except* for an upright walking hominin who lives worldwide.

Why were Old World monkeys so favored during the Late Miocene? While a detailed account is not necessary here (but see Turner and Maryanski 2008), it is important to highlight some key factors favorable for monkeys—but not apes. First, as the earth's climate cooled, the small size of monkeys became an advantage as they required less food and space. Second, since the Early Miocene, apes were dependent on fruits to meet their nutritional needs, while monkeys since the days of *Victoriapithecus* were also adapted to secondary foods, such as nuts and tuberous vegetables. Third, monkey and ape *life history characteristics* differed dramatically. Miocene hominoids had for a long time been in the "slow lane" with delayed maturation and reproduction, while monkeys had been in the "fast lane" with much faster maturation and reproduction. Thus, what was advantageous for apes—larger bodies, a dietary bias of energy-rich fruits, and slow maturation and reproduction—was now a disadvantage in a colder climate with far less resources. One long-standing hypothesis is that monkeys also gained a competitive dietary edge over hominoids by evolving an ability to digest unripe fruits instead of waiting for them to fully ripen (as apes and

[20] Only the huge teeth and jaws of *Gigantopithecus* have been recovered, but some maintain that this giant ape is the stem ancestor of the legendary Sasquatch, Yeti, and Bigfoot who are alleged to still roam the forests of Asia and the Pacific Northwest. The real significance of *Gigantopithecus*, of course, is the realization that such a hulking ape survived in parts of Vietnam, India, and China for over 9 million years.

humans must do today). In fact, some contemporary Old World monkeys (i.e., the colobines) have very specialized digestive systems with multichambered stomachs and a chemistry that enables them to ingest and effectively neutralize the toxic compounds in plants, leaves, seeds, and unripe fruits.[21] As apes were forest huggers and constrained by their dependence on fruit consumption, any disruption in their foraging habits would have dire consequences, especially when coupled with a slow maturation and reproduction rate. All in all, by the Late Miocene, ape lifeways were becoming very problematic. That Old World monkeys faced this challenge and thrived means they were in possession of survival strategies that apes lacked (Kay and Ungar 1997; Potts 2004; Pilbeam and Young 2004; Temerin and Cant 1983; Andrews 1981).

The Last Common Ancestor of Great Apes and Humans

What were the consequences of this massive die-off? Not all hominoid species went extinct, of course, or apes and humans would not be around today. But who lived and who died had revolutionary implications for later hominin evolution and, hence, for the evolution of the hominin brain. For as selection acted, it targeted the prolific quadrupedal apes with the "monkey-like" body plans (like *Proconsul*), whereas it cherry picked for survival apes with some degree of "lower branch" suspensory behavior—like the rare Early Miocene *Morotopithecus*. It is plausible that *Morotopithecus* is ancestral to the ape lineages that survived, as the early appearance of a suspensory ape in the Miocene fossil record means that selection had something to work on that could be modified (Nakatsukasa 2008; Gebo et al. 1997; Maclatchy et al. 2000; Young 2003; MacLatchy 2004). Yet why would *Morotopithecus* types survive and *Proconsul* types die out after millions of years of great reproductive success? More to the point: what endowments did the mother lineage of living apes and humans possess to avoid this slaughterhouse extinction? Answering this question is crucial if we wish to cast light on the nature of the hominin mind.

Let us start with the molecular clock showing that apes and humans are monophyletic with the following splitting estimates: the gibbon branched away between 21 and 19 million years ago, the orangutan between 15 and 12 (Mya), gorillas between 9 and 8 (Mya), and the common chimpanzee and hominins branched away between 7 and 5 million years ago. So the LCA of living apes and humans lived approximately about 20 million years ago (Chatterjee et al. 2009; Israfil et al. 2011; Bradley 2008; Gagneux and Varki 2001).

The next step is to use cladistics to isolate out the *derived* features or evolutionary novelties shared by living hominoids. As our goal this time is to reconstruct proto-hominin cognition, it is prudent to exclude the little gibbon ape as it branched away nearly 20 million years ago and seemingly underwent a dwarfing process, while the large-bodied gorilla, chimpanzee, orangutan, and proto-hominin were the same animal until about 15 million years ago. Our out-group lineage this time is their closest sister taxa—Old World monkeys. The question now is as follows: Since the split, what "evolutionary novelties" or derived traits do large-bodied apes (and humans) uniquely hold in common?

Hominoid Anatomy, Locomotion Patterns, and Social Networks

The skeletal structure of living apes (and humans) denotes a past adaptation to a positional repertoire of suspensory and vertical climbing behavior. This does not mean that the last common ancestor (LCA) was a specialized brachiator like the modern gibbon, only that it engaged in some

[21] The appearance in the fossil record of *Microcolobus*, an ancestral colobine about 10 million years ago, nicely fits the ripe-fruit-unripe-fruit hypothesis as apes began their accelerated decline during the late Miocene (Raaum et al. 2005).

degree of suspensory hanging with the arms held directly overhead (Pilbeam and Young 2004). This adaptation called for the peculiar ability to support oneself by the arms and shoulders (all monkeys require some other support). Second it called for strong finger flexion, for if the hand is to suspend the body like hanging fruit, it must be formed into a kind of hook. Third, the hand (instead of the foot in monkeys) must be capable of propelling the primate body through space. And fourth, using forelimb suspension alone for feeding or moving about required an extreme range of supination which made necessary a specialized and distinct wrist joint. All monkeys (even the so-called semibranchiators) have only 90 degrees of supination due to a joint pattern specialized for four-limbed quadrupedal locomotion (Lewis 1974: 158). Apes and humans have about 180 degrees of supination so that a body held by the prehensile hands can pivot at about the needed half circle for suspension (for discussions, see Oxnard 1963: 166; Corruccini and McHenry 1975; Gregory 1916; Napier 1963; Young 2003).

How might this unique locomotor pattern contribute to the expansion of the ape/proto-hominin neocortex? For openers, it requires precise hand-eye coordination, as the motor cortex must be able to react quickly to what the visual cortex "sees" and these corresponding brain signals must work together effectively. Hand-over-hand travel also required finer cortical motor control for precise delicate movements compared to the four-footed locomotion of *Proconsul* (the early ape) or *Victoriapithecus* (the early monkey) which is a relatively stereotyped and less flexible motion. It also places greater emphasis on mastering individually acquired navigational skills, especially for the proficient calculation of egocentric distance, requiring in turn greater memory for learning secondary depth cues. The loss of the substantial support of four limbs and reliance upon two limbs would also promote greater cortical proficiency to detect cues to surface features—that is, to temperature, smoothness, roughness, and stickiness along with what lies under the surface of objects. For example, a refined sense of texture would help to determine the support capacity of a tree branch.

We can also logically infer that weight suspension using the hands alone (that is allowing the arms to hang vertically above the head) in feeding or in a hand-over-hand locomotion would also intensify selection for expansion of association cortices for coordinating haptic (parietal lobe) and visual responses (occipital lobe). For, although the visual organ is dominant in hominoids, this sense alone would be insufficient to supply data about the immediate environment for vertical climbing or the gripping of tree limbs above the head. Thus, unlike monkeys (and humans) who use vision almost entirely for locomotion guidance, the peculiar nature of forelimb dominant locomotion would require compensation from other modalities, which, in turn, would lead to the cross-modal association of, for example, the somatosensory and visual cortices to prevent sensory disharmony.

But let us return to our main question: why would selection favor swinging apes over quadrupedal apes who had the inside track for millions of years? One reason is surely to be found in their distinctive foraging strategies. Both quadrupedal and suspensory apes were large-bodied frugivores, but the quadrupeds were "top-branch" crown feeders, while the swingers were "under-branch" borderline feeders. So if we assume that monkeys acquired the ability to digest unripe fruits, they could consume these fruits at the crown before they matured, leaving the quadrupedal apes literally dying of hunger. But neither monkeys nor quadrupedal apes could step on the fragile, swaying branches at the terminal ends to pick fruits. This gymnastic feat required the skillful interplay of a trapeze artist who could distribute its weight among several branches or suspend its body below or above the thin branches (Fleagle 1999).

Another advantage favoring suspensory apes was probably their "free spirit" social lifeways, and here, their acrobatic habits can help us to infer organizational patterns. A large-bodied ape feeding on scattered fruits at the margins is exploiting a very specialized niche that could never support large or tight-knit kinship groups. An earlier cladistic analysis done on contemporary ape social networks revealed that, compared to monkeys, living apes have maverick social networks with few kinship ties and many fluid weak ties (see Turner and Maryanski 2008; Maryanski and Turner 1992; Maryanski 1992, 1995). Since monkeys (and most social mammals) are organized around strong kinship bonds, what advantage is there to a weakly tied social structure, and how is it created? In monkey societies, adolescent males depart their natal group at puberty, and monkey females remain clustered in tight-knit matrilineal networks composed of mothers, sisters, daughters, and other female relations. In contrast, all hominoids evidence the rare pattern of dispersing at puberty both males and females (or only females in the case of chimpanzees). In conjunction with a promiscuous mating system (or a near one in the case of gorillas), this effectively shuts down both large foraging parties and blood-tied kinship networks. Thus, while female monkeys are the anchors of group cohesion and intergenerational continuity in monkey societies, the dispersal of ape females at puberty effectively bars monkey-style organizational arrangements. Thus, in a foraging niche with slim resources, there is a strength in weak ties over strong ties because individuals are not embedded in strong-tie cliques, giving apes the freedom to be self-reliant and to forage independently in space and, except for mother and dependent offspring, to be relatively unencumbered by relational obligations. But great apes do engage in "friendships," which puts a premium on memory because, unlike monkey societies with daily face-to-face interactions in enduring groups, apes evolved a "fission-fusion" system where individuals can forage alone when resources are scarce or join others when resources are rich, such as when seasonal trees are swollen with soft ripe fruits. In fact, tactical interactional skills would be essential to sustain a loose-knit sociality. This type of social structure, one built on weak ties rather than strong ties, requires a more inclusive and complex mind-set with the deft and versatile capacity to remember dispersed community members and to interact with them on an occasional basis. For example, in chimpanzee societies, male dyads meeting up often elicit "greeting rituals" if they have not been in propinquity for a time since their last interaction. Among the great apes, the orangutan is the least social hominoid and is probably the best prototype for the last common ancestor of the great apes. Orangutans are semi-solitary and seem to avoid most interactions with conspecifics except for adolescents who do get together occasionally for a "night out on the town." Gorillas are comparatively more social than orangutans, although adult females in a group only rarely interact with each other. Chimpanzee males are the most social apes, and they live in a widespread "fission-fusion" community structure, although scholars have long proposed that even gorillas may also live in a regional or community level of organization like that of chimpanzees (for general discussions, see Falk 2000; Relethford 2010; Stanford et al. 2013 and for a full listing of references, see Turner and Maryanski 2008: 28ff; Maryanski and Turner 1992; Maryanski 1987, 1995).

The Cognitive Capacities of the Great Apes

Hominoid brain expansion began with *Proconsul* and *Morotopithecus*—the Early Miocene apes with a brain volume of about 150 cm³ (which is about the size of baboons, the largest living monkeys). Ape cranial materials during the Middle and Late Miocene are rare, but two well-preserved specimens of the Late Miocene are those of *Sivapithecus* and *Dryopithecus*. Notably, *Dryopithecus* had enhanced association areas and a brain size between 300 and 350 cm³ which is within the range of living ape brains (Sherwood et al. 2008). In addition, both had a suspensory anatomy, a slow-lane *life history* profile in their physical development (essential for allowing the brain to grow larger), and a set of other advanced hominoid features. *Sivapithecus* has been floated as ancestral to orangutans, while *Dryopithecus* is a candidate for the LCA of gorillas, chimpanzees, and hominins (for discussions, see Begun 2007; Henke and Hardt 2007). Endocasts can tell us about brain size and shape, but for insights into Miocene ape cognition, we need to turn to living chimpanzees, gorillas, and orangutans and ask the following: What cerebral traits did hominoids acquire since the ape/monkey split in the Oligocene?

As measured by their ability to learn complex tasks, contemporary apes are more intelligent than monkeys. While larger species typically have larger brains and can learn more complicated tasks, the ape neocortex has a larger brain to body size ratio than expected when compared to other mammals.

The hominoid brain is essentially a model of the monkey brain except for the increased convoluted surface, which buries the surface of the neocortex in the fissures (Semendeferi and Damasio 2000). As cortical folding can harbor substantially more neuron densities in a limited space, this allows for more intricate connections among circuits. This trend is apparent in the ape (and human) motor cortex, which is more refined than monkeys seemingly because of the shift to forelimb dominant locomotion. Hominoid prehensile hands also evidence more cortically based precision for the gripping of objects, while the soft fingertip pads have enhanced raised ridges for enriched tactile perception (Holloway 1968; Napier and Napier 1985). During hominoid evolution, cortical tissue was also augmented in specialized association zones with complex multimodal regions for sensory convergence. And novel changes in neurochemistry and circuitry enhanced cell activity for memory, learning, and other higher cognitive functions (for discussions, see Sherwood and Hof 2007).

An emergent property of this cognitive leap was the appearance of a *social mind* with revolutionary facets. One component is *a sense of self.* The touchstone for self-recognition is the "looking glass" test of identifying who is in the mirror. Monkeys and apes have virtually the same visual equipment, so both "see" the mirror as a physical object, but they differ in their perception of what is reflected in the mirror. Monkeys always "see" another monkey, whereas great apes "see" their own image (deWaal et al. 2005). If an animal can recognize "self" as an object in the environment, to use G.H. Mead's words, it has the potential to have an abstract sense of self (Mead 1934). A sense of self is exceedingly rare, and, thus far, only great apes, dolphins, elephants, and humans are known to have a self-identify (Suddenforf and Collier-Baker 2009; Gallup 1998; de Waal et al.). Indeed, Suddendorf and Collier-Baker (2009: 1676) concluded that in the primate world, it is "only the descendants of a humanoid that probably lived between 13.8 and 18 Myr ago [that] have so far reliably demonstrated that they know who it is that… looks back at them when they look in the mirror." And if this cognitive capacity to recognize self in a mirror can enhance fitness, it can be selected upon to produce what humans reveal today: a series of identities that direct and guide their behaviors and emotional responses to others and social structures.

A self-identity, in turn, opens up a broader range of cognitive boundaries. First, knowledge of self is an enormous liberating force, promoting cognitive flexibility and intentional purposeful behavior. Second, as Mead (1934) emphasized, a self-consciousness entails an ability to be both subject and object to yourself. And this reflexivity is a special form of social consciousness because if you can take yourself as an object, you can reflect that others do as well and this enables you to assume the role of the other (Plotnik et al. 2006). Following Tomasello and Herrmann (2010: 3–8), Call and Tomasello (2007) great apes live in a world where they can form various types of social relationships, including ones based on "friendship." They can identify and engage in third party relationships, and they have the social intelligence to interpret the acts and goals of their social partners and form alliances and social maneuvers using a kind of "*Machiavellian* intelligence." Apes live in the here and now, but they can also recall things that happened in the past and anticipate or imagine things that might transpire in the future. In essence, Tomasello and Herrmann note: "Great apes…operate on their cognitive worlds in ways very similar to humans."

The final upshot of the hominoid mind was a cognitive leap to a new order of things—from visual representations of physical objects to their symbolic representations. Apes interact routinely with conspecifics by using symbolic gestures that are channeled through their visual-touch modalities (which are both under cortical control). According to Tomasello and Call (2007: 226), the "apes are learning or inventing many of their gestures, and they are choosing particular gestures for particular contexts, following up with other gestures if the first one does not work. Apes have social/communicative goals, and in their gestural communication… they are pursuing them flexibly" (and see, King 2004).

Abundant evidence also exists that linguistically trained apes have the sensory foundation to use symbols referentially to express their thoughts or to represent three-dimensional objects in space by using human sign language or a computer board of abstract lexigrams. African apes (and probably orangutans) also possess a humanlike auditory cortex that is primed to process and link a linguistic phonemic code with visual images and sounds. Apes cannot "talk" as their *vocal responses* remain under limbic control. But since their auditory perception is cortically based, they have the capacity when raised in a linguistic environment to learn *spontaneously* the grammar rules of human speech to the extent of comprehending complete sentences. Then, in response to the spoken message, they can either hand sign or use a computer board to punch out the correct series of abstract lexigrams for shared symbolic communication (for discussions, see Savage-Rumbaugh et al. 1993; Savage-Rumbaugh and Lewin 1994; Maryanski 1996, 1997). As great apes all possess these cognitive domains, they were surely inherited from their LCA and represent a huge suite of preadaptations that could be selected upon for the cognitive evolution of the hominin mind.

Summary

The Miocene epoch that began about 25 million years ago ended about 5 million years ago. It began with the branching away of Old World monkeys and apes from a last common anthropoid ancestor. The early monkeys like *Victoriapithecus* were scarce throughout most of the Miocene. Instead it was the heyday for apes—from the quadrupedal *Proconsul* to the swinging *Morotopithecus*. The apes dominated the forest landscape for most of the Miocene, but by the Late Miocene, their reign ended as "upper branch" apes went extinct leaving behind a few "under-branch" swinging apes on the hominoid family tree. The cladistic analysis of the LCA of great apes and hominins suggests a strong "founder effect" of nested phyletically based social and cognitive proclivities. The LCA population was predisposed toward (a) a macro-organizational structure built from weak ties, low-density, fission-fusion networks, high individualism, and rather low sociality when compared with monkey species and (b) a higher-order level of cognition that included self-awareness, comprehension of others' mental states, causal and logical reasoning, and capacities for symbolic representations.

After Old World monkeys took the lead in the playoffs, this trend accelerated, and monkeys today are now the overwhelmingly nonhuman primate success story, whether measured by number of species or by geographical location. Indeed, monkeys now inhabit savannas, forests, mountains, and even snow-covered terrains. But the hominoids still had an ace to play as eventually one of those swinging apes would give rise to the last common ancestor of chimpanzees and hominins.

The Evolution of Hominins and the Shift to a Terrestrial Niche

Chimpanzees and hominins shared a stem ancestor until the Late Miocene—some 7–6 million years ago. During the Early Pliocene about five million years ago, fossil beds in Africa chronicle the appearance of a number of hominin species with most placed in the genus—*Australopithecus*. As first cousins, ancestral chimpanzees and australopithecines closely resembled each other in physical appearance and brain size (375–450 cm^3) except that the australopiths had two standout features: smaller canines and, notably, an upright walking gait, which is *the* defining trait of a hominin. Both cousins coexisted for millions of years in a forest/woodland ecology, but, then, all the austropiths went extinct perhaps because of dwindling forest resources as the earth steadily grew colder with glaciers in the future. However, when selection favored the hominin line with a habitual upright walking gait, the stepping stones were laid for a future hominin to take up a wholly different lifestyle from chimpanzees (Ward et al. 2011; Hare 2011). About 2 ½ million years ago between the late Pliocene and Early Pleistocene (i.e., the Plio-Pleistocene), the *Homo* lineage appeared in Africa. Fossils of early *Homo* species reveal a gradual transition from their australopithecine ancestors toward a more humanlike appearance. And although early species of *Homo* were not much taller or heavier than the australopithecines, some *Homo* brains evidence an altered shaped skull with more modern frontal lobes and a cranial capacity in the range of 509–810 cm^3 (the modern human range is 1,200–1,600 cm^3) (Fleagle 1999; Turner and Maryanski 2008: 58ff; Stanford et al. 2013).

The next grade of hominin was *Homo erectus* who appeared about 1.8 million years ago. Here we have an ancestor with smaller jaws and teeth, larger body size (some individuals were nearly 6 feet tall), and a human anatomy from the neck down, although they were much more burly than modern humans. All signs point to *erectus* as on the road to humanity especially in brain size which grew closer to the human range with an average brain volume of 970 cm^3. This hominin was a true savanna dweller, with some erectus populations leaving Africa and moving into diverse ecosystems in Europe and Asia (rather reminiscent of the Middle Miocene apes that left Africa for parts of Eurasia). Thus, while the upright walking australopithecines evidenced relatively little change in brain size over millions of years, a change in ecological niche—from forest to open terrain—set into motion the adaptation of *Homo* to a new habitat, a shift that would eventually lead to the appearance of early modern *Homo sapiens* between 200,000 and 150,000 years ago (Fleagle et al. 2010; Jablonski et al. 2000).

A new environment always exerts selection pressures—and the more dissimilar the environment from the previous one, the more pronounced are selection pressures for change. Yet selection is conservative in the sense that adaptive responses to change are always guided by the existing facilities already present in a species' taxonomic repertoire. As G.L. Stebbins (1969, 1978: 65) has long emphasized, all sophisticated animals moving into a new ecology are already richly endowed with intricate programmed patterns that must be integrated with any new adaptation. Hence, "the mutations which are most likely to be accepted by selection…are those which individually have relatively slight effects on the phenotype." Thus, any modifications to the hominin line would have to involve a compromise between current selection forces and conservative phyletic traits which would limit the kinds of structural changes and elaborations possible.

Now, if we imagine the first hominins that sought to adapt to a parkland or savanna environment, we can assume that they took with them the neuroanatomy that had fostered their former adaptation to a forest habitat. What challenges did they face? One drawback was a dominant visual system that evolved for life in trees, whereas most ground-living mammals are olfactory dominant because it is automatically self-alerting and superior for detecting predators and prey at long distances by their enduring scents. In contrast, vision is inadequate for long distance perception in open terrain as it is not automatically alerting (as smell is) and often preoccupied with routine tasks, and it is useless after dark for detecting night predators, such as large prowling cats or snakes hiding in the grass. Yet it would take an act of evolutionary gymnastics at this late stage of hominin evolution to change a visually dominant primate into an olfactory dominant one. For as Stebbins (1969: 105) reminds us the *principle of conservation* states that:

> Once a unit of action has been assembled at a lower level of the hierarchy of organization and performs an essential function in the development of organisms at higher levels, mutations that might interfere with the activity of this unit are so strongly disadvantaged that they are rejected at the cellular level and never appear in the adult individual in which they occur.

A second problem for early *Homo* was their inherited social proclivities that would place constraints on the kinds of organizational arrangements possible. For once *Homo* confronted a dangerous world with large predators, with foods widely dispersed, and with the security of a leafy canopy out of sight, selection pressures were surely triggered for tighter-knit groups. But what kinds of social structures were possible for a "hang-loose" evolved ape? One well-known model has early *Homo* populations organized like baboon troops because these monkeys originally lived in propinquity with early hominins and thus would face similar selection pressures for food, sleeping sites, and protection from predators. If *Homo* populations were organized like contemporary baboons, group living would consist of large troops centered around hierarchical matrilineages with adult males positioned in military formation at the front and rear flanks of the troop as they moved across open terrain. To create these ranked matrifocal lineages, mothers and daughters must remain in spatial proximity as they do in monkey societies. However, as the cladistic analysis (briefly discussed earlier) on contemporary ape social networks concluded, a pattern of female-biased dispersal is a deep-seated phyletic trait. The *regularity hypothesis* used to test this assumption also confirmed that the modifications made by living descendants of the LCA evidence a strong systemic bias toward this structural trait, so that despite very different organizational arrangements, all living apes have social networks that are congruent with this ancestral footprint of female-biased dispersal (see Maryanski and Turner 1992; Turner and Maryanski 2008). Buttressing this finding is recent chemical data on early hominin dentition that indicates that they too had a transfer pattern of female-biased dispersal after puberty (see Callaway 2011; Copeland et al. 2011).

Another problem is that a baboon troop structure is just not realistic for large-bodied primates. It works well for baboons because they are comparatively small and can subsist on less nutritious foods. Hominins are twice the size of baboons with big brains that require large quantities of high-energy things to eat. So it is unlikely that a large cluster of hominins could forage together on a regular basis.

A second model is a chimpanzee organizational blueprint. As species usually build on the social structure that they inherit, it is far more likely that early *Homo* would start out with a social structure similar to their closest cousins who still share nearly 99% of their genetic material (Sherwood et al. 2008; Gagneux and Varki 2001). In fact, as Gagneux and Varki (2001: 2) pointed out, "if taxonomic classification were based solely upon genomic DNA sequence similarity, the nearly 99% identity of human, chimpanzee and bonobo genomes would require a reclassification of the latter two into the genus *Homo*." Chimpanzees (as discussed earlier) have a fission-fusion organization which is anchored in a macrolevel community organization. The ranging boundaries of a single community can be 8–80 square miles, with as many as 120 members and with locals sharing a "sense" of community. This intangible "commonwealth" is the only stable chimpanzee grouping (except for mother and dependent offspring) as promiscuous "bed hopping" between the sexes is the norm. On any given day, community members move about on their own, or join temporary clusters or "parties" which spontaneously gather for a few minutes, hours, or a day. Only rarely, if ever, does the entire community cluster in spatial propinquity. Yet community members all know the locals from strangers, and, indeed, a male stranger in a foreign community is usually attacked or killed. For this reason, males never leave their community ranges, whereas females depart after puberty to join another community. Thus, chimpanzee society is composed of loosely connected individuals whose members move about at will, often forge alone, hang out with a preferred close "friend," or join an impromptu gathering. Hence, community members are bound together by mostly weak ties and for males a few strong male friendship ties. Adult females (who move into a community) have weak ties with each other or what Jane Goodall calls a "neutral relationship" being neither friendly nor unfriendly (Goodall 1986: 17). Adult kinship ties exist, but they are limited to male siblings (if they are close in age) and between a mother and her adult son(s). While mothers and sons could theoretically mate and settle down to raise their offspring, thereby creating a unilineal kinship network with intergenerational continuity, the chimpanzee brain is wired for sexual avoidance between mothers and sons (seemingly a bioprogrammer to prevent inbreeding depression) (see Turner and Maryanski 2005). Thus, for the most part, chimpanzees have no extended relational anchors, giving adult individuals the freedom to move about and live apart within their community as they wish. And being self-reliant with few kinship ties, individuals can interact with a greater number of individuals. This ability to form mostly non-kinship-based weak or moderate ties takes a special kind of social mind, one able to envision "self" and 60 or more dispersed "selves" within a community. Indeed, for chimpanzees, there is a strength in weak ties over strong ties as it allows for the linking of a large, fluid population into a community form of organization, and in this sense, a chimpanzee community provides support for Mark Granovetter's "weak ties" thesis (Granovetter 1973; Maryanski 1992).

Now if humans' ape relatives have these proclivities, could they still be a foundational part of the hominin social mind? If so, how might selection have modified and expanded on such a fluid, relational structure? A detailed discussion of this transition is not possible here (see Turner and Maryanski 2008), but as hominins ventured further and further away from their forest home, a more tightly knit social structure would be crucial to survival. Yet if we reject the model of a baboon "female-bonded" society (as so-called in the literature) and if the ape system is simply not sufficiently structured, what alternatives are possible?

The best clues are found in the ethnographic record on well-studied hunting and gathering societies. For at least 99% of human history (or until garden farming was adopted about 10,000 years ago), humans "lived off the land" from coastal areas to the near deserts and frozen Arctic zones. Not surprisingly, hunter-gatherer societies share with chimpanzee societies many organizational features. Hunter-gatherers share a common foraging range and live in loosely organized bands of 30–100 individuals who move about in a yearly "seasonal round." They are also called "fission-fusion" societies because individuals within a shared "big band" region can freely join or leave a band depending on individual preferences and procurable resources. However, a hunter-gatherer society has a dramatically altered relational core than apes because it is centered around self-sufficient nuclear families who can forage apart from other families when resources are scarce and come together when resources are plentiful. Another consistent pattern is that the majority of hunter-gatherer societies (and most traditional societies) favor *patrilocal* residence and *female-biased dispersal*, although female exogamy is dressed up with invented cultural practices, such as a universal marriage rule, the touchstone for the creation of human kinship networks. Thus, the hallmark of a hunter-gatherer society is a loose-knit social structure and not a tight-knit troop. Even kinship and marriage norms are designed to accommodate a shifting collection of individuals. And, like chimpanzees, emphasis is placed on individual mobility, personal autonomy, and fluid networks. In short, hunter-gatherers do not display the high collectivism that is often imputed to human nature; they do, however, reveal a structure not evident among apes—relatively stable kinship groupings, which it appears, are what eventually allowed hominins to be sufficiently organized to survive open-country conditions (see Bailey and Aunger 1990 for a comparative relational analysis and Turner and Maryanski 2008 for a detailed discussion).

Finally, it is meaningful to ask the following: What other neurological changes in the hominin social mind were possible for an evolved forest ape who shifted to savanna terrain? For the most part, even the most intense selection forces could not have uprooted the basic hominoid legacy. Early *Homo* had not only benefited from nearly 65 million years of general primate evolution, but hominins are the end product of over 20 million years of ape evolution making it nearly impossible for large, radical mutations to have taken hold. In fact, as R.A. Fisher, the brilliant statistician, remarked, "the probability that individual mutations will contribute to evolution is in inverse correlation to the intensity of their effect on the developing phenotype" (quoted in Stebbins 1969: 104). Moreover, as Mendel discovered and as Fisher (1959: 16–17) has stressed, species are not passively "awaiting the next favorable mutation…" Instead species are already "abundantly supplied with heritable variation, prepared in advance for changes in all directions, and sensitively poised to respond to every kind of selective influence." Of course, many *small* random mutations (or point mutations) occurred during hominin evolution, but much of the burden was on natural selection to extend, elaborate, combine, or alter *already existing characteristics* of the hominin neuroanatomy. Selection could work successively on tail ends of bell-shaped distributions of existing traits that enhanced fitness and, in this way, alter the hominin genome. For example, our visual system is our paramount sense when it comes to the location, size, and shape of objects in space, and it is a major player in social communication. But in wide open terrain, hominins were faced with significant visual handicaps—not just after dark but even during the day because vision (unlike smell) is not self-alerting and always requires "on-guard" active attention and is often distracted. In trees, the vision and haptic senses had worked in close correspondence in many complex ways such as coordinating body movements as both were wired for voluntary neocortical control. But once hominins abandoned "swinging" from trees, their prehensile hands were no longer useful for guiding the travel of a biped or for alerting a biped of dangerous traveling conditions (e.g., a wobbly tree branch). However, the primate auditory cortex could be beefed up as an early warning system. While it was already under cortical control for the *perception of sounds*, the *vocal-auditory channel* was still under subcortical (or limbic) control, and so it could only emit mostly species-specific emotionally based vocalizations. Thus, in the *Homo* adaptive zone, selection acted by rewiring the inherited nature of this sensory arrangement by liberating the auditory/vocal channel from limbic control, placing it increasing under cortical control (by extending neuronets down to limbic centers), and, then, integrating it with the cortically based visual and haptic modalities (see Maryanski 1996 for a discussion). This control is not complete, as anyone who has seen someone screaming with emotion can attest, but auditory reposes in voice are mostly under cortical control (Deacon 2007).

While the evidence suggests that the auditory-vocal channel was originally rewired to assist the visual modality because it is (like smell) a self-alerting detection system that under cortical control could generate rational intended responses to environmental stimuli, this process set in motion far-reaching changes that would determine the direction of further hominin evolution. For once vocal production came under cortical control, it allowed for the creation of new voluntary sounds, especially since upright walking had already opened up the vocal tract. Then, with modifications to the hominin vocal apparatus that involved the tongue, larynx, pharynx, and related structures, more and more finely tuned and articulated sounds could be made. Hominin neuroanatomy also underwent extensive modifications that included augmentation of association tissue, greater complexity in multimodal regions, and more intricate interconnections among brain parts. Selection also usurped already existing zones on the left side of the brain, such as Broca's area in the frontal lobe (which is functionally unique in humans but cytoarchitecturally homologous in higher primates), to control the production of vocal utterances as speech is downloaded from the way the brain thinks, which, for the most part, is not in terms of words. Then, in addition to other neurological refinements, such as linking up of an already in place capacity among the great apes for symbolic representations (via the association cortices around the inferior parietal lobe), with the hominin extension of auditory sounds, much of the neurological work for speech had already been accomplished. Thus, spoken language is not the result of dramatic mutations nor is it a mysterious property of humans. It is the result of modifications and elaborations to an *already existing anthropoid sensory system* and the extension of symbolic and other cognitive abilities *already present* in our hominoid legacy. As Deacon (2007: 531) phrased it: "Generations of comparative neuroanatomists have failed to identify even one major novel brain structure in humans. This suggests that our special adaptations for language are the result of using previously evolved primate brain structures in new ways and in new combinations."

Thus, there was a "hominin adaptive zone" on the savanna long before there was an explosive growth of the neocortex in the ancestral *Homo* brain. What occurred then was an unfolding sequence of cranial enlargements, which is one of the hallmarks of primate evolution, but which culminated in human evolution. Given that selection acted to foster such striking brain development in human ancestors means that it must have greatly enhanced survival and reproductive success, despite the fact that a big brain is exceedingly costly to maintain. Hence, we should now ask the following: What dramatic events occurred to initiate the cognitive genus of *Homo sapiens?*

One key factor was that an adaptive shift to a new habitat may have been the trigger for *regular* use of standardized stone tools—mostly flake and choppers which first appear in the fossil record about 2.5 million years ago, giving us a baseline for the beginning of hominin technology. Thousands of early stone tools (known formally as the Oldowan tradition) have been discovered in East African sites. Microscopic examinations of the wear patterns on these assemblages along with experimental tryouts on their possible uses indicate they can be adapted for a variety of functions. By the time of

Homo erectus (circa 1.8 to about 250,000 years ago), a new, more advanced stone tool technology began to appear in the archaeological record (known as the Acheulean tradition). The hand ax was representative of this tradition, along with a variety of bifacial tool designs such as scrapers and cleavers that were used for specialized purposes—signaling the first cognitive leap in technological evolution. About 800,00 years ago, the archaeological record is conclusive that *Homo erectus* began to use fire seemingly for cooking meat, staying warm, seeing at night, and keeping predators at bay. Making specialized tools and using fire are hallmarks of human cognitive and social behavior allowing us to begin to piece together a little about how *Homo erectus* lived as they left remains of their material culture all over the African, Asian, and European continents. Their tool kits and occupation sites tell us that *erectus* populations relied upon a heavier flesh-based diet because these tools were utilized for the butchering of big game animals, thus signaling a shift from earlier hominins. Their tool assemblages are also highly uniform, indicating that the knowledge to produce them was passed along from generation to generation. For example, the design of the hand ax persisted as a core tool from 1.5 million years ago to 250,000 years ago, suggesting it was part of a cultural tradition and used in standardized ways. As food collectors, *erectus* populations naturally consumed plants, grains, and fruits depending on available resources, but it is surely the case that the increase in meat consumption is tied to the need for protein to support a taller body and a larger brain, which continued to expand. Indeed, the migration of animal herds may account, in part, for the dispersal of *erectus* all over the Old World. The material traces that *Homo erectus* left behind tell us that during their reign they underwent major cognitive changes tied to rewiring of the brain, and these led to organizational changes. In all likelihood, it was these evolving erectus populations that became the ancestral stock for the rise of later species of *Homo*. Yet the preserved fossils and artifacts of *erectus* do not tell us much about their social lifeways, aside from the fact that they were highly successful nomadic food collectors with a greater reliance on meat than earlier hominins. As they lived in open-ranging areas with many carnivores, they were probably organized into band level societies similar to the studied hunter-gatherer populations discussed earlier. And while their material assemblages point to distinctive cultural traditions with a steady increase in complexity and diversity, surely accompanied by cognitive development, we can only speculate about their cognitive behaviors. What we can document is the evolution of creative minds involved in the technological-based pursuit of stone tool technology.

By the Middle Pleistocene (circa 780,000–130,000 years ago), hominins appear more humanlike in facial features, and a cranial capacity near and perhaps overlapping with the size of modern humans, thus signaling the transition from *Homo erectus* to *Homo sapiens*. These descendants are usually labeled "advanced *Homo erectus*," "archaic *Homo sapiens*," or "*Homo heidelbergensis*" (to give them a species name). While their stone tool kits were similar to those of *Homo erectus*, some have creative variations, again representing technological-based intelligence. Fully modern-looking humans with big brains, rounded skulls, and true chins first appear in the fossil record between 200,000 and 150,000 years ago. They are surely the descendants of Middle Pleistocene archaic humans, but just where and when they evolved remains a mystery. Even more of an enigma is the discovery that the human genome has much less genetic variation than the chimpanzee genome. This suggests that early *Homo sapiens* underwent a tight "bottleneck" event and that all living humans may have originally descended from a tiny, ancestral gene pool—consisting of hundreds or at best 10,000 individuals. This means that there is actually relatively little variation in the *Homo sapiens* gene pool (Harpendig et al. 1998; Gagneux 2002; Kasserman et al. 1999; Kaessmann and Pääbo 2002; Adams et al. 2000). One implication of this finding is that, given the minimal genetic variability among humans (despite the 7 billion individuals now on earth), we should not be surprised that human beings when confronted with problems of survival work out relatively similar solutions. It also puts to rest any notions that *Homo sapiens* can be divided up into different races on the basis of illusionary qualitative distinctions among populations. For on most counts, there is more variation *within* human populations than between them.

Thus, goes the strange tale of how the hominin mind evolved in a mosaic pattern, although it reveals a unity of structure and function. First, the anthropoid sensory modalities set the stage for a primate tuned to the outside world and to voluntary, cortically based behavior. Then, after the split between monkeys and apes, their subsequent evolution took different paths, with apes dominating the forests for nearly 15 million years until monkeys gained the upper hand. The extinction of all quadrupedal apes that could not compete with monkeys followed, but the swinging apes survived, and one of those lineages became the LCA of the great apes, hominins, and humans. By the time of *Homo*, hominins were already equipped with all kinds of preadaptations on which selection could go to work, and coupled with a variety of point mutations, the hominin mind eventually emerged.

Yet the genius of the hominin mind rests not only on its amazing cognitive and intellectual potential but on its amazing social potential to cooperate with others—even strangers. As Tomasell and Herrmann (2010: 5) recently highlighted, "Virtually all of human's highest cognitive achievements are not the work of individuals acting alone but rather of individuals collaborating in groups… ending up with such things as complex social institutions structured by joint goals, division of labor, and communicative symbols." Thus, selection built on the elementary collaboration of weak tie networks found in great ape societies, and notably those of our closest African cousins, to fashion a strikingly novel *social mind* with the potential for macrolevel coordination—a sociality that allows for the integration of millions of individuals (see Machalek 1992). In addition, the hominin mind has an evolutionary novelty truly all its own—*selection favoring change itself.* Hominin evolution was a time of great climatic change, from warm periods through long glacial periods of cold and ice. The hominin brain is one that faced so many changes and endured such constant selection pressures that the long-term outcome was a structure built around creativity, variation in socialization, and the ability to adapt to a sweeping variety of new challenges. Thus, the hominin mind is wired for adaptability; it is not locked into very many hard-wired predispositions, and it not highly specialized but, instead, more generalized. The coevolution of culture with biology is what allows for this capacity—a truly unique capacity that along with our one-of-a-kind sociality is the true secret of the hominin mind.

References

Adams, E., Cooper, S., Thomson, G., & Parham, P. (2000). Common chimpanzees have greater diversity than humans at two of the three highly polymorphic MHC class I genes. *Immunogenetics, 51*, 410–424.

Andrews, P. (1981). Species diversity and diet in monkeys and apes during the miocene. In C. B. Stringer (Ed.), *Aspects of human evolution*. London: Taylor & Francis.

Andrews, P. (2007). The biogeography of hominid evolution. *Journal of Biogeography, 34*, 381–382.

Andrews, P., & Kelley, J. (2007). Middle miocene dispersals of apes. *Folia Primatol, 78*, 328–343.

Bailey, R., & Aunger, R. (1990). Humans as primates. The social relationships of Efe Pygmy men in comparative perspective. *International Journal of Primatology, 11*, 127–145.

Barton, R. A. (1998). Visual specialization and brain evolution in primates. *Proceedings of the Royal Society, London, B Biological Sciences, 265*, 1933–1937.

Begun, D. (2007). Fossil record of miocene hominoids. In W. Henke & T. Hardt (Eds.), *Handbook of paleoanthropology 1* (pp. 921–977). Berlin: Springer.

Benefit, B. R. (1999). *Victoriapithecus:* The key to old world monkey and catarrhine origins. *Evolutionary Anthropology, 7*, 155–174.

Benefit, B., & McCrossin, M. (1997). Earliest known old world monkey skull. *Nature, 388*, 368–371.

Bhatnagar, K. P., & Smith, T. D. (2007). The vomeronasal organ and its evolutionary loss in catarrhine primates. In J. Kaas & T. Preuss (Eds.), *Evolution of nervous systems: A comprehensive reference, V4* (pp. 41–148). Amsterdam: Elsevier.

Bloch, J., & Boyer, D. (2002). Grasping primate origins. *Science, 298*, 1606–1610.

Bloch, J., & Boyer, D. (2007). New skeletons of Paleocene-Eocene plesiadapiformes: A diversity of arboreal positional behaviors in early primates. In M. J. Ravosa & M. Dagosto (Eds.), *Primate origins: Adaptations and evolution* (pp. 535–581). New York: Springer.

Bradley, B. (2008). Reconstructing phylogenies and phenotypes: A molecular view of human evolution. *Journal of Anatomy, 212*, 337–353.

Call, J., & Tomasello, M. (2007). *The gestural communication of apes and monkeys*. Mahwah: Lawrance Erlbaum.

Callaway, E. (2011). Female australopiths seek brave new world. *Nature*, June 1, p. 1038.

Campbell, F., & Maffer, L. (1976). Contrast and spatial frequency. In R. Held & W. Richards (Eds.), *Recent progress in perception* (Scientific American). San Francisco: W.H. Freeman and Company.

Casagrande, V. A., & Khaytin, I. (2007). The evolution of parallel visual pathways in the brains of primates. In J. Kaas & T. Preuss (Eds.), *Evolution of nervous systems: A comprehensive reference, V4* (pp. 87–119). Amsterdam: Elsevier.

Chatterjee, H., Simon, H., Barnes, I., & Groves, C. (2009). Estimating the phylogeny and divergence times of primates using a supermatrix approach. *BMC Evolutionary Biology, 9*, 259–278.

Cheney, D. L., & Seyfarth, R. M. (1980). Vocal recognition in free-ranging vervet monkeys. *Animal Behavior, 28*, 739–751.

Coleman, M. (2009). What do primates hear? a meta-analysis of all known nonhuman primate behavioral audiograms. *International Journal of Primatology, 30*, 55–91.

Coleman, M., & Colbert, M. (2010). Correlation between auditory structures and hearing in non-human primates. *Journal of Morphology, 27*, 511–532.

Coleman, M., Kay, R., & Colbert, M. (2010). Auditory morphology and hearing sensitivity in fossil new world monkeys. *The Anatomical Record, 293*, 1711–1721.

Conroy, G. (1990). *Primate evolution*. New York: Methuen.

Copeland, S., Sponheimer, M., de Ruiter, D., Lee-Thorp, J., Codron, D., le Roux, P., Grimes, V., & Richards, M. (2011). Strontium isotope evidence for landscape use by early hominins. *Nature, 474*, 76–78.

Corruccini, R., Ciochon R., & McHenry H. (1975). Osteometric shape relationships in the wrist joint of some anthropoids. *Folia Primatologioa, 24*, 250–274.

Deacon, T. W. (2007). The evolution of language systems in the human brain. In J. Kaas & T. Preuss (Eds.), *Evolution of nervous systems: A comprehensive reference, V4* (pp. 529–547). Amsterdam: Elsevier.

Dean, M. C., & Leakey, M. (2004). Enamel and dentine development and the life history profile of *Victoriapithecus macinnesi* from maboko island, Kenya. *Annals of Anatomy, 186*, 405–412.

Deane, A. (2009). Early miocene catarrhine dietary behaviour: The influence of the red queen effect on incisor shape and curvature. *Journal of Human Evolution, 56*, 275–285.

deWaal, F. B., Dindo, M., Freeman, C. A., & Hall, M. (2005). The monkey in the mirror: Hardly a stranger. *Proceedings of the National Academy of Sciences, 102*, 1140–1146.

Dominy, N., Ross, C., & Smith, T. (2004). Evolution of the special senses in primates: Past, present and future. *The Anatomical Record Part A, 281 A*, 1078–1082.

Durkheim, E. ([1895] 1938). *The division of labor in society* (George Simpson, Trans.). New York: The Free Press.

Falk, D. (2000). *Primate diversity*. New York: W.W. Norton & Company.

Falk, D. (2007). Evolution of the primate brain. In W. Henke & T. Hardt (Eds.), *Handbook of paleoanthropology* (Vol. 1, pp. 1133–1162). Berlin: Springer.

Filler, A. (2007). Emergence and optimization of upright posture among hominiform hominoids and the evolutionary pathophysiology of back pain. *Neurosurg Focus, 23*, 1–6.

Fleagle, J. (1999). *Primate adaptation and evolution*. New York: Academic.

Fleagle, J., Shea, J., Grine, F., Baden, A., & Leakey, R. (2010). *Out of Africa* (Vol. 1). London: Springer.

Fortelius, M., & Hokkanen, A. (2001). The trophic context of hominoid occurrence in the later miocene of Western Eurasia: A primate-free view. In L. de bonis, G. Koufos, & P. Andrews (Eds.), *Hominoid evolution and climatic change in Europe* (pp. 19–47). Cambridge: Cambridge University Press.

Freides, D. (1973). Human information processing and sense modality: Cross-modal functions, information complexity, memory, and deficit. *Psychological Bulletin, 81*(5), 284–310.

Gagneux, P., & Varki, A. (2001). Genetic differences between humans and great apes. *Molecular Phylogenetics and Evolution, 18*, 2–13.

Gagneux, P. (2002). The genus pan: Population genetics of an endangered outgroup. *Trends in Genetics, 18*, 327–330.

Gallup, G. G. (1998). Self-awareness and the evolution of social intelligence. *Behavioral Processes, 42*, 239–247.

Gebo, D., Malit, N., & Nengo, I. (1997). New proconsuloid postcranials from the early Miocene of Kenya. *Primates, 50*, 311–319.

Gibson, J. J. (1962). Observations on active touch. *Psychological Review, 69*, 477–491.

Glaser, D. (2007). The evolution of the sweetness receptor in primates. In J. Kaas & T. Preuss (Eds.), *Evolution of nervous systems: A comprehensive reference* (Vol. 4, pp. 121–128). Amsterdam: Elsevier.

Glaser, D., Tinti, J. M., & Nofre, C. (1995). Evolution of the sweetness receptor in primates. 1. Why does alitame taste sweet in all prosimians and simians, and aspartame only in old world simians? *Chemical Senses, 20*, 573–584.

Glazko, G. V., & Nei, M. (2003). Estimation of divergence times for major lineages of primate species. *Molecular Biology and Evolution (Supplement), 20*, 424–434.
Glendenning, K. K., & Masterton, R. B. (1998). Comparative morphometry of mammalian central auditory systems: Variation in nuclei and form of the ascending system. *Brain, Behavior and Evolution, 51*, 59–89.
Goodall, J. (1986). *The chimpanzees of Gombe: Patterns of behavior*. Cambridge, MA: Harvard University Press.
Granovetter, M. (1973). The strength of weak ties. *American Journal of Sociology, 78*, 1360–1380.
Gregory, W. (1916). Studies on the evolution of the primates. *Bulletin of the American Museum of Natural History, 35*, 239–355.
Hackett, T. (2003). The comparative anatomy of the primate auditory cortex. In A. Ghazanfar (Ed.), *Primate audition: Behavior and neurobiology* (pp. 199–226). Boca Raton: CRC Press.
Hackett, T. (2008). Anatomical organization of the auditory cortex. *Journal of the American Academy of Audiology, 19*, 774–779.
Halloway, R. (1978). The relevance of endocasts for studying primate brain evolution. In C. R. Noback (Ed.), *Sensory systems of primates*. New York: Plenum.
Hamrick, M. (2001). Primate origins: evolutionary change in digital ray patterning and segmentation. *Journal of Human Evolution, 40*, 339–351.
Hare, B. (2011). From hominoid to hominid mind: What changed and why? *Annual Review of Anthropology, 40*, 293–309.
Harpendig, H., Gurven, M. R., et al. (1998). Genetic traces of ancient demography. *Proceedings of the National Academy of Sciences, 95*, 1961–1967.
Harrison, T., & Andrews, P. (2009). The anatomy and systematic position of the early miocene proconsulid from meswa bridge, Kenya. *Journal of Human Evolution, 56*, 479–496.
Heffner, R. (2004). Primate hearing from a mammalian perspective. *The Anatomical Record, 281A*, 1111–1122.
Heffner, H., & Heffner, R. (2008). High frequency hearing. In P. Dallos, D. Oertel, & R. Hoy (Eds.), *Handbook of the senses: Audition* (pp. 55–60). New York: Elsevier.
Heidegard, H., Beil, B., Hilbig, H., Call, J., & Bidmon, H.-J. (2009). Superior olivary complex organization and cytoarchitecture may be correlated with function and catarrhine primate phylogeny. *Brain Structure and Function, 213*, 489–497.
Hogg, C., Neveu, M., Stokkan, K.-A., Folkow, L., Cottrill, P., Douglas, R., Hunt, D., & Jeffery, G. (2011). Arctic reindeer extend their visual range into the ultraviolet. *The Journal of Experimental Biology, 214*, 2014–2019.
Holloway, R. (1968). The evolution of the primate brain: Some aspects of quantitative relations. *Brain Research, 7*, 121–172.
Hoover, K. (2010). Smell with inspiration: The evolutionary significance of olfaction. *Yearbook of Physical Anthropology, 53*, 63–74.
Israfil, H. S. M., Zehr, A. R., Mootnick, M. R., & Steiper, M. E. (2011). Unresolved molecular phylogenies of gibbons and siamangs (family: Hylobatidae) based on mitochondrial, Y-linked loci, and X linked loci indicate a rapid miocene radiation or sudden vicariance event. *Molecular Phylogenetics and Evolution, 58*, 447–455.
Jablonski, N. G., Whitfort, M. J., Roberts-Smith, N., & Qinqu, X. (2000). The influence of life history and diet on the distribution of catarrhine primates during the Pleistocene in eastern Asia. *Journal of Human Evolution, 39*, 131–157.
Jacobs, G. H. (1993). The distribution and nature of colour vision among the mammals. *Biological Reviews of Cambridge Philosophical Society, 68*, 413–471.
Jacobs, G. H. (2004). Photopigment variations and the evolution of anthropoid vision. In C. F. Ross & R. F. Kay (Eds.), *Anthropoid origins: New visions* (pp. 645–664). New York: Kluwer/Plenum.
Jacobs, G. H. (2007). The comparative biology of photopigments and color vision in primates. In J. Kaas & T. Preuss (Eds.), *Evolution of nervous systems: A comprehensive reference* (Vol. 4, pp. 80–85). Amsterdam: Elsevier.
Jacobs, G. H., & Deegan, J. F. (1999). Uniformity of colour vision in old world monkeys. *Proceedings of the Royal Society London B, 266*, 2023–2028.
Jacobs, G. H., & Williams, G. A. (2001). The prevalence of defective color vision in old world monkeys and apes. *Color Research and Application [supplement], 26*, 123–127.
Janečka, J., Miller, W., Pringle, T., Wiens, F., Zitzmann, A., Helgen, K., Springer, M., & Murphy, W. (2007). Molecular and genomic data identify the closest living relative of primates. *Science, 318*, 792–794.
Jeffers, R., & Lehiste, I. (1979). *Principles and methods of historical linguistics*. Cambridge, MA: MIT Press.
Jerison, H. J. (1973). *Evolution of the brain and behavior*. New York: Academic.
Jerison, H. J. (2007). What fossils tell us about the evolution of the neocortex. In J. Kaas (Ed.), *Evolution of nervous systems: A comprehensive reference* (Vol. 3, pp. 1–12). Amsterdam: Elsevier.
Kagaya, M., Ogihara, N., & Nakatsukasa, M. (2010). Is the clavicle of apes long? an investigation of clavicular length in relation to body mass and upper thoracic width. *International Journal of Primatology, 31*, 209–217.
Kass, J. (2007). The evolution of sensory and motor systems in primates. In J. Kaas & T. Preuss (Eds.), *Evolution of nervous systems: A comprehensive reference* (Vol. 4, pp. 35–57). Amsterdam: Elsevier.

Kass, J. (2008). The evolution of the complex sensory and motor system of the human brain. *Brain Research Bulletin, 75*, 384–390.

Kass, J., & Pons, T. P. (1988). The somatosensory system of primates. In H. D. Steklis & J. Erwin (Eds.), *Comparative primate biology, Vol. 4: Neurosciences* (pp. 421–468). New York: Alan R. Liss.

Kaessmann, H., & Pääbo, S. (2002). The genetic history of humans and the great apes. *Journal of Internal Medicine, 251*, 1–18.

Kasserman, H., Wiebe, V., & Pääbo, S. (1999). Extensive nuclear DNA sequence diversity among chimpanzees. *Science, 286*, 1159–1161.

Kay, R., Ungar, P. (1997). Dental evidence for diet in some Miocene Catarrhines with comments on the effects of phylogeny on the interpretation of adaptation. In D. Begun, C. Ward, & M. Rose (Eds.), *Function, phylogeny, and fossils: Miocene hominoid evolution and adaptation*. New York: Plenum.

Keller, G., & Barron, J. (1987). Paleodepht distribution of Neogene hiatuses. *Paleoceanography*, *2*, 697–713.

Keller, H. (1904). *The world I live in*. London: Hodder & Stoughton.

Kelley, J. (2002). Life-history evolution in miocene and extant apes. In N. Minugh-Purvis & K. J. McNamara (Eds.), *Human evolution through developmental change*. Baltimore: John Hopkins University Press.

Kelley, J., & Smith, T. M. (2003). Age at first molar emergence in early miocene *Afropithecus turkanensis* and life-history evolution in the *Hominoidea*. *Journal of Human Evolution, 44*, 307–329.

King, B. J. (2004). *Dynamic dance: Nonvocal communication in the African great apes*. Cambridge: Harvard University Press.

King, A., & Nelken, I. (2009). Unraveling the principles of auditory cortical processing: Can we learn from the visual system? *Nature Neuroscience, 12*, 698–701.

Kirk, C. (2004). Comparative morphology of the eye in primates. *The Anatomical Record Part A, 281A*, 1095–1103.

Kirk, C. (2006). Visual influences on primate encephalization. *Journal of Human Evolution, 51*, 76–90.

Krubitzer, L., & Kaas, J. (2005). The evolution of neocortex in mammals: How is phenotypic diversity generated. *Current Opinion in Neurobiology, 15*, 444–453.

Le Gros Clark, W. E. (1962). *The antecedents of man*. Chicago: Quadrangle Books.

Lederman, S. J., & Klatzky, R. L. (2009). Haptic perception: A tutorial. *Attention, Perception and Psychophysics, 71*, 1439–1459.

Lemelin, P., & Jungers, W. (2007). Body size and scaling of the hands and feet of prosimian primates. *American Journal of Physical Anthropology, 133*, 828–840.

Lewis, O. J. (1974). The wrist articulations of the anthropoidea. In F. Jenkins (Ed.), *Primate locomotion* (pp. 143–167). New York: Academic.

Lucas, P., Dominy, N., Riba-Hernandex, P., Stoner, K., Yamashita, N., Loría-Calderón, E., Petersen-Pereira, W., Rojas-Durán, Y., Salas-Pena, R., Solis-Madrigal, S., Osorio, D., & Darvell, B. (2003). Evolution and function of routine trichromatic vision in primates. *Evolution, 57*, 2636–2643.

Lyon, D. C. (2007). The evolution of visual cortex and visual systems. In J. Kaas & L. Krubitzer (Eds.), *Evolution of nervous systems: A comprehensive reference* (Vol. 3, pp. 267–306). Amsterdam: Elsevier.

Machalek, R. (1992). The evolution of macrosociety: Why are large societies rare? *Advances in Human Ecology, 1*, 33–64.

Maclatchy, L. (2004). The oldest ape. *Evolutionary Anthropology, 13*, 90–103.

MacLatchy, L., Gebo, D., Kityo, R., & Pilbeam, D. (2000). Postcranial functional morphology of *Morotopithecus bishopi,* with implications for the evolution of modern ape locomotion. *Journal of Human Evolution, 39*, 159–183.

Martin, R. D. (1990). *Primate origins and evolution: A phylogenetic construction*. London: Chapman & Hall.

Maryanski, A. R. (1987). African ape social structure: Is there strength in weak ties? *Social Networks, 9*, 191–215.

Maryanski, A. (1992). The last ancestor: An ecological network model on the origins of human sociality. *Advances in Human Ecology, 2*, 1–32.

Maryanski, A. (1995). African ape social networks: A blueprint for reconstructing early hominid social structure. In J. Steele & S. Shennan (Eds.), *Archaeology of human ancestry*. London: Routledge.

Maryanski, A. (1996). Was speech an evolutionary afterthought? In B. Velichkovsky & D. Rumbaugh (Eds.), *Communicating meaning: The evolution and development of language* (pp. 79–97). Mahwah: Lawrence Erlbaum.

Maryanski, A. (1997). Primate communication and the ecology of a language niche. In U. Segerstråle & P. Molnár (Eds.), *Nonverbal communication: Where nature meets culture* (pp. 191–209). Mahwah: Lawrence Erlbaum Associates.

Maryanski, A., & Turner, J. (1992). *The social cage: Human nature and the evolution of society*. Stanford: Stanford University Press.

Masterton, B., & Diamond, I. (1973). Hearing: Central neural mechanisms. In E. Carterette & M. Friedman (Eds.), *Handbook of perception no. 3, biology of perceptual systems* (pp. 408–448). New York: Academic.

Mead, G. H. (1934). *Mind, self and society*. Chicago: University of Chicago Press.

Meldrum, J. (2006). *Sasquatch: Legend meets science*. New York: Forge.

Miller, E. R., Benefit, B. R., McCrossin, M. L., Plavan, J. M., Leakey, M. G., El-Barkooky, A. N., Hamdan, M. A., Abdel Gawad, M. K., Hassan, S. M., & Simons, E. L. (2009). Systematics of early and middle miocene old world monkeys. *Journal of Human Evolution, 57*, 195–211.

Molnár, Z., Tavare, A., & Cheung, A. F. P. (2007). The origin of neocortex: Lessons from comparative embryology. In J. Kaas & L. Krubitzer (Eds.), *Evolution of nervous systems: A comprehensive reference* (Vol. 3, pp. 13–26). Amsterdam: Elsevier.

Nakatsukasa, M. (2008). Comparative study of moroto vertebral specimens. *Journal of Human Evolution, 55*, 581–588.

Napier, J. (1963). Brachiators and brachiators. In J. Napier & N. A. Barniat (Eds.), *The primates. Symposia of the Zoological Society of London* 10.

Napier, J. R., & Napier, P. H. (1985). *The natural history of the primates*. Cambridge, MA: MIT Press.

Nofre, C., Tinti, J. M., & Glaser, D. (1996). Evolution of the sweetness receptor in primates. 11. "Gustatory responses of non-human primates to nine compounds known to be sweet in man". *Chemical Senses, 21*, 747–762.

Norman, P., & Don Cameron, H. (1977). Cladistic methods in textual, linguistic, and phylogenetic analysis. *Systematic Zoology, 26*, 380–385.

Oxnard, C. E. (1963). Locomotor adaptation in the primates. In J. Napier, W. A. Barnicot (Eds.), *The Primates. Symposia of the Zoological Society of London* 10.

Pilbeam, D., & Young, N. (2004). Hominoid evolution: Synthesizing disparate data. *Comptes Rendus Palevol, 3*, 305–321.

Platnick, N., & Cameron, D. (1977). Cladistic methods in textual, linguistic, and phylogenetic analysis. *Systematic Zoology, 26*, 380–385.

Plotnik, J., de Waal, F., & Reiss. D. (2006) Self-recognition in an Asian elephant. *Proceedings of the National Academy of Sciences USA, 103*, 17053–17057.

Potts, R. (2004). Paleoenvironmental basis of cognitive evolution in great apes. *American Journal of Primatology, 62*, 209–228.

Preuschoft, S., & Preuschoft, H. (1994). Primate nonverbal communication: Our communication heritage. In W. Nöth (Ed.), *Origins of semiosis* (pp. 66–100). Berlin: Mouton de Gruyter.

Preuss, T. M. (2007). Primate brain evolution in phylogenetic context. In J. Kaas & T. Preuss (Eds.), *Evolution of nervous systems: A comprehensive reference* (Vol. 4, pp. 1–34). Amsterdam: Elsevier.

Purvis, D., White, L., Zheng, D., Andrews, T., & Riddle, D. (1996). Brain size, behavior and the allocation of neural space. In D. Magnusson (Ed.), *The lifespan development of individuals: Behavioral, neurobiological, and psychosocial perspectives*. Cambridge: Cambridge University Press.

Raaum, R., Kirstin, L., Sterner, N., Noviello, C., Stewart, C., & Disotell, T. (2005). Catarrhine primate divergence dates estimated from complete mitochondrial genomes: Concordance with fossil and nuclear DNA evidence. *Journal of Human Evolution, 48*, 237–257.

Radinsky, L. B. (1970). The fossil evidence of prosimian brain evolution. In C. Noback & W. Montagna (Eds.), *The primate brain*. New York: Appleton.

Radinsky, L. B. (1974). The fossil evidence of anthropoid brain evolution. *American Journal of Physical Anthropology, 41*, 15–28.

Rakic, P., & Kornack, D. (2001). Neocortical expansion and elaboration during primate evolution: A view from neuroembryology. In K. R. Gibson & D. Falk (Eds.), *Evolutionary anatomy of the primate cerebral cortex* (pp. 30–56). Cambridge: Cambridge University Press.

Rakic, P., & Kornack, D. (2007). The development and evolutionary expansion of the cerebral cortex in primates. In J. Kaas & T. Preuss (Eds.), *Evolution of nervous systems: A comprehensive reference* (Vol. 4, pp. 243–259). Amsterdam: Elsevier.

Relethford, J. (2010). *The human species*. Boston: McGraw Hill.

Ross, C. F., & Martin, R. D. (2007). The role of vision in the origin and evolution of primates. In J. Kaas & T. Preuss (Eds.), *Evolution of nervous systems: A comprehensive reference* (Vol. 4, pp. 59–76). Amsterdam: Elsevier.

Rossie, J. B. (2005). "Anatomy of the nasal cavity and paranasal sinuses in *aegyptopithecus* and early Miocene African catarrhines. *American Journal of Physical Anthropology, 126*(126), 250–267.

Rouquier, S., & Giorgi, D. (2007). The loss of olfactory receptor genes in human evolution. In J. Kaas & T. Preuss (Eds.), *Evolution of nervous systems: A comprehensive reference* (Vol. 4, pp. 129–139). Amsterdam: Elsevier.

Rowe, T., Macrini, T., & Luo, Zhe-Xi. (2011). Fossil evidence on origin of the mammalian brain. *Science, 332*, 955–957.

Savage-Rumbaugh, S., & Lewin, R. (1994). *Kanzi: The ape at the brink of the human mind*. New York: Wiley.

Savage-Rumbaugh, S. J. M., Seveik, J., Brakke, K., Williams, S. L., & Rumbaugh, D. (1993). Language comprehension in the ape and child. *Monographs of the society for research in child development* (Vol. 58, 3–4). Chicago: University of Chicago Press.

Seiffert, E., Perry, J., Simons, E., & Boyer, D. (2009). Convergent evolution of anthropoid-like adaptations in Eocene adapiform primates. *Nature, 461*, 118–121.

Semendeferi, K., & Damasio, H. (2000). The brain and its main anatomical subdivisions in living hominoids using magnetic resonance imaging. *Journal of Human Evolution, 38*, 317–332.

Sherwood, C. C., & Hof, P. R. (2007). The evolution of neuron types and cortical histology in apes and humans. In J. Kaas & T. Preuss (Eds.), *Evolution of nervous systems: A comprehensive reference* (Vol. 4, pp. 355–378). Amsterdam: Elsevier.

Sherwood, C., Subiaul, F., & Zawidzki, T. (2008). A natural history of the human mind: Tracing evolutionary changes in brain and cognition. *Journal of Anatomy, 212*, 426–454.

Silcox, M. T. (2007). Primate taxonomy, plesiadapiforms, and approaches to primate origins. In M. J. Ravosa & M. Dagosto (Eds.), *Primate origins: Adaptations and evolution* (pp. 143–178). New York: Plenum Press.

Silcox, M., Benham, A., & Block, J. (2010). Endocasts of microsyops (microsyopidae, primates) and the evolution of the brain in primitive primates. *Journal of Human Evolution, 58*, 505–521.

Simons, E. (1987). New faces of aegyptopithecus from the oligocene of Egypt. *Journal of Human Evolution, 16*, 273–289.

Smith, T. D., Bhatnagar, K. P., Shimp, K. L., et al. (2002). Histological definition of the vomeronasal organ in humans and chimpanzees with a comparison to other primates. *Anatomical Record, 265*, 176–192.

Snowdon, C. (1990). Language capacities of Non-human animals. *Yearbook of Physical Anthropology, 33*, 215–243.

Sousa, A., & Wood, B. (2007). The hominin fossil record and the emergence of the modern human central nervous system. In J. Kaas & T. Preuss (Eds.), *Evolution of nervous systems: A comprehensive reference* (Vol. 4, pp. 292–336). Amsterdam: Elsevier.

Stanford, C., Allen, J., & Antón, S. (2013). *Biological anthropology*. Boston: Pearson.

Stebbins, G. L. (1969). *The basis of progressive evolution*. Chapel Hill: University of North Carolina Press.

Stebbins, G. L. (1978). The dynamics of evolutionary change. In H. Evolution (Ed.), *Ediated by noel korn* (pp. 61–79). New York: Holt, Rinehart and Winston.

Stein, P., & Rowe, B. (2011). *Physical anthropology*. New York: McGraw Hill.

Suddenforf, T., & Collier-Baker, E. (2009). The evolution of primate visual self-recognition: Evidence of absence in lesser apes. *Proceedings of the Royal Society B: Biological Sciences, 276*, 1671–1677.

Taylor, M. M., Lederman, S. J., & Gibson, R. H. (1973). Tactual perception of texture. In E. Carterrette & M. Friedman (Eds.), *Handbook of perception* (Vol. 5). New York: Academic.

Temerin, L., & Cant, J. (1983). The evolutionary divergence of old world monkeys and apes. *American Naturalist, 122*, 335–351.

Tomasello, M., & Call, J. (2007). Ape gestures and the origins of language. In J. Call & M. Tomasello (Eds.), *The gestural communication of apes and monkeys* (pp. 221–239). Mahwah: Lawrance Erlbaum.

Tomasello, M., & Herrmann, e. (2010). Ape and human cognition: What's The difference? *Currrent Directions in Psychological Science, 19*, 3–8.

Turner, J., & Maryanski, A. (2005). *Incest: Origins of the taboo*. Boulder: Paradigm Press.

Turner, J., & Maryanski, A. (2008). *On the origin of societies by natural selection*. Boulder: Paradigm Publishers.

Von Senden, M. (1960). *Space and sight. The perception of space and shape in the congenitally blind before and after operation* (Peter Heath, Trans.). London: Methuen.

Vorobyev, M. (2004). Ecology and evolution of primate colour vision. *Clinical and Experimental Optometry, 87*, 231–238.

Walker, A. D., Falk, R. S., & Pickford, M. (1983). The skull of *Proconsul africanus: Reconstruction of cranial capacity. Nature, 305*, 525–527.

Ward, C. V. (1993). Torso morphology and locomotion in proconsul nyanzae. *American Journal of Physical Anthropology, 92*, 291–328.

Ward, C., Kimbel, W., & Johanson, D. (2011). Complete fourth metatarsal and arches in the foot of australopithecus afarensis. *Science, 331*, 750–753.

Whishaw, I. Q. (2003). Did a change in sensory control of skilled movements stimulate the evolution of the primate frontal cortex? *Behavioral and Brain Sciences, 146*, 31–41.

Williams, B., Kay, R., & Kirk, C. (2010). New perspectives on anthropoid origins. *Proceedings of the National Academy of Sciences, 107*, 4797–4804.

Wise, S. P. (2007). Evolution of ventral premotor cortex and the primate way of reaching. In J. Kaas & T. Preuss (Eds.), *Evolution of nervous systems: A comprehensive reference* (Vol. 4, pp. 157–165). Amsterdam: Elsevier.

Wolpoff, M. (1999). *Paleoanthropology*. Boston: McGraw-Hill.

Young, N. (2003). A reassessment of living hominoid postcranial variability: Implications for ape evolution. *Journal of Human Evolution, 45*, 441–464.

Zalmout, L., Sanders, W., MacLatchy, L., Gunnell, G., Al-Mufarreh, Y., Ali, M., Nasser, A.-A., Al-Masari, A., Al-Sobhi, S., Nadhra, A., Matari, A., Wilson, J., & Gingerich, P. (2010). New oligocene primate from Saudi Arabia and the divergence of apes and old world monkeys. *Nature, 466*, 360–365.

Zhang, J., & Webb, D. M. (2003). Evolutionary deterioration of the vomeronasal pheromone transduction pathway in catarrhine primates. *Proceedings of the National Academy of Sciences, 100*, 8337–8341.

Chapter 19
The Evolution of the Neurological Basis of Human Sociality

Jonathan H. Turner and Alexandra Maryanski

It is often assumed that humans are highly social by virtue of bioprogrammers lodged in the neurology of the brain. But, if we take a more evolutionary and comparative perspective, this common assumption can be called into question. Humans are, in essence, evolved apes and share a high percentage of their genes with the great apes—chimpanzees (common and bonobo), gorillas (lowland and highland), and orangutans. While the great apes do reveal propensities for empathy (De Waal 1996, 2009), for justice and fairness (De Waal 1991; Bronson and De Waal 2003; Brosnan et al. 2005), and reciprocity (De Waal 1989, 1996; De Waal and Bronson 2006), they do not evidence high solidarity, nor do they reveal permanent group structures (Maryanski 1986, 1987, 1992; Maryanski and Turner 1992). Rather, the larger community or home range of many square miles appears to be their natural social structure, with groups forming and disbanding in a fusion-fission process. Moreover, the great apes reveal virtually no inter-generational continuity in either their home range or group structures because, except for male chimpanzees, both males and females leave their natal communities at puberty—never to return. Add to this pattern of male and female transfer the high level of promiscuity among the great apes, where paternity is never known, anthropomorphic efforts to see "family" groupings among the great apes are overdrawn.

Even if we look at present-day humans when they interact, one of the striking characteristics of this interaction is how hard individuals work at making face-to-face encounters viable. There are elaborate greeting and closing rituals, often accompanied by elaborate (somewhat contrived) emotional displays, constant attention to face, nodding of heads to assure agreements and understandings, immediate apologies for any misunderstandings, rhythmic synchronization of body and talk to assume that everyone is on tract, and many other interpersonal processes that signal, to us at least, that this is not a natural process (Goffman 1958; Turner 2002, 2008, 2010; Collins 2004). Compare face-to-face interaction, however, with people's easy identification with larger social structures—school, community, nation—where they almost effortlessly mobilize emotionally charged symbols and intense commitments to these larger structures, even when they have not directly interacted with people. This larger macro structure is "more natural" to humans because it is closer to the home-range of the great apes and the common structure of the ancestors that we shared with the great apes.

Thus, humans are not so naturally social as is often assumed, if sociality means the formation of cohesive groupings by interactions among conspecifics. For humans to become more social, we argue, the hominin brain had to be significantly rewired to increase sociality at the level of face-to-face interaction; and this rewiring took millions of years of evolution. This enhanced sociality, however, did not involve installing direct bioprogrammers for forming groups, packs, troops, pods,

J.H. Turner (✉) • A. Maryanski
Department of Sociology, University of California, Riverside, CA 92521, USA
e-mail: jonathan.turner@ucr.edu; alexandra.maryanski@ucr.edu

D.D. Franks and J.H. Turner (eds.), *Handbook of Neurosociology*, Handbooks of Sociology and Social Research, DOI 10.1007/978-94-007-4473-8_19, © Springer Science+Business Media B.V. 2013

herds, and other biologically driven group formations evident among most other mammals, but rather, natural selection worked indirectly on subcortical areas of the primate brain to enhance hominins' and ultimately human's emotionality. Humans form groups by ramping up the emotional energy to create bonds of solidarity and commitments to group formations. And as J. H. Turner will outlined in Chapter 10 in this volume, enhanced emotionality, coupled with other behavioral propensities of the great apes, made humans more social and able to form tight-knit group structures—something that their great ape cousins cannot easily do. However, because humans form groups indirectly, via the arousal of emotions, the emotions can make human their grouping highly volatile. Emotions are always a double-edged sword because group solidarity depends upon positive emotions but the majority of primary emotions that humans have are more negative and, hence, potentially disruptive to social solidarity. Other mammals can allow these negative emotions to emerge because, in the end, the bioprogrammers that push for group formation override episodes of negative emotional arousal and the resulting episodes of intra-group conflict. For humans, such programmers do not exist, and hence, the burden falls upon culture (norms, beliefs, and values), sanctions, and moral emotions like *guilt* and *shame* to bring episodes of breached interaction back into some harmony Scheff 1988; Turner 2008, 2010).

Let us now backtrack and document in more detail why we make what may seem like outrageous assertions about human sociality. We will begin with a very brief overview of the evolution of primates; then, we will employ Alexandra Maryanski's (1986) cladistic analysis to document the low sociality of our distant ancestors. Next, we will turn to comparative neuroanatomy and document, with the limited data available, that human and great ape subcortical areas are very different, indicating that the human subcortex was under intense selection to enhance emotions over the last seven million years.

A Short History of Primate Evolution

Some 64 million years ago, a small rodent-like mammal climbed into the arboreal habitat to initiate the primate line, which is now divided into three branches: prosimians or pre-monkeys, monkeys, and apes. Current classifications add a fourth branch—*Homo*—and place humans or their direct ancestors into a separate lineage but we are, in reality, just another ape. Apes and monkeys were not dramatically different from each other 30 million years ago, except in their life history characteristics (Wolpoff 1999; Falk 2000), and for our purposes here, they coevolved for at least ten million years, beginning around 33 million years ago. Then, about 23 millions years ago, they differentiated in significant ways. Some of this differentiation was the result of monkeys' increased fitness relative to apes in the arboreal habitat. Monkeys began to occupy and dominate the verdant cores of trees, perhaps because they acquired the ability to eat unripe fruit, which is something that apes cannot do to this day (Andrews 1981, 1996). The result was for apes to be pushed to the terminal feeding areas of the arboreal habitat where there is not enough food or room to support large numbers of individuals permanently. In contrast, monkeys could support larger, more permanent groups in the core areas where food is more plentiful, and so they developed the social structure that we see today among all monkeys—a social structure revolving around generations of female matrilines, male dominance hierarchies, and cohesive troop organization. Females never leave their natal group and form dense kin networks within their natal group; males migrate at puberty to another group and begin the competition for dominance, while being replaced by males from other troops. Thus, monkey societies are built around well-structured local groups (Maryanski and Turner 1992; Turner and Maryanski 2008).

Apes may have had structures similar to those of monkeys at one time, but their niche in the arboreal habitat precluded large or stable groups. Selection worked to reduce strong ties so that individuals would not form large, permanent grouping. Selection favored animals that had weaker ties and that could be mobile as resource levels varied. Thus, for many million years, selection pushed for weaker ties among apes forced to live in the terminal feeding areas of the arboreal habitat (i.e., the tops of trees and the undersides of branches). The anatomy of apes, including their neuroanatomy, was altered by having to adapt to this set of niches, along a number of fronts: apes developed stronger arms, wrists, and hands than monkeys; apes acquired the capacity to brachiate (rotate their arm 360 degrees, something a monkey cannot do); apes developed somewhat better sensitivity and strength in their fingers; and apes became smarter than monkeys with larger brains. But, critical to our argument beyond morphological changes to ape anatomy was the transformation of their social morphology or structures. This structure was built around weak ties, mobility, individualism, and constant fusion-fission of groupings. If humans' more immediate ancestors had been monkeys, then the assumption that humans are naturally social and prone to group solidarity would be justified. But, humans are evolved apes and, to the extent that we have bioprogrammers from our early ape ancestors, these push humans to be more individualistic, mobile, and community-oriented. Thus, group cohesiveness would have to be achieved by a different route than hard-wire bioprogrammers for group formations among the ancestors of humans, once group cohesiveness had selective advantages.

Around ten million years ago, the forest of Africa began to recede and the great savannas began to open up. Many arboreal primates were forced from this habitat to the floor of the savanna, where predators were common. Monkeys did not have great difficulty making this transition because baboons and other large monkeys can live on the savanna today because they are well organized, in an almost militaristic fashion when they walk across open-country savanna. Apes, however, are not organized at the group level; and the result was a great extinction of ape species as the forest receded. No ape today, except humans, can live on the savanna. As selection worked on ape phenotypes and underlying genotypes, a number of strategies for survival in twentieth habitats emerged. For example in Asia, one was to make apes enormous (as tall as 8 ft) as a response to predation but the food requirements and the problems of releasing heat from such a large animal eventually doomed apes like *Gigantopithecus* to extinction some 1.6 millions years ago. Some apes may have become more hierarchical, since they reveal propensities for hierarchy in the natural habitats, but without the female matrilines to hold the troop together, this was a doomed strategy.

And so, the interesting question becomes this: How did the ancestors of humans beat the odds and survive in open-country savanna, where tight-knit group organization would have a selective advantage in food foraging and defense against predators? Before answering this question—which is at the core of the neurological story that we want to tell—let us firmly establish the fact that apes are indeed weak-tie animals and that, in fact, the common ancestor that humans have with apes was probably even more weak-tie than present-day apes, except for orangutans who are virtually solitary.

Alexandra Maryanski's Cladistic Analysis

Most research on primates has a behavioral bias, with emphasis on recording patterns of behavior among individuals. Sociology is the science of social structure, and so, emphasis is on the patterns of relationships that emerge and persist among con-specifics. This distinction between behavior and social structure is often lost on psychologists and ethologists who do most of the research on primates, but it is critical to understanding the social organization of apes. Alexandra Maryanski (1986) conducted a pioneering study when she re-coded all of the existing behavioral data from published studies on primates in terms of the network structure among con-specifics. That is, emphasis was on who formed social relations with whom, and to what degree of intensity and permanence? She created a simple scale: very weak or no ties; moderate ties; and strong ties. Strength of ties was coded by rates and durations of interaction, grooming, and other behaviors signaling a strong, moderate, weak, or null tie. She coded data on all species of apes, and select but representative species of monkeys (since there are so many species of monkeys compared to the handful of ape species). These data have been tabulated and published in a number of places (Maryanski 1987, 1992, 1993, 1995; Maryanski and Turner 1992; Turner and Maryanski 2005, 2008; Turner 2000), and so we will not reproduce the tabulations here. But the overall pattern is as described earlier. Ape societies reveal very few strong ties, mostly those associated with mother-infant bonding which is a pattern among all mammals. Virtually all other ties are weak or, at best moderate, with the exception of gibbon-siamangs where males and females form a lifetime bond (although these are very small Asian apes, way off the human line). Other exceptions include: male chimpanzees sometimes develop attachments to their mother, although they do not form a permanent group, a moderate-to-strong bond can exist for a lifetime as is evidenced by frequent visits; lead-silverback gorilla male harems of females can perhaps been seen as an exception, although these harems are more appearance than reality because females sneak off to have sexual relations with other males and use the silverback as a babysitter, and moreover, the tie is broken once females no longer have offspring; brothers among chimpanzees sometimes form moderate-to-strong bonds or even friendships with non-kin males. Otherwise, great ape societies do not form cohesive groups because:

1. Sexual relations among males and females are promiscuous, with paternity never known, even in gorilla harems; and with nothing like a family present.
2. At puberty, all females leave their natal group and community to be replaced by immigrating females from different communities who remain relative strangers to each other and do not form strong ties.
3. At puberty, all males leave the natal community, except chimpanzee males who remain in their natal community for their lifetime
4. Adult-male and adult-female relations do not exist beyond promiscuous sexual encounters.

As is evident, then, there is little basis for strong or permanent ties among the great apes, except the harems of gorillas and the ties of chimpanzee males to each other and their mothers. None of these lead to permanent groups, although the gorilla harem persists for a time but eventually breaks apart as offspring leave the natal group. Male chimpanzees will visit their mother for a time and hang out with brothers or other males, but these are not permanent groups, just prolonged encounters that are iterated but that do not form a cohesive group structure.

After recording the data, Maryanski (1986) performed a cladistic analysis in order to determine the nature of the social structure organizing the behaviors of the last common ancestor to humans and contemporary great apes. Cladistic analysis involves a reconstruction of a structure based upon the frequency of characteristics among related species. If a set of species all has the same characteristics, for example, it can be assumed that the last common ancestor of these species also revealed these characteristics. If a particular species in this related set does not reveal one of these characteristics but still reveals the other, then it can be assumed that this one characteristic evolved in response to particular selection pressures in a habitat or niche. For example, among all apes except for chimpanzee males, both males and females leave their natal communities at puberty. It is reasonable to assume that this one exception is the outcome of selection pressures operating in a particular niche that tied males to their natal community.

Another feature of cladistic analysis is to have a set of sister species as a comparison point to the species of interest. In this case, Maryanski did network analysis on representative species of monkeys, which all revealed the general pattern of female matrilines and male transfer at puberty, male dominance hierarchies, and tight-knit group structures. It can be assumed, therefore, that the last common ancestors of these species of monkeys also revealed these structural characteristics making for tighter-knit groups. Thus, the differences between monkeys and apes are real and are part of each's ancestral line because they each evidence distinctive patterns of network ties typical of their respective lineages.

With the calculations in place, Maryanski concluded that the last common ancestor of the great apes was probably most like the contemporary orangutan in its social structure, or in this case, the last common ancestor of apes and humans revealed an almost complete lack of social structure. The only strong tie among orangutans is mother-young bonding up to puberty, at which point males and females leave their mother's community forever. Otherwise, no strong ties exist among orangutans, a network structure (or lack thereof) that was probably typical of the last common ancestor. Such a weak structure would be highly adaptive in the terminal feeding areas of the arboreal habitat where food supplies were unpredictable and uneven, with individuals rather than groups moving about to find sufficient resources. But, if selection pressures suddenly shifted toward demands for more social organization, such weak-tie animals would be at an enormous disadvantage, and it all likelihood would go extinct.

And so, as the forest on the savanna receded and as apes were pushed onto the predator-ridden savanna, they were doomed; they simply did not have bioprogrammers for troop organization compared to monkeys. The result was the mass extinction of apes, except for the handful that exist today. Among the great apes, there are two subspecies of chimpanzees and gorillas, and one species among orangutans and humans. Of the hundreds, perhaps thousands of species of apes that once existed, this small handful is all that is left. There are about the same number of species among gibbons and siamangs, but these are very distant cousins and are not considered great apes.

The Weakness of Weak Ties

As various species of apes were forced to the floor of the savanna, they were forced to adapt to a terrestrial environment. Apes evidence a number of anatomical and behavioral propensities that make them less fit in such an environment. First, they are slow compared to other terrestrial animals, especially predators that could easily hunt them down. They cannot outrun predators because their bodies and limbs evolved in an arboreal habitat. Second, they are visually dominant, a trait that evolved, once again, as a means to adapt to an arboreal habitat, but the result was a decrease in their sense of smell. Thus, apes in the distant past could not easily smell predators; like humans and apes today when on the savanna, they must rely upon their acute sense of vision or less acute auditory system, but predators can hide behind bushes and trees, and thus sent is a much better defense mechanisms of detection against them. Third, apes are highly emotional; and when danger presents itself, they have trouble controlling emotional outbursts. Yet, a loud primate on the savanna floor was, no doubt, soon a dead primate. Indeed, apes will rant and wildly gesticulate when facing danger, thus drawing attention to themselves. Moreover, because apes have an individualistic streak, they will run and dance around as individuals, rather than forming a collective phalanx as do most terrestrial monkeys when confronting danger. Fourth, and most important, apes do not reveal natural propensities to form permanent groups for foraging and defense. Apes in their current environments will often defend their home range collectively, and so, there is some behavioral propensity to organize for defense, but these defense groups are not permanent. They form for patrols in forest habitats to make sure that other ape males do not enter their home range and disband, but these patrols do not represent a sentry group that persists for a long time. Thus, without bioprogrammers for more *enduring group structures* like most other mammals, and especially mammals subject to predation, apes could not consistently organize collective defenses, nor could they effectively organize hunting and foraging. True, male chimpanzees will sometimes coordinate a hunt through the use of non-verbal gestures, but they do not consistently do so on a daily basis in more enduring hunting bands. The more those early ape pioneers on the savanna were forced to move away from the edge of forests, the more this lack of tight-knit social structure at the level of the group became a handicap, dooming many species to extinction because they could not defend against predators, or effectively forge for food as a group.

For natural selection to increase group solidarity, it needs something to select on in the ape genome (via ape phenotypes), but the group solidarity evident among monkeys had been selected out over a 20 million year period, and thus, there was no existing behavioral propensity on which to select. Mutations could not occur with sufficient rapidity or focus to enhance "groupness," and so apes in open-country grasslands were unlikely to survive. Without strong ties among adults, and between adults and their offspring over longer periods of time, group solidarity cannot increase. The transfer of offspring at puberty broke the capacity for inter-generational groups; the immigration of unrelated and unfamiliar females from outside communities (and males as well for most apes) reduced the chances of permanent bonds among females and males, as is evident among present-day apes; and while migrating males may have begun to compete for dominance hierarchies, such hierarchies are not very effective when females reveal little solidarity. In contrast to apes, female monkeys form cohesive matrilines, and coupled with male hierarchies, this two-pronged phalanx represents a very effective fitness-enhancing strategy for terrestrial monkeys. But, for apes, hierarchy alone may have increased fitness to some degree and allowed some species to endure for a time, but without female solidarity and attachment to males in conjugal pairs, such quasi groups would not be highly fit under conditions of high predation. Males could easily find themselves trying to herd their harems of individualistic females, while at the same time defending the troop—not an easy task and not likely to increase fitness.

So, there is an obvious weakness to weak ties, but we should note that there is also a strength in weak ties that is evident today in human societies. There are very few macro societies among animals, and most are insect societies where millions of individuals can be organized for concerted action. Humans are a huge animal compared to an insect but we too can organize macro societies, albeit often somewhat precariously, because our ape ancestors did not form tight-knit groups and, in fact, were oriented to the larger, inclusive community more than the local group. We can live with relative strangers; we do not need to know everyone around us; we can play assigned roles (like insects); and so, the weakness of weak ties becomes a *strength* of weak ties in the evolution of macro-level societies among humans. In contrast, a monkey could never organize a macro society because they are oriented to the matrilines of females or the hierarchies among males in the *local* group. Yet, what was needed in the first movements of primates to the savanna was strong-tie groupings; and apes are singularly incapable of developing such groups, especially if Maryanski's cladistic reconstruction is correct. If our last common ancestor with present-day great apes was virtually solitary, then such an animal would have little chance of surviving in the predator-ridden savanna. Without the protection of the trees, even as temporary refuge, an individualistic ape would not be fit. Even a species of apes with some rudimentary group structure, such as hierarchy, or basic interpersonal skills like the capacities for empathy or reciprocity, would not be sufficiently fit.

Hominin or Hominid Evolution

Over-Emphasis on the Neocortex

Those species of animals on or near the human line are termed *hominids* or, more recently, *hominins*. Sometime about 7 million years ago, the descendants of the last common ancestor to humans and contemporary great apes split, with those species of apes forced to the savanna all going extinct, except for hominins, and with the ancestors of contemporary apes surviving by virtue of finding niches in primary and secondary forests (ironically their survival is now threatened by overpopulation and deforestation by humans). Somehow, the hominin ancestors of humans acquired the capacity to survive in more open country savanna conditions and, eventually, the ability to move to many new ecological niches outside of Africa. How did they do so?

The most common social science explanation is that the brain grew, language emerged, and culture evolved so that hominids could become better organized and capable of adapting to many diverse environments. The problem with this story is that the brain of hominins did *not grow dramatically* for millions of years, and so the language and culture that supposedly accompanying large brains could not have developed to anything near a human measure until, at best, 1.5 million years ago when *Homo erectus* emerged. Indeed, the brain of hominins does not increase beyond that evident among chimpanzees until 2.5 million years, and even then, the brain of *Homo habilis* was only an additional 125 cc greater than that among chimpanzees. And, clearly earlier hominins had been able to survive on the savanna with brains the same size or only slightly bigger than present-day chimpanzees. Even *homo erectus*, the presumed immediate hominin before *Homo sapiens*, had a median brain size that was less than 1,000 cc, the lower limit for humans, but it is possible and even likely that this sudden jump in brain size for *Homo erectus* indicates that something about a larger neocortex structure was fitness enhancing.

Still, long before this rapid increase in the overall size of hominin brain, older species of hominids had found a way to get better organized so as to form fitness-enhancing groups. How did this happen? Our answer is that the subcortical parts of the brain were subject to selection in ways that increased the ability to experience a wider range of emotions that, in turn, enhanced the few existing capacities of present–day apes for language facility, for empathy, for justice, for reciprocity, and for self-awareness *without* significantly increasing the size of the neocortex. The subcortex may have grown a bit, but this growth would not show up on endocasts in the same way that growth in the neo-cortex that envelops and surrounds the subcortical areas of the brain would. Research has, perhaps, spent too much time on the neocortex in explaining what is unique about humans; we argue that what really makes humans unique is *the increase in the capacity for emotions* and the use of emotions for forge stronger bonds among weak-tie apes.

Re-focusing on the Subcortical Areas of the Brain

Many of the behavioral capacities that make human interaction and patterns of social organization possible are evident in present-day apes and, in some cases, higher monkeys as well. But, without enhanced emotional capacities, these extant behavior propensities—that is, empathy, language facility, self-recognition, reciprocity, justice—would not be sufficiently powerful to cause stable groups to form among apes—now, or in the distant past millions of years ago. Only with enhanced emotional capacities that intersected with the neurological modules generating other behavioral abilities of all present-day apes (and hence the common ancestor of apes and humans) would produce an ape capable of forming stable and cohesive group structures.

Existing Capacities of the Last Common Ancestor to Apes and Humans

Visual Dominance and the Capacity for Language

All of the great apes can learn human language up to the same capacity as a 3 year old baby. Whether by sign language or by learning arbitrary signs on computer boards, apes can "speak" in sentences and interact with humans (see Rumbaugh and Savage-Rumbaugh 1990 and Savage-Rumbaugh et. al. 1988, 1993; Savage-Rumbaugh and Lewin 1994 for data that make this generalization indisputable).

There was much original resistance to this idea because language is one of the hallmarks of humans' designation as "Homo" and somehow above apes. If we think about it, for language to evolve, there had to be something for natural selection to select on that was common to all of the ape ancestors of humans. Mutations could not lead to the spread of language because if by some miracle a single ape could use language, whom would it speak to? If it tried, in fact, it might be ostracized as deviant and thus would probably be less fit and not reproduce itself. Thus, selection was not working so much on mutations as on the ends of Bell curves of traits that potentially could increase sociality. If traits and capacities evident on the tail end of the curve would enhance fitness, selection on these traits would make it more likely that those carrying this trait would reproduce, while those without the trait would be less likely to reproduce. As the gene pool shifted to the fitness enhancing end of the Bell curve, then the next Bell curve would be disproportionately composed to members of this enhanced trait, with the tail end of this curve being favored by selection. And, in this way, directional selection can work very rapidly, if a trait has fitness-enhancing value.

But, this begs the question of how apes had the capacity to use language in the first place. Some time ago, Norman Geschwind (1965a, b, c, 1985) and later with his student, Antonio Damasio (Geschwind and Damasio 1984), made the argument that during the switch from an olfactory-dominant but now arboreal mammal that had originally ascended into the arboreal habitat to a visually dominant primate, a considerable amount of rewiring of the association cortices connecting three major lobes in the neocortex was necessary. This rewiring had the consequence of making higher primates capable of language production and comprehension at a rudimentary level (Maryanski 1996; Maryanski et al. 1997; Maryanski and Turner 1992). This consequence was a purely chance effect of selection working to change the dominance hierarchy among the sense modalities of evolving primates.

The relative dominance of the parietal lobe (haptic sense modality), temporal lobe (auditory sense modality), and occipital lobe (visual modality) needed to be reordered. Vision had to replace olfaction as dominant, with haptic and auditory subordinated to vision. Thus, when an ape or human hears or feels something unusual, it will immediately turn its eyes and look at what it senses; and the visual modality will dominate what an ape or human does. Geschwind and Damasio argued that the areas around the inferior parietal lobe, where the temporal, occipital, and parietal lobes meet, are what regulate this shift to visual dominance. This shift allowed primates in general to be much more fit in an arboreal habitat where seeing one's way around a three dimensional environment is fitness enhancing compared to smelling, hearing, or feeling one's way around a three-dimensional arboreal habitat.

What has been the fate of olfaction, which is the dominant sense modality of most mammals, including the small mammal that initiated the primate order? The olfactory bulb resides in subcortical areas of the brain and thus is not directly under the control of the neocortex. With the more general shift to visual dominance, the olfactory bulb is dramatically reduced in apes and humans, and hence, it does not pose much power to override or disrupt vision, although we often will experience a flood of emotions with certain olfactory experiences.

Since all primates are visually dominant and since the shift to visual dominance is what wired in language capacities in higher apes, why cannot monkeys learn languages like apes? Thus, something more has to occur, above and beyond rewiring of the neocortex for visual sensory dominance, to make a primate capable of using language. This extra ingredient must be the increased brain size of the great apes. They developed larger brains because they were forced to live in more hazardous environments where one false step in the terminal feeding areas high in the trees would mean death by gravity; a smarter animal is more likely to be able to assess strength of branches, remember easy pathways through the trees that are safe, and generally assess what will allow movement high up in the arboreal habitat. The extra dose of intelligence, coupled with the shift to visual dominance, represented pre-adaptations for language capacity. They did not evolve for language; rather they evolved to increase fitness of primates in general and apes in particular for movement around the hazardous areas of the arboreal habitat.

As pre-adaptations, however, they are present and available to selection, if increased capacity for language, is fitness enhancing. And, since this language potential is present in all apes of a certain degree of intelligence, once selection begins to work on tail ends of the Bell curve distributing this potential, language could evolve. But for spoken language in the human measure to evolve, much more complex changes in the parietal lobe regulating muscles, lips, and other structures necessary for language production are necessary. It now appears that some of these structures have only been under selection for 200,000 years or less (Enard et al. 2002a, b), which would suggest that speech as we know it may be unique to humans. Thus, did this incipient potential for language simply "sit there" in the phenotype and underlying genotype of hominins for seven million years? As Turner has argued (2000), it did not; in fact, it was usurped very early in hominin evolution as a mechanism for creating a visually based, and emotionally charged, body language with its own phonemes and syntax; and this visually based language of emotions was *the first language* that evolved millions of years ago to increase nuanced interactions that led to emotionally charged social bonds among conspecifics that, in turn, increased group solidarity and hence fitness of early hominins. We will return to this argument shortly.

The Capacity to See Self

Another artifact of intelligence among mammals appears to be the capacity to recognize one's reflection in a mirror as an image of self (Gallup 1970, 1982; Gallup et al. 1995). Relatively few mammals can do so—at present, only humans, great apes, elephants, and dolphins are known to recognize themselves in a mirror, although other intelligent animals like whales may also have this capacity. Human interaction is mediated by individuals' respective identities, which are cognitive and emotional senses of who we are and what we should expect from others (Turner 2002, 2008, 2010). Indeed, much human interaction involves efforts to assure that others will acknowledge the self or identity that we present to others (Burke and Stets 2009). This dominance of self is, no doubt, the outcome of selection over millions of years as it worked on this capacity to recognize self that all great apes, and hence, all hominins possess. And with expansion of the neocortex, various types and layers of self could become evident—e.g., core identities, role identities, social identities, group identities, moral identities, etc. It is clear that only humans have elaborated many dimensions of identity that direct and channel the flow of all interpersonal activity, and all of these have large effects of the bonding among persons and the formation of group solidarities.

But, long before the neocortex began to expand, I would argue that images of self became intertwined with emotions, with hominins feeling a range of positive and negative emotions when their emerging sense of self was verified, or not verified. Indeed, eventually as moral emotions like *shame* and *guilt* evolved (see later section), this combination of self evaluation by feelings of *pride*, *shame*, and *guilt* represented a powerful force of social control that kept individuals in line and thereby promoted group solidarity. Thus, the moral self that gets so much play in recent years, may be one of the oldest senses of identity in the human clade. Again, more on this later.

Empathy

A number of investigators have argued that primates are good natured (De Waal 1996, 2009) because they can call out in themselves the responses of others, especially emotionally laden responses. There is considerable speculation about the role of mirror neurons (see chapters in this volume), which monkeys also have, in this capacity for empathy. We would invoked G. H. Mead's (1934) concept of "role-taking," which is to be preferred to the long-winded notion of "the theory of mind," to make this same point (see Also Franks' Chap. 4, in this volume). Apes surely can read the gestures of conspecifics and, to some degree, put themselves in the place of these others, and then respond sympathetically to them. And the more emotions attached to the gestures emitted by others, the more active will role-taking be and, hence, the more will sympathetic responses promote solidarity. Thus, once emotions are under some control, once their range and subtly has expanded, and once they become part of virtually all visual and, to a lesser extent, auditory or tactile behaviors of animals, the greater will be their solidarity. Given that the capacity for empathy or role-taking was present in the last common ancestor to apes and humans, enhancing emotionality would dramatically increase the range of role-taking among hominins; and as the nuance and subtly of role-taking increased, stronger bonds would increase among members of emerging grouping of hominins. But, this effect of role-taking depends, we believe, on the expansion of hominins emotional responses to each other, and so as selection expanded emotions, it gave more power to role-taking to promote bonds, thereby increasing selection on the capacity for both.

Reciprocity

For some time, researchers have argued that reciprocity is a universal human trait (Fiske 1991), and as evolutionary psychology emerged as a distinct sub-discipline, one of the first lines of argument was that natural selection had installed modules in the brain causing hominins and then humans to have behavioral propensities to return favors and otherwise give resources to others who have given resources to them (Cosmides 1989; Cosmides and Tooby 1992). The data on primates is suggestive but far from clear, but it too suggests that there is something like a sense of reciprocity in primates (De Waal and Bronson 2006) and, thus, in the last common ancestor to the great apes, hominins, and humans.

Reciprocity obviously has implications for group formation because, when individuals feel obligations to return favors or resources to others, the act of reciprocation increases the social bonds among individuals and, moreover, increases the likelihood that these bonds of reciprocity will endure over time. Thus, it is very likely that this module, wherever it is in the brain, could be selected upon if selection favored more enduring social bonds among hominins.

With enhanced emotionality, reciprocity gains even more traction to forge social relations. Individuals now feel a sense of *guilt* or even *shame* if they do not reciprocate, and a sense of *pride* and *happiness* if they do; and if others do not reciprocate, the *anger* leads individuals to negatively sanction others and bring them back into line. The end result would be enhanced solidarity as each act of reciprocity increased the flow of positive emotions in emerging group structures.

Fairness and Justice

As Sarah Bronson and Fans de Waal (2003) have documented for capuchin monkeys, primates appear to have a sense of fairness and justice in this sense: they compare their behavioral costs relative to rewards to the costs and rewards of others. Thus, for example, if a capuchin monkey in exchange with a trainer must give more effort for less reward or even the same effort for less reward compared to another capuchin that can be observed exchanging with a trainer, it will stop exchanging resources with its trainer until a greater or equal reward received by the other capuchin is offered. Clearly, there is some calculation about what is fair or just in terms of costs and rewards among monkeys; and it appears that the same propensity exists among the great apes, although the documentation is less definitive. The tragic assault several years ago in San Francisco by two chimpanzees who had been excluded from the "birthday party" of a third chimpanzee given by the latter's former caretakers documents how intense the negative emotions of those excluded from an exchange can be. For the two who were not allowed to participate, there was clearly a sense of injustice that aroused such intense emotions that they would attack adult humans, causing great harm and even death. (see Franks Chap. 15, in this volume.) Humans, of course, elaborate notions of justice into complex ideologies and beliefs, backed up by intense emotions.

One of the biggest problems of almost all groups engaged in concerted and coordinated activity is problem of "free-riding" (Hechter 1987; Olsen 1971/1967). When individuals contribute their fair share to joint activities, they experience positive emotions, whereas when some do not contribute but free-ride off the work of others, these others become angry because of the perceived injustice and become ready to sanction negatively those who try to free-ride, and among humans, they create norms to limit free-riding and monitoring systems to assure that these norms are followed. Michael Hechter emphasized these processes in a book titled *Principles of Group Solidarity* (1987), and this title makes the essential point, that group solidarity will increase when free-riding is limited. Thus, a module in the brain devoted to assessing justice and fairness, especially over contributions to group actions, will work to increase group solidarity; and this module could have been subject to selection during the course of hominin evolution. Moreover, as the range and intensity of emotions was also expanded, these emotions would give justice calculations some "teeth" and motivate individuals to monitor others and sanction them if necessary in order to assure that coordinated activities of groups are shared—which, in turn, would increase group solidarity.

In sum, then, there existed in the last common ancestor of present-day great apes and humans a series of behavioral propensities that could be subject to selection and that would be enhanced by increased emotionality. Each of these alone but, more importantly, all of them together would increase group solidarity among low-sociality apes. Some, like visual dominance and the emergence of the neurological wiring for language and the capacity to see self as an object were pre-adaptations—simple byproducts of selection for other outcomes. But, once present, they could be subject to further selection and thereby enhanced if they increased fitness. Others such as behavioral capacities for empathy, reciprocity, and justice/fairness may have been subject to selection early on to increase sociality or otherwise promote some cohesion. And, whatever their reason for being present, they could be subject to further selection if they enhanced group solidarity and cohesiveness.

We do not think, however, that these capacities alone could have converted low-sociality apes to highly organized animals capable of living full time on the open-country African savanna. These traits bought them some time, allowing some species of apes to hang together, perhaps at the edge of the forest and savanna, to survive until the power of enhanced emotions began to evolve by natural selection working on the subcortical areas of the brain to produce a highly emotional animal that became one of humans' hominin ancestors. Emotions themselves would increase solidarity, and enhanced emotions attached to existing behavioral capacities for empathy, self-evaluation, reciprocity, and justice would only increase their power to forge stronger and more permanent groups, and thereby allow some hominins to survive in even more open country savanna conditions. It is these hominins who that were on the direct line to *Homo sapiens sapiens*.

Rewiring the Hominin Brain for Enhanced Emotions

Rough but Suggestive Measurements of Key Brain Structures

Over a decade ago in the library stacks (in the old fashioned days before Google searchers) J. H. Turner stumbled upon the data summarized in Table 19.1; the original data set was more extensive, summarizing comparative measurements on the relative size, controlling for body size which is correlated with brain size, among monkeys, apes, and humans (Stephan 1983; Stephan and Andy 1969, 1977; Stephan et al. 1988). Only the data on apes and humans is presented in

Table 19.1 Relative size of brain components of apes and humans, compared to *Tenrecinae*

Brain component	Apes (Pongids)	Humans (Homo)
Neocortex	61.88	196.41
Diencephalon thalamus hypothalamus	8.57	14.76
Amygdala	1.85	4.48
Centromedial	1.06	2.52
Basolateral	2.45	6.02
Septum	2.16	5.45
Hippocampus	2.99	4.87
Transition coritices	2.38	4.43

Source: Data from Stephan (1983), Stephan and Andy (1969, 1977), and Eccles (1989)
Note: Numbers represent how many times larger than *Tenrecinae* each area of the brain is, with *Tenrecinae* representing a base of 1

Table 19.1. As part of the controls, the numbers represent how many times larger are various parts of the brain in apes and humans using as a base of 1 a simple mammal, *Tenrecinae* (a mammal that is probably very much like the one that originally ascended into the arboreal habitat to initiate the evolution of primates). Thus, the median size of ape brains, controlling for body size, is 61 times greater than that of *Tenrecinae*, whereas the human brain is 196 times larger (which makes the human brain three times larger than ape brains). This growth in the neocortex from that of the common ancestor of humans and apes is what gets most the press, because this growth allowed for language and culture although, as we noted earlier, language was already a behavioral capacity of apes and hence the last common ancestor.

What is more interesting to us in these data are the measurements on subcortical areas of the brain where emotions are ultimately generated. These measurements were not taken, of course, on all key centers of emotions, as we now understand them today, because this was not the focus of the study. Still, the measurements are revealing. For example, the amygdala is the ancient center (inherited from reptiles) of anger and fear (located in different parts of the amygdala), and the amygdala in humans is over two times larger than it is in apes. Why, then, would this have occurred? And, why would selection increase centers for anger and fear when positive emotions are what increase solidarity in animals without bioprogrammers for group cohesion? Much of the difference in the relative size of the amygdala is in the basolateral area, and interestingly, as Eccles (1989) notes, this area is devoted primarily to generating pleasure. Thus, slapped onto the amygdala of humans—that is, to the centers for anger and fear—are neuronets devoted to pleasure—a rather odd juxtaposition. Moreover, as is outlined in more detail in Chapter 10, the amygdala in humans has also been usurped for other functions, most having to do with operating as a conduit between the prefrontal cortex (the center of thought and decision making) and subcortical areas generating emotions. This additional work as a conduit also accounts for some of the size differences. Thus, natural selection was working to reduce dominance of fear and anger as emotional responses by adding centers for pleasure on ancient areas of the brain, while at the same time, enhancing the communication and, I suspect, cortical control of emotions by expanded neuronets connecting the cerebral cortex to the subcortical areas of the brain where emotions arise.

The other measurements are also interesting. Of particular note is the size of the human septum that, again, is over twice as large in humans as in apes. The septum is the source of sex drives that make intercourse pleasurable, and as field studies reveal, the great apes are highly promiscuous and enjoy sex immensely; and so we can ask: Why would humans who are only 15% on average larger than chimpanzees need more sex drive that our closest ancestor with whom we share 99% of our genes? One possible answer is that the additional size is related to other kinds of emotional pleasure, above and beyond the pleasure of the sex act itself. Perhaps emotions like "love" and other kinds of deeper attachments and commitments to sexual partners evident today in humans are a byproduct of the larger septum; and indeed, if such is the case, this larger septum may be the neurological basis of the nuclear family. Such attachments would be fitness enhancing for hominins who did not have nuclear families; any permanence in male–female bonding of sexual partners would increase local group cohesiveness and provide protection for vulnerable offspring.

All of the other structures listed in Table 19.1 have effects on emotions. For example, the hippocampus is where emotional memories are stored and probably where repressed memories are sent among humans. Part of using emotions to forge social bonds is to control emotions, and repression is one source of control (somewhat dysfunctional for individuals but, in the short run, functional for group solidarity); and so, to the extent that negative emotions needed to be pushed below consciousness to promote short-term positive relations, the larger size of the hippocampus—well over twice the size evident in the great apes—might signal this capacity of humans to push negative emotions below the level of consciousness. Also, since the transition cortices are what regulate working memory and what is available to working memory as the information of working memory is fed into, and pulled from, the hippocampus, it is again not surprising that these three cortices are also larger by a factor of 2. The thalamus and hypothalamus are also involved in emotions because they route sensory inputs to the relevant lobes in the neocortex, but equally important to subcortical emotion centers; and it is for this reason that emotional responses are often aroused before we become cognitively aware via a neo-cortical sensory center [primarily because this distanced traveled by impulse is a shorter, by fractions of a nanosecond, from, say, the thalamus to amygdala than from the thalamus to occipital lobe, thus generating fear or anger responses before a person can see what is causing the fear or anger response(LeDoux 1996)].

To be sure, these data are not definitive, only suggestive. They require an explanation, and our explanation emphasizes that natural selection was grabbing onto *any* area of the hominin brain that would generate emotions and increase its size (at one end of the Bell curve) in order to expand the level of emotional outputs of by the brain. At the same time, natural selection was also rewiring neuronets that connect and increase cortical control of these areas. Future research on the neurology of emotions, we believe, should focus on comparative anatomy, examining what emotion centers in, for example, chimpanzees do compared to their larger and more richly networked areas in humans. Such comparisons would give us more insight into how natural selection was working to make humans more emotional.

The First Language

If you ask students to watch soap operas, even those in another language with the sound turned off, they have little trouble following and telling you about the story line. Indeed, both of us have ruined many movies that we planned to watch at home by glancing up, ever-more often, at the movie being played on the airline's screens. By simply watching the "body language"—face as it reveals emotions, body countenance, body positioning, etc.—it is easy to follow what is going on, thus making it less interesting to watch the movie at home with the sound on. Social scientists tend to think of language as mostly auditory, but the more primal language among humans is the "language of emotions" (Turner 2000). Millions of years before auditory (spoken) language reached human proportions, hominins relied upon emotional phonemes signaled by face, body, and auditory inflections. These signals were strung together by a syntax, just as humans do today when two or more people interact, thereby making them a language because the signals and their ordering by a syntax produces common meanings. In fact, if we really want to know something about a person or situation, we rarely pay close attention to what is said; instead, we use our dominant and incredibly fine-tuned visual sense modality to read emotions.

As we emphasized earlier, the great apes are already wired for language capacity, and it seems unlikely that this capacity was simply left dormant for millions of years when selection was working to enhance sociality. Since emotions are the bonding force of all humans, and the language signaling emotions is so well developed in all cultures, it is far more reasonable to presume that this language evolved first, long before auditory language that required extensive neurological and physiological reworking of the parietal lobe in the brain and body structures around the face, lips, vocal track, tongue, larynx, etc. (Duchin 1990). Moreover, verbal language is more instrumental; it cannot communicate affect in the same way as face and body can. When people say that they are angry or hurt, for example, we look at their face and body to be sure that such is really the case. To say "I love you" means nothing unless the visual body language of emotions "says" so. Thus, since natural selection was working rapidly to grab onto whatever neurological or physical structure could enhance social bonds among weak-ties animals, it is not surprising that selection finally hit upon what turned out to be the solution faced by rapidly-going-extinct species of apes: enhance emotions and use the already built in neurological capacity for language in great apes to successively increase the emotional repertoire of hominins so that they could forge bonds by a language of affect that, more than any other force (save for dedicated bioprogrammers that hominins did *not* have), could increase sociality among weak-tied and non-group forming apes.

Indeed, it is now clear, as we mentioned earlier, that many of the structures that lead to finely articulated speech did not evolve until *Homo sapiens*, and it is unlikely that the language of emotions could have evolved at this late date. It must have *already been in place* when spoken language, which would obviously have great selective advantages, was piggy-backed to an ancient language of emotions that had been evolving for millions of years. We tend to think of body language as a supplement or adjunct to spoken language, but we would argue the opposite is the case: it is not only our primal language but also our primary language when it comes to forging strong ties, and this is what selection did: Take a weak-tie animal and strengthen its ties, networks, and hence groups through enhancing its behavioral capacities to feel and use a larger palate of emotions. Such is still the basic way in which groups form and hang together.

Sometimes, but not always, developmental sequences in infants mirror evolutionary sequences. Newborn infants can, with a couple of weeks, mirror facial expressions of emotions, such as a smile or frown, from a caretaker—months and for some functions a year before the brain begins wire up infants for auditory language comprehension. Since caretaker-infant bonding is critical to survival of big-brained and biologically vulnerable offspring, emotions are the easiest route to assure emotional charged bonds between infants and adults. Using emotions was a much easier and less cognitive route to forging critical bonds among primates that desperately needed bonding mechanism in the absence of bioprogrammers for sociality. Indeed, up to about the age of two, caretakers rely upon emotional expressions, including auditory ones, to figure out what their children need and want; and as they respond to their children they introduce words and the syntax of spoken language, but at the same time, they still use the language of emotions. Thus, the two languages become somewhat interwoven, but there is no doubt about which one came first, which one is primal, and in the end, which one is most important to affective ties of social solidarity. It is the language of emotions.

The Expansion of Hominins' and Humans Emotional Repertoire

The expansion of hominins' emotional capacities occurred over time, as initial natural selection on emotion centers generated more solidarity and, hence, fitness. With selection on the tail ends of the Bell curve, reliance on mutations would be less important than directional selection on existing phenotypes in subcortical areas of the brain and the underlying genotype. We believe that selection went through four phases: (1) increased cortical control of emotions by expansion of neuronets between neocortical and subcortical areas of the brain, (2) increased variations on hard-wired primary emotions, (3) mixing of two primary emotions into first-order elaborations of emotions, and (4) mixing of three primarily emotions into second-order elaborations. Each earlier phase set the stage for the next phase, with perhaps the last phase only reached with *Homo erectus* or even *Homo sapiens*.

Increased Cortical Control of Emotions

As mentioned earlier, the great apes do not have full cortical control over their emotions, or even moderate control of their emotions when aroused. This lack of control would be a liability on the predator-ridden savanna, especially for a poorly organized primate. Today, some species of monkeys that live in open country conditions march across their territories in virtual silence, so as to avoid attracting the interest of predators. But, for emotions to be controlled to this degree, additional neuronets from the prefrontal cortex to subcortical emotion centers become necessary; and there is ample evidence in humans that these nets are extensive (see Chapter 10). Indeed, much of the conversion of the amygdala into a switching station between neocortical and subcortical areas of the brains; and what we see in humans today, is simply the end point of evolution that began very early on the hominin line. Humans still "lose it" or "go postal" but not as much as one would think when we recognize how dramatically enhanced our behavioral capacities to experience emotions became during the course of hominin evolution. It is, in some ways, rather remarkable that humans have as much control over their emotions as they normally do.

Expansion of the Pallate of Primary Emotions

The great apes today evidence a palate of primary emotions, or emotions that are hard-wired into primate neurology. Most of these are inherited from reptiles (i.e., anger and fear) or early mammals (e.g., satisfaction, happiness, disappointment, sadness). There are discrete areas for some of these primary emotions—e.g., anterior cingulated gyrus for some aspects of happiness such as playfulness, posterior cingulate for sadness—but many emotions are the work of neurotransmitters and neuro-active peptides as they affect body systems. There needs to be more work on what modules, if any, are responsible for what have been hypothesized to be primary emotions (see Turner and Stets 2005: 15–16; Turner 2000: 68–69; Turner 2008: 4–5 for reviews). There is general consensus that *fear*, *anger*, *happiness*, and *sadness* are primary emotions because they appear on everyone's list; other hypothesized primarily emotions included *disgust* (Emde 1962; Izard 1992/1977; Ekman 1984), *surprise* (Plutchik 1980; Ekman 1984), *expectancy* (Panksepp 1982: Osgood 1966), and a few others. We will emphasize the four that everyone agrees upon—that is, fear, anger, happiness, and sadness.

One of the easiest routes for natural selection to take was to expand the range of variation in, and intensity of, these primary emotions. Perhaps greater control of emotions in general introduced key wiring of the brain to do so, but it is clear that humans can feel and express a range of primary emotions from low through middle to high intensity. In Table 19.2, we offer illustrations of this variation along four continua labeled *satisfaction-happiness*, *aversion-fear*, *aggression-anger*, and *disappointment-sadness* (Turner 2000, 2008). No new neuronets or modules would have to be created to increase variations; rather, the range of existing modules (on the Bell scurve) would simply be expanded as natural selection worked on tail ends of neuronets or modules—perhaps only increasing their size. By expanding the range of primary emotions, more nuanced emotions can be felt and expressed, thereby making social bonding more diverse and flexible.

Moreover, expanding primary emotions may have helped overcome a very large obstacle to using emotions to generate social solidarity: the fact that three of the four primary emotions are negative

Table 19.2 Variants of primary emotions

	Low intensity	Moderate intensity	High
Satisfaction-happiness	Content	Cheerful	Joy
	Sanguine	Buoyant	Bliss
	Serenity	Friendly	Rapture
	Gratified	Amiable	Jubilant
		Enjoyment	Gaiety
			Elation
			Delight
			Thrilled
			Exhilarated
Aversion-fear	Concern	Misgivings	Terror
	Hesitant	Trepidation	Horror
	Reluctance	Anxiety	High anxiety
	Shyness	Scared	
		Alarmed	
		Unnerved	
		Panic	
Assertion-anger	Annoyed	Displeased	Dislike
	Agitated	Frustrated	Loathing
	Irritated	Belligerent	Disgust
	Vexed	Contentious	Hate
	Perturbed	Hostility	Despise
	Nettled	Ire	Detest
	Rankled	Animosity	Hatred
	Piqued	Offended	Seething
		Consternation	Wrath
			Furious
			Inflamed
			Incensed
			Outrage
Disappointment-sadness	Discouraged	Dismayed	Sorrow
	Downcast	Disheartened	Heartsick
	Dispirited	Glum	Despondent
		Resigned	Anguished
		Gloomy	Crestfallen
		Woeful	
		Pained	
		Dejected	

Source: Data from Turner (Turner 1996a, b)

and, by themselves, do not promote solidarity. By allowing for low-intensity variants of anger, fear, and sadness, the power of these negative emotions to disrupt solidarities would be reduced; and indeed, such low intensity emission of these emotions may have encouraged efforts by others to respond to conspecifics who were experiencing negative emotions. By expanding the range of satisfaction-happiness, more hooks and options for using positive emotions to forge bonds of solidarity would have become available to hominins. And once selection took this route, additional ways of reducing the power of negative emotions to disrupt solidarity could evolve, while the available palate of positive emotions would expand. One of these paths that generated hooks and options was the

Table 19.3 First-order elaborations of primary emotions

Primary emotions	First-order elaborations
Satisfaction-happiness + *aversion-fear* →	Wonder, hopeful, relief, gratitude, pride, reverence
Satisfaction-happiness + *assertion-anger* →	Vengeance, appeased, calmed, soothed, relish, triumphant, bemused
Satisfaction-happiness + *disappointment-sadness* →	Nostalgia, yearning, hope
Aversion-fear	
Aversion-fear + *satisfaction-happiness* →	Awe, reverence, veneration
Aversion-fear + *assertion-anger* →	Revulsed, repulsed, antagonism, dislike, envy
Aversion-fear + *disappointment-sadness* →	Dread, wariness
Assertion-anger	
Assertion-anger + *satisfaction-happiness* →	Condescension, mollified, rudeness, placated, righteousness
Assertion-anger + *aversion-fear* →	Abhorrence, jealousy, suspiciousness
Assertion-anger + *disappointment-sadness* →	Bitterness, depression, betrayed
Disappointment-sadness	
Disappointment-sadness + *satisfaction-happiness* →	Acceptance, moroseness, solace, melancholy
Disappointment-sadness + *aversion-fear* →	Regret, forlornness, remorseful, misery
Disappointment-sadness + *assertion-anger* →	Aggrieved, discontent, dissatisfied, unfulfilled, boredom, grief, envy, sullenness

evolution of what J. H. Turner has term first-order elaborations, or mixes of two primary emotions (Turner 1996a, b, c, 1997, 1998, 1999, 2000, 2002, 2008, 2010) —an idea originally proposed by Robert Plutchik (1980) and pursued by many others.

The Evolution of First-Order Elaborations of Emotions

In Table 19.3, J. H. Turner's outline of *first-order elaborations* of the four primary emotions—that is, satisfaction-happiness, aversion-fear, aggression-anger, and disappointment-sadness—is presented (Turner 2000, 2008). A first-order elaboration is mixing (in some unknown neurological manner) of a greater amount of one primary emotion with a lesser amount of another primary emotion. This mixing, which is metaphorically like a color wheel (Plutchik 1980), dramatically expands to range, type, and intensity of emotions that individuals can experience, express, and interpret, as is evident in Table 19.3. First-order elaborations not only expand the emotional palate significantly, they also generate a higher ratio of more associative to disassociative emotions when compared to variations in primary emotions alone (see Table 19.2). For example, mixing a greater amount of happiness with fear produces emotions like *wonder*, *hopeful*, *relief*, *gratitude*, *pride*, and *reverence*—all of which can be associative. Or, even mixes of negative emotions like sadness and fear produce less disassociative emotions like *regret*, *forlorn*ment and *remorse*. Of course, such is not always the case; some very powerful first-order negative emotions like *vengeance* can come from mixing happiness with anger; and thus, using emotions to generate solidarity is always a double-edge sword.

Nonetheless, it is clear that this mixing through new neuronets connecting modules or simultaneous activation of modules dramatically expands the range and intensity of emotions that can be used to form strong bonds, to sanction, and to promote solidarity. Still, there can be little doubt that the obstacle posed by three of the four primary emotions being negative is not completely obviated. Emotions in either their primary or first-order form can still be very negative.

Table 19.4 The structure of second-order emotions: shame, guilt, and alienation

	Rank-ordering of constituent primary emotions		
Emotion	1	2	3
Shame	Disappointment-sadness (at self)	Assertion-anger (at self)	Aversion-fear (at consequences for self)
Guilt	Disappointment-sadness (at self)	Aversion-fear (at consequences for self)	Assertion-anger (at self)
Alienation	Disappointment-sadness (at self, others, situation)	Assertion-anger at others, situation)	Aversion-fear (at consequences for self)

Second-Order Elaborations of Emotions

Most of us think that our dog can express *shame* and *guilt*, but in fact, we are probably imputing these emotions to our dogs. The dog is typically expressing subordination to your anger at something that it has done. Even a much smarter animal like a chimpanzee is not expressing *shame* or *guilt* but, once again, subordination to its trainer. Yet, somehow, humans developed the behavioral capacity to experience *shame* and *guilt*, and these two emotions are perhaps the most important emotions of social control in human sociocultural formations. They facilitate social control because individuals who experience these emotions feel pain, sanction themselves, and become motivated to make amends, thereby reducing the monitoring costs and sanctioning activities of others. They drive people to conform to expectations and moral codes, and they thus promote more cohesive and enduring patterns of social solidarity.

But, how are these emotions generated? J. H. Turner's hypothesis is that they represent a further effort to reduce the disassociative power of the three negative emotions. By mixing fear, anger, and sadness together—again, in some unknown manner—individuals will experience either *shame* or *guilt*, depending upon the relative power of these three negative emotions. Table 19.4 outlines a view of the structure of shame and guilt (Turner 2000), as well as another second-order elaboration mixing the three negative primary emotions. The difference between *shame*, or feelings that self has not lived up to the expectations of others, and *guilt* where individuals feel that they have violated moral codes is the relative power of *anger*. In shame the second most powerful emotion after disappointment is anger, whereas in guilt the second most powerful emotion is fear, with anger at self having lesser intensity than is the case with shame. *Alienation* is much like shame in its basic structure but shifts the attribution of anger to others or situation, away from self, and thus produces a very different kind of emotion.

Selection may have worked toward such second-order emotional elaborations because they eliminate the problem of negative emotions disrupting solidarity and, in fact, transform them into emotions of social control—which would have been fitness enhancing millions of years ago. But here, we think that interaction with the neocortex becomes critical to understanding why we do not see such emotions in higher mammals; they depend upon a sufficiently large brain that can produce normative expectations and moral codes. Thus, perhaps only with *Homo erectus* could shame and guilt emerge because only with this hominin does the brain even approach the human measure. A 500 cc brain like that in *Homo habilis* is not, we believe, sufficiently large to allow for guilt because such animals did not and could not frame moral codes using arbitrary symbolic codes organized by a syntax. Shame could have evolved earlier, perhaps, because it is an emotion that is aroused with an animal who feels that it has not met the expectations of others that are immediately seen and felt during face-to-face ?; these do not need to be codified symbolically, but only sensed and felt.

Yet another obstacle, however, to experiencing shame and guilt is that the sense of self and identity has to be developed beyond mere recognition of one's reflection in a mirror. A shame- or guilt-experiencing animal must see its self as an object of evaluation by itself and others in relation to expectations and moral codes. Would a 500 cc brain be capable for having a cognitive conception of itself as a certain kind of being, deserving of responses from others? We think it more likely that a 900 cc brain like that evident in some very intelligent members of *Homo erectus* could have an "identity" that was on the line during interaction. With an identity, social control becomes more internal; individuals are motivated to have their identity verified in the eyes of others and to be very motivated to have others (and self) feel that they have met expectations and conformed to moral codes. Such an animal could build very cohesive and structured groups, and as a consequence, would have been very fit as *Homo erectus* clearly was (since this hominin migrated all over the globe).

One hypothesis might be that as natural selection was working to produce second-order emotional elaborations, fitness was increased only among those members of a species at the high end of the Bell curve of intelligence, where incipient identities might be forming and where some symbolic coding might be occurring. The result was a co-evolution with greater intelligence pushing forward selection on emotion centers for second-order emotional elaborations, and vice verse. Thus, the larger brain may have initially evolved not so much to make hominins culture-users and speakers but to give them more powerful emotions to *evaluate and regulate themselves* and to thereby form strong and solidarity-generating social bonds. Only as the neocortex reached human proportions could culture revolving around arbitrary signs denoting complex meanings evolve, but again, this type of culture still relied heavily on the more primary language—the language of emotions with guilt and shame as key mechanisms of social control.

Toward an Evolutionary Sociology of Neurosociology

The story outlined in this chapter is, of course, highly speculative. Yet, its subject matter must be at the heart of neurosociology. It is critical that neurosociology understand how the human brain evolved and what selection pressures were at work in rewiring the hominin and then human brain to produce a more social animal from weak-tied ancestors with few, if any, direct bioprogrammers for group or troop formation. We hesitate to call what we are doing evolutionary psychology because this approach is burdened by some very limiting assumptions. Rather, what is essential, we believe, is that neurosociology be committed to comparative neuroanatomy in which our closest primate relatives are used as a "distant mirror" to see what the brain and its behavioral outputs of the last common ancestors to humans and apes were like. Then, we can view differences in the brain of the great apes and humans as evidence of how natural selection changed the structure of the brain and how these changes altered the level of sociality among weak-tie hominins to ever-more social animals that eventually could form stronger ties and more cohesive groups, and indeed massive macro societies. This is what sociology can offer neurology: a story of *why* the human brain is wired differently in some critical areas, especially subcortical areas, compared to our closest primate relatives, one of whom is genetically very close to us and probably should be part of humans place on the primate family tree—that is, *Homo*. In understanding how this unique wiring of the brain evolved, we can better understand how it affects human behavior, interaction, and social organization; and this is what neurosociology gives sociology: new tools to understand the sociocultural domain of the social universe.

References

Andrews, P. (1981). Species diversity and diet in monkeys and apes during the Miocene. In C. B. Stringer (Ed.), *Aspects of human evolution*. London: Taylor & Francis.
Andrews, P. (1996). Palaeoecology and hominoid palaeoenvironments. *Biological Review, 71*, 257–300.
Bronson, S. F., & de Waal, F. B. M. (2003). Fair refusal by capuchin monkeys. *Nature, 439*, 128–140.
Brosnan, S. F., Schiff, H. C., & de Waal, F. B. M. (2005). Tolerance for inequity may increase with social closeness in chimpanzees. *Proceedings of the Royal Society of London, 272*, 253–58.

Burke, P. J., & Stets, J. E. (2009). *Identity theory*. New York: Oxford University Press.

Collins, R. (2004). *Interaction ritual chains*. Princeton: Princeton University Press.

Cosmides, L. (1989). The logic of social exchange: Has natural selection shaped how humans reason? *Cognition, 31*, 187–276.

Cosmides, L., & Tooby, J. (1992). Cognitive adaptations for social exchange. In J. H. Barkow, L. Cosmides, & J. Tooby (Eds.), *The adapted mind: Evolutionary psychology and the generation of culture*. New York: Oxford University Press.

de Waal, F. B. M. (1989). Food sharing and reciprocal obligations among chimpanzees. *Journal of Human Evolution, 18*, 433–459.

de Waal, F. B. M. (1991). The chimpanzee's sense of social regularity and its relation to the human sense of justice. *American Behavioral Scientist, 34*, 335–349.

de Waal, F. (1996). *Good natured: The origins of right and wrong in humans and other animals*. Cambridge: Harvard University Press.

de Waal, F. B. M., & Brosnan, S. F. (2006). Simple and complex reciprocity in primates. In P. Kappeler & C. P. van Schaik (Eds.), *Cooperation in primates and humans: Mechanisms and evolution* (pp. 85–106). Berlin: Springer.

de Waal, F. (2009). *The Age of empathy: Nature's lessons for a kinder society*. New York: Three Rivers Press.

Duchin, L. (1990). The evolution of articulate speech: Comparative anatomy of the oral cavity in Pan and Homo. *Journal of Human Evolution, 19*, 687–697.

Eccles, J. C. (1989). *Evolution of the brain: Creation of self*. London: Routledge.

Ekman, P. (1984). Expression and the nature of emotion. In K. Scherer & P. Edman (Eds.), *Approaches to emotion* (pp. 319–343). Hillsdale: Lawrence Erlbaum.

Emde, R. N. (1962). Level of meaning for infant emotions: A biosocial view. In W. A. Collins (Ed.), *Development of cognition, affect and social relations* (pp. 1–37). Hillsdale: Lawrence Erlbaum.

Enard, W. M., et al. (2002a). Molecular evolution of TOXP2, a gene involved in speech and language. *Nature, 418*, 869–872.

Enard, W. M., et al. (2002b). Intra-and interspecific variation in primate gene expression patterns. *Science, 296*, 340–342.

Falk, D. (2000). *Primate diversity*. New York: W. W. Norton.

Fiske, A. P. (1991). *Structures of social life*. New York: Free Press.

Gallup, G. G., Jr. (1970). Chimpanzees: Self-recognition. *Science, 167*, 88–87.

Gallup, G. G. (1982). Self-awareness and the emergence of mind in primates. *American Journal of Primatology, 2*, 237–248.

Gallup, G. G., Povinelii, D., Suarez, S., Anderson, J., Lethmate, J., & Menzel, E. (1995). Further reflections on self-recognition in prmates. *Animal Behavior, 50*, 1525–1532.

Geschwind, N. (1965a). Disconnection syndromes in animals and man, part I. *Brain, 88*, 237–294.

Geschwind, N. (1965b). Disconnection syndromes in animals and man, part II. *Brain, 88*, 585–644.

Geschwind, N. (1965c). Disconnection syndromes in animals and man. *Brain, 88*, 237–285.

Geschwind, N. (1985). Implications for evolution, genetics, and clinical syndromes. In S. Glick (Ed.), *Cerebral lateralization in non-human species*. New York: Academic.

Geschwind, N., & Damasio, A. (1984). The neural basis of language. *Annual Review of Neuroscience, 7*, 127–147.

Goffman, E. (1958). *The presentation of self in everyday life*. New York: Penguin.

Hechter, M. (1987). *Principles of group solidarity*. Berkeley: University of California Press.

Izard, C. E. (1992/1977). *Human emotions*. New York: Plenum Press.

LeDoux, J. E. (1996). *The emotional brain: The mysterious underpinnings of emotional life*. New York: Simon & Schuster.

Maryanski, A. (1986). African ape social structure: A comparative analysis. Ph.D. dissertation, University of California.

Maryanski, A. (1987). African ape social structure: Is there strength in weak ties? *Social Networks, 9*, 191–215.

Maryanski, A. (1992). The last ancestor: An ecological-network model on the origins of human sociality. *Advances in Human Ecology, 2*, 1–32.

Maryanski, A. (1993). The elementary forms of the first proto-human society: An ecological/social network approach. *Advances in Human Evolution, 2*, 215–241.

Maryanski, A. (1995). African ape social networks: A blueprint for reconstructing early hominid social structure. In J. Steele & S. Shennan (Eds.), *Archaeology of human ancestry* (pp. 67–90). London: Routledge.

Maryanski, A. (1996). Was speech an evolutionary afterthought? In B. Velichikovsky & D. Rumbaugh (Eds.), *Communicating meaning: The evolution and development of language*. Mahwah: Erlbaum.

Maryanski, A., & Turner, J. H. (1992). *The social cage: Human nature and the evolution of society*. Stanford: Stanford University Press.

Maryanski, A., Molnar, P., Segerstrale, U., & Velichikovsky, B. (1997). The social and biological foundations of human communication. In P. Weingart, S. Mitchell, P. Richerson, & S. Maasen (Eds.), *Human by nature*. Mahwah: Erlbaum.

Mead, G. H. (1934). *Mind, self, and society*. Chicago: University of Chicago Press.
Olsen, M. (1971/1967). *The logic of collective action*. Cambridge, MA: Harvard University Press.
Osgood, C. E. (1966). Dimensionality of the semantic space for communication via facial expressions. *Scandanavian Journal of Psychology, 7*, 1–30.
Panksepp, J. (1982). Toward a general psychobiological theory of emotions. *The Behavioral and Brain Sciences, 5*, 407–467.
Plutchik, R. (1980). *Emotion: A psychoevolutionary synthesis*. New York: Harper and Row.
Rumbaugh, D., & Savage-Rumbaugh, E. Sue. (1990). Chimpanzees: Competencies for language and numbers. In W. Stebbins & M. Berkley (Eds.), *Comparative perception* (Vol. 2). Wiley and Sons/Blackwell: New York/Cambridge.
Savage-Rumbaugh, S., & Lewin, R. (1994). *Kanzi: The ape at the brink of the human mind*. New York: Wiley.
Savage-Rumbaugh, S., Seveik, R., & Hopkins, W. (1988). Symbolic cross-model transfer in two species. *Child Development, 59*, 617–625.
Savage-Rumbaugh, S., Murphy, J., Seveik, J., Brakke, K., Williams, S. L., & Rumbaugh, D. (1993). *Language comprehension in the ape acnd Child* (Monographs of the Society for Research in Child Development, Vol. 58). Chicago: University of Chicago Press.
Scheff, T. J. (1988). Shame and conformity: The deference-emotion system. *American Sociological Review, 53*, 395–406.
Stephan, H. (1983). Evolutionary trends in limbic structures. *Neuroscience and Biobehavioral Review, 7*, 367–374.
Stephan, H., & Andy, O. J. (1969). Quantitative comparative neuroanatomy of primates: An attempt at phylogenetic interpretation. *Annals of the New York Academy of Science, 167*, 370–387.
Stephan, H., & Andy, O. J. (1977). Quantitative comparison of the amygdala in insectivores and primates. *Acta Antomica, 98*, 130–153.
Stephan, H., Baron, G., & Frahm, H. (1988). Comparative size of brains and brain components. In H. Steklis & J. Erwin (Eds.), *Neurosciences* (Vol. 4). New York: Alan Liss.
Turner, J. H. (1974). A cybernetic model of legal development. *Western Sociological Review, 5*, 3–16.
Turner, J. H. (1996a). The evolution of emotions in humans: A darwinian-durkheimian analysis. *Journal for the Theory of Social Behaviour, 26*, 1–34.
Turner, J. H. (1996b). Cognition, emotion, and interaction in the big-brained primate. In K. M. Kwan (Ed.), *Social processes and interpersonal relations*. Greenwich: JAI Press.
Turner, J. H. (1996c). Toward a general sociological theory of emotions. *Journal for the Theory of Social Behavior, 29*, 132–162.
Turner, J. H. (1997). The evolution of emotions: The nonverbal basis of human social organization. In U. Segerstrale & P. Molnar (Eds.), *Nonverbal communication: Where nature meets culture*. Hillsdale: Erlbaum.
Turner, J. H. (1998). The evolution of moral systems. *Critical Review, 11*, 211–232.
Turner, J. H. (1999). The neurology of emotions: Implications for sociological theories of interpersonal behavior. In D. Franks (Ed.), *The sociology of emotions*. Greenwich: JAI Press.
Turner, J. H. (2000). *On the origins of human emotions: A sociological inquiry into the evolution of human affect*. Stanford: Stanford University Press.
Turner, J. H. (2002). *Face to face: Toward a theory of interpersonal behavior*. Palo Alto: Stanford University Press.
Turner, J. H. (2008). *Human emotions: A sociological theory*. Oxford, UK: Routledge.
Turner, J. H. (2010). *Theoretical principles of sociology, volume 2 on microdynamics*. New York: Springer.
Turner, J. H., & Maryanski, A. (2005). *Incest: Origins of the taboo*. Boulder: Paradigm Press.
Turner, J. H., & Maryanski, A. (2008). *On the origins of societies by natural selection*. Boulder: Paradigm Press.
Turner, J. H., & Stets, J. E. (2005). *The sociology of emotions*. New York: Cambridge University Press.
Wolpoff, M. (1999). *Paleoanthropology*. Boston: McGraw-Hill.

Chapter 20
The Neurosociology of Reward Release, Repetition, and Social Emergence

Michael Hammond

Using more and more sophisticated technologies, scientists have begun to uncover a number of preconscious patterns in neurophysiology and brain activity related to general reward release in the human body. These patterns are associated with rewards such as various positive emotions and hedonistic pleasures. Sociologically, these reward dynamics are important because they point to opportunities to create special social structures that can trigger yet more reward release. Some of these structures can offer additional rewards on a mass basis to more than 50% of the population. Due in part to the nature of these different reward patterns, the costs to individuals and groups using these social creations also vary. These different costs played a key role in shaping the historical sequence in the emergence of these structures, thereby outlining one path of human social evolution.

This chapter focuses on three sets of these reward patterns, and three emergent structures offering additional rewards on a mass basis. The first section looks at an attachment interest that provides many rewards for highly repetitive interaction with a small variety of individuals and lesser repetition for a larger number of ties. This pattern helped to open the window for the emergence of religious gods and spirits throughout our history. The second section turns to a status interest that provides many rewards for a great variety of stimuli, even if these arousers are quite similar in status content. This pattern created a window for the emergence of elevated ascriptive inequality with the exodus from our context of origin as hunters and gatherers. The third section looks at consumption interests that offer extensive rewards for a wide variety of stimuli that are distinct from one another, and for serial novelty in arousers. This pattern sets the stage for the emergence late in our history of mass production technologies feeding a host of such interests.

Romantic Attachment and Beyond

In many species, there is an evolutionary advantage in encouraging individuals to spend a great deal of very repetitive time together in order to accomplish one task or another. Often, there is a selective advantage to ensuring that individuals do not just tolerate such repetitive interaction, but also that they seek out such repetition in some of their social ties. High repetition attachments come in all shapes and sizes, but the underlying reward dynamics are similar. Some means must be found to provide body rewards for spending a great deal of time with other individuals. The more the time

M. Hammond (✉)
Department of Sociology, University of Toronto, Toronto, ON, Canada, M5S 2J4
e-mail: michaelhammond@rogers.co

D.D. Franks and J.H. Turner (eds.), *Handbook of Neurosociology*, Handbooks of Sociology and Social Research, DOI 10.1007/978-94-007-4473-8_20, © Springer Science+Business Media B.V. 2013

together, the greater the total rewards must be. Classic studies on repetition encouraging neurohormones like vasopressin and oxytocin have highlighted one factor in providing such rewards in a number of mammalian species for activities such as child rearing and long-term mating (Donaldson and Young 2008). In regard to humans, neuroscientists have recently been using the latest technologies to look at these rewards in regard to our social ties. Some of the most striking studies have focused on one of the highest repetition reward responses, romantic infatuation. Just as LeDoux (1996) began his classic investigations of the neurophysiology of emotions with the study of fear, one of the most extreme emotions, neuroscientists have found it useful to use an extreme case of repetition rewards to uncover some of the basic dynamics of enticing humans into attachments.

Infatuation is a great example of a repetition incentive. Under its spell, even just thinking about another person can stimulate one reward spike after another. Normally, such a high rate of repetition would lead to reward suppression in which rewards are reduced as repetitions pile up. With infatuation, at least over the short term, this does not occur. Some qualities or characteristics can be used over and over as reward triggers, leading to something like a positive addiction to another person. These elevated infatuation rewards can only be sustained over the short term, but they are often crucial to setting the stage for strong longer term attachments (Fisher et al. 2006). With infatuation rewards, even small physical and personality differences among individuals can become the basis for very strong bonds in comparison to repetitive interaction with others outside the infatuation circle. Reinforced by such high repetition rewards, certain social ties are favored over others. This favoritism is a crucial bias at the heart of network formation in our social lives.

We do not have the neurotechnologies to study directly the pleasurable experiences of infatuation, but we can look for activity in the brain that appears to be associated with such rewards. What are some of the neurological markers for these romantic attachments? Using functional magnetic resonance imaging (fMRI) technologies, neuroscientists have demonstrated that there are a number of areas of the brain regularly involved in such intense ties (for a summary of these findings, see Marazziti 2009; Ortigue et al. 2010). Some like the caudate nucleus in the dorsal striatum and the nucleus accumbens in the ventral striatum are in the dopaminergic rich sub-cortical areas (Aron et al. 2005; Bartels and Zeki 2004; Fisher et al. 2006, 2010; Ortigue et al. 2007). High repetition addictions normally exert their power by prolonging the influence of dopamine on target neurons, and addiction-like infatuation appears to be no different in this respect. With ties to the limbic system, and roles in cognitive processing, decision making, and the evaluation of positive and negative reinforcers, the medial and lateral orbitofrontal cortices are also regularly activated in these attachment dynamics (Aron et al. 2005; Fisher et al. 2006, 2010), as are other parts of the limbic system such as the thalamus (Ortigue et al. 2007). With its role in memory and mental association, the hippocampus is also lit up, and brain areas related to anxiety and fear in the amygdala are dampened down (Aron et al. 2005; Bartels and Zeki 2004). The insula, another area related to mediating emotions, is also active, and the same is true for the anterior cingulate cortex, another brain area linked to emotions, as well as to reward anticipation and empathy (Aron et al. 2004; Bartels and Zeki 2004). This complexity is no accident. Like the status and consumption interests discussed later, attachment is too important an interest to rely on a single mechanism, and too important to be left entirely to individuals to figure out appropriate behavioral strategies. This complexity points to the long period in which natural selection worked on making highly repetitive interaction attractive in favoring some social ties over others. Furthermore, as we shall see below, the same areas play a role in fMRI brain studies of religious activities.

In looking at longer term strong ties, two groupings have emerged. The first is based on high repetition rewards for interaction that is relatively consistent over time. This pattern shares some neurophysiological aspects of the infatuation pattern, but also has some different elements indicating that there is much overlap but not perfect identity in different types of attachment (Beauregard et al. 2009; Fisher et al. 2002, 2006; Ortigue et al. 2010). Long-term attachments are costly in terms of time and effort, and these heavily favored ties are therefore limited in number. Like infatuation, these ties are also costly in terms of negative emotions, such as the anxiety and fear that comes with the possible loss of such bonds, and this too limits the number of these ties. However, strong attachments have a high tolerance for repetitive interaction over the long haul; and even with a lower average reward spike than infatuation bonds and even with some negative emotional costs, the total accumulation of rewards can be very high. As a result, these ties can provide profound comfort and support, as well as an intense sense of commitment and belonging.

The second type of strong attachments is based on bonds established with high interaction rates at one period in life by individuals who are parted by one factor or another, such as marriage or migration into another community. Common examples are biological relatives or close friends who are then separated, and thus have lower rates of regular interaction over time. These kinds of ties exhibit a higher average reward spike per interaction, because of the extra reward release that is triggered by the high situational contrast values when those separated for most of the time find short-term opportunities to get together. These attachments are also limited in number, but they are a key part of the inter-group extended tie networks that have been so important to the success of our species. Finally, for a social omnivore like us, there is substantial additional room for other ties with lower repetition rewards and hence on average, with lower reward totals, thereby filling out a tie network with a larger number of less strong bonds. As we shall see next, this space for lower repetition rewards would become important for the social creation of some additional strong ties to religious beings.

High repetition rewards in attachments do not mean that each and every interaction triggers a high reward spike. This response pattern would be unsustainable over the long term. High repetition rewards mean that the probability of there being some kind of positive reward with each interaction is higher for a small number of ties than the reward probability is for a far greater number of other ties. This difference is a key part of what makes some ties so very important and other ties less important to an individual. Such repetition rewards set the bar very high in marking out strong ties. As noted above, heavily favored ties are very rewarding, but also very costly, and hence must be limited in number by natural selection. One such mechanism would be to limit the reward pathways for high repetition (Hammond 2010); but whatever the mechanism involved in such capping of the number of strong ties, there is a consistent pattern across a wide variety of cultures, with an average of approximately 10 such bonds, and a common range from 5 to 15 (Dunbar 2010). However, is there another means to get some additional strong ties with less extreme repetition, and thereby to favor some extended ties over others? I would like to speculate that just such an opportunity to bias our interest is provided by natural selection in shaping attachment rewards. Understanding these repetition dynamics provides a new way to look at the social emergence of religion across our history as a vehicle to socially create additional attachment favoritism by generating high contrast arousers that can build strong bonds with fewer repetitions.

Religion and Attachment

The study of religious experiences has been another active arena for neuroscientists, and the neuroanatomical overlap between brain areas altered by interpersonal attachment dynamics and by religious experience has been a robust finding in this research. For instance, Schjodt and his colleagues (2008) used fMRI to study the impact of prayers to a god, and traced increased activity in the caudate nucleus and the nucleus accumbens, the same dopaminergic areas that lit up in fMRI interpersonal attachment studies. Using the same technology, Beauregard and Paquette (2006) also traced similar changes in the caudate nucleus. They also showed changes in the insula and anterior cingulate, once again brain areas related to emotions and reward anticipation that also showed up in fMRI studies of strong ties. Using single photon emission computed tomography (SPECT),

Newberg et al. (2001, 2003) showed the same kind of changes in the medial and lateral orbitofrontal cortices, the amygdala, the thalamus, and the cingulate gyrus that were seen in attachment studies. (For a summary, see McNamara 2009.) Altogether, it appears clear that in one way or another, religion is able to piggyback on the reward circuitry for strong attachments. With such piggybacking, religion is able to offer individuals additional comfort and support, as well as an additional sense of belonging and solidarity.

However, there is a potential problem in such an overlap. As we have seen, repetition favoritism in attachments is costly, and only a few ties can be shaped in this way. If religion piggybacks on attachment reward wiring, might it not interfere with or in another way compromise some of those personal ties? The numbers do not show such a result. Individuals with ties to religious beings do not have on average fewer strong personal ties, and vice versa, those without religion do not have on average more strong personal bonds. By some means, the problems in this overlap are minimized, if not completely set aside. Of course, there could be, in popular terms, a god gene or god module in the brain, and that could mean that the fMRI results simply show common pathways to different ultimate rewards. However, postulating such god things in the human body raises many problems in evolutionary biology that go far beyond the scope of this paper; and there is much simpler explanation for the general lack of confusing overlap based on another part of the dynamics of strong attachments.

Contrast sensitivity is a part of any interest that routinely has to assess a variety of stimuli. Neuroscientists have studied such sensitivity in a number of sensory systems, such as vision, hearing, and food assessment (Carrasco 2006; Hillier and Miller 1991; Rolls et al. 1986). The same pattern occurs again and again. The greater the contrast values in an arouser, the greater the response in the system. We do not have at present the non-invasive tools to study this pattern in human interests such as attachment. However, there is no reason not to assume that in addition to repetition sensitivity, reward systems for our attachments are also sensitized by natural selection with reward pathways to respond to contrast values among potential arousers. For instance, with these pathways, contrasts in physical attractiveness or personality characteristics can become one basis for different reward responses in attachment choice. Once again, the rule is the more striking the contrasts, the more initially appealing are the stimuli. Of course, for most individuals, it is the repetition rewards that are the most crucial to tie strength over the long term, and not the contrast values of qualities of the individuals in a bond. After all, the evolutionary logic in favor of high repetition rewards is to increase the likelihood that virtually all individuals will be able to form some strong ties based on the reward totals made possible by such a reward pattern. Since there are relatively few high contrast differences among all individuals in a population, such differences normally cannot be the primary basis for tie strength over time.

As with other human interests, there is a great deal of flexibility in using such contrast sensitivity in tie formation. This flexibility opens a special window for a species with our extensive cognitive talents. What about exploiting the reward space for ties with lower repetition rates than for the strongest personal ties, but still with room for some regular interaction over the long term? Normally, the lower repetition rates would mean lower reward totals, except in a case where the arouser had an extreme high contrast value. Such elevated contrasts could trigger a more elevated reward spike on average. Over time, with such heightened contrasts providing reward bonuses, interaction that is more periodic and shorter in length could still produce significant reward totals. Normal human beings are unlikely to provide such extreme contrasts for one another in the long run. There is simply not that much natural variation among individuals. But what about using super-human beings with all kinds of supernatural qualities that normal humans could never fully match? If such beings could be found, they would provide a rich vein of additional reward release by taking advantage of the basic contrast sensitivity of an attachment interest.

How to create such a reward window? Use our extraordinary cognitive capacities and imagination to create gods or spirits that have all kinds of remarkable qualities that normal humans simply do not possess. These beings do not have to be really real; but interaction with them, or interaction with other humans in the presence of such beings, can offer a special intensity of key body rewards, and feel every bit as real as ties to other humans. Pump contrasts up to extreme levels, so that fewer interactions can trigger lots of rewards over the same period of time; and bonds between gods and humans do not have to compete with high repetition reward ties between humans in order to extract major reward accumulations. Once again, the reward release pattern is not that each and every religious action produces a high spike, but rather that on average, such actions will have a higher spike. It is then possible to get major reward totals with fewer interactions, and at the same time, not interfere with strong bonds based on higher repetition rewards. I label these special social creations as enhancements because they encourage individuals to take advantage of body reward rules to get additional rewarding release that otherwise would not be available on a regular basis over the long term (Hammond 2006, 2010). The additional rewards are greater than individuals would receive for an equivalent number of lower repetition ties to other individuals. These bonus rewards would help to provide a greater package of additional attachment benefits, and also would help to fuel an increased commitment by individuals to social structures that can consistently provide such arousers.

In the pursuit of these additional ties, costs for individuals and groups become an important issue. Remember that these imaginary additions do not result in stronger ties in terms of total reward release over the long term. They simply have the same potential strength with less repetitive interaction. Individuals are therefore not going to wish to pay more in total for such additional ties in terms of time and effort expended. Our imagination is one ideal cost saver. If we had to actually pay out 10 times, or 100 times, or 1,000 times more costs for ties to beings 10 or 100 or 1,000 times more powerful or more knowledgeable than any other humans, and if we only got the same reward total as with high repetition reward personal ties, then few would be interested in such beings. However, since these beings are imaginary, some potential costs can be significantly reduced as part of the enhancement process. Collected together in a social construction we label as religion, these supernatural beings can be made regularly available on mass basis, that is, to at least 50% and perhaps to all members of the population.

Of course, even when enhanced, there are very real costs in terms of time and effort for such strong additional ties. These additions can also have costs in terms of negative emotional states among believers who, for instance, can be anxious or fearful about their relationship to these creations. Thus, as with strong personal ties, the strongest of these enhanced attachments must also be limited in number. However, even the smallest of human populations with the most limited of technologies have been able to create these special structures and make them available over the long term on a mass basis. Enhancements emerging later in our history would have some very different costs, and hence would not be as universal.

Altogether, by using contrast sensitive neural pathways to trigger additional rewards, humans can build additional strong ties to gods and spirits while still minimizing possible interference with interpersonal ties. With this neuroanatomical symmetry, these bonus ties use much of the same reward systems, and therefore, it is no wonder that these religious bonds can feel every bit as real as ties to other human beings. In addition, given this first overlap, it is not surprising that there are so many other affinities, such as in the anthropomorphic language of these sacred bonds, in the anthropomorphic representation of these beings, and in the parallels between types of personal attachments and types of attachments to the gods (Kirkpatrick and Granquist 2008). All these overlaps point to some means for piggybacking rewards without too much interference among the different kinds of bonds. We shall probably never know if religion continued to emerge historically as a by-product of other adaptations early in our prehistory, or if such religious piggybacking began as a by-product, and then become an adaptation itself favoring individuals having more contrast sensitivity with some additional rewards set aside for higher contrast and lower repetition arousers. The sociological consequences are much the same. If there is such a set aside, it might not be for religion per se, but rather for a high contrast religion-like creation, such as society itself. This was of course Durkheim's classic argument (Hammond 2003).

Inequality and Rewards

Creating status distinctions is another area of social life with a great deal of repetitive interaction. What kind of body reward release pattern might be related to such repetition? Due to ethical limitations on research with human subjects, some of the most sociologically important studies on the neurophysiology of these rewards for status activities have been done on non-human primates. For instance, in many primate species, natural differences among individuals are magnified into a status hierarchy through a long series of status competitions. In looking for the underlying mechanisms in such a magnification of differences, the neuroscientist Michael McGuire and his colleagues undertook a classic series of experiments measuring whole blood serotonin levels in male vervet monkeys in order to demonstrate that the neurotransmitter serotonin played a key role in the reward process for individual primates successful in repetitive status interactions (Brammer et al. 1994). They focused on serotonin because many studies in humans and non-humans had indicated that the regular administration of serotonin producing drugs, or of serotonin reuptake inhibitors, had significant and positive personality and behavioral effects on individuals. Normally, serotonin levels increase or decrease in relationship to the number of positive signals the individual receives from others in the group. The more positive signals, the higher is the serotonin level, and the fewer the positive signals, the lower the level. These scientists were able to shape the acquisition or loss of dominant status by manipulating serotonin levels with the systematic use of serotonin reuptake inhibitors. To do this, they removed some individual primates from a group and altered serotonin levels in those individuals. The scientists then reinserted the neurochemically altered individuals back into the population, and thereby bidirectionally altered their relative status position both upwards and downwards. Previously dominant individuals with experimentally lowered serotonin levels assumed subordinate positions, and previously subordinate individuals with raised levels assumed more dominant positions. They concluded that dominant or subordinate status affects the body chemistry of serotonin, and that conversely that altering serotonin levels influences status attainment. These altered status roles also demonstrated that all the primates in the experiments have the reward release capacities for elevated status, even though in normal circumstances only a few would reap the benefits of those rewards (Madsen 1994). There are many ongoing controversies about the causal role of neurotransmitters and hormones in triggering rewards for successful status interaction, but the details of these controversies are not at issue here. What is important is that in any status sensitive species, there have evolved elaborate mechanisms in the body rewarding individuals for repeated success in status competition, and that these mechanisms are not limited to a few individuals in a wider population.

One of the most striking aspects of neurophysiological experiments like those of McGuire and others is to see how magnified status distinctions can appear and persist even when only a small number of non-human primates are interacting in ordinary circumstances over the long term. The same process occurs in the wild, not only for species like the vervet monkey but also for other primates such as the chimpanzee. This pattern is only possible if status rewards can be triggered by a large number of interactions with the same individual, and also by a large number of similar interactions with other individuals. That is, status interactions between dominant and subordinate individuals can be repeated time and time again, and body rewards continue to flow to dominant individuals. As a result, there do not have to be many individuals in a population in order to fuel status expansion for a few. The same subordinates can provide a large number of reinforcement opportunities because the repetition rewards do not significantly dampen the impact of recurring status interaction.

In regard to human status interaction, neuroscientists do not allow themselves to conduct the kind of extreme experiments that McGuire and others have used on non-human primates. There are many potential negative outcomes for individuals in experiments using strong inequality, and scientists working with humans use more gentle interventions. For instance, in some neurological fMRI research, status interactions are carefully limited in computer games (Izuma et al. 2008; Zink et al. 2008). These experiments make it clear how status sensitive humans are, and delineate some of the underlying dynamics in the neural processing of social hierarchy in humans, such as role of the striatum in status related activities. As we saw earlier, the striatum is one of dopaminergic rich sub-cortical areas also active in attachment dynamics. These fMRI studies indicate that there are distinct reward circuits activated by attention to social rank. Martina Ly and her colleagues (2011) have focused attention on the ventral striatum by demonstrating that an individual's subjective socioeconomic status differentially influences activity in this area of the brain when processing status information. Similarly, another fMRI study has demonstrated that there are distinct neural representations for status in the human inferior parietal cortex (Chiao et al. 2009). There is also some evidence that brain dopamine receptor density in the striatum correlates with social status (Martinez et al. 2010). This of course raises the important question of whether changes in social status can be correlated with changes in this receptor density. There are also some intriguing indicators of a common dopaminergic neural currency for social and monetary rewards (Saxe and Haushofer 2008; Tobler et al. 2007). Such a common currency would represent an important extension of the fMRI studies of Read Montague and his colleagues on dopamine, expectations, and the evaluation of alternative actions (Montague and Berns 2002, 2006). In another line of neurochemical research led by Allan Mazur and his colleagues, there has been some very important human research on testosterone and success in competition in many different social arenas (Mazur and Booth 1998; Mazur 2005; Booth et al. 2006). Once again, the testosterone related dominance studies have raised many questions about causality and rewards in regard to neurotransmitters and hormones, but these controversies are not at issue here. What is important is that these fMRI and neurochemical studies demonstrate clearly that humans are a status sensitive species right down to a neurophysiological level, and that such a sensitivity is only possible if natural selection favored it over prolonged lengths of time in the evolutionary origin of our species.

However, with so many limitations on human experimentation, the neurotechnologies available at the present time to study status interactions are not likely to penetrate the dynamics of elevated inequality, the area with so many implications for the social sciences. Thus, to push further in looking for such implications in brain research, we must turn again to the primate studies and try to tease out other aspects of that research for the analysis of human inequality. One way to extend that analysis is to return to the role of repetition rewards in shaping behavior. This has been a fruitful path for analysis of attachment interests, and perhaps the same might be true for the study of human status dynamics.

As noted above, both in the laboratory and in the field, some non-human primate studies have demonstrated clearly the existence of extensive rewards for a great quantity of quite similar status interactions among even a small number of individuals. The result is that a micro social scale can produce permanent and magnified status distinctions. However, when we look at the anthropological evidence for human populations who persisted over time at such small scales, it appears to be the case that there are not high repetition rewards for status interaction with the same individual in the same general circumstances. All human populations exhibit some status differentiation; but if there were an equivalent repetition reward pattern in humans, then even micro human populations should exhibit again and again the same elevated status dynamics of other primates with a high repetition tolerance. Instead, anthropology shows that it is generally impossible to sustain magnified status distinctions over the long term with only a small number of individuals. As Boehm (1999) and others have demonstrated, these micro populations produce occasional status spikes; but normally there is successful resistance to any attempts to make these spikes permanent. It is only with larger numbers that status differentiation becomes regularly elevated, and resistance in defense of more modest distinctions becomes more ineffective in dampening status inflation over the long term.

This is intriguing sociologically, and suggests something about the probable reward wiring for status in humans. There appear to be no selective advantages to evolving new status rewards for the new primate species of human beings; and therefore, it is most likely that much the same basic body rewards exist in human beings as in other status sensitive primates (Madsen 1994). What seem to be different in humans are the reward release rules, and in particular, the rules for repetition rewards. As we have seen, in some other primates, there are extensive reward totals for repetitive status interaction with any specific individual; but in humans, there appear to be more limited rewards for such repetition. In our context of origin with its micro-populations, there was a limited variety of other individuals to provide status stimuli. With limited repetition rewards for status interaction with any one individual, and with a limited variety of other individuals, there simply were not the regular conditions to provide the reinforcements necessary to fuel high status positions on a long-term basis. This inequality capping had a number of advantages for such populations, particularly in regard to the high degree of economic cooperation necessary for survival in a hunting-gathering subsistence context. Elevated inequality makes long-term economic cooperation in food provisioning very problematical. For instance, male-male cooperation in activities like collective hunting, as well as female-female cooperation in other economic endeavors, is very difficult to mix with such inequality.

However, uniquely among primates, these more limited repetition rewards for status interaction with any specific human being appear to exist alongside additional reward release for a high total of quite similar status interactions with a great variety of individuals (Hammond 2008). The second part of this reward mix also had a number of advantages in our context of origin. If different ecological conditions made possible a slightly greater population concentrations, then individuals would be able to respond to the increased variety of status reinforcements that such small differences in density made possible. Furthermore, there were also some exceptionally rare instances in our pre-horticultural history when especially favorable ecological conditions made possible much larger populations and more permanent settlements over the long term. The regular expansion of social inequality in these extraordinary circumstances points again to the probable existence of status reward rules wired to be responsive to a large variety of similar reinforcements.

There were also occasional but important situations when populations were temporarily expanded in periodic gatherings of groups normally dispersed throughout the rest of the year. These gatherings were very useful for the exchange of a wide range of goods and services, marriage partners, new ideas, technological changes, etc. (Johnson and Earle 2000; Powell et al. 2010). Some evolutionary theorists even speculate that these gatherings were a key part of the social basis for cognitive and technological innovations as early in human history as the Upper Paleolithic (Richerson and Boyd 2005; Heinrich 2009). There were also temporary groupings for more violent activities such as warfare between groups (Keeley 1996). Given our status reward release rules, these many different temporary assemblies created a few special opportunities for status differentiation fueled by the additional variety of individuals. Such temporary variety could provide a surge in reinforcing status arousers that encouraged a few to pursue more elevated positions in the general activities of these assemblies, such as special food provisioning opportunities in collective hunts, extraordinary religious rituals, and combat. In these conditions, a few individuals would receive a double bonus in reward release, with variety points for the temporarily increased numbers and contrast points for the special quality of such occasional situations.

In these special conditions, some individuals would have to respond quickly to such occasional and temporary, but still important situations. A variety and novelty sensitive reward release pattern reacting to an expanded albeit temporary range of similar arousers is ideal for such a rapid response. If more individually distinct reinforcements were necessary to trigger additional reward release, then the costs to status seekers in terms of time and effort would be too great, and the status interactions develop too slowly, to provide these short-term reward spikes. Of course, such special situations were generally unsustainable in our context of origin. Before too long, individuals returned to smaller groups and more mundane activities; and the status dampening impact of limited repetition rewards for status interaction with any specific individual and a limited variety of individuals came into play again.

In these micro populations, another consequence of reward wiring for similar variety is a preference for multiple status distinctions that increase the possibility of using the same individuals more often as status reinforcers for different categories. With such small numbers, reinforcement can come from the same individuals focusing on a specific distinction, or from the same individuals focusing on a number of distinctions. There is less repetition and more total variety of reinforcement from the second alternative. However, creating such multiple distinctions means a trade-off among individuals. Those successful in one status arena must also be willing to support those successful in other arenas. I recognize your distinction if you recognize mine. Without such reciprocity, there are simply not the bodies available to trigger sufficient reward release to make the costs of multiple status acquisition worthwhile. However, this reciprocity makes high contrast distinctions even more problematical in such small populations. How are a few going to achieve high contrast differentiation if simultaneously they must spend substantial time and effort supporting the status distinctions of many others? The more binary distinctions seen in other status sensitive primates simply will not work as effectively with human reward wiring and micro populations.

Altogether, in comparison to other primates, human status wiring seems to have a Goldilocks principle of not too little and not too much. If there are limited rewards for repetition of the same stimuli and for variety, then there is going to be very little inequality, as in the case of the bonobo. If there are rewards for both repetition and variety, then there is going to be extensive inequality, as with the vervet monkey and the chimpanzee. With limited rewards for repetition, but more rewards for variety, the human case is not as clear cut. This is probably one reason why neurophysiological studies of human status interaction have more inconsistency than studies in other primates. The human reward mix normally produces an in-between moderate inequality total in the micro populations marking most of our history. However, this moderation is predominant only as long as the dynamic is constrained by the demography of our context of origin.

The Exodus and the Inequality Explosion

With the exodus from our context of origin, populations become permanently larger in area after area around the world. With reward release rules keyed to the increased variety of stimuli that such permanent settlements provided, inequality began to expand, and then to explode as social scale grew again and again. This expansion was an indirect consequence of a mix of status reward release rules forged by natural selection in a very different context. With these enlarged populations, the dampening impact of a limited repetition release for status interaction with any single individual was eroded, and the variety release rule become more and more important in triggering more and more rewards for status distinctions. Thus, the demographic changes of the exodus opened a reward triggering window that before had only been available on an occasional and temporary basis. This produced a dramatic reversal of fortune in which reward rules began to have the opposite effect in comparison to their earlier role. The expansion restricting rules on one context became fuel for permanent inequality inflation in other circumstances. In the face of such a reversal, the general resistance to elevated inequality that was so successful earlier in our history became less and less effective. We are still caught up today in the consequences of this often tragic development.

One of the most striking developments in this inequality explosion was the emergence of high contrast ascriptive status distinctions in which one group of individuals sets itself apart from another group on the basis of different qualities attributed to the two groups but not actually tested on a fair and open basis, even though the two groups regularly interact with one another. This is something not emerging from the status reward rules of other primates. It is one thing to see highly repetitive status competition with body rewards magnifying performance differences among individuals. This occurs in many primate species. We also see that pattern after the exodus with the emergence of such roles as the "big man" who may have hundreds of different individuals providing highly similar but still very attractive status reinforcers. However, elevated ascriptive differentiation applied to whole groups of regularly interacting individuals within a population, such that even those groups very subordinate in some dimensions are very dominant in other dimensions, is something very human.

As the McGuire experiments on serotonin manipulation and status alteration in vervet monkeys demonstrated, the body rewards are there in all the individual primates for elevated status positions, even if in more normal circumstances, only a few would be able to take real advantage of these rewards. There is no reason to assume that the human situation is any different. The existence of this reward reservoir opens a window for a mass social creation that could make elevated distinctions and status rewards widely available within a population. With low repetition rewards for any single stimulus, the first part of any such reward package must be to expand the number of reinforcing stimuli available. The post-exodus population surge provided just such an incentive on a long-term basis. The elevated population concentration meant that a wide variety of status reinforcers were now available on a mass basis. Even when divided up into sub-populations, there was still enough variety of stimuli to trigger major rewards as one group of individuals set themselves up as dominant over others. As populations continue to grow, multiple distinctions become possible as more and more sub-groups would still have the critical mass to trigger additional rewards.

Secondly, to further reward release through ascription, there must be some mass basis to take advantage of the contrast sensitivity of status reward wiring. In all primates wired in one way or another for major status rewards, there is a reward rule to provide higher release totals for higher contrast distinctions. As we have seen in looking at contrast sensitivity in attachment, this is the general rule in any strong interest that has to assess a variety of arousers. In the human case, real performance differences might create a few such distinctions among individuals in a population. However, natural differences among individuals in a large group are much greater than the differences between such large groups. How then to provide elevated contrasts on a mass basis when real differences among large groups of individuals in a population are either non-existent, or comparatively small?

Just as in the case of religion, the solution is to use our remarkable cognitive capacities and imagination to construct enhanced arousers in a cost effective manner. Create the extraordinary out of the ordinary. Exaggerate differences by taking natural disparities in one area, such as gender and physical strength, and attributing all kinds of other important distinctions on the basis of that initial difference. Then, on the basis of that attribution, exclude whole groups from even competing in one or more status arenas. Or, even better, simply make up major distinctions that exclude whole groups from different status competitions. Use gender, race, ethnicity, eye color, hair color, or whatever for this ascription. Then, in a series of repetitive status interactions, use those increased contrasts to trigger additional body rewards from our preconscious reward release rules, and use these extra rewards as fuel to maintain this social construction. As with religion, these distinctions do not have to be truly real in order to serve as reward triggers. Once again, natural selection leaves open a window of status rewards for differences that do not exist naturally, just as attachment rewards are there for bonds to beings who do not exist outside of our belief in them. Tie these imaginary status creations to a large population base providing a great variety of reinforcers, and enhanced ascriptive inequality explodes.

To reap these additional rewards, there are some real additional costs in terms of time, effort and risk in these new status creations. For many individuals, these costs and benefits can be kept somewhat in line because of the cost savings in using our exceptional cognitive skills to generate these inflated status distinctions in the first place. However, as with enhanced attachments, since there are some real costs, the number of such distinctions used by an individual must be limited. Of course, it is crucial to structure elevated ascription so that the groups claiming dominance do not have to prove in fair and open competition that they actually possess all the outstanding qualities they profess to have. Any such performance competition would be enormously costly and might expose the fragile or even non-existent basis for such ascriptive inflation. Also, as another means of cost control, it is useful to tie this expanded inequality and religion together in order to let them reinforce each other. Religion could provide some powerful justifications for these imaginary ascriptive distinctions, and some of these distinctions between humans could have the same hyper-elevation marking the distinctions between humans and gods. Furthermore, linking religion and inequality could provide an additional sense of social support, comfort, and belonging. When all of this is linked together, elevated mass status distinctions emerge as one of the most striking enhancements to appear with the exodus.

As with religion, this second historical enhancement provides bonus rewards that otherwise would be unavailable on a mass scale. With this enhancement, special rewards that were available for a few on an occasional and temporary basis in our context of origin become available for many on a regular and permanent basis. Elevated ascription provides a double bonus because it allows many to take advantage of the similar variety reward rule in a human status interest, as well as the elevated contrast reward rule. Given these potential rewards, it is no wonder that after the exodus, few were able to resist the appeal of such a social construction. This widespread adoption of elevated ascription on a mass scale was very useful to the much smaller elites at the top of status pyramids of wealth and political power. If individuals embrace one status enhancement, it is that much harder to oppose the very idea of inflated differentiation, such that mass ascription handicaps long-term effective resistance to other enhancements that create these elites.

The linkage of such special status distinctions to religion is also important in terms of reward repetition rules. With an inequality component that is highly tolerant of similar variety, religious activities could be more appealing more often. In fact, with the exodus, it is striking that religious ceremonies tend to occur with more regularity. For instance, among traditional foraging populations like the Kung, the key ceremony of the trance dance occurs quite irregularly, and on average only about once a month. Similarly, the practice of time consuming daily rituals or regularly scheduled assemblies is not common (Dunbar 2010). This spacing would make sense in terms of a religion reliant mainly on higher contrast stimuli and lower repetition rates in comparison to personal ties. Adding mass high contrast inequality could change the spacing dynamic dramatically. By intertwining religion with patriarchy, social class, ethnicity, or any other inflated status distinction, more repetitive ritual actions could be partly fueled by the inequality reward surge for higher repetition in elevated ascription. If linked together, the rule would be the greater the mass ascriptive inflation, the greater the average time and effort likely to be devoted to religious rituals by believers. Of course, this would occur most often when religion and mass ascription are the only two enhancements available to most individuals. As we shall see in the next section, if a third enhancement is available, then other possibilities emerge.

From the point of view of individuals in a group using elevated ascriptive distinctions, regular interaction between individuals in the same status group added to regular interaction with subordinate groups could create a treasure trove of stimuli that all share similar status content. All of this could have many repetitive aspects, but as we have seen, in humans and some other primates, such similarity is still given many rewards. Arousers that are more distinct from one another would be even more appealing, but such distinctiveness is not necessary for major reward release. This is one reason why stereotypical thinking is so appealing for so many individuals embracing this brand of elevated inequality. Stereotypes can provide a sea of quite similar interaction patterns. Formalizing such relationships is also very appealing to those using such ascriptive distinctions. Formal arrangements increase the likelihood of a similarity of outcome in providing a steady variety of reinforcers. The move levels in such formal structures, the more opportunities there are to provide a range of status interactions with a similar outcome.

With our variety biased status reward rules, the social creation of ascriptive inequality began in our context of origin as individuals sought additional status distinctions with a collective basis that did not require too many individual demonstrations of special performance capabilities. For instance, the anthropological record is replete with examples of the appearance of such ascription in gender distinctions. The male-male cooperation required to create part of this gender ascription probably carried a selective advantage in strengthening cooperation for other activities like collective hunting, and perhaps even for inter-group warfare. However, the expansion of such ascription was severely dampened by the micro-populations of that era. Any long-term magnification of ascriptive inequality ran into the dampening blanket provided by small populations with rules limiting the rewards for the repetition of any specific status stimuli. Under such a blanket, ascription remained comparatively light on a regular basis, and spiked only on a temporary and occasional basis. The exodus stripped away that constraint. A small percentage of individuals could then begin to accumulate massive political and economic inequality that also provided access to other goods and services triggering rewards from other interests. Simultaneously, a mass of individuals could be offered some additional status rewards with permanent elevated ascriptive contrasts and somewhat limited costs. In the face of such an offering, and without an alternative new enhancement package on the horizon, long-term resistance to general status inflation began to erode. It seems that given our status reward release rules, the modest distinctions of our earliest history could not compete with the enhanced status inflation made possible by the exodus.

Of course, there are even greater costs for those unable to benefit much, or at all, from the social construction of elevated inequality. Such inequality not only requires post-exodus population concentrations, but it also burdens many with having to provide massive reinforcements to others for these social creations. Given these radically unequal burdens, coercion and resistance will be a part of these social worlds. The pursuit of status inflation also creates constant conflict between individuals and groups competing for high status positions, between powerful elites and non-elites with only some ascriptive inequality, and between non-elites with different claims to ascriptive status. Handicapped by our status reward release rules, resistance either fails in the short term, or succeeds only to eventually produce another version of inflated inequality. It is therefore not at all surprising that during the longest part of human history, even when there was a basic understanding of plant cultivation that could support a growing population, most individuals favored fissioning and migration as a means to limit population concentration and the emergence of the heightened inequality accompanying this demographic change.

Only when the most favorable ecological niches in many parts of the world were filled by fissioning and migration did more and more individuals begin to settle down into permanent settlements and try to cope as best they could with a Pandora's box of problems tied to this change, such as status inflation. A few shot to the top of the status pyramids accompanying the exodus, but most had to make do in one way or another. Embracing ascriptive status inflation was one such vehicle for many trying to make something good out of a difficult situation. Only a few can remain consistently dominant as social scale grows and populations divide and then sub-divide into more and more status categories, each of which has to have winners and losers. The result is that most individuals have to cope by mixing the more appealing reward and cost packages of some ascriptive distinctions in their favor with a greater number of the less appealing packages of other distinctions not in their favor. This mass trade-off opens a window for enhancement substitution when a new enhancement opportunity begins to emerge.

The Great Substitution

Neuroscientists have also been active in studying other reward release rules for many other interests. One line of research has focused on the rewards for consuming things such as food. For instance, in a classic series of experiments, Edmund Rolls and his colleagues were able to use microelectrodes on specific neurons in the lateral hypothalamus of monkeys in order to study directly the firing rates of neurons devoted to a food interest (Rolls et al. 1986, 1999). This is another example of neuroscientists using techniques that are not normally allowed in similar research on fellow human beings. With such interventions, Rolls and his team were able to demonstrate clearly the preconscious impact of the repeated presentation of the same attractive food stimuli and the decay of neural responses to such repetition. They were also able to demonstrate the role of additional arousers with lesser or greater diversity in extending the interest of the primates in continuing to eat. One consistent finding was that additional foods that were only similar, but not really distinct, in comparison to other foods had a limited impact in such interest extensions. That is, this variety interest has a high degree of contrast sensitivity. Furthermore, unlike a status interest, the repetition of even high contrast foods did not trigger all that much in the way of additional rewards, and interest soon turned to yet more variety. Finally, they were able to show through diminishing neural firing spikes, the declining marginal utility that always accompanied the accumulation of multiple arousers. Once again, as with attachment and status interests, this complexity highlights an interest that is too important to rely on a single rule, and too important to be left entirely to individuals to discover the correct behavioral patterns.

Following up on the research of Rolls but using a somewhat different approach focusing on the trade-offs between favorites and variety in food preferences, behavioral psychologists have demonstrated that in non-human primates like the tufted capuchin, there are more rewards for extended variety in food consumption than for the repetition of a few high contrast favorites (Addessi et al. 2010; Santos and Chen 2009; Chen et al. 2006). Favorites are nice, but variety is the winner. Similarly, in their study of hedonic hotspots like the nucleus accumbens and the ventral pallidum with their brain circuits attaching pleasure and desire to foods, neuroscientists such as Kent Berridge and his colleagues have shown the same response pattern in humans, as well as highlighting the pleasurable impact of novelty in such consumption interests (Berridge et al. 2010; Smith et al. 2010. See also Tobler et al. (2007) for recent neurological research on declining marginal utility in humans). The evolutionary logic behind this pattern is that the selective advantage went to individuals who, whenever possible, would avoid too much repetition or similarity in consumption. All of these findings are important sociologically because they point to yet another opportunity for humans to use enhancements to exploit further the ancient neural reward release rules that are a part of our evolutionary past.

These findings are also important because they offer enhancement windows that can serve as a partial or even complete substitution for classic enhancements such as elevated ascribed inequality and religion. Although it is imperative for humans to find some enhancements, there is no such imperative to find any particular enhancement. Specific enhancements can be piggybacked, or they can be set aside if other enhancements seem to offer more reward triggers for the same or even lesser costs. For instance, as we have seen, the emergence of such ascriptive inequality was one of the few means available to large numbers of individuals to try to cope with the many difficult consequences of the exodus from our context of origin. Even though the costs were high, there were additional status rewards made available on a mass basis. Without new alternative enhancements offering many individuals other packages of bonus rewards, there was little chance of successful long-term resistance to this post-exodus emergence. Any new alternative enhancements would require both alternative body reward sources and alternative social structures that could pump up those rewards. Let us look at consumption interests in terms of providing one basis for such a substitution.

The first two parts of this article tried to demonstrate how attachment and status interests tended historically to focus on body rewards triggered by a limited number of high contrast distinctions. These are the favorites. However, as the neurosciences have shown, we also have interests that are preconsciously wired to prefer extensive diversity rather than a few favorites. For these interests, it is extended diversity with more limited contrasts on average that is most often the winner in terms of triggering body rewards. This consumption diversity is therefore somewhat different than status wiring in terms of its reward release rules. Status variety is very receptive to a large number of quite similar stimuli. For the big man, having a few hundred individuals recognize his special position is clearly attractive, even if this recognition has many repetitive elements. The total body rewards available from stockpiling a wide variety of such partial repetitiveness is great, and will entice some individuals to risk their very lives in order to ascend to such positions. For the consumer, having a few hundred varieties of bread available is not going to have anything near the same impact. There are some rewards for similar stimuli, but there must be more distinct contrasts among the individual food stuffs available in order to pile up a great total of body rewards. With status, reward wiring favors a few high contrast distinctions based on a great number of quite similar reinforcers. With food, this reward accumulation pattern is not as effective. The consumption interest favors an alternative strategy of piling up a greater variety of stimuli that are more distinct from one another across a greater number of food categories.

Furthermore, a consumption diversity interest is often better able to take advantage of the body reward bonuses for novel arousers. Using fMRI technologies, neuroscientists have focused on the so-called "novelty center", the substantia nigra and ventral tegmental area of the midbrain, which are linked to both cognitive information processing of hippocampus and the emotional information processing of the amygdala (Bunzeck and Duzel 2006). Not surprisingly, there seems to be a positive feedback loop between dopamine and novelty (Lisman and Grace 2005). Most human reward systems have an automatic set aside of some rewards in case attractive novel stimuli become available (Berns et al. 2001). That is, there is novelty default in which some additional rewards are made available for new stimuli. As the novelty wears off, partial habituation occurs, rewards decrease, and this creates a reservoir for responding to future novelty. Once again, there is contrast sensitivity in which the substantia nigra and ventral tegmental area become more activated by greater degrees of novelty (Bunzeck and Duzel 2006). For many high contrast enhancements such as religion and ascription, it is difficult to make too much use of this novelty bonus. After all, such high contrast stimuli most often offer just the opposite of novelty. These are eternal creations whose strength is found in their non-changing qualities. From such elevated distances, novelty is more often seen as a threat than as a further reward bonus.

Serial Novelty

Consumption diversity does not generally offer such eternal high contrasts, but it can offer more novelty bonuses. In fact, it can offer serial novelty with a string of novel arousers stretching over a lifetime. The reward bonuses from such a sequence can provide very attractive reward totals to a large number of individuals. It is this combination of enduring variety and serial novelty that offers a consumption alternative to classic high contrast enhancements. This combination normally has a material bias because presumably, material goods and services do not object if they are left behind for a new product. In contrast, strong personal ties create all kinds of difficulties when one is set aside for another. Attachment has a novelty set aside, but these bonds normally have greater costs than material stimuli in substituting arousers that can reap the benefits of novelty. Religion and serial novelty often have a difficult relationship, since most believers look to their faith for its continuity and enduring qualities and not as a source of new things. Similarly, elevated ascriptive inequality on a mass scale is most often based not on regular novelty, but rather on fixed differences between groups in a population. Long-term serial novelty in status acquisition is only possible for a small percentage of any population. Status ladders providing distinctly novel upward steps over a lifetime are simply too expensive to be made available on a mass basis.

As with other interests, in order to exploit fully the reward bonuses of consumption interests, individuals are going to have to find some means to enhance arousers. How to create new enhancement windows for these interests? Use the lessons from other enhancements. First, as with status enhancements, make available on a regular basis to the consumption interests a degree of diversity and novelty that was only available on an occasional and temporary basis in our context of origin. Second, as with attachment and status enhancements, find the means to provide such additional variety and novelty without requiring too many additional costs to access such rewarding packages.

If religion and ascriptive inequality were two social creations for high contrast and limited novelty reward bonuses, what social creation could provide reward triggers having lower contrast values mixed with higher diversity and novelty? The answer, of course, is found in mass production technologies capable of churning out a wide variety of goods and services to be consumed by a large percentage of the population. This mass production not only offers a vast surge in variety and novelty for all kinds of interests to all kinds of individuals, but it does so without requiring the same surge in time and effort devoted to such production. As we have seen, declining marginal utility is an important part of consumption interests. This means that for instance, a ten-fold increase in attractive stimuli will not produce a ten-fold increase in rewards, but rather a smaller increase. Therefore, an increase in arouser variety that has an identical increase in costs to access those arousers will not be very appealing. There have to be some enhancements that provide increased rewards without a proportionate increase in costs. Technologies can be one such vehicle. Indeed, the new enhancements can be so productive that they can even offer the promise of reducing some costs for many individuals in comparison to the personal costs of other enhancements such as ascriptive inequality. Altogether, such production offers up a partial or even complete substitute for other enhancements like religion and ascriptive inequality. Given these offers, it is no wonder that the emergence of this third enhancement vehicle ignited fierce opposition from defenders of these other classic enhancements.

Why a substitution of enhancements? One reason is that inflated ascription and the mass production of goods and services are more incompatible with one another than was the classic enhancement combination of religion and ascription. Mass production requires long-term economic growth in order to make it deeply appealing, and intense ascription gets in the way of such growth in a number of ways. For instance, such ascription limits social mobility of individuals and groups, keeps many marginal to the economic marketplace, impedes the geographical exchanges of goods and services, and resists technological change and innovation. Elevated inequality for some and technology heavy mass production are not as incompatible. Indeed, such production expands somewhat the size of the economic elite in comparison to the hyper-concentration of income in pre-industrial cultures. However, elevated inequality for many cannot be as easily spread throughout a population using vehicles like ascription. At the very least, the extreme economic inequality of large scale pre-industrial cultures must be partially eroded, and ascriptive inequality must be severely curtailed if mass production structures are going to offer more and more individuals an enhanced variety of goods and services.

Similarly, serial novelty can be very appealing for many interests; but as we have seen, it is often not very compatible with many classic high contrast enhancements offering eternal truths about life and social status. The more that a culture relies on serial novelty to top up reward totals, the harder it is to also rely so heavily on these traditional enhancements. Once again, to embrace serial novelty, these classics must be at least partially set aside by a large number of individuals in a population in favor of other enhancement packages. The classics need not disappear, but they must play a more marginal role on average as a host of individuals seek out other means to trigger bonus rewards from the body. In contrast, the use of technological enhancements can have the opposite impact on interests such as curiosity. Humans are a curious species, and neuroscientists have demonstrated in fMRI studies a role for both the dopaminergic reward system and the information comprehension and anticipation areas in the brain, the same areas we see active in other interests such as attachment (Kang et al. 2009). In most individuals, curiosity is not as strong as other basic interests such as attachment or food consumption. However, a curiosity interest is very sensitive to novelty (Loewenstein 2007). With expanding technologies for communication, information, and entertainment, curiosity can be exploited again and again on a mass basis for reward triggering through serial novelty, and this interest too can play an active role in fabricating substitutes for other traditional enhancements. Similarly, such technologies can allow many to develop their social networks in novel ways that trigger many rewards from an attachment interest, without relying as heavily on religious arousers to extend this interest.

Without such alternatives for enhancement substitution, individuals will not be highly attracted to these newer social structures because in such a change, they would lose many rewards available from the older historical enhancements. Only as alternatives becomes more and more widespread do other enhancement packages come to erode more and more. For instance, in terms of these dynamics, the long-term inequality trend would be a gradual replacement beginning with a little enhanced production and a lot of ascriptive inequality. As mass consumption spreads, elevated ascriptive distinctions can diminish. These divisions do not necessarily disappear, but they erode in efficacy. But why give up elevated ascriptive differences, and by extension elevated rewards, the very divisions the world had feasted on for so long, unless something else was being offered in that change? Why accept moderate inequality, and by extension moderate rewards, without such an exchange? Given our ancient reward release rules for various interests, successful long-term resistance to such classic inequality will only occur as new alternative enhancements are made available to more and more individuals. This produces an inequality split in a population embracing the new enhancements. A few continue to maximize inflated inequality in economic and political power positions, and many others substitute alternative enhancements and embrace a more moderate inequality package.

If the enhancement model is correct, the key issue for many individuals is not about any specific enhancement, but rather about enhancement bundling or substitution. This means that there could be a tipping point in regard to the waxing and waning of inequality. Initially, in enhancement substitution, there must be both new technologies and shrinking inequality in order to entice many to come over to the new consumption enhancement. But at some point, if elevated consumption levels can be maintained by other means, it is possible for some kinds of inequality to grow again. For instance, if others are gaining more in economic wealth, such a relative loss is not necessarily a key issue for those individuals where reward bonuses have shifted more to consumption itself, as long as they still have sufficient economic resources to fuel their consumption. This is the tipping point at which populations can split even more in terms of inequality as older alliances in anti-inequality politics between middle and lower income groups begin to unravel. The greater the increases in productivity, by technology or any other means, the sooner this tipping point comes. Similarly, the more that credit is available to extend consumption, the sooner this tipping point comes, but only as long as the credit extension can be maintained.

Of course, all technologies are not enhancements. For instance, if new food production technologies lead to a population increase, then the net result for any individual can be the same food provisioning per person. If the population surge is even greater than the technological change, the net result could be less per person in a Malthusian scenario. Or, new technologies might lead to increased production per person, but most of this increase could be directed to a small elite, and have little impact on the population as a whole. Furthermore, just having a few technological changes is not going to provide the mass enhancement packages that could challenge other classic reward enhancements. A little additional production is not going to go far in challenging such high contrast creations. As a result, the innovations required to exploit this third reward package from our evolutionary past are staggering in their breadth. While some costs might be reduced, such as personal costs rooted in elevated ascriptive inequality, the social creation of a sea of more distinct and novel stimuli for multiple interests on a mass scale over the long term is in other ways enormously costly in total, both to individuals and groups, locally and globally, as well as to the earth itself. No wonder this enhancement option only appeared late in our history.

Individuals using extreme consumption diversity and serial novelty on a mass scale to trigger bonus rewards are no more a representation of "true" human nature than individuals using elevated ascriptive inequality for such rewards. Because of their alternative cost packages, the first is preferable in some ways, but not because it is somehow closer to what being human really is. Both are based on opportunities to exploit ancient reward release patterns forged by natural selection in conditions very different than our context of origin. Both therefore represent one type of social evolution emerging from the mixture of ancient rules and new contexts. Given this mixture, both have instability built into them. For instance, since there is nothing inherently natural about elevated inequality on a mass scale, there will always be opposition and conflict surrounding any specific manifestation of this bonus structure, even though such efforts will not be successful in the long run unless a new alternative enhancement package can be offered. Similarly, since there is nothing natural about serial novelty on a mass scale, there will always be dissatisfaction for some based on the erosion of any short-term success in creating such novelty, and dissatisfaction for others based on the desperate hope that there is some long-term alternative to such self-eroding dynamics. Furthermore, just as the demands for mass elevated inequality constrain the structural options of societies pursuing that enhancement path, the demands of mass diversity and novelty limit structural options. Once committed to such alternative bonus packages, there are not that many structural alternatives for regularly creating these packages on a long-term basis. That is one reason behind the structural convergence on such cultures, just as there was convergence in the history of mass inequality structures.

Conclusion

This article focuses on three preconscious reward release patterns outlined by neuroscientific studies of human and other primates. The attachment pattern provides many rewards for highly repetitive interaction with a small variety of individuals and less repetitive interaction with a larger number of individuals. The status pattern offers many rewards for a great variety of stimuli, even if these arousers are on average quite similar. The consumption pattern provides additional rewards for a variety of stimuli, but only if they are more diverse or even novel on average. Each pattern of reward release has contributed to the emergence of social structures that can provide additional reward packages on a mass basis. There is a sequence in the emergence of these reward enhancing structures because of the different costs in these social creations. This is one reason for the historical sequence of religion, elevated ascriptive inequality, and mass production in the social evolution of enhancements.

References

Addessi, E., Mancini, A., Crescimbene, L., Ariely, D., & Visalberghi, E. (2010). How to spend a token? Trade-offs between food variety and food preferences in tufted capuchin monkeys(Cebus apella). *Behavioral Processes, 83*(3), 267–275.

Aron, A., Fisher, H. E., Mashek, D. J., Strong, G., Li, H.-F., & Brown, L. L. (2005). Reward, motivation and emotion systems associated with early-stage intense romantic love. *Journal of Neurophysiology, 94*, 327–337.

Bartels, A., & Zeki, S. (2004). The neural correlates of maternal and romantic love. *NeuroImage, 21*, 1155–1166.

Beauregard, M., & Paquette, V. (2006). Neural correlates of a mystical experience in Carmelite nuns. *Neuroscience Letters, 405*, 186–190.

Beauregard, M., Paquette, V., Courtemanche, J., Paquette, V. S., & St-Pierre, E. L. (2009). The neural basis of unconditional love. *Psychiatry Research, 172*, 93–98.

Berns, G. S., McClure, S. M., Pagnoni, G., & Montague, P. R. (2001). Predictability modulates human brain response to reward. *Journal of Neuroscience, 21*(8), 2793–2798.

Berridge, K., Ho, C. Y., Richard, J. M., & Feliceantonio, A. G. (2010). The tempted brain eats: Pleasure and desire circuits in obesity and eating disorders. *Brain Research, 1350*, 43–64.

Boehm, C. (1999). *Hierarchy in the forest: The evolution of egalitarian behavior*. Cambridge, MA: Harvard University Press.
Booth, A., Granger, D., Mazur, A., & Kivlighan, K. (2006). Testosterone and social behavior. *Social Forces, 85*(1), 167–191.
Brammer, G., Raleigh, M., & McGuire, M. (1994). Neurotransmitters and social status. In L. Ellis (Ed.), *Social stratification and socioeconomic inequality* (pp. 75–91). Westport: Praeger.
Bunzeck, N., & Duzel, E. (2006). Absolute coding of stimulus novelty in the human substantia nigra/VTA. *Neuron, 51*(3), 369–379.
Carrasco, M. (2006). Covert attention increases contrast sensitivity: Psychophysical, neurophysiological, and neuroimaging studies. In S. Martinez-Conde, S. L. Macknik, L. M. Martinez, J. M. Alonso, & P. U. Tse (Eds.), *Progress in brain research* (pp. 33–70). New York: Elsevier.
Cassidy, J., & Shaver, P. (Eds.). (2008). *Handbook of attachment*. New York: Guilford Press.
Chen, M. K., Lakshminaryanan, V., & Santos, L. (2006). How basic are behavioral biases? Evidence from capuchin monkey trading behavior. *Journal of Political Economy, 114*(3), 517–537.
Chiao, J., Harada, T., Oby, E., Li, Z., Parrish, T., & Bridge, D. (2009). Neural representations of social status hierarchy in human inferior parietal cortex. *Neuropsychologia, 47*, 354–363.
deHaan, M., & Gunnar, M. (2009). *Handbook of developmental social neuroscience*. New York: Guilford Press.
Donaldson, Z. R., & Young, L. J. (2008). Oxytocin, vasopressin, and the neurogenetics of sociality. *Science, 332*(5903), 900–904.
Dunbar, R. (2010). *How many friends does one need? Dunbar's number and other evolutionary quirks*. London: Faber.
Edwards, D. (2006). Competition and testosterone. *Hormones and Behavior, 50*, 681–683.
Elliott, R., Friston, K. J., & Dolan, R. J. (2000). Dissociable neural responses in human reward systems. *Journal of Neuroscience, 20*(16), 6159–6165.
Ellison, P. T., & Gray, P. B. (Eds.). (2009). *Endocrinology of social relationships*. Cambridge, MA: Harvard University Press.
Fehr, E., & Camerer, C. F. (2007). Social neuroeconomics: The neural circuitry of social preferences. *Trends in Cognitive Sciences, 11*(10), 419–427.
Fisher, H., Brown, L. L., Aron, A., Strong, G., Mashek, D., Aron, A., Mashek, D., Li, H., & Brown, L. L. (2002). Defining the brain systems of lust, romantic attraction, and attachment. *Archives of Sexual Behavior, 31*(5), 413–419.
Fisher, H., Brown, L. L., Aron, A., Strong, G., Mashek, D., Aron, A., & Brown, L. L. (2006). Romantic love: A mammalian brain system for mate choice. *Philosophical Transactions of the Royal Society of London Biological Sciences, 361*(1476), 2173–2186.
Fisher, H., Brown, L. L., Aron, A., Strong, G., & Mashek, D. (2010). Reward, addiction, and emotion regulation systems associated with rejection in love. *Journal of Neurophysiology, 104*, 51–60.
Glimcher, P., Camerer, C., Fehr, E., & Poldrack, R. (Eds.). (2009). *Neuroeconomics: Decision making and the brain*. London: Elsevier.
Hammond, M. (2003). The enhancement imperative: The evolutionary neurophysiology of Durkheimian solidarity. *Sociological Theory, 21*(4), 359–374.
Hammond, M. (2006). Evolutionary theory and emotions. In J. Stets & J. Turner (Eds.), *Handbook of the sociology of emotions* (pp. 368–385). New York: Springer.
Hammond, M. (2008). *Reversal of fortune: How evolutionary adaptations for limiting inequality become fuel for inflated inequality*. Annual Meeting of the American Sociological Association, Boston.
Hammond, M. (2010). *The inequality explosion, the great substitution, and social evolution*. Annual Meeting of the American Sociological Association, Atlanta.
Henrich, J. (2009). The evolution of innovation-enhancing institutions. In M. J. O'Brien & S. Shennan (Eds.), *Innovations in cultural systems: Contributions from evolutionary anthropology* (pp. 99–120). Cambridge: MIT Press.
Hillier, D., & Miller, J. (1991). Auditory contrast sensitivity in normal and hearing impaired listeners. *Journal of the Acoustical Society of America, 89*(4B), 1938.
Izuma, K., Saito, D., & Sadato, N. (2008). Processing of social and monetary rewards in the human striatum. *Neuron, 58*(2), 284–294.
Johnson, A. W., & Earle, T. (2000). *The evolution of human societies*. Stanford: Stanford University Press.
Kang, M. J., Hsu, M., Krajbich, I., Loewenstein, G., McClure, S., Wang, J., & Camerer, C. (2009). The wick in the candle of learning: Epistemic curiosity activates reward circuitry and enhances memory. *Psychological Science, 20*(8), 963–973.
Keeley, L. (1996). *War before civilization*. New York: Oxford University Press.
Kirkpatrick, L., & Granquist, P. (2008). Attachment and religious representations of behavior. In J. Cassidy & P. Shaver (Eds.), *Handbook of attachment* (pp. 906–933). New York: Guilford Press.
Kringelbach, M., & Berridge, K. (Eds.). (2010). *Pleasures of the brain*. Oxford: Oxford University Press.
LeDoux, J. (1996). *The emotional brain*. New York: Simon and Shuster.

Lisman, J., & Grace, A. A. (2005). The hippocampal-VTA loop. *Neuron, 46*(5), 703–713.
Loewenstein, G. (2007). The psychology of curiosity. In G. Loewenstein (Ed.), *Exotic preferences: Behavioral economics and human motivation* (pp. 121–180). New York: Oxford University Press.
Ly, M., Haynes, M., Barter, J., Weinberger, D., & Zink, C. (2011). Subjective socioeconomic status predicts human ventral striatal responses to social status information. *Current Biology, 21*(9), 794–797.
Madsen, D. (1994). Serotonin and social rank among human males. In R. Masters & M. McGuire (Eds.), *The neurotransmitter revolution: Serotonin, social behavior, and the law* (pp. 146–158). Carbondale: Southern Illinois University Press.
Marazziti, D. (2009). Neurobiology and hormonal aspects of romantic attachments. In M. de Haan & M. R. Gunnar (Eds.), *Handbook of developmental social neuroscience* (pp. 265–280). New York: Guilford Press.
Martinez, D., Orlowska, D., Narendran, R., Slifstein, M., Liu, F., Kumar, D., Broft, A., Van Heertum, R., & Kleber, H. D. (2010). Dopamine type 2/3 receptor availability in the striatum and social status in human volunteers. *Biological Psychiatry, 67*(3), 275–278.
Masters, R., & McGuire, M. (1994). *The neurotransmitter revolution: Serotonin, social behavior, and the law*. Carbondale: Southern Illinois University Press.
Mazur, A. (2005). *Biosociology of dominance and deference*. Lanham: Rowman and Littlefield.
Mazur, A., & Booth, A. (1998). Testosterone and dominance in men. *The Behavioral and Brain Sciences, 21*(3), 353–363.
McNamara, P. (2009). *The neuroscience of religious experience*. Cambridge: Cambridge University Press.
Montague, R., & Berns, G. (2002). Neural economics and the biological substrates of valuation. *Neuron, 36*(2), 265–284.
Montague, R., Berns, G., King-Casas, B., & Cohen, J. (2006). Imaging valuation models in human choice. *Annual Review of Neuroscience, 29*, 417–448.
Newberg, A., Alavi, A., Baime, M., Pourdehnad, M., Santanna, J., & d'Aquili, E. (2001). The measurement of regional cerebral blood flow during the complex cognitive task of meditation: A preliminary SPECT study. *Psychiatry Research: Neuroimaging, 106*, 113–122.
Newberg, A., Pourdehnad, M., & d'Aquili, D. (2003). Cerebral blood flow during meditative prayer. *Perceptual and Motor Skills, 97*, 625–630.
O'Brien, M. J., & Shennan, S. (2010). *Innovations in cultural systems: Contributions from evolutionary anthropology*. Cambridge: MIT Press.
Ortigue, S., Bianchi-Demicheli, F., Hamilton, A., & Grafton, S. (2007). The neural basis of love as a subliminal prime: An event-related fMRI study. *Journal of Cognitive Neuroscience, 19*, 1218–1230.
Ortigue, S., Bianchi-Demicheli, F., Hamilton, A., Grafton, S., Bianchi-Demicheli, F., Patel, N., Frum, C., & Lewis, J. (2010). Neuroimaging of love: FMRI meta-analysis evidence toward new perspectives in sexual medicine. *The Journal of Sexual Medicine, 7*, 3541–3552.
Powell, A., Shennan, S. J., & Thomas, M. G. (2010). Demography and variation in the accumulation of culturally inherited skills. In M. J. O'Brien & S. Shennan (Eds.), *Innovations in cultural systems* (pp. 137–160). Cambridge, MA: MIT Press.
Richerson, P. J., & Boyd, R. (2005). *Not by genes alone: How culture transformed human evolution*. Chicago: University of Chicago Press.
Rolls, E., Murzi, E., Yaxley, S., Thrope, S., & Simpson, S. (1986). Sensory specific satiety: Food-specific reduction in responsiveness of ventral forebrain neurons after feeding in the monkey. *Brain Research, 368*, 79–86.
Rolls, E., Murzi, E., Yaxley, S., Thrope, S., & Simpson, S. (1999). *The brain and emotion*. Oxford: Oxford University Press.
Sanchez, R., Parkin, J. C., Chen, J. Y., & Gray, P. G. (2009). Oxytocin, vasopressin, and human social behavior. In P. Ellison & P. Gray (Eds.), *Endocrinology of social relationships* (pp. 319–339). Cambridge, MA: Harvard University Press.
Santos, L., & Keith Chen, M. (2009). The evolution of rational and irrational economic behavior: Evidence and insight from a non-human primate species. In P. Glimcher, C. Camerer, E. Fehr, & R. Poldrack (Eds.), *Neuroeconomics: Decision making and the brain* (pp. 81–94). London: Elsevier.
Saxe, R., & Haushofer, J. (2008). For love or money: A common neural currency for social and monetary reward. *Neuron, 58*(2), 164–165.
Schjodt, U., Stodkilde-Jorgenson, H., Geertz, A., & Roepstorff, A. (2008). Rewarding prayers. *Neuroscience Letters, 443*, 165–168.
Smith, K. S., Mahler, S. V., Pecina, S., & Berridge, K. (2010). Hedonic hot spots: Generating sensory pleasure in the brain. In M. Kringelbach & K. Berridge (Eds.), *Pleasures of the brain* (pp. 27–49). Oxford: Oxford University Press.
Tobler, P., Fletcher, P., Bullmore, E., & Schultz, W. (2007). Learning-related human brain activations reflecting individual finances. *Neuron, 54*(1), 167–175.
Young, L., & Wang, Z. (2004). The neurobiology of pair bonding. *Nature Neuroscience, 7*(10), 1048–1054.
Zink, C., Tong, Y., Chen, Q., Bassett, D., Stein, J., & Mayer-Lindenberg, A. (2008). Know your place: Neural processing of social hierarchy in humans. *Neuron, 58*(2), 273–283.

Part IV
The Neurology of Social Issues and Problems

Chapter 21
Persistent Inequality: A Neurosociological Perspective

Jeff Davis

Introduction

In light of the significant advances in social neuroscience, it is past the time for sociologists to consider the brain as an important sociological phenomenon. A sizeable body of social neuroscience research illuminates the sensitivity of neurobiological processes to the social environment (Cacioppo and Patrick 2008; Insel and Fernald 2004; McEwen and Gianaros 2010). Comparative analyses of human and non-human species suggest nervous systems evolved under strong social selection (Robinson et al. 2008). Complex sociological dynamics such as rituals (Graybiel 2008), location in an organizational hierarchy (Creel and Sands 2003), opportunities to increase relative social standing (Burmeister et al. 2005), social exclusion and ostracism (Eisenberger et al. 2003; Masten et al. 2009; Slavich et al. 2010), the learning and transmission of cultural information (Hakeem et al. 2009), morality and distributive justice (Fehr and Camerer 2007; Greene and Paxton 2009; Koenigs et al. 2007) have left large imprints on the evolutionary path of the human brain (Kenrick ref).

This chapter advances a neurosociological model of a highly complex sociological process – persistent inequality. Numerous studies of persistent inequality have been compiled. As noted by many ethnographers, individuals and families who tend to get trapped in lower socioeconomic status exhibit preferences and behaviors which value short-term outcomes. Short-sightedness inexorably leads to behaviors, such as high sexual activity and aggression, which lock an individual and their offspring into lower socioeconomic status (Schwalbe et al. 2000). The central thesis of the model states that persistent lower status is ultimately a result of a high degree of environmental uncertainty induced by subjugated social status. Almost by definition, members of a subjugated social status experience environmental conditions which preclude the exercise of control. In many ways, the environment is comprised of harsh, unpredictable stressors such as financial insecurity, episodic violence, instability in social relationships, and risk of acute stress-related illnesses. Perhaps most unpredictable is one's chances of upward social and economic mobility. As I argue, environmental uncertainty has significant consequences for three areas of the brain which are directly involved in the behaviors observed by ethnographers – the prefrontal cortex (PFC), the anterior cingulate cortex (ACC), and the vertebrate social behavior network (VSBN). Activities in these three areas are

This project was supported by a grant from the U.S. National Institutes of Health, National Center on Minority Health and Health Disparities, Research Infrastructure at Minority Institutions program (Grant Number: 1 P20 MD003942-01).

J. Davis(✉)
Department of sociology, California State University, Long Beach, CA 90842, USA
e-mail: jeffrey.davis@csulb.edu

D.D. Franks and J.H. Turner (eds.), *Handbook of Neurosociology*, Handbooks of Sociology and Social Research, DOI 10.1007/978-94-007-4473-8_21, © Springer Science+Business Media B.V. 2013

delicately balanced in order to regulate decision-making and behavior. Exposure to environmental uncertainty disrupts the regulatory process resulting in a strengthening of short-term preferences and increases in reproductive and aggressive behaviors.

Mechanisms of Persistent Inequality

Figure 21.1 shows a basic model of inequality developed by Tilly (1998) and elaborated throughout this chapter. Inequality is organized around social categories which are used to sort individuals and to make decisions about the allocation of material and cultural resources. Categorical inequality refers to the unequal allocation of resources across social categories, primarily as the result of systemic discrimination and social closure within local social and economic organizations. Tilly (1998) identified four major mechanisms which embed categorical inequality into the fabric of a society: exploitation, opportunity hoarding, emulation, and adaptation. Figure 21.1 shows only the unique contributions of each mechanism to the persistence of categorical inequality, but there exists a dynamic interdependency among them as well.

Exploitation is "...where powerful, connected people command resources from which they draw significantly increased returns by coordinating efforts of outsiders whom they exclude from the full value added of that effort" (Tilly 1998, p. 74). Opportunity hoarding is the control and confinement of a valuable resource (or a value producing resource) to members of a category. Both mechanisms originate in the local cultural and institutional environment. Cultural norms about relative social standing and the formal rules sustaining local social hierarchies shape how individuals view others. These views are often carried into social and economic organizations and influence decision-making processes within them.

Consequently, categorical inequality becomes manifest in a variety of organizational processes such as discrimination in hiring and promotion. The highest degrees of categorical inequality are found where the internal structures of social and economic organizations are most reliant upon the categorization schemes endorsed by local culture and institutions.

Exploitation and opportunity hoarding are the primary, proximate causes of categorical inequality. However, the intensity of their effects on the installation of categorical inequality is a result of the two mechanisms of emulation and adaptation. According to Tilly (1998), emulation generalizes categorical inequality across individuals and organizations within a society. Menchaca (1995) provides an excellent example of emulation in her study of the development of Mexican-Anglo relations in southern California since the Mexican-American War. As Anglos expropriated land from Mexican families during the war period, many turned to wealthiest of Anglo landowners for advice on the most effective methods for controlling Mexican laborers. Consequently, the exploitative forms of labor control used by large landowners were readily copied by small landowners (Menchaca 1995). The spread of exploitation practices was one of the major factors in emergence and persistence of what Menchaca (1995) refers to as "social apartness"– Mexican-Anglo segregation in southern California.

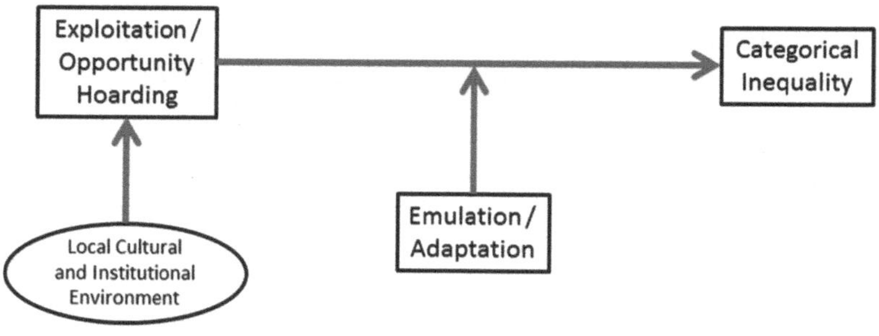

Fig. 21.1 Tilly's (1998) basic model of durable inequality

Whereas emulation causes a system of categorical inequality to spread, adaptation locks categorical inequality into the relationships between groups. It is broadly defined as the "...invention of procedures which ease day-to-day interaction" (Tilly 1998, 2005). However, these solutions have some undesirable, unintended consequences for the persistence of inequality. Consider some classic ethnographic studies of youth growing up in conditions of poverty. Sociologists frequently observe among these youth a pessimistic view of their chances of mainstream success (Anderson 1978, 1999; MacLeod 1995; Whyte 1943; Willis 1977). Their pessimism derived from a realistic assessment of a long family history of struggles and the poor quality of the local social and economic environment. One group of youth in MacLeod's (1995) study, for example, concluded that striving for upward mobility was a futile effort because several family members had previously failed in their efforts. Moreover, good jobs were simply scarce. In response to objectively poor environments, young adults engage in "dropping out behavior" (Schwalbe et al. 2000), withdrawing from activities prescribed by normative society such as school and the mainstream labor market. From the individual's perspective, dropping out might lead to a number of positive outcomes such as closer social ties with peers, higher self-esteem, and a general perception that mainstream life is relatively unimportant for well-being (Schwalbe et al. 2000). Unfortunately, these are most often short-term benefits with the costs of steep, negative trade-offs for future well-being (Day et al. 1999).

A Neurosociological Model of Persistent Inequality

A neurosociological perspective on Tilly's model is not new. Massey (2007) was probably the first to do so. He argued that while categorical inequality is fundamentally the result of an interaction between institutional environments and market conditions, categorization is generally an expression of the universal tendency of humans to use heuristics as a means of understanding environments (Massey 2007). The capacity for categorization evolved in response to energetic constraints (Massey 2007), which have actually shaped numerous aspects of brain functioning (Niven and Laughlin 2008). Categorization of the social and natural world is a method of conserving energy for cognitive processing of stimuli (Massey 2007). In some important ways, the use of heuristics can yield an accurate view of the social and physical environment and greatly aid decision-making (Gigerenzer and Brighton 2009). However, it can also easily lead to maladaptive outcomes – such as prejudice. Whether it actually leads to prejudice is mostly a function of conditions of the institutional environment of the individual (Massey 2007).

Whereas Massey (2007) addressed the most central concept in Tilly's model, my focus is on the two mechanisms of emulation and adaptation. Figure 21.2 illustrates the model. The main consequence of categorical inequality is the experience of greater environmental uncertainty in the lives of subjugated groups. For these groups, uncertainty arises in everyday decisions and social interactions. The poor quality of local job markets creates unpredictable changes in employment opportunities, instability and distrust among coworkers, and vulnerability to abusive employers (Bourdieu and Accardo 1999). In turn, economic vulnerability creates distrust and uncertainty in social interaction (Bourdieu and Accardo 1999). Simple social exchanges within such environments have sensitive tipping points, turning either friendly or hostile in a somewhat chaotic manner. For example, whether an interaction turns friendly or hostile often depends upon the individuals' ability to correctly interpret highly nuanced signals (e.g., a nod of the head, a glance in one's direction) (Anderson 1990; Gambetta and Hamill 2005; O'Donnell and Sharpe 2000). The wrong interpretation could result in disastrous consequences.

Efforts to overcome the social vulnerability and unpredictability associated with lower social status often involve high-risk behaviors. For young men, both issues are typically resolved by expending the only viable resource they have – physical talent – in order to build a local reputation. Unfortunately, this involves frequent challenges from other young men, often in the form of violent confrontations (Anderson 1999; Barker 2005; Bourgois 2002; Carroll et al. 2003; Emler and Reicher 1995;

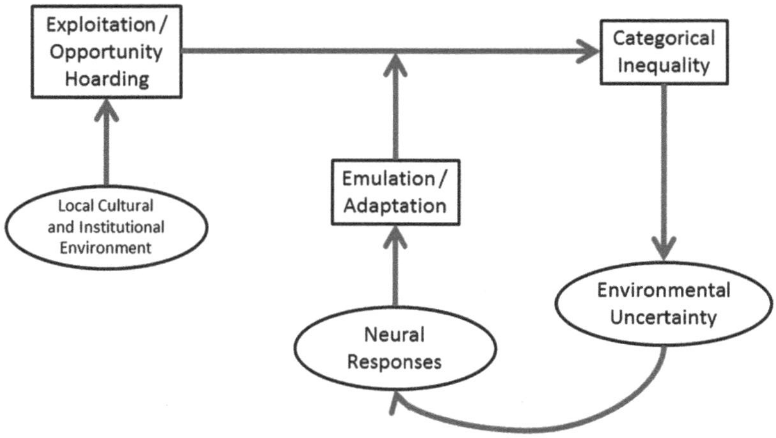

Fig. 21.2 A Neurosociological model of persistent inequality. The primary effect of inequality is to increase an individual's uncertainty about the current and future socioeconomic environments. In turn, uncertainty triggers neural processes associated with short-term decision making behaviors

O'Donnell and Sharpe 2000). For young women, child birth is sometimes perceived as a means of bringing stability and purpose to an otherwise chaotic existence (Edin and Kefelas 2005). Social support from peers can increase as well as social standing (Anderson 1999). However, young unwed mothers risk losing steady support from family members. It seems more likely that the mother becomes stigmatized for having a child out-of-wedlock (Edin and Lein 1997). For both men and women, vulnerability and uncertainty in social interactions can be reduced copying the behaviors of others who seem least vulnerable in the environment. One can copy behaviors prescribed by local codes of conduct and imitate vernacular in order to "fit in" and find a degree of comfortable acceptance from others (Anderson 1999). However, copying the behaviors of those who are perceived to be least vulnerable might also mean developing a closer relationship with peers who routinely engage in high-risk behaviors. Stronger ties with such peers will ensure low social mobility over the long run (Whyte 1943; Willis 1977).

Neuroscientists have observed that life in an unpredictable environment impacts neural functioning much like any other significant form of stress (Arnsten 2009; Munhoz et al. 2006; Rosen and Donley 2006; Wingfield and Kitaysky 2002)(McEwen also). There is also evidence suggesting that living in a poor quality environment can trigger neural activity associated with stronger preferences for short-term rewards, attention to the immediate environment, neglect of long-term costs of actions, and higher levels of aggressive behaviors (Manuck et al. 2003, 2005). Behavioral studies have linked environmental uncertainty to a stronger tendency toward imitation and social learning (Ball-Rokeach 1973; Coussi-Korbel and Fragaszy 1995; LaLand 2004; Salancik and Pfeffer 1978). Exposure to unpredictable environments, especially during childhood, has also been linked with a higher risk of teenage parenthood and higher fertility over the life span (Belsky et al. 2010; Brumbach et al. 2009; Chisholm 1999; Ellis et al. 2009), but many questions remain about the underlying neurobiological mechanisms (Maeda and Tsukamura 2006; Matsuwaki et al. 2003).

The model in Fig. 21.2 proposes that environmental uncertainty initiates what neuroscientists refer to as a "bottom-up" behavioral control process (Arnsten 2009; Yu and Dayan 2002). In a non-stress situation, behaviors are primarily regulated by a "top-down" process where higher-level cognitive control functions inhibit behavioral impulses from the affective centers of the brain (e.g., the limbic system). However, exposure to stressors such as uncertainty tends to increase activity in the limbic system and simultaneously impairs higher-level cognitive control (Arnsten 2009; Yu and Dayan 2005).

The Predictive Brain

To understand the stressful impact of uncertainty it is first necessary to understand the evolutionary value of predictions about the future. Current theory holds that prediction (or predictive learning) evolved because it conferred distinct speed advantages in the location of vital energy resources and predator avoidance (Shettleworth 1998; Watson 2008). Prediction is an act of associative learning where two or more events are temporally and/or spatially distant (Shettleworth 1998). It requires a means of transduction of environmental stimuli and a system of memory. Evolution has apparently found diverse ways to develop these traits. While vertebrates depend upon various elements of cognition to make predictions, simple organisms have evolved non-cognitive means of predictive learning. In experimental studies, researchers were able to select for associative learning traits in a population of Escherichia coli (E. coli). Subsequent generations evolved the capacity to accurately predict the location of food resources (Tagkopoulos et al. 2008). Researchers have also found that predictive capacity develops early in some vertebrate species. Species of frogs, for example, are to detect environmental signals associated with future risk of predation while still in the embryonic stages of development (Warkentin and Caldwell 2009). These studies and others demonstrate that organisms with the ability to accurately predict the temporal distribution of energy resources and predation risk have a clear advantage in survival.

Predictions about the future are also used in reproductive decisions. In various species, the timing of sexual maturation and subsequent reproduction are highly sensitive to an organism's perceptions about future environmental conditions (Davis and Werre 2008). Here also, the organism relies upon environmental cues to make predictions about the future and acts accordingly. If the organism predicts higher quality environmental conditions in the future, reproduction is delayed until that time when conditions improve. If the future is perceived to be in a worse condition than the present, then sexual development is accelerated and reproductive effort is generally increased (Wasser and Place 2001).

Empirical evidence for this pattern of behavior abounds for non-human species. For example, there is a large literature documenting some rather sophisticated methods of reproductive control, depending upon current and likely future states of the environment. One method is diapause, a neuro-physiological mechanism which allows an organism to suspend fertilization or embryonic growth until environmental conditions improve (Renfree and Shaw 2000). Evidence from studies of human reproductive decision-making and behavior is growing and the patterns resemble those found among other species. Collectively, these studies suggest that predictions about the future influence human reproductive biology. Wasser and Place (2001), for example, argued that the close association between the quality of local social environments and reproductive outcomes implies a process they refer to as reproductive filtering. The reproductive filtering hypothesis states that reproduction will be suppressed in a stressful environment if the individual perceives improvements in future conditions (Wasser and Place 2001). Their analyses of women undergoing fertility treatment shows that treatment of women in stressful social conditions (e.g., poor quality of social support, hostility in social relationships) was less successful for women who believed that conditions would improve in the future (Wasser and Place 2001). Longitudinal studies have also found higher levels of sexual and reproductive activity among women who perceive no improvement or worse environmental conditions in the future (Belsky et al. 1991; Brumbach et al. 2009; Chisholm et al. 2005; Ellis et al. 2009).

Lastly, prediction plays a central role in facilitating an organism's efforts to modify the environment – an important activity which can significantly enhance long-term survivability and reproductive success (Odling-Smee et al. 2003). Activities related to environmental modification are referred to as niche construction (Odling-Smee et al. 2003). The ultimate goal of niche construction is to exercise control over fluctuations in the local environment. Specifically, niche construction stabilizes the flow of local energy resources necessary for survival and reproduction. It can effectively dampen stochastic fluctuations in resource levels occurring over time. Predictive learning allows an organism to detect and evaluate environmental cues which indicate the how the quality of the local environment might

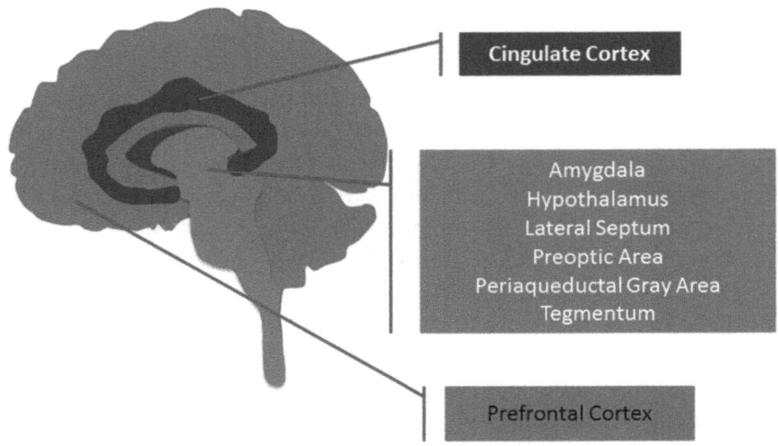

Fig. 21.3 Neuroanatomy of the prefrontal cortex, cingulate cortex, and structures of the vertebrate social behavior network

change over time. In a very real sense, predictive learning helps the organism efficiently allocate efforts such that it avoids wasting precious time and energy on modifying an environment that will yield low returns in the future (Odling-Smee et al. 2003).

Observations from behavioral studies lead to the conclusion that prediction is a major mechanism in the regulation of behavior and the accuracy of a prediction can have significant consequences for the fitness of the organism and its offspring. A behavior with such adaptive importance is bound to have many biological traces of strong selection. According to a large body of neuroscience research, the human brain is aptly equipped for predictive learning (Bubic et al. 2010) and is constantly making predictions to construct representations of the future (Bar 2009). Much of this activity takes place within the prefrontal cortex (Bar 2009; Bechara et al. 1999), although other structures also play crucial roles such as the anterior cingulate cortex (Allman et al. 2001; Walton et al. 2007), the hippocampus (Eichenbaum and Fortin 2009) and the cerebellum (O'Reilly et al. 2008; Szpunar et al. 2007). Working together, various components keep the individual in a continuous state of anticipation and readiness to adapt to environmental challenges (Bar 2009). For my purposes, I focus on the interactions between three areas: the dorsomedial prefrontal cortex (dmPFC), the anterior cingulate cortex (ACC), and the vertebrate social behavior network (VSBN). Figure 21.3 shows the locations and some of the sub-components of each structure.

The prefrontal cortex (PFC) is that part of the brain comprising the anterior area of the frontal cortex. The PFC enables working memory – the process of storing information temporarily, retrieving information from long-term memory, manipulating it and then selectively discarding information deemed irrelevant to the task at hand (Bledowski et al. 2009). The PFC is also responsible for the selection of appropriate social behaviors (Arnsten 2009; Kouneiher et al. 2009) and general decision-making (Fellows and Farah 2007; Osaka et al. 2007; Rougier et al. 2005; Wallis 2007). Substrates of the PFC have distinct functions for decision-making and behavior. Many of these functions have been uncovered by studies of behavior among individuals who have suffered damage to specific parts of the brain. Of all substrates, the dorsomedial PFC (dmPFC) seems to play the most significant role in forming predictions, assessing prediction errors, and resolving conflicts in information (Modirrousta and Fellows 2008; Rushworth et al. 2007). The dmPFC reacts particularly strong to environmental cues which have reliably signaled a future reward in past experiences (Ishikawa et al. 2008).

The anterior cingulate cortex (ACC) has numerous roles in cognition, behavior, and physiological functioning (Allman et al. 2001; Devinsky et al. 1995). Neuropsychological research on phenomena such as cognitive dissonance reveals that the ACC plays a crucial role in the detection and monitoring of cognitive conflict (Carter and van Veen 2007; van Veen and Carter 2006; van Veen et al. 2001). Consequently, researchers have also found that the dorsal ACC (dACC) is also directly involved in attitude changes intended to rapidly reduce cognitive conflict (van Veen et al. 2009). There seems to be broad agreement that ACC seems to play an important role in adaptive decision-making – that is, decision-making under uncertainty (Allman et al. 2001; Kennerley et al. 2006). However, the exact nature of the role is unclear. Recent computational models have proposed that the dACC is involved in the assessment of prediction error. Specifically, it facilitates the estimation of the likely degree of error in predictions (Brown and Braver 2007). However, other research casts doubt on this role. Kennerley et al. (2006) suggest the actual role of the dACC is to weigh the opportunity costs of a future action against the history of previous rewards for the action and alternative actions.

The vertebrate social behavior network (VSBN) is composed of several structures distributed across two broad areas – the basal forebrain and the midbrain (Goodson 2005; Newman 1999). The specific areas are as follows: (within the basal forebrain) extended medial amygdala; preoptic area; anterior hypothalamus; ventromedial hypothalamus; lateral septum; (within the midbrain) periaqueductal gray area; and tegmentum (Goodson 2005). Structures within the basal forebrain have long been associated with socio-emotional experiences (Morgane et al. 2005). In mammals these structures have been associated with a range of behaviors related to reproduction and aggression. For example, the lateral septum is a central structure in the regulation of aggressive behavior. Depending upon the level of certain hormones, the lateral septum directs aggressive behavior toward others who pose a danger to self or offspring (Lee and Gammie 2009; Singewald et al. 2011). Recent work is illuminating the role of the VSBN in prediction and decision-making. For example, the midbrain is the origin of dopaminergic neurons (Iversen 2010). Activity in these neurons increases significantly in response to the expectation of a reward (Schultz et al. 2002). Some research has also found increased activity in response to uncertainty about the probability of receiving a reward (Fiorillo et al. 2003).

The PFC, ACC, and VSBN are interconnected through numerous projections of neural networks (Salzman and Fusi 2010). The projections are complex in terms of number of neurons involved, the size of neural networks, the diversity of hormones transmitted, the role of particular dendritic receptors, and the actions of secondary chemical messengers such as cyclic adenosine monophosphate (cAMP). A full review of projections is well-beyond the scope of this chapter and the summary offered here is narrowly focused on the neuroscience of decision-making. These limitations notwithstanding, three general conclusions can be drawn from the literature. First, dopamine is one of the major neurotransmitters involved in prediction and decision-making (Daw et al. 2006; Morris et al. 2006; Schultz et al. 2002). Second, the ACC acts as a mediator between the PFC and VSBN structures (Etkin et al. 2006; Miller and Cohen 2001). It has both cognitive and affective elements to assist communication between the PFC and VSBN structures (Devinsky et al. 1995).

Third, it is not the prediction per se that is source of behavioral change, rather, it is prediction error (Bubic et al. 2010; Schultz and Dickinson 2000). Prediction errors occur when the actual outcome of a process is either worse than the predicted outcome (a negative prediction error) or better than the predicted outcome (a positive prediction error)(Dayan and Abbott 2001; Montague et al. 1996; Niv et al. 2005). A prediction error is encoded in the brain via changes in the levels of dopamine in the reward centers of the brain (e.g., the midbrain) and in substrates of the frontal cortex (e.g., the orbitofrontal cortex, ventromedial prefrontal cortex) (Bayer and Glimcher 2005). When an individual experiences a negative prediction error, dopamine levels drop below normal baseline levels in the midbrain and prefrontal cortex. This will ultimately lead to the discontinuation (or extinction) of a behavioral pattern (Schultz and Dickinson 2000). A positive prediction error results in an increase in dopamine levels above baseline and has the effect of reinforcing a behavioral pattern (Schultz and Dickinson 2000). It should also be noted that a spike in dopamine above baseline levels enhances attention to the immediate environment, almost to the complete neglect of distal environments (Hills 2006).

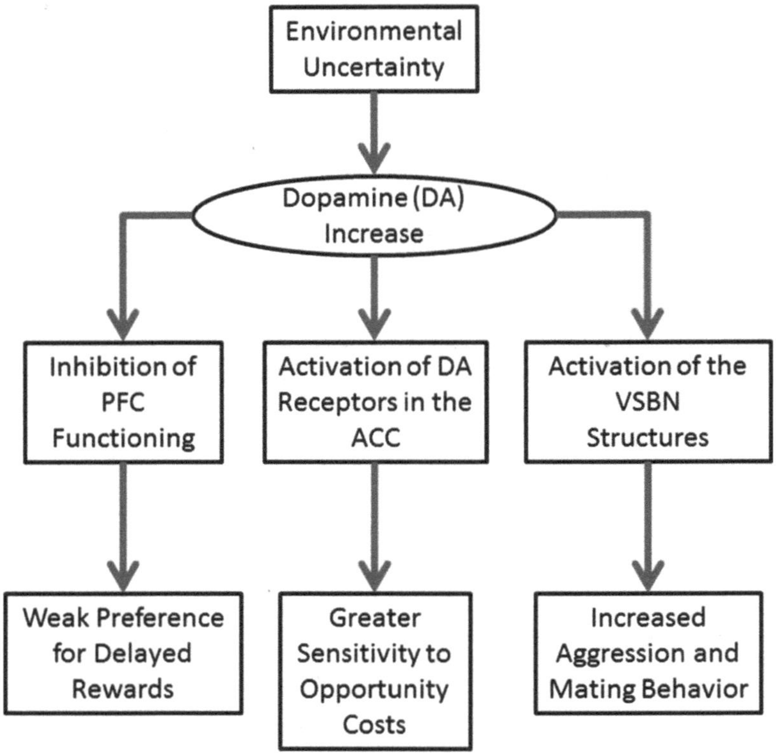

Fig. 21.4 Model of the neural responses to environmental uncertainty

Brain and Behavior Under Environmental Uncertainty

In a variable but predictable environment, decision-making is a "top-down" process. Behavior is rule-based and the influence of autonomic responses to sensory stimuli is mitigated (Miller and Cohen 2001). Neuronal activities in the PFC and ACC regulate behavioral inputs from the VSBN. Environmental uncertainty, because it is a stressful condition, inverts this process (Arnsten 2009). Figure 21.4 is a representation of one pathway by which exposure to unpredictable environments impacts the PFC, ACC, and VSBN. The model assumes that exposure is chronic rather than acute. The nervous systems of various species can readily handle acute forms of stress (Sapolsky 2004). In fact, research suggests that nervous systems evolved under conditions of acute stress exposure (Munhoz et al. 2006; Sapolsky 2004). Prolonged exposure has detrimental effects in terms of two major consequences. First, stress-response hormones can become "fixed" at high circulating levels under chronic stress exposure (Creel and Sands 2003; Munhoz et al. 2006; Rodrigues et al. 2009). Second, prolonged exposure can set in motion a vicious cycle between stress exposure and stress response, with stress-related neuronal processes becoming increasingly sensitive to each additional exposure over time (see Cacioppo and Patrick 2008 for a detailed discussion).

Figure 21.4 shows that environmental uncertainty, in the form of unpredictable stressors and uncertain threats to well-being, triggers an increase in the activity of dopaminergic neurons and dendritic receptors for dopamine. The midbrain is the primary origin of dopaminergic neurons (Bayer and Glimcher 2005) but large populations of neurons with dopamine receptors are also located in PFC (Arnsten 2009) and ACC (Assadi et al. 2009). Under environmental uncertainty, activity of dopaminergic neurons in the midbrain increases above baseline levels (Fiorillo et al. 2003). The net effect of this is to focus attention on the immediate environment (Hills 2006). The effects of increased levels of dopamine in other areas of the VSBN produce behavioral patterns of the same commonly found in sociological studies of persistent inequality. For example, the medial preoptic area controls the release of gonadotropin, a reproductive hormone in males and females (Dominguez and Hull 2005). While exposure to stress typically lowers the circulating level of gonadotropin (Creel and Sands 2003), high levels of dopamine in the preoptic area have been linked to increased sexual activity in males (Hull et al. 2004). In fact, it seems that the medial preoptic area plays a critical mediating role in the relationship between stress and release of reproductive hormones (Williamson and Viau 2008).

Researchers have also found extensive dopamine receptors, particularly D5 receptors, in the anterior hypothalamus. This is significant because the anterior hypothalamus has been associated with the control of aggression. Higher levels of dopamine in the anterior hypothalamus excite aggressive behavior in studies on mammalian species (Schwartzer and Melloni 2010). Perhaps more significant, other studies have shown that above-baseline levels of dopamine in the anterior hypothalamus are also trigger intense male-male aggression in defense of mating relationships (Gobrogge et al. 2007). This resonates strongly with the observations of researchers such as Anderson (1999), who have documented hyper-violent male-male conflicts over sexual opportunities with females within the context of deprived and unpredictable environments.

Another common observation in studies of subjugated groups is the propensity to devalue behavioral choices with delayed payoffs. Adhering to mainstream prescriptions for success, e.g. going to college, is often perceived as a having little value. Note also that the devaluation is based on observations of (1) how others failed to find any significant degree of success while pursuing mainstream goals and (2) local opportunities, which do not require advance formal training, yet still allow one to obtain a standard living comparable to everyone else in the environment (see Anderson 1999; MacLeod 1995, and Willis 1977 for examples). Perhaps this decision-making is the result of the impact of dopamine on the ACC. An increase in dopamine in the ACC generally increases the willingness of an individual to invest large amounts of effort in the pursuit of high-risk, high-value rewards (Schweimer and Hauber 2006). However, the value of the action is a function of opportunity costs, where opportunity costs rise with the temporal delay of payoffs (Niv et al. 2007). It seems that an individual's preference is stronger for actions with lower opportunity costs (Niv et al. 2007).

The effect of dopamine on the functioning of the PFC is similar to the complex effects commonly found in other biological processes. There really can be too much of a good thing. In a review of literature, Arsten (2009) describes the relationship between dopamine levels and functioning of the PFC as an inverted U-shape. When dopamine levels are too low, PFC functioning is impaired. This condition has been associated with cognitive disorders such as attention deficit/hyperactivity disorder (Cook et al. 1995) and delinquency (Guo et al. 2008). Increases in dopamine to normal levels produce significant improvements in working memory, thereby increasing attention and learning capacity (Hills 2006). However, studies of mammalian species show that supranormal levels of dopamine actually impair working memory (Vijayraghavan et al. 2007; Zahrt et al. 1997) through suppression of other molecular processes necessary to maintain a high level of functioning in the PFC (Arnsten 2009).

Given the importance of the dmPFC to prediction and decision-making, the detrimental effect of supranormal levels of dopamine on working memory might explain the commonly observed relationship between subjugated social status and short-sighted decisions involved in adaptation to categorical inequality. Recent findings, for example, suggest that supranormal levels of dopamine can cause an individual to steeply discount the value of temporally delayed rewards (Pine et al. 2010). This is consistent with the findings on the behavioral effects of high levels of dopamine in the ACC (cited above).

Toward Neurosociology of Prediction

The origins of emulation and adaptation (as well as exploitation and opportunity) can be difficult to discern from purely structural models of persistent inequality. Structural explanations fall short of explaining the microfoundations of inequality. Without identifying mechanisms at these levels, the value of sociological analysis of inequality will be restricted to its descriptive richness. However, a neurosociological perspective can bring important mechanisms to light and, as a result, explain core problems in research on persistent inequality.

There are at least three ways neurosociology can contribute to the development of a more powerful structural theory of inequality. First, it can yield a viable theory of social action under inequality. Even now, there is no widely accepted theory of social action within sociology. Rational choice has been roundly successful but it is limited by a lack of regard for the problem of uncertainty. Its assumptions simply do not allow for it. Yet, uncertainty abounds in social decision-making and it is clearly one major vector by which environmental conditions generate dramatic, seemingly irrational changes in behavior (Gigerenzer 2008; Klein 2009). In response to the shortcomings of rational choice theory, social scientists have turned increasingly toward the concept of adaptive rationality (Haselton et al. 2009; Kenrick et al. 2009; Vanberg 2002). At its core, a model of adaptive rationality is still informed by some of the major premises of rational choice theory. However, it recognizes that behavior can be program-based as well as calculative depending upon past, current and predicted states of the environment (Vanberg 2002). Program-based behaviors are evolved, atavistic behaviors strongly favored by natural selection to enhance survivability and reproductive success (Vanberg 2002).

Theoretical models of adaptive rationality suggest that as environmental change becomes less predictable, program-based behaviors become increasingly expressed (Boyd and Richerson 2002). In unpredictable environmental conditions, calculative decision-making strategies such as trial-and-error learning can become too costly to pursue. That is, the costs of an error are likely to outweigh the benefits of the trial (Boyd and Richerson 2002). Decisions become increasingly restricted to program-based strategies such as social learning and genetically prescribed preferences (Boyd and Richerson 2002). In essence, then, to the degree that categorical inequality generates environmental uncertainty for a subjugated group the more common program-based behaviors will be among its members.

Second, a neurosociology of prediction can theorize more precisely the relationship between the environment generated by structural inequality and individual behavior. I argue that the best model for understanding the role of prediction is the threshold trait model. In its simplest form, threshold trait models predict the expression of phenotypic dimorphism in a population of individuals of the same species (Roff 1996). The model focuses on the interaction of three factors: (1) an underlying genetic liability which can express one of two phenotypes; (2) a physiological system of monitoring the environment; and (3) a change in environmental conditions. Liability is normally distributed in the population. The monitoring system is acutely attuned to the environment. It transduces environmental signals and stores that information in memory. Thus, it allows the individual to track changes in the environment and make predictions about future environmental states. The environment fluctuates between two states and the fitness of a phenotype depends upon the state of the environment (Roff 1996). The interaction between the factors can be summarized as follows. When a predictive cue from the environment is detected by the monitoring system it, in turn, triggers the expression of a phenotype in anticipation of environmental change. The phenotype expressed is the one with higher fitness in the new environment (Roff 1996).

More complex models of threshold traits account for the fact that natural selection has conserved many traits which buffer an organism's development against environmental changes (Dworkin 2005) and that phenotypes expressed often incur costs to future fitness (Roff 1996). Overall, however, models of threshold traits suggest some important considerations for understanding the relationship between environmental conditions and individual behavior. First, threshold models propose the relationship is non-linear. Second, environmental induction of phenotypic variability within a population is likely to result in a bimodal distribution rather than a standard normal distribution – which is the one usually tested in sociological research. Third, threshold models also propose that understanding the non-linear impact of the environment and the resulting distribution of traits requires close attention to the mechanisms the individual relies upon to track and predict changes in the local environment. Without specification of this mechanism and its properties, a model of the relationship between environmental conditions and individual-level behavior is severely limited.

Third, a neurosociology of prediction can offer a more powerful interpretation of emulation and adaptation to subjugated social status and their inherently detrimental outcomes. In summary, I argue that emulation and adaptation are not simply meant to ease day-to-day interaction. Rather, they are primary methods for rapidly building a predictable cultural niche. Unfortunately, the emphasis on the rapidity of niche development incurs large long-term costs.

Copying the behaviors of others in the local environment can quickly bring about a stable niche in a variety of ways. At the most basic level, emulation encourages perceptions of common traits and interests among individuals sharing a common social space (Diijksterhuis 2005). From this point, individuals can develop stronger bonds of trust, long-term friendships, and greatly extend social networks (Diijksterhuis 2005). Each of these outcomes has the benefit of minimizing social threats and minimizing the amount of effort needed to access material resources. Likewise, adaptation through the creation of local cultural norms can quickly promote an order and predictability in social interactions, even among strangers (see Anderson 1999 for example).

Problems arise because rapidly achieving control over the social environment will often require the employment of social strategies which are inherently detrimental over the long run. Two such strategies are fear mongering and the threat of violence. Fear tactics and violence have been used to "encourage" quick conformity within a population (Caplin 2003) or bring about radical cultural and institutional change within a specific community (Klein 2007). In their studies of subjugated groups, ethnographers have frequently noted the use of violence and threats of violence as social currency. Some individuals build reputations for violence and, in effect, are able to control interactions with others from a distance (see Anderson 1999; MacLeod 1995; Willis 1977). Though able to quickly create a predictable environment, violence and fear stimulate some of the same neurobiological processes generated by uncertainty (Rosen and Donley 2006). Higher-level cognitive functioning is likewise compromised. Behavior becomes rule-based. Consequently, the predictable environment created is inherently unsustainable over the long run.

Conclusions

Neurosociology clearly has the promise to illuminate the cognitive and neurobiological mechanisms of social behavior but this is not its sole contribution to sociology. Rather, neurosociology brings our attention to the fact that the individual is intricately dependent upon sociological processes at multiple levels of biological functioning and that this is the major outcome of the operation of sociological forces over evolutionary time (Franks 2010; TenHouten 1997). But a new perspective cannot advance without solid empirical research programs in place. These research programs must be able to answer existing questions in the sociological literature and yield novel insights and directions for the future. A focus on prediction and predictive learning can help neurosociology achieve these goals. Because sociological forces have so strongly shaped the evolution of the human brain, how an individual predicts the future is also likely to be a reflection of social selection dynamics. That is, just as the predictive behaviors of some animals are tuned to cues in the physical environment the predictive behaviors of humans are probably tuned to specific sociological cues such as relative social status and the formation of coalitions. This is at least suggested by some researchers (Kenrick et al. 2003).

A measure of predictions about the future would rest squarely on such an argument and provide a solid empirical dimension to neurosociology.

In my view, neurosociology greatly expands the breadth and depth of sociology as a discipline. Since it is mechanism-based, it can contribute greatly to the development of powerful explanatory theories within sociology. Measures built on neurosociological theories will undoubtedly bring even greater methodological integration across disciplines. I predict a bright future for neurosociology.

References

Allman, J. M., Hakeem, A., Erwin, J. M., Nimchinsky, E., & Hof, P. (2001). The anterior cingulate cortex. *Annals of the New York Academy of Sciences, 935*, 107–117.

Anderson, E. (1978). *A place on the corner*. Chicago: University of Chicago Press.

Anderson, E. (1990). *Streetwise: Race, class, and change in an urban community*. Chicago: University of Chicago Press.

Anderson, E. (1999). *Code of the street: Decency, violence, and the moral life of the inner city* (Vol. 1). New York: W. W. Norton and Company.

Arnsten, A. F. T. (2009). Stress signalling pathways that impair prefrontal cortex structure and function. *Nature Reviews Neuroscience, 10*, 410–422.

Assadi, S. M., Yncel, M., & Pantelis, C. (2009). Dopamine modulates neural networks involved in effort-based decision-making. *Neuroscience and Biobehavioral Reviews, 33*, 383–393.

Ball-Rokeach, S. J. (1973). From pervasive ambiguity to a definition of the situation. *Sociometry, 36*, 378–389.

Bar, M. (2009). Predictions: A universal principle in the operation of the human brain. *Philosophical Transactions of the Royal Society B: Biological Sciences, 364*, 1181–1182.

Barker, G. T. (2005). *Dying to be men: Youth and masculinity and social exclusion*. Oxford: Routledge.

Bayer, H. M., & Glimcher, P. W. (2005). Midbrain dopamine neurons endcode a quantitative reward prediction error signal. *Neuron, 47*, 129–141.

Bechara, A., Damasio, H., Damasio, A. R., & Lee, G. P. (1999). Different contributions of the human amygdala and ventromedial prefrontal cortex to decision-making. *Journal of Neuroscience, 19*, 5473–5481.

Belsky, J., Steinberg, L. D., & Draper, P. D. (1991). Childhood experience, interpersonal development and reproductive strategy: An evolutionary theory of socialization. *Child Development, 62*, 647–670.

Belsky, J., Steinberg, L., Houts, R. M., & Halpern-Felsher, B. L. (2010). The development of reproductive strategy in females: Early maternal harshness –>earlier menarche –>increased sexual risk taking. *Developmental Psychology, 46*, 120–128.

Bledowski, C., Rahm, B., & Rowe, J. B. (2009). What "works" in working memory? Separate systems for selection and updating of critical information. *The Journal of Neuroscience, 29*, 13735–13741.

Bourdieu, P., & Accardo, A. (1999). *The weight of the world: Social suffering in contemporary society*. Stanford: Stanford University Press.

Bourgois, P. (2002). *In search of respect: Selling crack in El Barrio*. Cambridge: Cambridge University Press.

Boyd, R., & Richerson, P. J. (2002). Norms and bounded rationality. In G. Gigerenzer & R. Selten (Eds.), *Bounded rationality: The adaptive toolbox* (pp. 281–296). Cambridge, MA: The MIT Press.

Brown, J. W., & Braver, T. S. (2007). Risk prediction and aversion by anterior cingulate cortex. *Cognitive, Affective, & Behavioral Neuroscience, 7*, 266–277.

Brumbach, B., Figueredo, A., & Ellis, B. (2009). Effects of harsh and unpredictable environments in adolescence on development of life history strategies. *Human Nature, 20*, 25–51.

Bubic, A., Von Cramon, D. Y., & Schubotz, R. I. (2010). Prediction, cognition and the brain. *Frontiers in Human Neuroscience, 4*:25, doi: 10.3389/fnhum.2010.00025.

Burmeister, S. S., Jarvis, E. D., & Russell, F. (2005). Rapid behavioral and genomic responses to social opportunity. *Public Library of Science Biology, 3*, e363–e371.

Cacioppo, J. T., & Patrick, W. (2008). *Loneliness: Human nature and the need for social connection*. New York: W. W. Norton.

Caplin, A. (2003). Fear as a policy instrument. In G. F. Loewenstein, D. Read, & R. Baumeister (Eds.), *Time and decision: Economic and psychological perspectives on intertemporal choice* (pp. 441–458). New York: Russell Sage.

Carroll, A., Green, S., Houghton, S., & Wood, R. (2003). Reputation enhancement and involvement in delinquency among high school students. *International Journal of Disability, Development and Education, 50*, 253–273.

Carter, C. S., & van Veen, V. (2007). Anterior cingulate cortex and conflict detection: An update of theory and data. *Cognitive, Affective, & Behavioral Neuroscience, 7*, 367–379.

Chisholm, J. S. (1999). *Death, hope, and sex: Steps to an evolutionary ecology of mind and morality*. Cambridge: Cambridge University Press.

Chisholm, J. S., Quinlivan, J. A., Petersen, R. W., & Coall, D. A. (2005). Early stress predicts age at menarche and first birth, adult attachment, and expected lifespan. *Human Nature, 16*, 233–265.

Cook, E. H., Jr., Stein, M. A., Krasowski, M. D., Cox, N. J., Olkon, D. M., Kieffer, J. E., & Leventhal, B. L. (1995). Association of attention-deficit disorder and the dopamine transporter gene. *The American Journal of Human Genetics, 56*, 993–998.

Coussi-Korbel, S., & Fragaszy, D. M. (1995). On the relation between social dynamics and social learning. *Animal Behaviour, 50*, 1441–1453.

Creel, S., & Sands, J. L. (2003). Is social stress a consequence of subordination or a cost of dominance? In F. B. M. de Waal & P. L. Tyack (Eds.), *Animal social complexity: Intelligence, culture and individual societies* (pp. 153–169). Cambridge, MA: Harvard University.

Davis, J., & Werre, D. (2008). A longitudinal study of the effects of uncertainty on reproductive behaviors. *Human Nature, 19*, 426–452.

Daw, N. D., O'Doherty, J. P., Dayan, P., Seymour, B., & Dolan, R. J. (2006). Cortical substrates for exploratory decisions in humans. *Nature, 441*, 876–879.

Day, S., Papataxiarchis, E., & Stewart, M. (1999). Consider the lilies of the field. In S. Day, E. Papataxiarchis, & M. Stewart (Eds.), *Lilies of the field: Marginal people who live for the moment* (pp. 1–25). Boulder: Westview Press.

Dayan, P., & Abbott, L. F. (2001). *Theoretical neuroscience: Computational and mathematical modeling of neural systems*. Cambridge, MA: MIT Press.

Devinsky, O., Morrell, M. J., & Vogt, B. A. (1995). Contributions of anterior cingulate cortex to behaviour. *Brain, 118*(Pt 1), 279–306.

Diijksterhuis, Ap. (2005). Why we are social animals: The high road to imitation as social glue. In S. Hurley & N. Chater (Eds.), *Perspectives on imitation: From neuroscience to social science* (Imitation, human development, and culture, Vol. 2, pp. 207–220). Cambridge, MA: MIT Press.

Dominguez, J. M., & Hull, E. M. (2005). Dopamine, the medial preoptic area, and male sexual behavior. *Physiology and Behavior, 86*, 356–368.

Dworkin, I. (2005). Canalization, cryptic variation, and developmental buffering: A critical examination and analytical perspective. In B. Hallgrimsson & B. K. Hall (Eds.), *Variation: A central concept in biology* (pp. 131–158). Amsterdam: Elsevier.

Edin, K., & Kefelas, M. (2005). *Promises I can keep: Why poor women put motherhood before marriage*. Berkeley: University of California Press.

Edin, K., & Lein, L. (1997). *Making ends meet: How single mothers survive welfare and low wage work*. New York: Russell Sage.

Eichenbaum, H., & Fortin, N. J. (2009). The neurobiology of memory based predictions. *Philosophical Transactions of the Royal Society B: Biological Sciences, 364*, 1183–1191.

Eisenberger, N. I., Lieberman, M. D., & Williams, K. D. (2003). Does rejection hurt? An FMRI study of social exclusion. *Science, 302*, 290–292.

Ellis, B., Figueredo, A., Brumbach, B., & Schlomer, G. (2009). Fundamental dimensions of environmental risk. *Human Nature, 20*, 204–268.

Emler, N., & Reicher, S. (1995). *Adolescence and delinquency: The collective management of reputation*. Oxford: Blackwell Publishing.

Etkin, A., Egner, T., Peraza, D. M., Kandel, E. R., & Hirsch, J. (2006). Resolving emotional conflict: A role for the rostral anterior cingulate cortex in modulating activity in the amygdala. *Neuron, 51*, 871–882.

Fehr, E., & Camerer, C. F. (2007). Social neuroeconomics: The neural circuitry of social preferences. *Trends in Cognitive Sciences, 11*, 419–427.

Fellows, L. K., & Farah, M. J. (2007). The role of ventromedial prefrontal cortex in decision making: Judgment under uncertainty or judgment Per Se? *Cerebral Cortex, 17*, 2669–2674.

Fiorillo, C. D., Tobler, P. N., & Schultz, W. (2003). Discrete coding of reward probability and uncertainty by dopamine neurons. *Science, 299*, 1898–1902.

Franks, D. D. (2010). *Neurosociology: The nexus between neuroscience and social psychology*. New York: Springer.

Gambetta, D., & Hamill, H. (2005). *Streetwise: How taxi drivers establish their customers' trustworthiness*. New York: Russell Sage.

Gigerenzer, G. (2008). *Rationality for mortals: How people cope with uncertainty*. New York: Cambridge University Press.

Gigerenzer, G., & Brighton, H. (2009). Homo heuristicus: Why biased minds make better inferences. *Trends in Cognitive Sciences, 1*, 107–143.

Gobrogge, K. L., Liu, Y., Jia, X., & Wang, Z. (2007). Anterior hypothalamic neural activation and neurochemical associations with aggression in pair-bonded male prairie voles. *The Journal of Comparative Neurology, 502*, 1109–1122.

Goodson, J. L. (2005). The vertebrate social behavior network: Evolutionary themes and variations. *Hormones and Behavior, 48*, 11–22.
Graybiel, A. M. (2008). Habits, rituals, and the evaluative brain. *Annual Review of Neuroscience, 31*, 359–387.
Greene, J. D., & Paxton, J. M. (2009). Patterns of neural activity associated with honest and dishonest moral decisions. *Proceedings of the National Academy of Sciences, 106*, 12506–12511.
Guo, G., Roettger, M. E., & Cai, T. (2008). The integration of genetic propensities into social-control models of delinquency and violence among male youths. *American Sociological Review, 73*, 543–568.
Hakeem, A. Y., Sherwood, C. C., Bonar, C. J., Butti, C., Hof, P. R., & Allman, J. M. (2009). Von economo neurons in the elephant brain. *The Anatomical Record: Advances in Integrative Anatomy and Evolutionary Biology, 292*, 242–248.
Haselton, M. G., Bryant, G. A., Wilke, A., Frederick, D. A., Galperin, A., Frankenhuis, W. E., & Moore, T. (2009). Adaptive rationality: An evolutionary perspective on cognitive bias. *Social Cognition, 27*, 733–763.
Hills, T. T. (2006). Animal foraging and the evolution of goal-directed cognition. *Cognitive Science, 30*, 3–41.
Hull, E. M., Muschamp, J. W., & Sato, S. (2004). Dopamine and serotonin: Influences on male sexual behavior. *Physiology and Behavior, 83*, 291–307.
Insel, T. R., & Fernald, R. D. (2004). How the brain processes social information: Searching for the social brain. *Annual Review of Neuroscience, 27*, 722.
Ishikawa, A., Ambroggi, F., Nicola, S. M., & Fields, H. L. (2008). Dorsomedial prefrontal cortex contribution to behavioral and nucleus accumbens neuronal responses to incentive cues. *The Journal of Neuroscience, 28*, 5088–5098.
Iversen, L. L. (2010). *Dopamine handbook*. Oxford/New York: Oxford University Press.
Kennerley, S. W., Walton, M. E., Behrens, T. E. J., Buckley, M. J., & Rushworth, M. F. S. (2006). Optimal decision making and the anterior cingulate cortex. *Nature Neuroscience, 9*, 940–947.
Kenrick, D. T., Li, N. P., & Butner, J. (2003). Dynamical evolutionary psychology: Individual decision rules and emergent social norms. *Psychological Review, 110*, 3–28.
Kenrick, D. T., Griskevicius, V., Sundie, J. M., Li, N. P., Li, Y. J., & Neuberg, S. L. (2009). Deep rationality: The evolutionary economics of decision making. *Social Cognition, 27*, 764–785.
Klein, N. (2007). *The shock doctrine: The rise of disaster capitalism*. New York: Metropolitan Books.
Klein, G. (2009). *Streetlights and shadows: Searching for the keys to adaptive decision making*. Cambridge, MA: MIT Press.
Koenigs, M., Young, L., Adolphs, R., Tranel, D., Cushman, F., Hauser, M., & Damasio, A. (2007). Damage to the prefrontal cortex increases utilitarian moral judgements. *Nature, 446*, 908–911.
Kouneiher, F., Charron, S., & Koechlin, E. (2009). Motivation and cognitive control in the human prefrontal cortex. *Nature Neuroscience, 12*, 939–945.
LaLand, K. N. (2004). Social learning strategies. *Learning & Behavior, 32*, 4–14.
Lee, G., & Gammie, S. C. (2009). GABAA receptor signaling in the lateral septum regulates maternal aggression in mice. *Behavioral Neuroscience, 123*, 1169–1177.
MacLeod, J. (1995). *Ain't no Makin' it: Leveled aspirations in a low income neighborhood*. Boulder: Westview Press.
Maeda, K.-I., & Tsukamura, H. (2006). The impact of stress on reproduction: Are glucocorticoids inhibitory or protective to gonadotropin secretion? *Endocrinology, 147*, 1085–1086.
Manuck, S. B., Flory, J. D., Muldoon, M. F., & Ferrell, R. E. (2003). A neurobiology of intertemporal choice. In G. Loewenstein, D. Read, & R. F. Baumeister (Eds.), *Time and decision: Economic and psychological perspectives on intertemporal choice* (pp. 139–172). New York: Russell Sage.
Manuck, S. B., Bleil, M. E., Petersen, K. L., Flory, J. D., John Mann, J., Ferrell, R. E., & Muldoon, M. F. (2005). The socio-economic status of communities predicts variation in brain serotonergic responsivity. *Psychological Medicine, 35*, 528.
Massey, D. S. (2007). *Categorically unequal: The American stratification system*. New York: Russell Sage.
Masten, C. L., Eisenberger, N. I., Borofsky, L. A., Pfeifer, J. H., McNealy, K., Mazziotta, J. C., & Dapretto, M. (2009). Neural correlates of social exclusion during adolescence: Understanding the distress of peer rejection. *Social Cognitive and Affective Neuroscience, 4*, 143–157.
Matsuwaki, T., Watanabe, E., Suzuki, M., Yamanouchi, K., & Nishihara, M. (2003). Glucocorticoid maintains pulsatile secretion of luteinizing hormone under infectious stress condition. *Endocrinology, 144*, 3477–3482.
McEwen, B. S., & Gianaros, P. J. (2010). Central role of the brain in stress and adaptation: Links to socioeconomic status, health, and disease. *Annals of the New York Academy of Sciences, 1186*, 190–222.
Menchaca, M. (1995). *Mexican outsiders: A community history of marginalization and discrimination in california*. Austin: University of Texas Press.
Miller, E. K., & Cohen, J. D. (2001). An integrative theory of prefrontal cortex function. *Annual Review of Neuroscience, 24*, 167–202.

Modirrousta, M., & Fellows, L. K. (2008). Dorsal medial prefrontal cortex plays a necessary role in rapid error prediction in humans. *The Journal of Neuroscience, 28*, 14000–14005.
Montague, P. R., Dayan, P., & Sejnowski, T. J. (1996). A framework for mesencephalic dopamine systems based on predictive Hebbian learning. *Journal of Neuroscience, 16*, 1936–1947.
Morgane, P. J., Galler, J. R., & Mokler, D. J. (2005). A review of systems and networks of the limbic forebrain/limbic midbrain. *Progress in Neurobiology, 75*, 143–160.
Morris, G., Nevet, A., Arkadir, D., Vaadia, E., & Bergman, H. (2006). Midbrain dopamine neurons encode decisions for future action. *Nature Neuroscience, 9*, 1057–1063.
Munhoz, C. D., Lepsch, L. B., Kawamoto, E. M., Malta, M. B., Lima Lde, S., Avellar, M. C., Sapolsky, R. M., & Scavone, C. (2006). Chronic unpredictable stress exacerbates lipopolysaccharide-induced activation of nuclear factor-kappaB in the frontal cortex and hippocampus via glucocorticoid secretion. *Journal of Neuroscience, 26*, 3813–3820.
Newman, S. W. (1999). The medial extended amygdala in male reproductive behavior: A node in the mammalian social behavior network. *Annals of the New York Academy of Sciences, 877*, 242–257.
Niv, Y., Duff, M., & Dayan, P. (2005). Dopamine, uncertainty and TD learning. *Behavioral and Brain Functions, 1*, 6.
Niv, Y., Daw, N. D., Joel, D., & Dayan, P. (2007). Tonic dopamine: Opportunity costs and the control of response vigor. *Psychopharmacology, 191*, 507–520.
Niven, J. E., & Laughlin, S. B. (2008). Energy limitation as a selective pressure on the evolution of sensory systems. *Journal of Experimental Biology, 211*, 1792–1804.
O'Donnell, M., & Sharpe, S. (2000). *Uncertain masculinities: Youth, ethnicity, and class in contemporary Britain.* London: Routledge.
O'Reilly, J. X., Marsel Mesulam, M., & Nobre, A. C. (2008). The cerebellum predicts the timing of perceptual events. *Journal of Neuroscience, 28*, 2252–2260.
Odling-Smee, J. F., LaLand, K. N., & Feldman, M. W. (2003). *Niche construction: The neglected process in evolution.* Princeton: Princeton University Press.
Osaka, N., Logie, R. H., & D'Esposito, M. (2007). *The cognitive neuroscience of working memory.* Oxford/New York: Oxford University Press.
Pine, A., Shiner, T., Seymour, B., & Dolan, R. J. (2010). Dopamine, time, and impulsivity in humans. *Journal of Neuroscience, 30*, 8888–8896.
Renfree, M. B., & Shaw, G. (2000). Diapause. *Annual Review of Physiology, 62*, 353–375.
Robinson, G. E., Fernald, R. D., & Clayton, D. F. (2008). Genes and social behavior. *Science, 322*, 896–900.
Rodrigues, S. M., LeDoux, J. E., & Sapolsky, R. M. (2009). The influence of stress hormones on fear circuitry. *Annual Review of Neuroscience, 32*, 289–313.
Roff, D. A. (1996). The evolution of threshold traits in animals. *The Quarterly Review of Biology, 71*, 3–35.
Rosen, J. B., & Donley, M. P. (2006). Animal studies of amygdala function in fear and uncertainty: Relevance to human research. *Biological Psychology, 73*, 49–60.
Rougier, N. P., Noelle, D. C., Braver, T. S., Cohen, J. D., & O'Reilly, R. C. (2005). Prefrontal cortex and flexible cognitive control: Rules without symbols. *Proceedings of the National Academy of Sciences of the United States of America, 102*, 7338–7343.
Rushworth, M. F., Buckley, M. J., Behrens, T. E., Walton, M. E., & Bannerman, D. M. (2007). Functional organization of the medial frontal cortex. *Current Opinion in Neurobiology, 17*, 220–227.
Salancik, G. R., & Pfeffer, J. (1978). Uncertainty, secrecy, and the choice of similar others. *Social Psychology Quarterly, 41*, 246–255.
Salzman, C. D., & Fusi, S. (2010). Emotion, cognition, and mental state representation in amygdala and prefrontal cortex. *Annual Review of Neuroscience, 33*, 173–202.
Sapolsky, R. M. (2004). *Why zebras don't get ulcers/Robert M. Sapolsky.* New York: Times Books.
Schultz, W., & Dickinson, A. (2000). Neuronal coding of prediction errors. *Annual Review of Neuroscience, 23*, 473–500.
Schultz, W., Dayan, P., & Montague, P. R. (2002). A neural substrate of prediction and reward. In J. Cacioppo, G. G. Berntson, R. Adolphs, C. S. Carter, R. J. Davidson, M. K. McClintock, B. S. McEwen, M. J. Meaney, D. L. Schacter, E. M. Sternberg, S. S. Suomi, & S. E. Taylor (Eds.), *Foundations in social neuroscience* (pp. 541–554). Cambridge, MA: MIT.
Schwalbe, M., Godwin, S., Holden, D., Schrock, D., Thompson, S., & Wolkomb, M. (2000). Generic processes in the reproduction of inequality: An interactionist analysis. *Social Forces, 79*, 419–452.
Schwartzer, J. J., & Melloni, R. H., Jr. (2010). Anterior hypothalamic dopamine D2 receptors modulate adolescent anabolic/androgenic steroid-induced offensive aggression in the Syrian hamster. *Behavioural Pharmacology, 21*, 314–322.
Schweimer, J., & Hauber, W. (2006). Dopamine D1 receptors in the anterior cingulate cortex regulate effort-based decision making. *Learning & Memory, 13*, 777–782.

Shettleworth, S. J. (1998). *Cognition, evolution, and behavior*. Oxford: Oxford University Press.
Singewald, G. M., Rjabokon, A., Singewald, N., & Ebner, K. (2011). The modulatory role of the lateral septum on neuroendocrine and behavioral stress responses. *Neuropsychopharmacology, 36*, 793–804.
Slavich, G. M., Way, B. M., Eisenberger, N. I., & Taylor, S. E. (2010). Neural sensitivity to social rejection is associated with inflammatory responses to social stress. *Proceedings of the National Academy of Sciences of the United States of America, 107*, 14817–14822.
Szpunar, K. K., Watson, J. M., & McDermott, K. B. (2007). Neural substrates of envisioning the future. *Proceedings of the National Academy of Sciences of the United States of America, 104*, 642–647.
Tagkopoulos, I., Liu, Y. C., & Tavazoie, S. (2008). Predictive behavior within microbial genetic networks. *Science, 320*, 1313–1317.
TenHouten, W. (1997). Neurosociology. *Journal of Social and Evolutionary Systems, 20*, 7–37.
Tilly, C. (1998). *Durable inequality*. Berkeley: University of California Press.
Tilly, C. (2005). *Identities, boundaries and social ties*. Boulder: University of Colorado Press.
van Veen, V., & Carter, C. S. (2006). Error detection, correction, and prevention in the brain: A brief review of data and theories. *Clinical EEG and Neuroscience, 37*, 330–335.
van Veen, V., Cohen, J. D., Botvinick, M. M., Stenger, V. A., & Carter, C. S. (2001). Anterior cingulate cortex, conflict monitoring, and levels of processing. *NeuroImage, 14*, 1302–1308.
van Veen, V., Krug, M. K., Schooler, J. W., & Carter, C. S. (2009). Neural activity predicts attitude change in cognitive dissonance. *Nature Neuroscience, 12*, 1469–1474.
Vanberg, V. J. (2002). Rational choice vs. Program-based behavior: Alternative theoretical approaches and their relevance for the study of institutions. *Rationality and Society, 14*, 7–54.
Vijayraghavan, S., Wang, M., Birnbaum, S. G., Williams, G. V., & Arnsten, A. F. T. (2007). Inverted-U dopamine D1 receptor actions on prefrontal neurons engaged in working memory. *Nature Neuroscience, 10*, 376–384.
Wallis, J. D. (2007). Orbitofrontal cortex and its contribution to decision-making. *Annual Review of Neuroscience, 30*, 31–56.
Walton, M. E., Croxson, P. L., Behrens, T. E. J., Kennerley, S. W., & Rushworth, M. F. S. (2007). Adaptive decision making and value in the anterior cingulate cortex. *NeuroImage, 36*, T142–T154.
Warkentin, K. M., & Caldwell, M. S. (2009). Assessing risk: Embryos, information, and escape hatching. In R. Dukas & J. M. Ratcliffe (Eds.), *Cognitive ecology II* (pp. 177–200). Chicago: University of Chicago Press.
Wasser, S. K., & Place, N. J. (2001). Reproductive filtering and the social environment. In P. T. Ellison (Ed.), *Reproductive ecology and human evolution* (pp. 137–158). Chicago: Aldine De Gruyter.
Watson, K. K. (2008). Evolution, risk, and neural representation. *Annals of the New York Academy of Sciences, 1128*, 8–12.
Whyte, W. F. (1943). *Streetcorner society: The social structure of an Italian slum*. Chicago: University of Chicago Press.
Williamson, M., & Viau, V. (2008). Selective contributions of the medial preoptic nucleus to testosterone-dependant regulation of the paraventricular nucleus of the hypothalamus and the HPA axis. *American Journal of Physiology – Regulatory, Integrative and Comparative Physiology, 295*, R1020–R1030.
Willis, P. (1977). *Learning to labor: How working class kids get working class jobs*. New York: Columbia University Press.
Wingfield, J. C., & Kitaysky, A. S. (2002). Endocrine responses to unpredictable environmental events: Stress or anti-stress hormones? *Integrative and Comparative Biology, 42*, 600–609.
Yu, A. J., & Dayan, P. (2002). Acetylcholine in cortical inference. *Neural Networks, 15*, 719–730.
Yu, A., & Dayan, P. (2005). Uncertainty, neuromodulation, and attention. *Neuron, 46*, 681–692.
Zahrt, J., Taylor, J. R., Mathew, R. G., & Arnsten, A. F. T. (1997). Supranormal stimulation of D1 dopamine receptors in the rodent prefrontal cortex impairs spatial working memory performance. *The Journal of Neuroscience, 17*, 8528–8535.

Chapter 22
The Neurobiology of Stereotyping and Prejudice

Todd D. Nelson

In 1922, journalist Walter Lippmann used the term "stereotype" to describe a human tendency to think of people in similar terms (e.g. having similar characteristics) based on a feature they shared. Famously, he said that people have "pictures in our heads" (p.3) of the world outside, and these act like templates into which we try to fit and organize the often disorganized, confusing information we get from the world. Since then, social scientists have borrowed Lippman's "stereotyping" term, and have been investigating that human proclivity for grouping people together based on a shared feature (e.g. skin color) and arriving at conclusions about everyone in that group that often apply to only a fraction of those individuals (Jussim et al. 2009). *A stereotype, therefore, is a belief about the link between a characteristic and a group.* For example, a common stereotype about older persons is that they are "grumpy." Of course, this may indeed describe *some* older persons, it certainly is not an appropriate descriptor for *all* older persons, which is why it is a stereotype. Inexorably tied to the stereotype concept is how this cognitive grouping habit influences the way we feel about the people we have grouped.

Typically, people like to think of themselves in positive ways, as above average, and as having a number of positive attributes (Greenberg et al. 1982; Taylor 1989). If we think of ourselves as better than others, and we assume others are "all the same" in a group, then we should tend to compare ourselves favorably relative to stereotyped groups. Indeed, this is the case (Allport 1954; Fiske 1998; Nelson 2006). The feelings that are usually evoked regarding the stereotyped group tend to be negative. We do not like others who are different, and who, we believe, do not share the same values as us. This negative feeling is prejudice[1] (Nelson 2006, 2009).

Over the decades, researchers have learned much about the factors that lead to the origin, maintenance, and reduction of stereotypes and prejudice (Nelson 2009). The goal of this chapter is to highlight how our understanding of stereotyping and prejudice has been influenced by the application of neuroscience and neuroimaging techniques which might allow us better insight into what

[1] A traditional way of thinking about stereotypes, prejudice, and discrimination is to think of them as an analog to the three components that make up an attitude: cognitive, affective, and behavioral components. Thus, the stereotype represents the belief about a group, prejudice represents the negative feeling toward the group members, and discrimination represents the negative behavior toward the group members. However, definitions of prejudice have become varied over the years, and many researchers have conceptualized prejudice as an attitude itself. That is, this approach combines the cognitive component of a stereotype (these people have this characteristic) with the evaluative component of prejudice ("I don't like" this person). This chapter will maintain the traditional view of prejudice as a 'negative feeling toward stereotyped group members.'

T.D. Nelson (✉)
Department of Psychology, California State University,
Stanislaus, CA 95382, USA
e-mail: tnelson@athena.csustan.edu

D.D. Franks and J.H. Turner (eds.), *Handbook of Neurosociology*, Handbooks of Sociology and Social Research, DOI 10.1007/978-94-007-4473-8_22, © Springer Science+Business Media B.V. 2013

parts of the brain are more active when engaged in cognitive activity concerning stereotyped outgroups. The central question: can neuroimaging pictures *of* our heads help us understand Lippmann's pictures *in* our heads?

Changing Methods

As with any scientific field, scientific advances in psychology progress as the sophistication of one's research methods allow. Sometimes, speculation and theory are not borne out empirically for decades, when the methods to test those suppositions have been invented/conceived. For example, many of William James' (1890)speculations about self, consciousness, and other aspects of psychology were not able to be tested (and empirically verified) until nearly a century later (Buck 1990).

Though integrating social science and neuroscience approaches to understanding behavior is an idea that is at least 20 years old (e.g., Cacioppo 1994), it has only been in the last 10 years that advances in modern neuroimaging methods have become sophisticated enough to allow us the ability to have a better idea of what is happening in the brain when we are perceiving others (Eberhardt 2005; Phelps and Thomas 2003).

Social Cognitive Neuroscience

As neuroimaging techniques matured, researchers began to see that a new, multi-level analysis could be conducted regarding human behavior, thought, and feeling. Social psychologists are interested in understanding social behavior. Cognitive psychologists want to understand information-processing in the brain, and neuroscientists want to understand the neural substrates to human cognition. In their influential 2001 paper, Ochsner and Lieberman proposed that these three approaches to understanding human thought and behavior be merged together into an interdisciplinary field called "social cognitive neuroscience (SCN)." The goal of this approach is to develop and empirically test new theories that refer to all three levels of analysis. In so doing, social psychologists can, for example, make use of the vast cognitive neuroscience database to explore questions about social phenomena. For example, are different parts of the brain activated when one employs different cognitive heuristics or does one part of the brain handle those varied (but conceptually similar) cognitive tasks? Either answer would be interesting, and help researchers further refine their existing theories concerning social psychological concepts.

Perhaps the most popular neuroimaging method in this new interdisciplinary research is functional magnetic resonance imaging (fMRI) (Ochsner and Lieberman 2001; Phelps and Thomas 2003). Like positron emission tomography (PET) scans, fMRI scans involve measuring blood flow to parts of the brain. However, unlike PET scans, fMRI scans are much more precise, and can yield a much more detailed image and location of activity levels of the brain. Both PET and fMRI assess blood flow to the brain, because when the brain is called upon to do a task, it requires more blood. fMRI assesses the amount of blood volume, blood flow, and oxygen use, referred to collectively as the blood oxygenated level dependent (BOLD) signal. The MRI machine is essentially a very powerful magnet, and when an individual lies inside the machine, and the MRI is turned on, the protons in their brain become aligned. Then, a short radio wave is sent through the chamber, knocking the protons out of alignment, and the protons become realigned. Parts of the brain that are more active, with greater concentrations of oxygenated blood send out a stronger signal. The fMRI measures the re-alignment, and is this able to pinpoint (within about 3–6 mm) which part of the brain is most active at a given moment (Cabeza and Nyberg 2000; Watson 2011). Below, we shall examine the relevant studies using fMRI to examine prejudice. How does the brain react when we perceive or think about people from stereotyped outgroups? What insights about social perception, intergroup interaction, and prejudice can we obtain when we understand the neural substrates of such cognitive activity?

The Amygdala, Fear, and Prejudice

Our brains are hard-wired to categorize things, and this process is rather automatic (Gardner 1985). This is highly adaptive, because it allows us to quickly perceive and think about a given stimulus, and move on to devote one's attention and thought to other things that may require more elaborate cognition. This works well, and it is fairly accurate… when dealing with inanimate objects. However, when we begin to apply this automatic categorization to *people*, we run into problems. That is, the accuracy of those automatic categorizations drops, because people are much more complex compared to inanimate objects. Nevertheless, our brains persist on automatically categorizing people. Research shows we tend to automatically categorize people along three dimensions: race, gender, and age. This tendency is so pervasive that researchers have called it "basic" or "primitive" categorization (Brewer 1988; Fiske and Neuberg 1990).

When we categorize people, we also automatically place them into an *ingroup* or *outgroup*. An ingroup is any group of which the perceiver is a member, and the outgroups are those groups to which the perceiver does not belong (Brewer 1979). There are many ingroups for each person. For example, some ingroups for your author would be: men, people who grew up in Minnesota, people who wear glasses, professors, parents, diet mountain dew aficionados, and Macintosh computer users. Research shows that whichever ingroup is salient in the perceiver's mind will affect what behaviors the perceiver feels free to do, and also what his/her expectations are for the behavior of those he/she is with (Allport 1954). So, when I am in a meeting with my fellow professors, there are certain behaviors, jokes, etc. that if enacted or uttered, would be considered strange and unprofessional. However, if one was out for drinks with one's friends, those same behaviors or jokes might arouse uproarious approval from one's friends.

Outgroup members are perceived to share many of the same features, characteristics, values and goals. We think that outgroup members "are all alike," and from that, stereotypes are born. We think of people in our ingroups as unique as fingerprints. This is referred to as the "*outgroup homogeneity effect*" (Hamilton 1976; Ostrom and Sedikides 1992). Because we tend to have more interactions with ingroup members (based on shared interests, values, goals, etc.), we feel comfortable with them, because we feel like we know their motives, expectations, and we therefore know how to behave with them. On the other hand, about outgroup members, we know relatively little, and, as mentioned above, that lack of knowledge leads to assumptions and sweeping evaluations (stereotypes) about the outgroup (that they "are all alike") so that we can at least feel like we know what to expect from them, and to try to abate the anxiety and indeed, fear (and prejudice) we may feel about the outgroup members (Taylor 1981). Providing further support for this perception of outgroups, recent research shows that the more we perceive outgroups as more heterogeneous, the less likely one will feel prejudice against those outgroup members (Brauer and Er-rafiy 2011).

We have seen that one reason why people may find themselves avoiding outgroup members, and developing negative prejudices about particular outgroups is that they have an underlying *fear* of that outgroup (Allport 1954; Jones 1997; Mackie et al. 2009). In the limbic system of the brain, the amygdala is responsible for the emotion of fear (Anderson and Phelps 2000; LeDoux 1996) and anger (Whalen et al. 2001). So, one might expect elevated activity levels in the amygdala when one perceives or thinks about a member of an outgroup. Indeed, Harris and Fiske (2006) found that extreme outgroup members provoke particularly strong activation in the amygdala. Using fMRIs, Phelps et al. (2000) found that Whites who displayed an outgroup bias against Blacks on the Implicit Association Test

(IAT, Greenwald et al. 1998) were much more likely to show a stronger amygdala activation when they viewed photos of Blacks than when they viewed photos of White individuals. A study by Hart et al. (2000) used fMRIs to examine amygdala response of Whites and Blacks to photos of White and Black persons. For all participants, the researchers found equally elevated amygdala responses (stronger BOLD signal) when first presented with ingroup and outgroup faces. On subsequent presentations of the stimuli faces, Hart et al. found greater amygdala activity in White and Black participants looking at an outgroup face. Hart and his colleagues suggest these results indicate that the amygdala response is affected by the relationship of the participant to the stimulus face. Specifically, it appears that during the first presentation of the ingroup and outgroup faces, participants began to habituate to ingroup faces. So, by the time the faces were presented in a subsequent trial, participants no longer showed elevated amygdala activity to those ingroup faces they had grown accustomed to, but their fear response remained for those outgroup faces.

Interestingly, research by Lieberman et al., (2005) showed the same pattern for Whites, but *not* for Black perceivers. That is, Black participants also showed elevated amygdala activity when they perceived photos of other Blacks compared to when they viewed photos of other whites. Lieberman and his colleagues suggest that their results indicate that the amygdala activation differences people show when viewing photos of Blacks vs. Whites do not reflect the novelty of the other race, but rather cultural learning about the other races. It also appears to be the case that the response of the amygdala can be modulated depending on the cognitive processing goals of the perceiver. Wheeler and Fiske (2005) conducted a study in which white participants were exposed to white and black faces and were asked to do one of three processing tasks: individuate the faces (decide whether the person liked vegetables), ascertain whether there was a dot on the target faces (visual search) or indicate whether the target face was older or younger than 21 years of age (social categorization). Data indicated greater amygdala response to black faces over white faces only in the social categorization condition. Over time, the way that researchers have interpreted the elevated amygdala activity has changed. Though this early research linked increased amygdala activity to fear, most current researchers view the spike in amygdala activity to outgroup faces as indicating the strength or intensity of an individual's reaction to the outgroup (Amodio and Lieberman 2009). It should also be remembered that although increased amygdala activity in these experiments may indicate an anger response, some data suggests that the amygdala reacts more strongly when one feels fear than when one feels anger (Whalen et al. 2001).

Research by Maddox and his colleagues (Maddox and Chase 2004; Maddox and Gray 2002) has demonstrated that people perceive light-skinned Blacks more favorably than dark-skinned Blacks, and this effect holds true for both Black and White perceivers (Dixon and Maddox 2005). To examine the neural correlates of these findings about skin tone and prejudice, Ronquillo and colleagues (2007) exposed White subjects to an fMRI scan while they perceived unfamiliar Black and White faces. Replicating previous research, their results showed that there was greater amygdala activity when subjects viewed a Black face, compared to when they viewed a White face. However, the data also showed that the amygdala activity to the Black face was the same whether it was a dark-skinned Black face or a light-skinned Black face. It appears that the amygdala only broadly and bluntly responds to what it sees as "outgroup face" and does not make any distinction among specific subtypes within that outgroup. Whether a light or dark skinned outgroup face is presented, the amygdala reacted the same.

Cross-Race Identification Bias

Because we have more experience with people in our ingroups compared to those in outgroups, it turns out that we therefore have a better ability to distinguish faces of people in our own racial group compared to those in other racial groups. This well-established finding has become known as the "cross-race identification bias" (Malpass and Kravitz 1969; Meissner and Brigham 2001; Platz and Hosch 1988). Research indicates that the brain devotes more cortical resources to the categorization of faces from our own race compared to those of other races (Feng et al. 2011). In one of the first studies to examine the neural systems related to same/other race facial memory and identification, Golby et al. (2001) exposed Black and White men to photos of faces of other Black and White individuals, and recorded their task performance while in an fMRI scanner. Results indicated superior memory for same-race faces, as predicted. Golby and his colleagues suggest that the differences in same vs. different race recognition and brain activation could be due to differential experience with members of one's own race over encounters with those of other races. That differential experience from one's lifetime of experience does not appear to be fixed however. Research by Tanaka and Pierce (2009) indicates that when participants were trained to individuate members from another race, they demonstrated an enhanced recognition of novel faces and evoked increased activation of the N250 component, associated with expert perception and recognition.

fMRI data showed that memory differences for both groups was related to activation of the left fusiform cortex and right parahippocampal and hippocampal areas. The authors suggest that the fusiform areas are therefore implicated in generating the cross-race identification bias. A problem with the Golby et al. study is that we cannot definitively conclude that the fusiform area is specialized for facial recognition (Phelps 2001). Previous research has shown, for example, that right, not left fusiform activation leads to good facial recognition (Kanwisher 2000; Kuskowski and Pardo 1999) and another study found that inverted faces strongly activate the fusiform area (Kanwisher et al. 1998). Moreover, some recent research shows that the superiority of same-face recognition can be eliminated simply by threatening the distinctiveness of one's own racial ingroup (Wilson and Hugenberg 2010). Does this mean that the activation of the fusiform area (by facial stimuli) can be suppressed by an overriding social concern? Clearly, much more research is needed before we can be sure about the neural correlates of the cross-race identification bias.

Event-Related Potentials (ERP) and Prejudice

Another popular method of assessing neural activity during cognitive processing is the use of event-related potentials (ERP) (for reviews, see Ito et al. 2006; Kubota and Ito 2009). ERPs show the averaged electrical activity of groups of neurons that fire when one views a stimulus or engages in some cognitive activity. The ERP is measured in the amplitude of the electrical waveform (indicating the amount of cognitive effort used during the activity) and the latency of the response (indicating the point at which the cognitive process was completed (Kubota and Ito 2009)). One nice advantage of the ERP method is that it can very specifically isolate onset and cessation of a cognitive event, even down to milliseconds. Below, we examine how researchers have used ERP to enhance our understanding of the neural underpinnings of cognitive activity related to person perception (Bartholow and Dickter 2007; Ito 2011) in general and stereotyping and prejudice against others (Ito et al. 2007) in particular.

Much research using ERPs has been devoted to understanding the neural underpinnings of initial social categorization. If we can understand when this takes place, and what brain components are responsible for this categorization, we will have a richer understanding of the beginning processes that underlie potential prejudice and stereotyping. ERP studies have demonstrated that we automatically categorize others along "primitive categories" (race, age, gender) within milliseconds of perceiving the other person (Ito and Bartholow 2009; Ito and Cacioppo 2000; Kubota and Ito 2007). Specifically, people process gender cues in the P200 component approximately 180 ms after viewing the target (Mouchetant-Rostaing et al. 2000), while racial cues tend to be processed in both the P200 and the N100 components within 122 ms (Correll et al. 2006; Willadsen-Jensen and Ito 2006).

Interestingly, in one study, even when attention was directed toward more specific social judgments (i.e. personality assessments), the processing of race was not affected (Ito and Urland 2005).

Research using ERPs has also demonstrated that the P300 component may show greater activation when one encounters a stimulus that does not fit one's expectations (Bartholow et al. 2001). Ito and her colleagues (2004) found evidence that the P300 component shows greater activation in Whites who, on a paper and pencil measure of explicit racism – (the Modern Racism Scale (McConahay 1986)) – showed prejudice against Blacks. Though more research is needed to support this research, it could therefore be the case that the 500 ms latency of the P300 activation represents the early processing of stereotypes and prejudice in the social perception process which begins with the social categorization of faces discussed earlier.

Implicit Stereotyping and Prejudice

For over 30 years, social and cognitive psychologists have made an important distinction between cognitive processes that are intentional vs. unintentional (Devine 1989; Graf and Schacter 1985; Kihlstrom 1990; Uleman and Bargh 1989). While different researchers label these distinctions differently (e.g. unconscious vs. conscious, or automatic vs. controlled), it is likely these can be subsumed under the broader distinction of implicit vs. explicit cognitive processes (Abelson 1994; Greenwald and Banaji 1995). Explicit cognitive activity are those cognitive processes about which the individual is consciously aware and can control. Implicit cognitive processes are those about which the individual is not aware and cannot control. Research has shown that one factor that influences whether one will stereotype another is if that person has been previously exposed to information that they are not aware of (Banaji and Greenwald 1994). In other words, suppose you are exposed to some biasing information about an outgroup. Later, you are tested about your attitudes toward that outgroup. At the time of testing, you were unaware of, and could not recall the prior exposure to that biasing information. Nevertheless, to the degree that your explicit judgments and behavior at the time of testing show a bias (relative to a control group who were not exposed to earlier biasing information), this would imply that an "implicit stereotyping" process was occurring (Greenwald and Banaji 1995).

We have reviewed evidence that thus far suggests an automatic amygdala activation to outgroup faces, especially among whites perceiving black faces. Cunningham and his colleagues (2004) wondered whether this automatic response might be attenuated by more deliberative cognitive control processes that are engaged later in the social perception process. They exposed some participants to black and white faces for 30 ms (a level below conscious perception), and found automatic amygdala activation, stronger for black than white faces. This difference was stronger for those who showed stronger race bias on the IAT. However, among those participants who were exposed to black and white faces for 525 ms (at conscious perception), there were no amygdala activation differences among those perceiving black vs. white faces. However, Cunningham et al. did find significant activation of the dorsolateral prefrontal cortex (DLPFC) and anterior cingulate (ACC) areas associated with inhibition, conflict and control. The prefrontal cortex activation was strongest among those Whites showing the highest anti-Black IAT scores, which Cunningham et al. interpret as corresponding to more material that needs to be controlled/inhibited. The researchers suggest that participants in this condition were seeking to control unwanted prejudicial responses to the black faces. These data suggest that it is indeed possible to control automatic negative reactions to outgroups. The inhibition function for the DLPFC was supported by Richeson et al. (2003) and Knutson et al. (2007). Importantly, data from Richeson et al. (2003) and Amodio et al. (2007) show that the ACC is specifically implicated in the detection of the need for cognitive control, and detecting potential conflict between explicit and implicit attitudes, respectively.

Conclusion

Since the advent of modern social neuroscience research on prejudice and stereotyping nearly two decades ago, considerable advances have been made in our understanding of the neural substrates that seem to be linked to certain processes and brain structures involved in the timecourse of perceiving and later thinking about outgroup members. We know that the perception and categorization of other-race faces is a rather automatic process that occurs very early (often within 130 ms), and involves the amygdala. Though initially believed to reflect a fear response to the outgroup face, researchers now think the amygdala response is an indicator of strength of response, and not valence. Another fascinating finding is the reciprocal influence of explicit cognition on the control of both conscious processes designed to not engage in stereotyping and also significantly attenuates amygdala responses early in social perception. While neuroimaging can implicate particular neural structures that appear to be specifically activated while engaging in a particular cognitive activity, it is important to remember that most neuroimaging studies point to *multiple* structures being involved in social perception, at various points in time, when perceiving outgroups. Moreover, certain structures are not only activated, for example, at the earliest phases of social perception, but often are activated at multiple points in the social perception process, depending on the perceiver, the target, and the perceiver's cognitive goals (Ito 2010; Ito and Bartholow 2009). Though we now have over 20 years of neuroimaging studies on stereotyping and prejudice, our understanding of the complex ways that brain structures interact in social perception is still quite rudimentary. Much more continuing research is needed to flesh out the many questions that remain. It is an exciting time for researchers in social neuroscience involved in addressing prejudice and stereotyping, as the more we learn about the parts of the brain that are involved in prejudice, the better equipped we are to understand how and when such prejudice can be eliminated prior to its influence on behavior, and other people.

References

Abelson, R. P. (1994). A personal perspective on social cognition. In P. G. Devine, D. L. Hamilton, & T. M. Ostrom (Eds.), *Social cognition: Impact on social psychology*. San Diego: Academic.

Allport, G. W. (1954). The nature of prejudice. Reading, MA: Addison-Wesley.

Amodio, D. M., & Lieberman, M. D. (2009). Pictures in our heads: Contributions of fMRI to the study of prejudice and stereotyping. In T. D. Nelson (Ed.), *Handbook of prejudice, stereotyping and discrimination* (pp. 347–365). New York: Psychology Press.

Amodio, D. M., Harmon-Jones, E., & Devine, P. G. (2007). Mechanisms for the regulation of intergroup responses: Insights from a social neuroscience approach. In E. Harmon-Jones & P. Winkielman (Eds.), *Fundamentals of social neuroscience* (pp. 353–375). New York: Guilford.

Anderson, A. K., & Phelps, E. A. (2000). Expression without recognition: Contributions of the human amygdala to emotional communication. *Psychological Science, 11*(2), 106–111.

Banaji, M., & Greenwald, A. G. (1994). Implicit stereotypes and prejudice. In M. P. Zanna & J. M. Olson (Eds.), *The psychology of prejudice: The Ontario symposium* (Vol. 7, pp. 55–76). Hillsdale: Erlbaum.

Bartholow, B. D., & Dickter, C. L. (2007). Social cognitive neuroscience of person perception: A selective review focused on the event-related brain potential. In E. Harmon-Jones & P. Winkielman (Eds.), *Social neuroscience: Integrating biological and psychological explanations of social behavior* (pp. 376–400). New York: Guilford.

Bartholow, B. D., Fabiani, M., Gratton, G., & Bettencourt, B. A. (2001). A psychophysiological examination of cognitive processing of affective responses to social expectancy violations. *Psychological Science, 12*, 197–204.

Brauer, M., & Er-rafiy, A. (2011). Increasing perceived variability reduces prejudice and discrimination. *Journal of Experimental Social Psychology, 47*, 871–881.

Brewer, M. B. (1979). In-group bias in the minimal intergroup situation: A cognitive-motivational analysis. *Psychological Bulletin, 86*, 307–324.

Brewer, M. B. (1988). A dual-process model of impression formation. In T. K. Srull & R. S. Wyer (Eds.), *Advances in social cognition* (Vol. 1, pp. 1–36). Hillsdale: Erlbaum.

Buck, R. (1990). William James, the principles of psychology, and experimental psychology. *Personality and Social Psychology Bulletin, 16*(4), 612–625.
Cabeza, R., & Nyberg, L. (2000). Imaging cognition II: An empirical review of 275 PET and fMRI studies. *Journal of Cognitive Neuroscience, 12*(1), 1–47.
Cacioppo, J. T. (1994). Social neuroscience: Autonomic, neuroendocrine, and immune responses to stress. *Psychophysiology, 31*(2), 113–128.
Correll, J., Urland, G. R., & Ito, T. A. (2006). Event-related potentials and the decision to shoot: The role of threat perception and cognitive control. *Journal of Experimental Social Psychology, 42*, 120–128.
Cunningham, W. A., Johnson, M. K., Raye, C. L., Gatenby, J. C., Gore, J. C., & Banaji, M. R. (2004). Separable neural components in the processing of Black and White faces. *Psychological Science, 15*(12), 806–813.
Devine, P. G. (1989). Stereotypes and prejudice: Their automatic and controlled components. *Journal of Personality and Social Psychology, 56*, 5–18.
Dixon, T. L., & Maddox, K. B. (2005). Skin tone, crime news, and social reality judgments: Priming the stereotype of the dark and dangerous black criminal. *Journal of Applied Social Psychology, 35*(8), 1555–1570.
Eberhardt, J. L. (2005). Imaging race. *American Psychologist, 60*(2), 181–190.
Feng, L., Liu, J., Li, J., Li, L., Ge, L., Tian, J., & Lee, K. (2011). The other face of the other-race effect: An fMRI investigation of the other-race face categorization advantage. *Neuropsychologia, 49*(13), 3739–3749.
Fiske, S. T. (1998). Stereotyping, prejudice, and discrimination. In D. T. Gilbert, S. T. Fiske, & G. Lindzey (Eds.), *The Handbook of Social Psychology* (pp. 357–411). New York: McGraw-Hill.
Fiske, S. T., & Neuberg, S. L. (1990). A continuum of impression formation, from category-based to individuating processes: Influences of information and motivation on attention and interpretation. In M. P. Zanna (Ed.), *Advances in experimental social psychology* (Vol. 23, pp. 1–74). New York: Academic.
Gardner, H. (1985). *The mind's new science: A history of the cognitive revolution*. New York: Basic Books.
Golby, A. J., Gabrieli, J. D. E., Chiao, J. Y., & Eberhardt, J. L. (2001). Differential responses in the fusiform region to same-race and other-race faces. *Nature Neuroscience, 4*(8), 845–850.
Graf, P., & Schacter, D. (1985). Implicit and explicit memory for new associations in normal and amnesic subjects. *Journal of Experimental Psychology: Learning, Memory, and Cognition, 11*, 501–518.
Greenberg, J., Pyszczynski, T. A., & Solomon, S. (1982). The self-serving attributional bias: Beyond self-presentation. *Journal of Experimental Social Psychology, 18*(1), 56–67.
Greenwald, A. G., & Banaji, M. (1995). Implicit social cognition: Attitudes, self- esteem, and stereotypes. *Psychological Review, 102*, 4–27.
Greenwald, A. G., McGhee, D. E., & Schwartz, J. L. K. (1998). Measuring individual differences in implicit cognition: The Implicit Association Test. *Journal of Personality and Social Psychology, 74*(6), 1464–1480.
Hamilton, D. L. (1976). Cognitive biases in the perception of social groups. In J. S. Carroll & J. W. Payne (Eds.), *Cognition and social behavior* (pp. 81–93). Hillsdale: Erlbaum.
Harris, L. T., & Fiske, S. T. (2006). Dehumanizing the lowest of the low: Neuroimaging responses to extreme outgroups. *Psychological Science, 17*(10), 847–853.
Hart, A. J., Whalen, P. J., Shin, L. M., McInerney, S. C., Fischer, H., & Rauch, S. L. (2000). Differential response in the human amygdala to racial outgroup vs. ingroup face stimlui. *Neuroreport: An International Journal for the Rapid Communication of Research in Neuroscience, 11*(11), 2351–2355.
Ito, T. A., & Cacioppo, J. (2000). Electrophysiological evidencce of implicit and explicit categorization processes. *Journal of Experimental Social Psychology, 36*, 660–676.
Ito, T. A. (2010). Reflections on social neuroscience. *Social Cognition, 28*(6), 686–694.
Ito, T. A. (2011). Perceiving social category information from faces: Using ERP's to study person perception. In A. Todorov, S. T. Fiske, & D. A. Prentice (Eds.), *Social neuroscience: Toward understanding the underpinnings of the social mind* (pp. 85–100). New York: Oxford University Press.
Ito, T. A., & Bartholow, B. D. (2009). The neural correlates of race. *Trends in Cognitive Sciences, 13*(12), 524–531.
Ito, T. A., & Urland, G. R. (2005). The influence of processing objectives on the perception of faces: An ERP study of race and gender perception. *Cognitive, Affective, and Behavioral Science, 5*, 21–36.
Ito, T. A., Thompson, E., & Cacioppo, J. T. (2004). Tracking the timecourse of social perception: The effects of racial cues on event-related brain potentials. *Personality and Social Psychology Bulletin, 30*(10), 1267–1280.
Ito, T. A., Urland, G. R., Willadsen-Jensen, E., & Correll, J. (2006). The social neuroscience of stereotyping and prejudice: Using event-related brain potentials to study social perception. In J. T. Cacioppo, P. S. Visser, & C. L. Pickett (Eds.), *Social neuroscience: People thinking about thinking people* (pp. 189–212). Cambridge, MA: MIT Press.
Ito, T. A., Willadsen-Jensen, E., & Correll, J. (2007). Social neuroscience and social perception: New perspectives on categorization, prejudice, and stereotyping. In E. Harmon-Jones & P. Winkielman (Eds.), *Social neuroscience: Integrating biological and psychological explanations of social behavior* (pp. 401–421). New York: Guilford.
James, W. (1890). *The principles of psychology*. New York: Holt.
Jones, J. (1997). *Prejudice and racism* (2nd ed.). New York: McGraw-Hill.

Jussim, L., Cain, T. R., Crawford, J. T., Harber, K., & Cohen, F. (2009). The unbearable accuracy of stereotypes. In T. D. Nelson (Ed.), *Handbook of prejudice, stereotyping, and discrimination* (pp. 199–227). New York: Taylor & Francis.

Kanwisher, N. (2000). Domain specificity in face perception. *Nature Neuroscience, 3*, 759–763.

Kanwisher, N., Tong, F., & Nakayama, K. (1998). The effect of face inversion on the human fusiform face area. *Cognition, 68*, B1–B11.

Kihlstrom, J. (1990). The psychological unconscious. In L. A. Pervin (Ed.), *Handbook of personality: Theory and research* (pp. 445–464). New York: Guilford Press.

Knutson, K. M., Mah, L., Manly, C. F., & Grafman, J. (2007). Neural correlates of automatic beliefs about gender and race. *Human Brain Mapping, 28*, 915–930.

Kubota, J. T., & Ito, T. A. (2007). Multiple cues in social perception: The timecourse of processing race and facial expression. *Journal of Experimental Social Psychology, 43*, 738–752.

Kubota, J. T., & Ito, T. A. (2009). You were always on my mind: Event-related potentials inform impression formation research. In T. D. Nelson (Ed.), *Handbook of stereotyping, prejudice, and discrimination* (pp. 333–345). New York: Psychology Press.

Kuskowski, M. A., & Pardo, J. V. (1999). The role of the fusiform gyrus in successful encoding of face stimuli. *NeuroImage, 9*(6), 599–610.

LeDoux, J. E. (1996). *The emotional brain.* New York: Simon & Schuster.

Lieberman, M. D., Hariri, A., Jarcho, J. M., Eisenberger, N. I., & Bookheimer, S. Y. (2005). An fMRI investigation of race-related amygdala activity in African- American and Caucasian-American individuals. *Nature Neuroscience, 8*(6), 720–722.

Lippmann, W. (1922). *Public opinion.* New York: Harcourt, Brace, Jovanovich.

Mackie, D. M., Maitner, A. T., & Smith, E. R. (2009). Intergroup emotions theory. In T. D. Nelson (Ed.), *Handbook of prejudice, stereotyping, and discrimination* (pp. 285–308). New York: Psychology Press.

Maddox, K. B., & Chase, S. G. (2004). Manipulating subcategory salience: Exploring the link between skin tone and perception of Blacks. *European Journal of Social Psychology, 34*(5), 533–546.

Maddox, K. B., & Gray, S. A. (2002). Cognitive representations of Black Americans: Reexploring the role of skin tone. *Personality and Social Psychology Bulletin, 28*(2), 250–259.

Malpass, R. S., & Kravitz, J. (1969). Recognition for faces of own and other race. *Journal of Personality and Social Psychology, 13*, 330–334.

McConahay, J. B. (1986). Modern racism, ambivalence and the modern racism scale. In J. F. Dovidio & S. L. Gaertner (Eds.), *Prejudice, discrimination andracism* (pp. 91–125). New York: Academic Press.

Meissner, C. A., & Brigham, J. C. (2001). 30 years of investigating the own-race bias in memory for faces: A meta-analytic review. *Psychology, Public Policy, and Law, 7*, 3–35.

Mouchetant-Rostaing, Y., Giard, M., Delpuech, C., Echallier, J., & Pernier, J. (2000). Early signs of visual categorization for biological and non-biological stimuli in humans. *Neuroreport: For Rapid Communication of Neuroscience Research, 11*(11), 2521–2525.

Nelson, T. D. (2006). The Psychology of prejudice (2nd Ed.). New York: Allyn & Bacon.

Nelson, T. D. (Ed.). (2009). *Handbook of prejudice, stereotyping, and discrimination.* New York: Taylor & Francis.

Ochsner, K. N., & Lieberman, M. D. (2001). The emergence of social cognitive neuroscience. *American Psychologist, 56*(9), 717–734.

Ostrom, T. M., & Sedikides, C. (1992). Outgroup homogeneity effects in natural and minimal groups. *Psychological Bulletin, 112*, 536–552.

Phelps, E. A. (2001). Faces and races in the brain. *Nature Neuroscience, 4*(8), 775–776.

Phelps, E. A., & Thomas, L. (2003). Race, behavior, and the brain: The role of neuroimaging in understanding complex social behaviors. *Political Psychology, 24*(4), 747–758.

Phelps, E. A., O'Connor, K. J., Cunningham, W. A., Funayama, E. S., Gatenby, J. C., Gore, C., & Banaji, M. (2000). Performance on indirect measures of race evaluation predicts amygdala activation. *Journal of Cognitive Neuroscience, 12*(5), 729–738.

Platz, S. J., & Hosch, H. M. (1988). Cross-racial/ethnic eyewitness identification: A field study. *Journal of Applied Social Psychology, 18*, 972–984.

Richeson, J. A., Baird, A. A., Gordon, H. L., Heatherton, T. F., Wyland, C. L., Trawalter, S., & Shelton, J. N. (2003). An fMRI investigation of the impact of interracial contact on executive function. *Nature Neuroscience, 6*, 1323–1328.

Ronquillo, J., Denson, T. F., Lickel, B., Lu, Z., Nandy, A., & Maddox, K. B. (2007). The effects of skin tone on race-related amygdala activity: An fMRI investigation. *Social Cognitive and Affective Neuroscience, 2*(1), 39–44.

Tanaka, J. W., & Pierce, L. A. (2009). The neural plasticity of other-race face recognition. *Cognitive, Affective, & Behavioral Neuroscience, 9*, 122–131.

Taylor, S. E. (1981). A categorization approach to stereotyping. In D. L. Hamilton (Ed.), *Cognitive processes in stereotyping and intergroup behavior* (pp. 88–114). Hillsdale: Erlbaum.

Taylor, S. E. (1989). *Positive Illusions: Creative self-deception in the healthy mind.* New York: Basic Books.

Uleman, J., & Bargh, J. (Eds.). (1989). *Unintended thought*. New York: Guilford Press.
Watson, S. (2011). *How fMRI works*. http://health.howstuffworks.com/medicine/tests-treatment/fmri1.htm
Whalen, P. J., Shin, L. M., Mcinerney, S. C., Fischer, H., Wright, C. I., & Rauch, S. L. (2001). A functional MRI study of human amygdala responses to facial expressions of fear versus anger. *Emotion, 1*(1), 70–83.
Wheeler, M. E., & Fiske, S. T. (2005). Controlling racial prejudice and stereotyping: Social cognitive goals affect amygdala and stereotype activation. *Psychological Science, 16*, 56–63.
Willadsen-Jensen, E. C., & Ito, T. A. (2006). Ambiguity and the timecourse of racial perception. *Social Cognition, 24*, 580–606.
Wilson, J. P., & Hugenberg, K. (2010). When under threat, we all look the same: Distinctiveness threat induces ingroup homogeneity in face memory. *Journal of Experimental Social Psychology, 46*, 1004–1010.

Chapter 23
Dominance, Violence, and the Neurohormonal Nexus

Allan Mazur

Introduction

Sociologist Randall Collins emphasizes that in face-to-face interaction, people rarely act violently, that virtually everyone is reluctant to physically damage another person. The reason is that violent confrontation is fraught with tension and fear, which act as a protective emotional barrier against inflicting harm or being harmed. When violence does occur, tension and fear usually ensure that attacks are brief and incompetent. Terrified shooters in a gun duel are unlikely to hit a target; clashing gang members are more bluff and bluster than lethal attackers. This picture completely contradicts the portrayal in action movies, where violence is perpetrated easily and efficiently, often over extended periods, usually free of anxiety (Collins 2008; Mazur 2009a). Still, obviously, violence does occasionally occur, and we may explain much of it as extreme outcomes of person-to-person dominance contests and their underlying neuroendocrinology.

Warfare is outside this inquiry. My concern is physical violence between people who are in proximity, who recognize one another as individuals and could talk to (or shout at) each other. Often this action is illegal and in the extreme is murder. Illegal killing occurs for diverse reasons including drug marketing, robbery, jealousy, mental derangement, religious or ideological commitment, and cash payment. Here I focus on what criminologist Jack Katz (1988) calls the most numerous type of criminal homicide, the impassioned killing of someone for what the killer regards as a good moralistic reason, perhaps the defense of his family, his property, or his good name.

Usually these murders of passion occur without premeditation. The episodes Katz had in mind develop quickly, occurring without thought of legal consequences. Often the killers are surprised by the unintended fatality, regarding the outcome as an accident. Reflecting their lack of forethought, many killers do not attempt to escape, or do so ineptly. Police make arrests in roughly 80% of homicide cases – usually within a day of the crime – compared to arrests in about 25% of robberies and 15% of burglaries.

Death is not necessarily the desired end point of an impassioned attack so much as hurting or physically punishing the victim. Whether an attack ends as a criminal homicide or an aggravated assault may be incidental, depending on such chance factors as the time to reach an emergency room, the quality of medical service, whether a gun was used, whether the falling victim's head hit concrete,

A. Mazur (✉)
Maxwell School of Citizenship and Public Affairs, Syracuse University,
Syracuse, NY 13244, USA
e-mail: amazur@syr.edu

D.D. Franks and J.H. Turner (eds.), *Handbook of Neurosociology*, Handbooks of Sociology and Social Research, DOI 10.1007/978-94-007-4473-8_23, © Springer Science+Business Media B.V. 2013

and so on. There is little reason to think such killings differ much from impassioned attacks whose victims survived. Murders are better documented than non-lethal assaults and therefore more amendable to analysis.

Victims are usually relatives, friends, or at least acquainted with their killers. Criminologists have repeatedly replicated Marvin Wolfgang's (1958) classic study of homicide, showing that fatal aggression between men is usually precipitated by a trivial altercation, perhaps an insult, curse, or jostling. This is followed by an escalation of hostile verbal actions that may look no different than many nonfatal arguments.

The Difference Between Aggression and Dominance

Numerous animal experiments, especially on rodents, show that raising testosterone increases aggressiveness. In interpreting this work, it is important to distinguish aggressive behavior from dominance behavior. An individual will be said to act *aggressively* if its apparent intent is to inflict physical injury on a member of its species. An individual acts *dominantly* if its apparent intent is to achieve or maintain high status – i.e., to obtain power, influence, or valued prerogatives – over a conspecific. Rodents typically dominate aggressively, but that is not true among the higher primates.

The distinction between aggression and dominance is particularly important for humans, because we normally assert our dominance without violence. Sports, spelling bees, elections, criticism, competitions for promotion, and academic jousting all involve contests for domination without intending physical injury. We understand that there are different motivations for dominance and violent aggression, though they may sometimes work concurrently.

Until fairly recently, researchers thought that young men with high testosterone were especially aggressive, rather like male rodents. This picture remains in the public mind, including images of body builders on anabolic steroids (chemically similar to testosterone) being prone to violent "road rages." It fact, empirical research shows little if any *direct* linkage between testosterone and physical aggressiveness. However, we shall see that testosterone may play an *indirect* role in interpersonal violence (Mazur 2008).

A common view among today's researchers is that testosterone is related primarily to dominant behavior among men, not to aggression as such (Mazur and Booth 1998; Archer 2006). Nearly all animal studies once interpreted as linking testosterone to aggression may as easily be interpreted as linking testosterone with dominance. On theoretical grounds, dominating mechanisms – whether aggressive or nonaggressive in form – would confer an evolutionary advantage in helping an individual acquire valued resources, especially in competition for mates. This is not simply a matter of a dominant man taking what he wants; women regard men who *look* dominant as attractive.

A Primer on Testosterone

Testosterone is the primary *androgen*, a class of steroid hormones that develop and maintain masculine features. Although testosterone is made in the adrenal cortex and ovary of females, it is produced in far greater amounts by the Leydig cells of the testis. Testosterone in men is secreted into the bloodstream in spurts, so measured levels can change considerably within a few minutes. The hormone has a circadian rhythm in both sexes, highest and most variable in the morning, lower and more stable during the afternoon. Testosterone is highly diluted in the blood, requiring sophisticated measurement techniques. Early studies on humans required blood samples, but most research today measures testosterone from saliva samples.

Many effects that we explain today by testosterone deficiency were obtained since ancient times by castration of men and animals, which was practiced not only to prevent fertility but also to prevent the development of secondary sexual characteristics, produce docility, reduce sex drive, and – in butchered animals – to produce fatter, more tender meat. Castrating a male chick, for example, makes its adult flesh more edible, and the capon fails to develop the rooster's head furnishings (red comb and wattles – markers of reproductive competence), does not crow or court hens, and does not fight other cocks. In Asia, eunuchs were presumed to be safe harem guards because of their lack of both interest and ability to copulate. Male sopranos and contraltos, emasculated to maintain their prepubescent voice range, were prominent in the opera and church music of seventeenth and eighteenth century Europe.

Our modern understanding began in the 1930s after the isolation and identification of testosterone. Reminiscent of the Curies' heroic extraction of minute amounts of radium from a ton of pitchblende, Fred Koch and his coworkers mashed tons of bull testicles to fractionate ounces of material sufficiently pure to make the combs of capons grow bright red. (Another researcher distilled 25,000 l of policemen's urine to obtain 15 mg of the androgen androsterone.) Chemical synthesis followed quickly, enabling experimenters to replace or enhance testosterone in animal subjects and human patients. An example is the classic study of hen peck-orders by Allee et al. (1939), who injected testosterone into low-ranking hens. These females became aggressive, and each rose in her status hierarchy, some to the top position. Furthermore, their comb size increased (a male characteristic), egg-laying was suppressed, some began crowing (rare in hens), and a few began courting other hens.

Testosterone Works Differently Perinatally, at Puberty, and in Adulthood

It is now clear that testosterone affects human males importantly but differently at three stages of life: perinatally (in utero and shortly after birth), during puberty, and in adulthood. This chapter focuses on the adult stage, but a brief review of earlier effects is worthwhile.

The mammalian fetus of both XX and XY individuals begins with undifferentiated sexual parts. A gene on the Y chromosome causes the asexual gonads to develop as testes; lacking this gene the gonads become ovaries. The sex chromosomes have little more to do with sex differentiation which hereafter is driven by hormones produced in the now sex-specific gonads. The testes produce testosterone during gestation, and production peaks again a month or two after birth, then declines by 6 months of age to the low range seen in later childhood. Testosterone and other testicular secretions cause the external genitalia to form into penis and scrotum rather than clitoris and labia, and internal ducts take the male form. The central nervous system is masculinized. The general rule, somewhat simplified, is that early exposure to greater amounts of testosterone will produce more male characteristics and fewer female characteristics, while less exposure to testosterone will produce the reverse. Perinatal manipulation of animal subjects, and developmental abnormalities among humans, show convincingly that even genetic females will show male forms if dosed early enough with testosterone, and genetic males will show female forms if deprived of the hormone.

Perinatal testosterone exposure affects behavior in a number of animal species. For example, young male rhesus monkeys normally engage in more threats and rough-and-tumble play than do females, but when testosterone is administered to pregnant monkeys, their pseudo-hermaphroditic female offspring exhibit male-type play behavior. Furthermore, by limiting testosterone administration to the later part of gestation, female offspring are produced who exhibit male-type play but retain female appearing genitals, showing that behavioral masculinization is independent of genital masculinization.

Many perinatal hormone effects are regarded as *organizing* the architecture of the body and brain, and the distribution of hormone receptors, into a relatively male-like configuration. When male testosterone increases later in life, it *activates* these preexisting structures. Thus, behaviors derive from the interaction of long-term organizational *and* shorter-term activation effects.

The testes greatly increase production of testosterone at puberty, elevating prepubescent serum levels ten or more times. This promotes growth of the penis, larynx (and deeper voice), muscles, beard and body hair, and sex interest. Boys who are hypogonadal or castrated before puberty do not experience these changes, but they can be induced by testosterone replacement therapy.

It is tempting to assume the testosterone surge in adolescent boys to be the cause of their seemingly heightened aggressiveness or pugnacity at that age. However, research has not verified a causal link between the hormonal and behavioral changes. First of all, physical aggression in boys does *not* generally rise during adolescence (Tremblay 2000). Actually, it is littler boys who are most prone to physically assault their peers, though they lack the muscle or weapons to do much damage. The well-documented rise in boys' antisocial behavior with puberty is due mainly to nonviolent delinquency such as vandalism and status violations (Rutter et al. 1998).

Attempts to evaluate the contribution of testosterone to adolescent social behavior have produced mixed findings. These are difficult studies to conduct because investigators must untangle the direct effect testosterone from other physical changes in the boy's body at puberty, which affect how people respond to him. Also, there are important *social* changes during the early teen years – entry into high school, taking a job, prolonged absence from parents, more dependence on peer approval – any of which may affect behavior independently of hormonal effects.

The empirical picture now emerging is that the adolescent rise in testosterone does *not* lead simply and directly to increased antisocial behavior among teenage boys. But once the social context is taken into account, we *do* see an effect of testosterone on dominance *in a direction consistent with the behavior of peers*. This was most dramatically shown when Rowe et al. (2004) considered whether boys did, or did not, have deviant peers. Boys with high testosterone committed a large number of "conduct disorders" (usually non-violent antisocial actions like lying or breaking in) *if* they had delinquent peers, but there was no testosterone-disorder relationship among boys without delinquent peers. As if in mirror image, boys with high testosterone were more likely to be chosen by other children as team leaders if they did *not* have delinquent peers, but there was no testosterone-leader relationship among boys with delinquent peers.

The primary lesson of this research is that one cannot assess the effect of hormones on behavior without taking into account the social context. An adolescent's behavior is importantly affected by relationships with parents and peers. Rising testosterone might have a different effect on a boy in a delinquent gang than on a member of the Boy Scouts (Booth et al. 2006).

Among American males, after testosterone peaks in the late teens and early twenties, it usually declines slowly with age (Mazur 2009b). The hormone decline among middle-age men does not reliably occur in non-industrial societies. In the United States, it seems a consequence of American men getting fatter as they get older. Among 1,880 Air Force veterans who participated in four medical examinations from 1982 to 1992, testosterone declined only among men who gained more than 10% in body fat. For men with slighter increases in fat, testosterone remained essentially level. For men who lost fat over the decade, testosterone actually increased (Mazur 1998).

By the end of teenage, the physical form of a boy has changed into that of a man so testosterone no longer influences behavior through major reorganization of the body. However, the level of testosterone circulating in the bloodstream at any moment may affect dominating behavior by activating receptors in organs or the nervous system. The remainder of this chapter focuses on such post-adolescent effects.

Reciprocal Causation

There is considerable evidence from a variety of settings that in men, circulating testosterone is correlated with dominant behavior (Burnham 2007; Carré et al. 2011). Most impressive are recent studies linking, with some qualification, endogenous increases (or decreases) in testosterone with increased (decreased) readiness to compete further (Carré et al. 2009; Mehta and Josephs 2006). However, correlation does not imply causation, and the question remains: Is high testosterone a *cause* of dominant and antisocial behavior? This could be answered with double-blind experiments, comparing the behavior of men whose testosterone was raised pharmaceutically with a control group receiving a placebo. If dominant actions increased under the testosterone treatment, that would implicate the hormone as a cause of the behavior. Such controlled experimentation has barely begun and not yet established an unequivocal conclusion (Zethraeus et al. 2009; Eisenegger et al. 2010; Zak et al. 2009).

If there is a link between testosterone and dominance, primate studies suggest it is reciprocal. Not only does testosterone affect dominance, but changes in dominance behavior or in social status cause changes in testosterone level. We have strong evidence on this "reverse" effect in humans. By now there have been many reports of testosterone changes in young men during athletic events as well as other competitive situations that are not physically taxing (Carré 2009; Maner et al. 2008; Mehta and Josephs 2008; Stanton et al. 2009b). Again with some qualification, male testosterone varies in predicable ways both before and after competitive matches.

First, testosterone rises shortly before their matches, as if in anticipation of the competition. This pre-competition boost may promote dominant behavior, increasing the chance victory, as occurs in rodents but has not yet been demonstrated in humans.

Second, for one or two hours after the match, testosterone of winners is usually high relative to that of losers. The win/loss effect even works vicariously, with supporters of one side or another in an athletic contest or an election showing corresponding changes in testosterone depending on whether or not "their" side won or lost (Bernhardt et al. 1998; Stanton et al. 2009a, b; Carré and Putnam 2010).

Thus, the pattern of testosterone fluctuations appears in nonphysical as well as physical competition, and in response to symbolic challenges and status changes among men. (Whether similar changes occur among women is unresolved.)

In theory, a man who has experienced a recent rise in testosterone, perhaps from a prior victory or a symbolic elevation in status, should be unusually assertive and may challenge someone of relatively high status in a contest for dominance. If the winner experiences rising testosterone as a result of his victory, this should sustain or increase his assertiveness and his display of dominant signs such as erect posture, sauntering or striding gait, and direct eye contact with others. Thus bolstered, the dominator may seek out new challenges and is primed to win them. The feedback between high testosterone and dominant demeanor may explain the momentum often associated with strings of triumphs. Success begets a high testosterone response, which begets more dominant behavior, which begets more success.

The loser experiences a drop in testosterone, reducing his assertiveness and increasing his display of deferential signs such as stooped posture, smiling, or eye aversion. Faced with a new dominance encounter, he is now at a psychic and physiological disadvantage. One defeat begets another because the loser becomes increasingly likely to retreat or submit. This may be an adaptive response, saving losers from further losses and possible damage.

Dominance Contests

Does testosterone play a role in daily challenges to status, either from strangers or from people well known to us? Like all primates, humans in face-to-face groups form themselves into fairly consistent dominance/status hierarchies so that higher-ranked members have more power, influence, and valued prerogatives than lower-ranked ones. Ranks are allocated either *cooperatively*, by consensus of those involved, or *competitively*, when there is disagreement over who should outrank whom.

To appreciate a person's decision to compete or cooperate, visualize two individuals (Ego and Alter) meeting for the first time. If their interaction is very brief or casual, the notion of ranking may never arise. However, in more extended or serious meetings, each will size up the other and gain some sense of their relative standings. If Ego thinks that Alter's status does or should exceed his own, he may defer to Alter without any dispute. In human terms, Ego may believe that Alter belongs in the higher rank, that Alter deserves it, that Alter could easily take it if Ego resisted, or that Alter would be more competent in the duties of high rank. In any case, ranks are allocated quickly and cooperatively. If Ego and Alter do not agree on their relative standings, then they may either break off the interaction or vie for the contested rank.

Ego's decision to compete or to comply will also depend on his motivation to dominate, which seems related to his testosterone level (among other factors). A man who has experienced a recent rise in testosterone, perhaps from a victory or a symbolic elevation in status, will be unusually assertive and may challenge someone of relatively high status. If both Ego and Alter decide to compete, their relative ranks are then determined by the outcome of one or more short dominance contests between them.

Nonhuman primates are commonly observed to establish and maintain their status hierarchies through a series of short face-to-face competitions between members of the group. Some competitions involve fierce combat; others are mild, as when one animal is obviously the more powerful and assertive or the other appears fearful. In such cases, a simple stare by the powerful animal, followed by the fearful animal's eye aversion or by its yielding something of value (perhaps food or a sitting place), may suffice. Sometimes a single contest is all that is needed to allocate ranks or to verify a preexisting rank relationship, but often the outcome is settled only after a series of contests.

A psychophysiological mechanism operating across this range of competition is the manipulation of stress levels (Mazur 2005). An exchange of threats or attacks is seen as an attempt by each individual to "outstress" or intimidate the other by inducing fear, anxiety, or other discomfort. Stress is experienced as both a feeling of discomfort and a syndrome of neurological responses. The individual who outstresses his adversary is the winner.

The model becomes clearer if we consider a concrete example. Consider two strangers, Ego and Alter, whose eyes meet, by chance, across a room. Let us say that one of the strangers, Ego, decides to hold the stare. The chance eye contact now becomes a dominance encounter. Ego's stare makes Alter uncomfortable. Alter may then avert his eyes, thus relieving his discomfort while, in effect, surrendering, or he may stare back, making Ego uncomfortable in return. In the latter case, the stare-down would continue, with each individual attempting to outstress the other until finally one person succumbed to the discomfort (and the challenger) by averting his eyes. The matter thus settled, the yielder usually avoids further eye contact, though the winner may occasionally look at the loser as if to verify his victory.

In this example, Ego's stare is assumed to elicit feelings of stress in Alter. Alter's eye aversion is assumed to relieve his own felt stress. Staring – the stress-inducing behavior – is a dominant sign associated with high status. Eye aversion is a deferential sign associated with low status. In other words, a dominant act (staring) elicits stress in the recipient; a submissive act (eye aversion) relieves stress in the actor. It is a central assumption of this model that most dominant and deferential acts work this way, inducing or relieving stress, respectively. These acts are the means whereby the adversaries wage their stress contest, each aiming "darts" at the other. Finally, when the stress is too great for one, he switches from dominant to deferential actions, thereby relieving his stress and simultaneously signaling his acceptance of the lower rank.

Within minutes of this outcome, we assume Ego (the loser) experiences a drop in testosterone, reducing his assertiveness, diminishing his propensity to display the dominant actions associated with high status, and increasing his display of such submissive signs as stooped posture, smiling, or eye aversion. Faced with a new dominance encounter, Ego is more likely than before to retreat or submit. On the other side, Alter, the winner, experiences the opposite effects: rising testosterone, increased assertiveness, and a display of dominant signs such as erect posture, sauntering or striding gait, and direct eye contact with others. Alter may seek out new dominance encounters and is bolstered to win them. This feedback between high (or low) testosterone and dominant (or submissive) demeanor would help to explain the momentum often associated with strings of triumphs or defeats: success begets a high testosterone response which begets more dominant behavior which begets more success

The Neurohormonal Nexus

Social interactions or other stimuli that are processed in the brain can affect the release of testosterone from the gonads. Some of the major steps are well known. Especially important are two small structures, the *hypothalamus* and *pituitary*, located at the base of all vertebrate brains. When signals from elsewhere in the brain reach the hypothalamus, it releases chemical factors that stimulate the nearby pituitary to release luteinizing hormone and follicle stimulating hormone, which flow through the blood until they reach receptors in the gonads, initiating the synthesis and release of testosterone. This is time consuming, on the order of seconds to minutes. The brain can also communicate almost instantly to distant organs via electric signals through the nervous system.

Besides signaling from the brain to other organs, reciprocal effects require that the other organs send signals back to the brain (or elsewhere in the body). Usually this is via hormones, which flow with the blood circulation from the organ of origination until they reach specialized receptor proteins, located on the surface or inside of certain cells. Each hormone binds to a particularly shaped receptor, like a key fitting into a lock to open the door. A hormone produces its effect only when it (or its metabolite) binds to its matching receptor.

Once testosterone enters the central nervous system, it can act directly onto androgen receptors or first metabolize to estradiol, which binds to estrogen receptors. These receptors are widespread in the central nervous system, so probably testosterone acts along multiple pathways and not in a simple, direct manner (Sarkey et al. 2008). Just what goes on inside the brain is little known and barely discernible but is the subject of active research.

The technique of functional magnetic resonance imaging (fMRI) allows investigators to "look into" the brains of human subjects while they are shown various stimulus materials or interact with other subjects, often communicating via computer linkage. The fMRI images show which parts of the brain have increased blood flow, indicating where increased neuronal activity occurs while subjects look at the stimuli. The technique requires that subjects remain immobile in the scanner for extended periods, so simulating a dominance contest requires considerable ingenuity on the part of investigators. One group showed subjects pictures of both angry and happy faces, after dosing them with testosterone (versus a placebo). If we accept that the angry faces (compared to the happy faces) simulate a dominance contest, then this work shows that during competition, testosterone activates a region of the orbitofrontal cortex, which is implicated in impulse control, as well as the amygdala and hypothalamus, both subcortical structures of the "limbic system" (Hermans et al. 2008). These structures are rich in androgen and estrogen receptors. Subjects in competitive games also show activity in the amygdala and orbitofrontal cortex (Carré et al. 2011). Possibly the orbitofrontal cortex mediates the effect of testosterone on dominant behavior (Mehta and Beer 2009). Studies along similar lines diverge is detailed findings but are consistent in showing the involvement of multiple brain structures, nearly always including the amygdala (Derntl et al. 2009; Stanton et al. 2009a, b; van Wingen et al. 2010). The amygdala is generally associated with emotional reactions including fear and in that broad sense is consistent with the model of dominance contests as involving an exchange of stressors.

Murder

In everyday life, dominance contests are based on the subtle manipulation of psychological and physiological stress, not on causing or even threatening physical harm to one's adversary (Mazur 2005). Most adult humans experience little violence, perhaps, as suggested at the outset of this chapter, because virtually everyone is reluctant to physically damage another person – because the tension and fear that accompany violent confrontation act as emotional barriers against inflicting harm or being harmed. But occasionally face-to-face competition escalates to a violent stage not originally intended or foreseen.

While a killing is rarely the outcome of a violent dominance contest, I focus on murder because it represents an unambiguous endpoint, has good statistics, has been much studied by criminologists, and is so distressing a phenomenon. In the U.S., killers and their victims are disproportionately young adult males. There is a large racial disparity in murder rates. For example, in New York City from 2002 through 2005 there were 1,662 reported murders. Killers were 93% male, 76% between ages 18 and 40, and 61% black (compared to a black population of 25%). Victims were 82% male, 69% between 18 and 40, and 60% black (McGinty 2006). The reasons for these patterns are not fully understood, but I propose that testosterone plays an indirect role.

Often the killer feels provoked by the victim, being the target of an insult, the butt of a joke, or an object of humiliation. "From the killer's perspective, the victim either teases, dares, defies, or pursues the killer… That the killer feels compelled to respond to a fundamental challenge to his worth is indicated as well by the frequent presence and the role of an audience" (Katz 1988: 20).

Extreme anger (rage) heightens the potential lethality of competition by shifting the adversaries' intentions from dominating to damaging the opposition. The red face of anger, with its raised voice and universally recognized facial gestures and body postures, likely entails particular neurophysiological actions. Alcohol may intensify the normal neurophysiological effect of rage.

Nisbett and Cohen (1996) attribute the historically high violence in the American South, compared to the North, to its "culture of honor" whereby Southern men, when challenged by insults to themselves or their families, are required to defend themselves as virtuous warriors or else lose face. Apparently as a result, Southern men are unusually alert to possible insults, reacting dominantly – sometimes violently – to speech or actions that might not be perceived as injurious in other cultures.

Leaving aside the particular historic roots of the South, there may be a general hypersensitivity to insult in *any* subculture that is (or once was) organized around young men who are unconstrained by traditional community agents of social control, as often occurs in frontier communities, gangs, among vagabonds or bohemians, and after breakdowns in the social fabric following wars or natural disasters. When young men place special emphasis on protecting their reputations, and they are not restrained from doing so, dominance contests become ubiquitous, the hallmark of male-to-male interaction.

The leading student of street behavior in America's inner cities, sociologist Elijah Anderson (1994: 88–89), vividly portrays the importance of dominance contests and their constant presence for poor young black men:

> (M)ost youths have…internalized the code of the streets…, which chiefly (has) to do with interpersonal communication…, (including) facial expressions, gait, and verbal expressions – all of which are geared mainly to deterring aggression….
>
> Even so, there are no guarantees against challenges, because there are always people looking for a fight to increase their share of respect – of "juice," as it is sometimes called on the street. Moreover, if a person is assaulted, it is important, not only in the eyes of his opponent but also in the eyes of his "running buddies," for him to avenge himself. Otherwise he risks being "tried" (challenged) or "moved on" by any number of others. To maintain his honor he must show he is not someone to be "messed with" or "dissed."
>
> …The craving for respect that results gives people thin skins. Shows of deference by others can be highly soothing, contributing to a sense of security, comfort, self-confidence, and self-respect…. Hence one must be ever vigilant against the transgressions of others or even appearing as if transgressions will be tolerated. Among

> young people, whose sense of self-esteem is particularly vulnerable, there is an especially heightened concern with being disrespected. Many inner-city young men in particular crave respect to such a degree that they will risk their lives to attain and maintain it.

We know that testosterone rises in men awaiting a contest, regardless of the eventual outcome of that contest. Generalizing to the street, hormone levels should be elevated in young men who are constantly vigilant against challenges to their reputations. Testosterone is also affected by the outcome of the contest, so persistent losers might be hormonally depressed, but most men – those with mixed outcomes or better – are expected to have elevated testosterone, thus hormonally primed to react assertively to real or perceived insult. But is it in fact true that testosterone levels are especially high among young black men living in poor neighborhoods?

The question has been addressed in two large epidemiological studies. One, based on 4,462 male U.S. Army veterans, answered in the affirmative. Younger black men who were poorly educated – those veterans most likely to be participants in inner-city honor cultures – did as a group have very high testosterone; among older men or well-educated men, race made no difference in hormone levels (Mazur 2005). However a second study, based on different sample of 1,637 men drawn from the U.S. population, showed testosterone of young, poorly educated black men to be unexceptional, thus contradicting the first study (Mazur 2009b). Like so many questions about neuroendocrinology and behavior, we do not have a clear answer.

Acknowledgment I appreciate the advice of Christoph Eisenegger.

References

Allee, W., Collias, N., & Lutherman, C. (1939). Modification of the social order in flocks of hens by the injection of testosterone propionate. *Physiological Zoology, 12*, 412–440.

Anderson, E. (1994). The code of the streets. *Atlantic Monthly, 5*, 81–94.

Archer, J. (2006). Testosterone and human behavior. *Neuroscience and Biobehavioral Reviews, 30*, 319–345.

Bernhardt, P., Dabbs, J., Fielden, J., & Lutter, C. (1998). Testosterone changes during vicarious experiences of winning and losing among fans at sporting events. *Physiology and Behavior, 65*, 59–62.

Booth, A., Granger, D., Mazur, A., & Kivlighan, K. (2006). Testosterone and social behavior. *Social Forces, 85*, 167–192.

Burnham, T. (2007). High-testosterone men reject low ultimatum game offers. *Proceeding Biological Science, 274*(1623), 2327–2330.

Carré, J. M. (2009). No place like home: Testosterone responses to victory depend on game location. *American Journal of Human Biology, 21*(3), 392–394.

Carré, J. M., & Putnam, S. K. (2010). Watching a previous victory produces an increase in testosterone among elite hockey players. *Psychoneuroendocrinology, 35*(3), 475–479.

Carré, J. M., Putnam, S. K., & McCormick, C. M. (2009). Testosterone responses to competition predict future aggressive behaviour at a cost to reward in men. *Psychoneuroendocrinology, 34*(4), 561–570.

Carré, J., McCormick, C., & Hariri, A. (2011). The social neuroendocrinology of human aggression. *Psychoneuroendocrinology, 36*(7), 935–944.

Collins, R. (2008). *Violence: A micro-sociological theory of violence*. Princeton: Princeton University Press.

Derntl, B., Windischberger, C., Robinson, S., Kryspin-Exner, I., Gur, R., Moser, E., & Habel, U. (2009). Amygdala activity to fear and anger in healthy young males is associated with testosterone. *Psychoneuroendocrinology, 34*, 687–693.

Eisenegger, C., Naef, M., Snozzi, R., Heinrichs, M., & Fehr, E. (2010). Prejudice and truth about the effect of testosterone on human bargaining behaviour. *Nature, 463*, 356–359.

Hermans, E., Ramsey, N., & van Honk, J. (2008). Exogenous testosterone enhances responsiveness to social threat in the neural circuitry of social aggression in humans. *Biological Psychiatry, 63*, 263–270.

Katz, J. (1988). *Seductions of crime*. New York: Basic Books.

Maner, J. K., Miller, S. L., Schmidt, N. B., & Eckel, L. A. (2008). Submitting to defeat: Social anxiety, dominance threat, and decrements in testosterone. *Psychological Science, 19*(8), 764–768.

Mazur, A. (1998). Aging and testosterone. *Science, 279*, 305–306.

Mazur, A. (2005). *Biosociology of dominance and deference*. New York: Rowman & Littlefield.

Mazur, A. (2008). Testosterone and violence among young men. In T. Walsh & K. Beaver (Eds.), *Biosocial criminology* (pp. 190–204). New York: Routledge.
Mazur, A. (2009a). A hormonal interpretation of Collins's micro-sociological theory of violence. *The Journal for the Theory of Social Behaviour, 39*, 434–447.
Mazur, A. (2009b). The age-testosterone relationship in black, white and Mexican-American men, and reasons for ethnic differences. *The Aging Male, 12*(2), 66–76.
Mazur, A., & Booth, A. (1998). Testosterone and dominance in men. *The Behavioral and Brain Sciences, 21*, 353–363.
McGinity, J. (2006, April 28). New York killers, and those killed, by Numbers. *New York Times*.
Mehta, P., & Beer, J. (2009). Neural mechanisms of the testosterone-aggression relation: The role of orbitofrontal cortex. *Journal of Cognitive Neuroscience, 22*, 2357–2368.
Mehta, P. H., & Josephs, R. A. (2006). Testosterone change after losing predicts the decision to compete again. *Hormones and Behavior, 50*, 684–692.
Mehta, P. H., & Josephs, R. A. (2008). Testosterone and cortisol jointly regulate dominance: Evidence for a dual-hormone hypothesis. *Hormones and Behavior, 58*, 898–906.
Nisbett, R. & Cohen, D. (1996). *Culture of Honor*. Boulder CO: Westview Press.
Rowe, R., Maughan, B., Worthman, C., Costello, E., & Angold, A. (2004). Testosterone, antisocial behavior, and social dominance in boys. *Biological Psychiatry, 55*, 546–552.
Rutter, J., Giller, H., & Hagell, A. (1998). *Antisocial behavior by young people*. New York: Cambridge University Press.
Sarkey, S., Azcoitia, I., Garcia-Segura, L. L., Garcia-Ovejero, D., & DonCarlos, L. (2008). Classical androgen receptors in non-classical sites in the brain. *Hormones and Behavior, 53*, 753–764.
Stanton, S., Wirth, M., Waugh, C., & Schultheiss, O. (2009a). Endogenous testosterone levels are associated with amygdala and ventromedial prefrontal cortex responses to anger faces in men but not women. *Biological Psychology, 81*, 118–122.
Stanton, S., Beehner, J., Saini, E., Kuhn, C., & Labar, K. (2009b). Dominance, politics, and physiology: Voters' testosterone changes on the night of the 2008 United States presidential election. *PLoS One, 4*(10), e7543.
Tremblay, R. (2000). The development of aggressive behaviour during childhood. *International Journal of Behavior and Development, 24*, 129–141.
Van Wingen, G. C., Mattern, R., Verkes, J. B., & Fernandex, F. (2010). Testosterone reduces amygdala-orbitofrontal cortex coupling. *Psychoneuroendocrinology, 35*, 105–113.
Wolfgang, M. (1958). *Patterns of criminal behavior*. Philadelphia: University of Pennsylvania.
Zak, P. J., Kurzban, R., Ahmadi, S., Swerdloff, R. S., Park, J., Efremidze, L., Redwine, K., Morgan, K., & Matzner, W. (2009). Testosterone administration decreases generosity in the ultimatum game. *PLoS One, 4*, e8330.
Zethraeus, N., Kocoska-Maras, L., Ellingsen, T., von Schoultz, B., Hirschberg, A., & Johannesson, M. (2009). A randomized trial of the effect of estrogen and testosterone on economic behavior. *Proceedings of the National Academy of Sciences, 106*, 6535–6538.

Chapter 24
Comprehending the Neurological Substratum of Paraverbal Communications: The Invention of SplitSpec Technology

Stanford W. Gregory Jr. and Will Kalkhoff

Introduction to SplitSpec Technology

An earlier book chapter (Gregory 1999) reviewed the specific background of our investigations into the social significance and neurology of the lower speech frequency in human communication. This lower speech frequency, termed the *paraverbal* frequency (beneath .5 kHz), was found to convey important nonverbal social information, and the previously published chapter outlined the methodology and results of numerous experiments showing how the paraverbal signal differs from the verbal; specifically, how it acts as an elemental mechanism of social status accommodation and social convergence between conversation partners. The present chapter will continue from where the last chapter left off and aims first to merge our past results and observations with a theoretical account making use of pertinent findings from physical anthropology, cognitive psychology, and neurology. With the theoretical background established, we will then review how our research led to the development of a technological innovation called "SplitSpec Technology," and how this innovation will fit into the future of human electronic communications.

In order to understand how SplitSpec Technology fits with modern electronic communications, we will explain the rationale behind this technology by highlighting and describing the milestones in its development. The view that there is something more too human communications than just the verbal word has been clear in human consciousness for centuries, yet it is still often assumed, mainly in the Western tradition, that the *primary* conveyance of communicated meaning is the "word," or the consciously controlled verbal portion of speech (Sarles 1975). It is only within the last half century that linguists, psycholinguists, sociolinguists and others have concentrated on the important nonverbal elements of human communications. For example, in the 1960s and 1970s, social psychological textbooks (e.g., Sampson 1971) began to address the subject of nonverbal communication and suggested that it comprised only four types of behaviors: kinesics (facial, body and limb cues), proxemics (spatial cues), tactility (touch cues), and paralinguistics (tonal speech cues). These four types have since been expanded to include a much broader array of nonverbal communications conveyed rhythmically by the body, olfactorally through smell, and paraverbally through aural communications conveyed by speech. It is the latter which comprises our interest and has given birth to SplitSpec Technology.

S.W. Gregory Jr. Ph.D. (✉) • W. Kalkhoff, Ph.D.
Department of Sociology, Kent State University, Kent, OH 44241, USA
e-mail: sgregory@kent.edu

D.D. Franks and J.H. Turner (eds.), *Handbook of Neurosociology*, Handbooks of Sociology and Social Research, DOI 10.1007/978-94-007-4473-8_24, © Springer Science+Business Media B.V. 2013

Theoretical Development

Laterality and Evolution

Modern humans increasingly make use of distally conveyed, wired or wireless, electronic communications that have replaced more traditional ambient space, proximal communications. Early hominids relied upon the latter, which made exclusive use of anatomical organs to support oral, nasal, audio, tactile and visual behaviors. The contralateral neural wiring of functions involved with hominid communications (and mammalian communications in general) is ancient and undoubtedly predates these species, for this feature appears to be present even in the precursors to fishes (McManus 2002; Shubin 2008). In these animals, stimuli often indicate a predator – something from which to flee – hence sensory input on one side of the body, whether hearing, vision, or mechanoreception (water currents), causes a reflex that contracts muscles on the contralateral (opposite side) side of the body, and as a result of this muscle contraction, the "prefish" (and its fish descendants) bends away from the perceived danger (H. Thewissen, personal communication, February 21, 2011).[1] These systems evolved to sustain the general survival of the species, and the evolution of these organs was directly accompanied by a cognitive and neuronal counterpart. In a way, the fish (and the prefish) really is "in us," as Shubin (2008) has shown.

The aural ability in mammals is quite complex, and it is doubtful that the present neural wiring is the same as when the system evolved. But, as noted above, it has always been important for mammals and earlier forms to determine the direction of a sound, and this is accomplished by comparing the timing and intensity of a signal arriving at the left and right ears. That is, if the left ear hears the signal first and more intensely than the right, then its source is most likely from the left. In addition, the nerve cells involved with this function in the brain are very finely tuned to support it (Thewissen, personal communication, February 21, 2011).

There is another lesser known function associated with contralateral wiring in mammals that tends to complicate the aural function. Neural wiring for the aural function is different from visual wiring over the optic chiasm[2] in that the complete aural signal from a stimulus is routed to *both* hemispheres. This seemingly odd neural routing is actually highly beneficial: it provides the organism with ability to activate its contralateral motor system for possible defense against predation, while not having to depend upon hemispheric neural systems on the side being attacked. This wiring system thus appears to comply with the conventional and quite ancient evolutionary model of species survival.

However, when primates, and possibly other species, appear on the scene, the contralateral "issue" becomes more complex. Broca's nineteenth century discovery of left cerebral lateralization of language more sharply emphasized the asymmetrical nature of brain function. Not only does the brain show asymmetry by its contralateral motor/sensory function, but other specific functions,

---

[1]Professor Hans Thewissen is professor of anatomy at the Northeastern Ohio Universities College of Medicine, and his specialized research interest is in marine mammals. The lateral lines in fish have a precursor inserted within the aural system of mammals – the lateral lines are spiraled and condensed thus forming the mammalian cochlea (Partridge 1982). The contralateral reflex, discussed above, is called C-start in fishes reflecting the "C" shape of the fish after the defensive reflex is affected. A marine, snake predator has made use of this reflex in pursuing its prey by first circling the fish, then activating the fish's C-start reflex with a slight aquatic perturbation in such a way to cause the fish to directly swim into the snake's mouth, thus turning the fish's defensive strategy into the snake's offensive strategy http://www.livescience.com/5503-snake-tricks-fish-swim-mouth.htlm.

[2]This difference in neural wiring has undoubtedly developed owing to the major difference in somatic position of (1) the eyes perceiving an apportioned but partially shared field of view as mimicked by the neural function of the optic chiasm, and (2) the ears perceiving a shared ambient air space but an apportioned proximity of aural space as mimicked by the neural function of the aural pathways (Geffen et al. 1971).

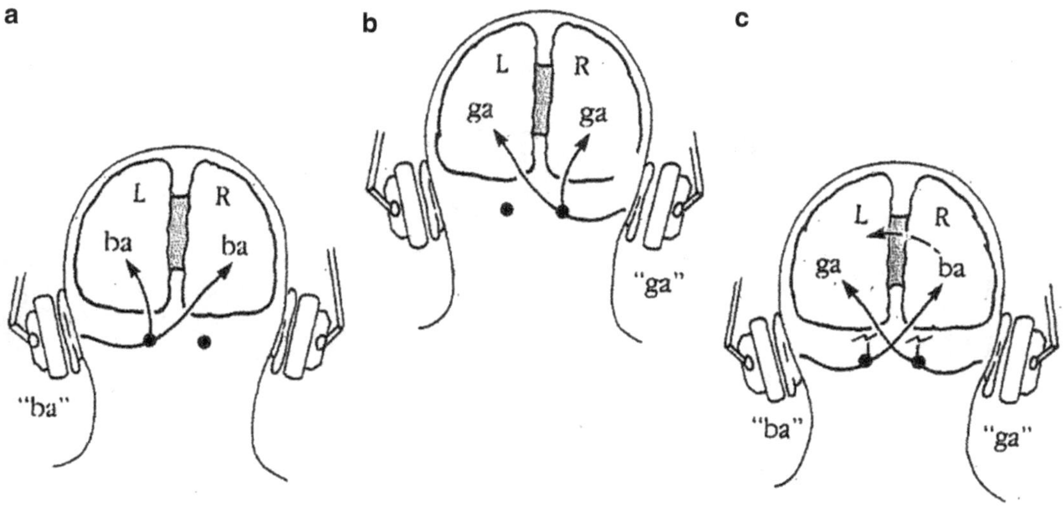

Fig. 24.1 Dochotic listening pathways for normal patients

including verbal language and other social behaviors, are lateralized as well.[3] For example, early on it was found that contralateral projections from ear to brain are stronger than ipsilateral (same side) pathways (Rosenzweig 1951). In other words, the ipsilateral signal can be suppressed in favor of the contralateral signal. Years later, Kimura's (1967) dichotic listening studies revealed that when *different* speech stimuli were input via electronic headsets *simultaneously* to each ear, participants most often reported the signal input to the right ear and thus the left hemisphere (see Fig. 24.1c). This occurs because when two different stimuli are presented simultaneously to each ear, the difference in strength of the pathways is exaggerated so that information sent along the ipsilateral route is suppressed. Thus the right ear more efficiently routes its stimulus contralaterally to the left aural cortex due to ipsilateral signal suppression to the right aural cortex. Though there is no difference between the two ears in a person's ability to detect or identify stimuli presented one at a time, when there is *competition* between the ears, as in the presentation of speech stimuli dichotically, the contralateral route is strengthened over the ipsilateral (Kimura 1961; Springer and Deutsch 1981). This effect has been shown in an exaggerated form with split-brain patients whose cerebral commissures (i.e., bands of fibers connecting the two brain hemispheres) have been severed (Milner et al. 1968), but whose *subcortical pathways* have been left intact (See Fig. 24.2). The effect is exaggerated because conflicting signals (i.e., different speech stimuli input simultaneously) passed from the right hemisphere over the cerebral commissure pathways in normal persons were eliminated in commissurotomy patients, allowing an unhampered contralateral signal to be routed to the left hemisphere (compare Fig. 24.1c with Fig. 24.2c). The right ear advantage first documented in Kimura's (1967) research applies mostly to speech sounds, for when nonverbal stimuli such as environmental cues (e.g., trees falling, dogs barking) were presented dichotically to the ears of normal participants, the *left* ear appeared to have an advantage (Curry 1967). The general point in this connection is that the hemisphere receiving a stimulus directly has an advantage over the hemisphere that receives identical information indirectly by way of the cerebral commissures. An important conclusion, particularly relevant for the purposes of our theory, is stated by Springer

[3]This left hemisphere lateralization for conspecific vocalizations has been found in Japanese macaques (Zoloth et al. 1979; May et al. 1988) as well as rhesus monkeys (Hauser et al. 1998).

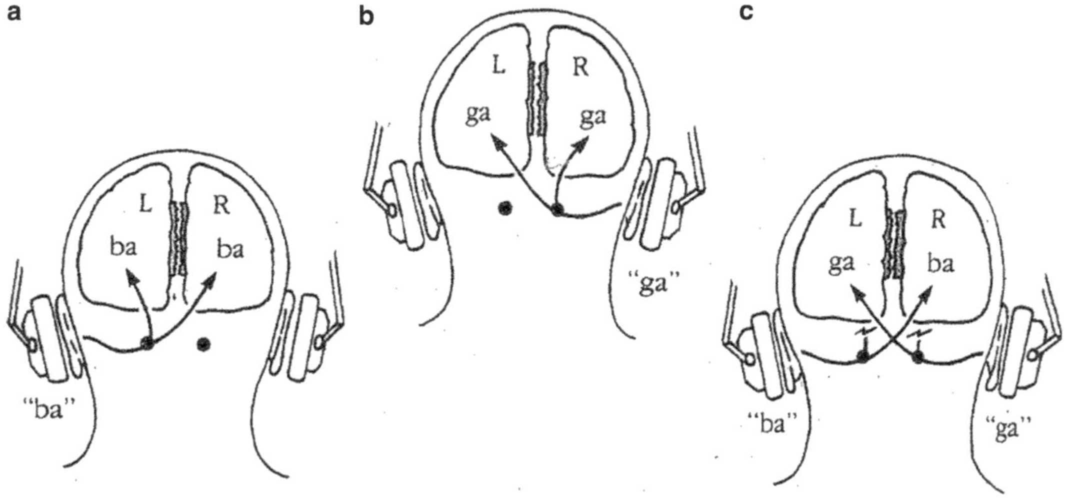

Fig. 24.2 Dichotic listening pathways for split-brain patients

and Deutsch (1981:64): "The condition where information is presented directly to the hemisphere *specialized for a specific function* would be expected to produce better performance – that is, more accurate or faster responding – than one in which information goes first to the other half of the brain [italics added]." These findings show that neural pathways have evolved to satisfy highly specialized purposes and have been lateralized for discerning different sources and types of information.

The social functions of the neural pathways described above have evolved, for the most part, in the pre-modern circumstances of social life, and various social "tools" have accommodated to the neural substrata supporting them. Gestural communications probably came first (Darwin 1872; Hewes 1973; Lieberman 1975), including many of the now-termed "nonverbal" types of communication. Early vestiges of modern language came next, followed by writing and then printing. Modern media such as wired forms of distal communications came much later, and these have now moved on to include wireless forms. All of these developments came with slight or sometimes vast alterations to basic face-to-face communications that early hominids had become accustomed to using. It appears that we will not be going back exclusively to our basic social roots of face-to-face communications in the near future, so we must make the best use of what we have in our neural toolbox. As noted above, early on we used our two ears for indicating the location and features of stimuli, and this applied to social relations as well, such as the much discussed "cocktail party effect" (Cherry 1957) wherein those with "normal" hearing can discern speech conveyed from a conversation partner amidst a cacophony of ambient speech. Yet while our hearing system and its neural substrate have adapted to normal social and environmental contexts, these systems are not necessarily most advantageously suited to modern electronic communications environments.

It is possible with modern electronic communications to actually alter the way communications signals are neurally processed by the brain. With the headset/microphone, the directional attribution function of the ears is unnecessary, but the contralateral pathways can be made useful for the predominantly social information conveyed via modern headset/microphone devices. The headset/microphone device transforms normal speech into an electronic signal. This signal can be adapted so input via neural pathways is more efficiently formatted for cognitive processing. This alteration of the normal speech signal could not have been developed without electronic communications, and though the hominid hearing system has served our species well, an advanced manipulation of the electronic signal input to the ears can lead to an enhanced cognitive processing that humans have not yet experienced. SplitSpec Technology offers a way of putting our ancient contralateral laterality to work in the enhancement of our communications.

Putting Contralateral Pathways to Work

From the above discussion, we suggest that the human aural system has evolved a neuronal and cognitive function that served early hominids and their more modern homo sapiens counterparts well in past contexts of habitation. However, technology has changed the habitation environment significantly since the introduction of modern communications technology. The modern environment is replete with social contexts that prohibit traditional, proximal, face-to-face interaction. Much of human communication interaction is carried on distally via wired or wireless speech or text technology. Some of the changes in social communications brought about by advanced technology have been detrimental, from serviceable, smaller issues associated with faulty technology to much more profound problems concerning elimination of affect in communications. However, the benefits of distal communications in modern societies have probably outweighed the disadvantages, and the prominent fact prevails that modern society is extraordinarily dependent on advanced electronic communications.

In the speaking and hearing realm, most technological innovations have emphasized traditional patterns of speaking, hearing and listening, and they comprise an array of devices, including sound amplification systems, stereo systems, Dolby sound systems, digital noise elimination, voice simulation, hearing aid systems for the hearing impaired, etc. There is scant evidence however that these innovations have actually changed neuronal and cognitive patterns pertaining to how humans comprehend speech sounds. Previous research, as summarized in the earlier chapter cited above (Gregory 1999), noted how the lower voice frequency transmits paraverbal, less consciously controlled affective information, and how the higher voice frequency transmits verbal, more consciously controlled and less affective information. We have also found in more recent research that the paraverbal signal is critical in forming perceptions of social status and that this signal is processed by the right cerebral hemisphere. Finally, in our most recent research we found that if the speech signal is filtered so that the paraverbal signal is input dichotically to the left ear (and thus routed contralaterally to the right cerebral hemisphere) while the higher speech frequency is routed to the right ear (and thus the left cerebral hemisphere), *experimental participants can complete an interactive task more accurately and more rapidly than if participants hear the normal monaural signal without filtration or dichotic input*. In short, under the former "enhanced" condition, participants can do their job better than if they are hearing the electronic communications without enhancement as in the later condition.

Under modern conditions, some distal communications systems require headsets and attached microphones. In these cases, ambient sound is attenuated or eliminated altogether, and an electronically reproduced signal is transmitted directly from a source (i.e., a microphone) to a receiver (i.e., a headset). This synthesized electronic system offers an opportunity for altering the voice signal to produce a more efficient neural routing. By introducing this altered listening system to the headset communications environment, it is possible to improve the efficiency of the means whereby the human brain comprehends and acts upon speech information. As noted above, the signal input to the left ear is routed contralaterally to the right cerebral via subcortical pathways. We have found that the right hemisphere is more efficient in processing the paraverbal frequency while the left hemisphere is more efficient in processing the verbal frequency. Filtering the signal so it is more efficiently routed for processing will thus improve efficiency of cognitive processing, thereby leading to more accuracy and timeliness of subsequent sensory/motor behaviors.

The evolved contralateral system originally introduced to defend against predation can now be synthetically adapted to enhance humans' comprehension of the speech signal and thus improve their behavioral efficiency. Based on our knowledge of neuronal processing of the aural signal, we submit that when dichotically filtered speech arrives at the ears – wherein the paraverbal signal is input to the left ear and the verbal signal is input to the right ear – competition between the two simultaneously arriving signals causes a more robust contralateral signal than ipsilateral signal to be passed to the respective hemispheres, i.e. the ipsilateral signal is *suppressed* and the contralateral is *permitted*. When the contralateral signal arrives at its appropriate neural destination, it is processed thus causing the appropriate motor responses. In the interim, because the signal is sent contralaterally and directly to the appropriate hemisphere it overwhelms the competitive signal arriving over the cerebral commissures (see Fig. 24.1c); and because an otherwise ipsilaterally contending signal is suppressed, the signal is more efficiently routed than if it were sent via normal, nonfiltered, monaural pathways. This dichotically enhanced condition can be likened to the increased contralateral transmission efficiency when the corpus callosum is severed as with split-brain patients (see Fig. 24.2c, where no signal is passed over the commissures), in that contending signals are eliminated.

On the contrary, in the monaural non-filtered listening condition various neural impediments arise. In reference to Fig. 24.1a, when a speech signal "ba" is input to the left ear, a participant responds to hearing "ba," because the left verbal hemisphere has received verbal information ipsilaterally; and in reference to Fig. 24.1b, when "ga" is input to the right ear and is sent contralaterally to the left verbal hemisphere the participant reports hearing "ga" because the left hemisphere has again received the appropriate verbal signal. However, in the dichotic condition with an *unfiltered competing signal being passed simultaneously to both ears* as in Fig. 24.1c, participants' reports are *most often* "ga" because this signal is received contralaterally in the appropriate left hemisphere and the ipsilaterally sent signal to the left hemisphere "ba" is suppressed to the left hemisphere but is allowed to enter the left via the commissures (see Fig. 24.1c). As noted in Kimura's (1967) research, in this condition participants show "a small right ear advantage" (Springer and Deutsch 1981:67), but when the same protocol is applied to split-brain patients, the right ear advantage is "highly exaggerated" (Springer and Deutsch 1981:69). This exaggeration is due to the severance of the commissures (see Fig. 24.2c) which prevents impedance from pathways over the commissures as in Fig. 24.1c with normal participants' dichotic listening.

As noted above, the neural pathways for the dichotic condition for normal participants hearing a monaural signal, differs from participants with a split-brain condition; however, when the signal is filtered and administered dichotically to normal participants, the result is strikingly similar to the condition for split-brain patients, and this is because the filtered dichotic condition contralaterally inputs only the *appropriate* signals to the respective hemispheres, thus impeding potency of interference from transmissions over the commissures as seen in Fig. 24.1c. This technologically conditioned neuronal system allows the dichotically filtered verbal and paraverbal signals to be neurally processed in a much more efficient manner thus leading to more accurate and rapid motor sensory responses.

In a number of published articles, we have presented experiments leading to such a system, and the subsequent portion of this chapter will review research milestones leading up to production of SplitSpec Technology.

To summarize:

1. Modern communications are increasingly accomplished via electronic media, which eliminates traditional face-to-face communication.
2. Dichotic listening studies have shown that neural routing for aural communications is conditioned by pathways which favor the subcortical contralateral over the subcortical ipsilateral pathways when a competing signal is simultaneously input to the ears.
3. Experiments on split-brain patients have shown an exaggerated effect in the increased strength of contralateral pathways.
4. Extant literatures in a number of disciplines attribute verbal stimuli functions predominantly to the left cerebral hemisphere, and nonverbal stimuli functions to the right hemisphere.

5. When information is routed directly to the appropriate hemisphere specialized for a specific function, performance is more accurate and timely than when information is routed to the appropriate hemisphere indirectly.
6. A system that routes speech information directly to the appropriate hemisphere via a dichotically filtered system can enhance accuracy and speed of human communications.
7. In modern electrically transmitted communications, the verbal/paraverbal filtered dichotically input signal is more cognitively efficient than the monaurally transmitted signal.

Milestones on the Road to SplitSpec Technology

The Paraverbal Milestone of Social Status Communications

As reported in the earlier review chapter (Gregory 1999), persons engaged in speech interaction mutually converge their paraverbal frequencies, and this convergence is conditioned by "social status accommodation," which is a process discussed by Howard Giles in a number of works in communications accommodation theory, or CAT (Giles and Smith 1979, Giles et al. 1987; Giles and Coupland 1991). Giles' work with CAT was adapted to our work on paraverbal frequencies (Gregory and Webster 1996), and the general notion in this connection is that lower status conversation partners adapt their paraverbal frequencies to the higher status partner, thereby leading to communications convergence. This type of communications accommodation is more specifically described in Gregory and Webster (1996), where a social ranking of celebrity guests on *Larry King Live* was produced using Fast Fourier Transform (FFT) analysis on guests' speech and that of Larry King. Specifically, the rank ordering was constructed by comparing paraverbal frequency spectra variation of Mr. King's guests over the duration of their interviews vis-à-vis the temporal variation of Mr. King's own paraverbal frequencies. If a guest showed a relatively higher level of variation over time than Mr. King, then the guest was adapting more to Mr. King than vice versa. Conversely, if Mr. King show a relatively higher level of variation over time than his guest, then he was adapting more to the guest than vice versa. Statistics derived from these calculations showed that some guests' spectra showed higher variation than others, which was also the case with Mr. King. From these data we were able to create a rank ordering of guests and Mr. King, essentially revealing that Mr. King accommodated to higher status guests, while lower status guests accommodated to Mr. King (see Table 1 in Gregory and Webster 1996 for the complete rank ordering). Finally, a panel of 600 students rated all the guests as well as Mr. King, and the results showed a strong correlation between student ratings of social status and ratings from the paraverbal spectral analysis.

Gregory and Webster (1996) led to several other studies focusing on asymmetrical power relations (Gregory et al. 1993, 1997, 2001; Gregory 1994; Gregory and Gallagher 2002; Kalkhoff and Gregory 2008). Gregory and Gallagher (2002) showed that paraverbal analysis could accurately predict outcomes of past U.S. presidential elections. Making use of altered methods gathered from previous paraverbal studies of social status differences, Gregory and Gallagher (2002) used data gathered from televised U.S. presidential debates from 1960 to 2000, where each candidate's paraverbal signal was analyzed and statistically processed to reveal "winners" of the debates for a particular election. The paraverbal statistic was compared with the popular vote and other statistics, and results showed a significantly accurate prediction for each of the 8 elections.[4]

[4]Though former Vice President Gore won the popular vote, he did not become president as the Supreme Court judgment caused George W. Bush to take the office of president.

This method applied to elections from 1960 to 2000 was replicated for the three most current presidential debates for the election in 2008, featuring Senator Obama versus Senator McCain (Kalkhoff and Gregory 2008). Results from this paraverbal analysis showed that Senator McCain exhibited more paraverbal dominance than Senator Obama during the beginning and middle parts of each debate, while Senator Obama exhibited more dominance toward the end of each debate. Senator Obama's "rope-a-dope" strategy, as his debate style has been dubbed, served him well, as Gallup trends indicated greater support for his candidacy and superior debate performance than McCain. The paraverbal analysis picked up on this strategy and revealed that *how much* dominance is exercised overall in the debates may be less consequential than *when* it is exercised.

Along with the social status milestone leading to SplitSpec Technology we have found that the paraverbal frequency not only carries a social status marker, but is also related to conversation quality. Elimination of the paraverbal frequency from speech of conversation partners reduces the quality of conversation as perceived by research participants who listen to dyadic conversations between other participants (Gregory et al. 1997).

The Paraverbal Milestone of Communications Quality

As noted above, previous studies have shown that conversation partners accommodate and converge paraverbal frequencies of their speech. These observations led us to question why convergence occurs, and in answering this question we postulated that accommodation and convergence must have something to do with enhancing the communications quality of conversation. In considering the postulate of quality we set up several experiments. The earliest experiment (Gregory et al. 1993) compared paraverbal results from analysis of six interviews between the same interviewer and six different partners along with assessments from sets of participants who listened to and evaluated the six interviews. The object of this research was to determine the role of the paraverbal in structuring the interview. The six dyadic interviews were analyzed using the FFT method described in the earlier chapter (Gregory 1999). In addition, a scale made up of semantic differential items (Osgood et al. 1957; Osgood 1964; Heise 1970; Miller 1977) was devised as a means of assessing judgments of the interviews by independent raters. For each of the six interviews, an instrument consisting of 34 semantic differential (SD) items was used to assess meanings for three separate points of focus: interviewer, interviewee, and interview itself. Excerpts of the interviews were aired to six different audiences of 32–35 participants. Each of the groups of raters heard one interview and were asked to complete the SD instrument three times, once for each point of focus (see Table 2 in Gregory et al. 1993 for details). Results from factor analysis of the SD instrument produced three factors classified as "evaluation," "potency," and "affect." The potency factor stood out from the rest, and so this factor was used in further comparisons with scores produced from the separate paraverbal analysis. The Pearson correlation coefficient for participant assessments of the interviews' potency factor and the paraverbal analysis scores was 0.85 ($p=0.016$). This struck us as a particularly strong association between very diverse ways of assessing interactional elements of the interviews. Results from this study led to a further refinement of our methodology to establish a metric of quality in conversations.

Following from Gregory et al. (1993), the next study of paraverbal affects on interaction quality (Gregory et al. 1997) involved *altering* the acoustic signal between interacting partners. A total of 60 participants in dyadic pairs were set apart in two different rooms (labeled A and B), and they communicated via headsets while accomplishing an interactive task. The speech signal from each partner was transmitted through a "Y" cable to both a dual-channel electronic high/low-pass filter and also a stereo tape recorder that captured the conversation in its unfiltered entirety. Switch settings on the filter were set to produce three experimental conditions: high pass (above .5 kHz), low pass (below .1 kHz), and unaltered. Several other groups of participants listened to samples of conversations from the three conditions, and they recorded their assessments on a semantic differential (SD) instrument similar to the one used in Gregory et al. (1993) and described above. The aired samples were from the *unaltered taperecorded conversations* from *all* experimental conditions. Results from participants' assessments of conversation excerpts using the SD instrument revealed that they perceived more positive social/evaluative worth, i.e. quality, in the full and low-pass groups as opposed to high-pass groups. These findings show strong evidence that elimination of the paraverbal signal (in the high-pass group) diminishes the social/evaluative quality of communication.

The final study in connection with paraverbal quality assessments compared the relative importance of the paraverbal channel over the *visual* channel. In other words, is the presence of the paraverbal channel or the visual channel more important, for example, when it comes to understanding instructions? Or are they both equally important? We discuss that question next.

Milestones in the Primacy of the Paraverbal Channel

The aim of Gregory et al. (2001) was to determine the potency of the paraverbal channel in comparison with the presence or absence of other communicative channels, in particular the visual channel. The study tested three hypotheses. First, when dyadic conversation partners engage in an interactive task using both visual and aural channels,[5] they will show less paraverbal accommodation than conversation partners who accomplish an identical task using the aural channel alone. Second, when *one* partner's paraverbal speech is filtered out during an interactive task such that *only* the higher (verbal) tones are allowed to "pass," it will produce an aural response in the non-filtered partner's speech that stimulates a proportionally decreased visual response by the paraverbally filtered partner. Finally, independent raters who watch and listen to the conversation partners complete the interactive task via a split-screen monitor will express less socio-emotional attention, less positive evaluation, and less positive affect than raters who make assessments of partners conversing via the aural channel alone.

Results from this study showed support for the first hypothesis, as inclusion of the visual channel clearly resulted in less paraverbal accommodation. In other words, paraverbal accommodation is less necessary with addition of the visual channel. As to the second hypothesis, the operational definition for visual attention was defined by the number of times partners would look up at their partner from the task materials on the desk in front of them; this was termed a "visual check". As noted above, only one partner's speech in the filtered condition was, in fact, filtered, thus it was possible to compare the visual checks and other observations between partners. Analyses of the data revealed a significant difference in visual checks only between filtered and unfiltered partners in the high-pass group (paraverbal is filtered out). This result means that the partner who received unfiltered speech (including the paraverbal signal) looked up *significantly less* than the partner who received filtered speech (without the paraverbal signal), and the latter looked up about as much as both partners in the "control" group, neither of whose speech was filtered. Importantly, whether or not filtering occurred was unknown to both partners. The partner who unknowingly received a filtered signal also unknowingly *returned* a complete signal. The returned complete signal includes an intact paraverbal frequency which is anomalous in that it has not been accommodatively encoded by normal interaction. That is, the partner who receives a filtered signal (yet returns a complete signal)

[5]The aural system in this study used the same headset microphone system as in previous research projects, and the visual system made use of closed circuit television monitors with cameras installed so participants could interact visually with one another.

has no acoustic information on which to base an appropriately encoded paraverbal response. And looking at the situation from the other direction, attempts by the partner who receives the complete signal to accommodate via the paraverbal signal are returned with a perceivably anomalous and anxiety-provoking paraverbal response from the partner who receives an incomplete, filtered signal.

The unusual paraverbal signal encoded and returned by the partner who receives the filtered signal is not anomalous because it is random. On the contrary, there is a strong chance that the returned signal may convey inappropriate, non-negotiated dominant social status content, and it would follow that this erroneously encoded signal is stress-provoking for the partner who receives the full signal, leading to decreased engagement – to a condition where "… cognitive resources are shifted more toward the self and less toward the partner" (Patterson 1995:10). In other words, the partner who receives the filtered signal absent paraverbal feedback defaults to a dominant paraverbal signal in return. Justification for this assumption is found in NG and Bradac's (1993) extensive review of the literature on effects from powerful and powerless vocal styles (see in particular pages 24–29). Based on a number of studies, Ng and Bradac (1993:27) claim that the high-power style is associated with "… communicator competence, status, dynamism, and attractiveness". With these rather rewarding outcomes, it is likely that the choice of paraverbal encoding by the partner who receives a paraverbally depleted signal would convey a higher, rather than a lower, power aspect.

To confirm this claim, a separate test was performed. As noted above, a previous study (Gregory and Webster 1996) used factor analysis to compare the paraverbal signals of Larry King and his celebrity guests. In that study, we found that more dominant speakers have higher loadings on the first extracted factor because their paraverbal signals show less variance. More dominant speakers are accommodated to by their conversation partner, and thus they change relatively little in the paraverbal range. Conversely, more deferent speakers accommodate more to their conversation partner, and so their paraverbal signals show more variance, which leads them to have higher loadings on the *second* (rather than the first) extracted factor. Reusing this technique for the primacy study, Gregory et al. (2001) indeed found a significant difference between filtered and unfiltered partners for the audio/visual, high-pass condition. As expected, partners who received a filtered signal had higher loadings on the first extracted factor (the dominance factor) than partners who received a complete signal. And again, we believe this occurs because *partners who receive a filtered signal obviously have no way to adapt to their partner's paraverbally depleted speech, thus the signal they send appears as a dominant signal insofar as it shows less variance.*

Finally, as to the third hypothesis in Gregory et al. (2001), findings supported the hypothesis and reveal a definitive affective shift occurring when the visual channel is added to the vocal channel during task interaction. With the addition of the visual channel, the evaluators' task becomes less of an interesting, social, and expressive endeavor and more of a defined, clear, and thus instrumental task.

Gregory et al. (2001) provided definite evidence of the significance, if not the primacy, of the paraverbal channel in the expression of social information that appears to be critical to interactive communication. Next we report on further investigations into the neural tracking of the paraverbal signal and build toward a determination of how this tracking can be manipulated to serve human communication needs.

The Neural Milestones of the Paraverbal Signal

In the early 2000s, though we had had success in discovering the communicative significance of the paraverbal signal and in showing its influence in supporting social status accommodation and convergence, we had not definitively explicated its cerebral connection. From our previous research we had only gained some hints about the cerebral location of paraverbal processing. And while at the time there was ample evidence in the cognitive science and psycholinguistic literatures concerning location of nonverbal affective communications, there had been no solid evidence pointing to the cerebral location of paraverbal frequency processing.

In striving to better target the neural location of the paraverbal signal, we began an investigation using the hints offered from our past research as a starting point. One hint came from our incorporation of early studies about the lower frequencies conducted by other investigators (Starkweather 1956a, b; Kramer 1964; Bradshaw and Nettleton 1983; Lieberman 1984). A second hint came from our own as well as others studies of laterality (Sperry 1959; Gazzaniga 1970, 1985, 2000). There was ample indication from these hints that the seat of paraverbal function is in the right cerebral hemisphere. Also, the early dichotic listening research by Kimura (1967, 1973) offered us a method for locating the paraverbal signal's neural source and for actually manipulating the paraverbal signal during behavioral tasks. Armed with this information, we spearheaded two separate projects, the first (Gregory et al. 2009) aimed at targeting the neural location and specific function of the paraverbal signal, and the second (Kalkhoff et al. 2009) aimed at specifically determining the behavioral effects from experimental manipulation of the paraverbal signal during a simulated task performance.

The Paraverbally Enhanced Communications Neural Location Project

The dichotic communications enhanced neural location project was a laboratory experiment and validation study using dichotic listening methods (Kramer 1964, 1967, 1973a, b; Springer and Deutsch 1981) to explore possible communication benefits of adapting the dichotic method by inputting the lower frequency paraverbal signal into the left ears of participants and the higher verbal signal into their right ears.[6] We termed this arrangement the "enhanced condition" because we surmised that listening would be enhanced when (1) the paraverbal signal is input to the left ear and then routed contralaterally (see above) to the right cerebral hemisphere, while (2) the verbal signal is simultaneously input to the left ear and then routed contralaterally to the right hemisphere. We compared this condition with a reversed dichotic pattern called the "confounded condition" (the opposite of the enhanced dichotic condition) and also a "control" condition involving no dichotic filtering. If the enhanced arrangement enabled participants to complete an interactive task in a more efficient, accurate, and timely manner (i.e., compared to the confounded and control conditions), then we could conclude that dichotically targeting the right hemisphere for the paraverbal signal and the left hemisphere for the verbal signal maximizes efficiency of neural routing of the aural signal; and thus confirms our having located the appropriate cerebral location for maximal processing efficiency. Finally, to corroborate our initial findings we constructed another experiment wherein participants evaluated unfiltered audio/visual samples of the three conditions described above using a semantic differential (SD) instrument as in our earlier experiments. In this experiment, we surmise if participants objectively evaluate the enhanced condition more positively than the other conditions, then the most appropriate neural targeting has been performed.

Participants in the first part of this experiment completed an interactive task from separate rooms over an audio/visual connection. Unlike our past experiments, participants' speech was input *dichotically* in accordance with the three experimental conditions described above. The second

[6] All prospective participants were administered the Oldfield Handedness Inventory (1970). Only right-handed participants could participate in this research, as there is a higher probability that left-handers may have reversed cerebral lateralization than right- handed participants. This practice was done in order to prevent a possible confound, and is generally the practice in dichotic listening research.

part was a validation study where we aired A/V samples of the recorded dyadic conversations (without audio filtering *regardless of condition*) to groups of participant evaluators using the standard SD instrument described above.

The dependent variables for the first part of the experiment were task completion time and task accuracy. For task completion time, results revealed statistically significant differences in the expected directions between the enhanced and confounded conditions (11.5 min versus 13.8 min, respectively) as well as between the enhanced and control conditions (11.5 min versus 14.2 min respectively). In both cases, participants in the enhanced conditions completed their task in significantly less time; about 20% faster, for example, in the enhanced condition versus the control condition. Though we did not find a significant difference between the control and confounded conditions, the relatively low mean task completion time for the enhanced condition meets the postulated assertion for this experiment. However, we did not expect the mean task completion time for the control condition to be greater than the completion time for the confounded condition, though this result does not lessen the importance of the predicted result for the enhanced condition.[7]

For task accuracy, results revealed statistically significant differences between the enhanced and control conditions (14.7 correct versus 13.9 correct) as well as between the confounded and control conditions (14.7 correct versus 13.9 correct), though we did not find a difference between the enhanced and confounded conditions.[8]

In the second part of this study, samples from audio/video taped participant interactions were aired to groups of "judges" who rated performances using the SD instrument described above. Factor analyses of the SD data indicated that only factor scores derived from the first factor, "affect" produced a significant result using "condition" as the independent variable. This factor included SD items such as comfortable/uncomfortable, friendly/unfriendly, relaxed/tense, pleasant/unpleasant, secure/insecure, sociable/unsociable, happy/sad, like/dislike. Also, post-hoc *t*-tests showed significant differences in the expected direction between the enhanced and confounded conditions as well as between the enhanced and control conditions. In both cases, raters perceived more positive affect among interactants in the enhanced condition. The confounded condition was not significantly different from the control condition. The result for the "affect" factor validated our theoretical stance that the paraverbal signal is critical for the communication of affect, and thus for the general positive, sociable evaluation of interaction.

The Paraverbally Enhanced Communications Driving Simulation Project

The second, most recent project adapted our paraverbal experimental protocol to a behavioral context of simulated driving where participants received turn-by-turn driving instructions and completed cognitive tasks while interacting with an experimenter via an audio communications system. Participants were randomly assigned to one of two conditions: an enhanced condition wherein the paraverbal signal was routed to the left ear (with right cerebral processing) and the verbal signal was routed to the right ear (with left cerebral processing), or a control group wherein participants heard unfiltered communications in both ears.

---

[7]No significant differences were found between the task completion time means for the confounded and control conditions. This was an unanticipated finding and we offer a possible explanation for this result. Though our theory establishes that routing the signal to the appropriate cerebral location enhances timely performance, the dichotic, contralateral, routing manipulation of the high and low-pass signals to the *inappropriate* hemispheres may involve a neuronal high/low-pass packaging of individual frequency bands, making them more efficient for rerouting than the raw, unfiltered monaural control signal.

[8]This result may be caused by a similar neural packaging affect as noted in note 7 above.

Before participating in the actual experiment, right-handed participants accomplished a preliminary driving exercise in order to acquaint them with the driving simulator.[9] When the experiment began, participants were instructed to follow driving instructions given by an experimenter who was located in an adjacent room (e.g., "Turn right at the next stop sign and proceed to the next traffic light then..."). Furthermore, as the participant attempted to operate the simulator in accordance with the driving directions, the participant was asked by the experimenter to repeat a series of digits in the forward direction (e.g., 5–3–8–2) and later in the digits backward direction (e.g., 2–8–5–3). Digits-forward creates attentional demands while the digits-backward imposes spatial demands, including visiospatial imaging processes.

The most important result from this experiment is that the participants' odds of experiencing a simulator cessation (i.e., crashing) were significantly lower in the enhanced condition compared to the control condition. Specifically, only five out of 28 participants (17.9%) crashed in the enhanced condition, while 14 out of 31 participants (45.2%) crashed in the control condition. Thus there were 64% fewer crashes in the enhanced condition versus the control condition. Expressed in terms of "odds," participants in the control condition were almost six times more likely to crash than participants in the enhanced condition. It is evident crash risk was significantly lower for participants who communicated with the administrator using a dichotically enhanced communication system.

Discussion

In this chapter, we presented a theoretical orientation to our research, and our most recent studies. We also described results from the program of studies that laid a foundation for development of SplitSpec Technology. In 2007, and in partnership with the Kent State University Office of Technology Transfer, we filed an application to the U.S. Patent Office for a device and method based on our research. Later in 2009 we had our ideas actualized by contracting a circuit designer to develop a prototype product that accomplishes the dichotic filtered alteration of the acoustic signal, and we have named this device "SplitSpec Technology".

In 2010, we made contact with several members of the Multi-Sensory Research Team in the Army Research Laboratory for Translational Neuroscience at Aberdeen Proving Ground (Aberdeen, Maryland). There we made a presentation of our work, and members of the APG Team are now working with us to create a neuroscience research project aimed at further testing and developing the neural connections associated with the effects of SplitSpec Technology. This research will involve use of laboratory and ambulatory electroencephalographic (EEG) technology to explore the effects of paraverbal enhancement on the behavior of Army personnel in laboratory and field settings. Finally, it is our hope that SplitSpec Technology will eventually be applied to a multitude of applications that lead to fuller and more complete communications between persons interacting via electronic media.

References

Bradshaw, J. L., & Nettleton, N. C. (1983). *Human cerebral asymmetry*. Englewood Cliffs: Prentice-Hall.

Cherry, C. (1957). *On human communication*. Cambridge: The MIT Press.

Curry, F. K. (1967). A comparison of left-handed and right-handed subjects on verbal and non-verbal dichotic listening tasks. *Cortex, 3*, 343–352.

[9]The driving simulator (Simulator Systems International S-3300 Modular Driving simulator) consisted of simulated driving equipment, i.e. wheel, pedals, etc., including consoles with the necessary audio/visual effects.

Darwin, C. (1872). *The expression of emotions in man and animals*. London: John Murray.
Gazzaniga, M. S. (1970). *The bisected brain*. New York: Appleton.
Gazzaniga, M. S. (1985). *The social brain*. New York: Basic Books.
Gazzaniga, M. S. (2000). Cerebral specialization and interhemispheric communication – Does the corpus callosum enable the human condition? *Brain, 123*, 1293–1326.
Geffen, G., Bradshaw, J. L., & Wallace, G. (1971). Interhemispheric effects on reaction time to verbal and nonverbal visual stimuli. *Journal of Experimental Psychology, 87*, 415–422.
Giles, H., & Coupland, N. (1991). *Language: Contexts and consequences*. Pacific Grove: Brooks/Cole.
Giles, H., Smith, P. M., & Patricia, M. (1979). Accommodation theory: Optimal levels of convergence. In G. Howard, N. Robert, & R. St. Clair (Eds.), *Language and social psychology* (pp. 45–65). Baltimore: University Park Press.
Giles, H., Mulac, A., Bradac, J. J., & Johnson, P. (1987). Speech accommodation theory: The next decade and beyond. In *Communication yearbook* (Vol. 10, pp. 13–48). Newbury Park: Sage.
Gregory, S. W., Jr. (1994). Sounds of power and deference: Acoustic analysis of macro social constraints on micro interaction. *Sociological Perspectives, 37*, 497–526.
Gregory, S. W., Jr. (1999). Navigating the sound stream of human social interaction. In D. D. Franks & T. S. Smith (Eds.), *Mind, brain, and society* (pp. 247–285). Stamford: JAI Press.
Gregory, S. W., Jr., & Gallagher, T. J. (2002). Spectral analysis of candidates' nonverbal vocal communication: Predicting U.S. Presidential election outcomes. *Social Psychology Quarterly, 65*, 298–308.
Gregory, S. W., Jr., & Webster, S. (1996). A nonverbal signal in voices of interview partners effectively predicts communication accommodation and social status perceptions. *Journal of Personality and Social Psychology, 70*, 1231–1240.
Gregory, S. W., Jr., Webster, S. W., & Huang, G. (1993). Voice pitch and amplitude convergence as a metric of quality in dyadic interviews. *Language and Communication, 13*, 195–217.
Gregory, S. W., Jr., Dagan, K. A., & Webster, S. (1997). Evaluating the relation of vocal accommodation in conversation partners' fundamental frequencies to perceptions of communication quality. *Journal of Nonverbal Behavior, 21*, 23–43.
Gregory, S. W., Jr., Green, B. E., Carrothers, R. M., Dagan, K. A., & Webster, S. W. (2001). Verifying the primary of voice fundamental frequency in social status accommodation. *Language and Communication, 21*, 37–60.
Gregory, S. W., Jr., Kalkhoff, W., Harkness, S. K., & Paull, J. L. (2009). Targeted high and low speech frequency bands to right and left ears respectively improve task performance and perceived sociability in dyadic conversations. *Laterality: Asymetries of Body, Brain and Cognition, 14*, 423–440.
Hauser, M. D., Agnetta, B., & Perez, C. (1998). Orienting asymmetries in rhesus monkeys: The effects of time domain changes on acoustic perception. *Animal Behavior, 56*, 41–47.
Heise, D. R. (1970). The semantic differential and attitude research. In G. F. Summers (Ed.), *Attitude measurement*. Chicago: Rand McNally.
Hewes, G. W. (1973). Primate communication and the gestural origin of language. *Current Anthropology, 14*, 5–24.
Kalkhoff, W., & Gregory, S. W., Jr. (2008). Beyond the issues: Nonverbal vocal communication, power rituals, and 'Rope-A-Dopes' in the 2008 presidential debates. *Current Research in Social Psychology, 14*, 39–51.
Kalkhoff, W., Gregory, S. W., Jr., & Melamed, D. (2009). Effects of dichotically enhanced electronic communication on crash risk and performance during simulated driving. *Perceptual and Motor Skills, 108*, 449–464.
Kimura, D. (1961). Cerebral dominance and the perception of verbal stimuli. *Canadian Journal of Psychology, 15*, 166–171.
Kimura, D. (1967). Functional asymmetry of the brain in dichotic listening. *Cortex, 3*, 163–178.
Kimura, D. (1973a). The asymmetry of the human brain. *Scientific American, 228*, 70–78.
Kimura, D. (1973b). Manual activity during speaking. I. Right handers. *Neuropsychologia, 11*, 45–50.
Kramer, E. (1964). Elimination of verbal cues in judgments of emotion from voice. *Journal of Abnormal and Social Psychology, 68*, 390–396.
Lieberman, P. (1975). *On the origins of language: An introduction to the evolution of human speech*. New York: Macmillan Publishing Co., Inc.
Lieberman, P. (1984). *The biology and evolution of language*. Cambridge: Harvard University Press.
May, B., Moody, D. B., & Stebbins, W. C. (1988). The significant features of Japanese macaque Coo sounds. *Animal Behavior, 36*, 1432–1444.
McManus, C. (2002). *Right hand, left hand: The origins of asymmetry in brains, bodies, atoms and cultures*. London: Weidenfeld & Nicolson.
Miller, D. C. (1977). *Handbook of research design and social measurement*. New York: Mckay.
Milner, B., Taylor, L., & Sperry, R. W. (1968). Lateralized suppression of dichotically presented digits after commissural section in man. *Science, 161*, 184–185.
Ng, S. H., & Bradac, J. J. (1993). *Power in language: Verbal communication and social influence*. Newbury Park: Sage.

Oldfield, R. C. (1970). The assessment and analysis of handedness: The Edinburgh inventory. *Neuropsychologia, 9*, 97–113.
Osgood, C. E. (1964). Semantic differential technique in the comparative study of cultures. *American Anthropologist, 66*, 171–200.
Osgood, C. E., Suci, G. J., & Tannenbaum, P. H. (1957). *The measurement of meaning*. Urbana: University of Illinois Press.
Partridge, B. L. (1982). The structure and function of fish schools. *Scientific American, 246*, 114–123.
Patterson, M. L. (1995). A parallel process model of nonverbal communication. *Journal of Nonverbal Behavior, 19*, 3–29.
Rosenzweig, M. R. (1951). Representations of the two ears at the auditory cortex. *American Journal of Physiology, 167*, 147–158.
Sampson, E. E. (1971). *Social psychology and contemporary society*. New York: Wiley.
Sarles, H. B. (1975). A human ethological approach to communication: Ideas in transit around the Cartesian impasse. In A. Kendon, R. M. Harris, & M. R. Key (Eds.), *Organization of behavior in face-to-face interaction*. The Hague: Mouton Publishers.
Shubin, N. (2008). *Your inner fish: A journey into the 3.5 billion history of the human body*. New York: Pantheon Books.
Sperry, R. W. (1959). The growth of nerve circuits. *Scientific American, 201*, 68–75.
Springer, S. P., & Deutsch, G. (1981). *Left brain right brain*. New York: W. H. Freeman and Company.
Starkweather, J. A. (1956a). Content-free speech as a source of information about the speaker. *Journal of Abnormal and Social Psychology, 52*, 394–402.
Starkweather, J. A. (1956b). The communication value of content-free speech. *The American Journal of Psychology, 69*, 121–123.
Zoloth, S. R., Petersen, M. R., Beecher, M. D., Green, S., Marler, P., Moody, D. B., & Stebbins, W. (1979). Species-specific perceptual processing of vocal sounds by monkeys. *Science, 204*, 870–873.

Chapter 25
A Neurosociology of Mental Health

Anne F. Eisenberg

Introduction

Over the past 40 years increasing numbers of Americans have been diagnosed with, and treated for, serious and chronic mental illness (not including those with some form of dementia) resulting in greater numbers of people on permanent disability. According to Whittaker (2010: 220), prior to 1980 it is estimated that about 150,000 children were taking Ritalin and that "today, perhaps 3.5 million American children take a stimulant for ADHD." Additionally, improved health care for the baby boomer generation has resulted in American adults living longer and more likely to develop some form of dementia. The Alzheimer's Association website identifies that the annual total of new cases of Alzheimer's is projected to double by 2050. Specifically, from 2000 to 2010 there was a 10% increase in new cases of Alzheimer's; by 2030 there will be 50% more cases than in 2000; and by 2050 it is projected that there will be 130% more diagnosed cases of Alzheimer's than there were in 2000. Finally, "in 1987, there were 1.25 million people receiving an SSI [monthly Supplemental Security Income] or SSDI [Social Security Disability Insurance] payment because they were disabled by mental illness, or 1 in every 184 Americans" (Whitaker 2010: 6). He goes on to note that in 2007 the "number of disabled mentally ill on the SSI and SSDI rolls soared to 3.97 million… [and] the disability rate was 1 in every 76 Americans. That is more than double the rate in 1987, and six times the rate in 1955." (Whitaker 2010: 7)

More generally, mental illness now spans the entire life course starting with children being diagnosed with ADHD or even depression to disorders such as bipolar and schizophrenia appearing in young adult years, and dementia in the aged. The cost of caring for American's mental illness will overburden existing health care systems and further strain economic systems. As with the origins of sociology in the nineteenth century providing a new and unique perspective in understanding the impact of capitalism on human society, so too can neurosociology provide a new, unique, and compelling perspective in addressing the challenges wrought by the impending mental illness crisis in the United States.

A.F. Eisenberg (✉)
Department of Sociology, New York State University,
Geneseo, NY 14454, USA
e-mail: eisenber@geneseo.edu

D.D. Franks and J.H. Turner (eds.), *Handbook of Neurosociology*, Handbooks of Sociology and Social Research, DOI 10.1007/978-94-007-4473-8_25,

Mental Health, Mental Illness and Neurosociology

We can define mental health as "how we think, feel and act as we cope with life. It also helps determine how we handle stress, relate to others and make choices. Like physical health, mental health is important at every stage of life, from childhood and adolescence through adulthood." (See Medline Plus 2011) This definition implies that the more able we are to manage our lives, the more mentally healthy we are. Thus, mental illness is when "medical conditions… disrupt a person's thinking, feeling, mood, ability to relate to others and daily functioning. Just as diabetes is a disorder of the pancreas, mental illnesses are medical conditions that often result in a diminished capacity for coping with the ordinary demands of life." (National Alliance on Mental Illness) Andreasan (2001: 4) states that "mental illnesses are often ignored, misunderstood, or stigmatized" despite the fact that, as she claims, they are "among the most common diseases that afflict human beings." Ratey (2001: 337) argues that there is a biological basis for every mental disorder whereby "a person's emotional state is an important therapeutic consideration, of course, because that is the means by which the brain communicates to itself qualitative aspects of experience." Scientists distinguish between the brain and the mind with the brain defined as the physical organ while the mind is "the product of activity occurring in the brain at the molecular, cellular, and anatomical levels." (Andreasan 2001: 27) Thus, every mental illness can be described in terms of a deficit with some aspect of the brain – the physical organ – resulting in a diminished capacity for managing everyday life.

Andreasan (2001: 172) identifies the four stages "of medical progress in understanding and conquering almost all kinds of biomedical diseases" as being, first, to isolate a syndrome, which "refers to seeing a pattern in symptoms and their changes over time that suggests that their cluster may define a specific disease." In the second stage we identify its pathophysiology which "refers to the mechanisms by which a disease arises and… continues to progress." (Andreasan 2001: 173) The third stage is to find "a treatment to reverse the pathophysiology," which she admits may occur before it is actually understood, and the final stage is to find "a way to prevent the pathophysiology from arising." (172–173) The general process of understanding how the brain might be damaged and not correctly functioning starts with the display of symptoms noticed either by the patient or by those close to him or her. For the psychiatrist and therapist, this process starts with a simple question such as "How do you feel?" along with follow-up questions to more specifically understand how the patient perceives and interprets the world around him or her. Responses to these questions serve as indicators of the possible location of damage to, or within, the brain. When first meeting with a patient, the neurologist literally tests for neurologic localization through a series of exams focusing on mental status (particularly memory), cranial nerves, motor function, sensory function, coordination, and reflexes (Goldberg 2004; Ratey 2001; Cozolino 2006). Thus, mental illness is typically studied from a neurological, psychological, or pharmacological/clinical basis.

The traditional sociological approach to studying mental illness focuses on three broad areas – the social antecedents and social contexts in which mental illness might develop; the impact of mental illness on interpersonal and group dynamics; and institutional responses to mental illness (Aneshensel and Phelan 1999; Cockerham 2010; Rogers and Pilgrim 2010; Tausig et al. 2003). Recent sociological research of mental illness includes examining beliefs and perceptions about mental illness, social stressors contributing to mental illness, service usage, and systemic/institutional issues in providing mental health care (Avison 2010; Zhang et al. 2011; Hochhausen et al. 2011; Uneo 2010; Lindrooth et al. 2005). While this research does examine social contexts and antecedents to mental illness, such as stress and one's environment, it does not consider the connection between the social, the biological and the neurological nor the implications for treatment of specific mental illnesses/ disorders.

There are two relatively new approaches to studying the link between the social and the physiological – social neuroscience and biosociology. Cacioppo and Berntson (2002: 3) state that social neuroscience "addresses fundamental questions about the mind and its dynamic interactions with the biological systems of the brain and the social world in which it resides… [by studying] the relationship between neural and social processes…." The basis of social neuroscience is that humans are social animals that construct meaning and that social behavior can impact the brain as much as the brain is known to impact social behavior. While much of the work in social neuroscience focuses on the impact of biological/neurological and cognitive factors on intra- and interpersonal behaviors, an increasing number of researchers are also examining how the social affects the biological. Examples of the latter research includes Knox and Uvnäs-Moberg's (2002) study on social isolation and heart disease and Cohen et al.'s (2002) study of the connection between social ties and the common cold. Interestingly, social neuroscientists are biologists, psychologists, neurologists, and anthropologists. Despite the emphasis on the social, sociologists are not part of these developments.

The second new approach to studying the link between the social and the physiological is promulgated by sociologists. Biosociology "covers a wide range of topics, from microsociological to macrosociological, with the unifying feature being an acknowledgement of the role of biology in human social life." (Hopcroft 2011: 3) The first biosociologist was Frenchman Maurice Halbwachs who, in writing about aphasia in 1935, argued that it occurs due to a significant change in individuals' connections to others. Building on Durkheim's ideas about the role groups play in the learning process, he contended that "it is in an absent or disordered relationship to that group that the causes of any individual mental disorders should be sought, instead of launching into unverifiable conjectures as to the state of an individual's brain." (Marcel and Mucchielli 2010: 143) TenHouten is the first American to connect the sociological with the neurological in 1973 with further developments beginning in the 1980s and 1990s (Franks and Smith 1999; Franks 2006; TenHouten 1996; Smith and Stevens 1996; Mazur 1985). This area of research now includes examining the intersection of social behavior (such as religiosity, crime, status attainment) with emotions, evolution, genetics, and hormones (among others – see Hopcroft (2011) for a bibliography). The majority of biosociological research examines how some aspect of our biology affects social behavior. Neurosociology, as originally coined by TenHouten and subsequently further developed by Franks, focuses on social interactions as the unit of analysis and examines how social interactions shape brain function and thus how we perceive and interpret the world.

I argue that the neurosociological perspective provides insight to understanding and treating mental illness. There are two ways that social interactions can shape the brain. First, recent research clearly indicates that mirror neurons are responsible for managing social interactions. Thus, one can assume that the quality and quantity of our social interactions will affect the degree to which mirror neurons are activated. Neural activation levels determine the degree and complexity of the neural networks that develop in the brain. The more developed are our neural networks the "healthier" are our brains. Changes in one's interactions may likely lead to changes in neural pathways and neurochemical production associated with the mirror neurons. The second way that social interactions may shape the brain centers on the non-recursivity of social and physical phenomenon. The majority of medical models concerning mental illness focus, as does any scientific model, on one causal pathway in which the biological affects the behavioral and social (such as dopamine's potential effect on impulse control in children in explaining ADHD). The neurosociological perspective seeks to complement already accepted medical and neurological understanding of mental illness by asking how social interactions effect neural health and neurochemical production, which then effects individuals' behaviors. More importantly, the neurosociological perspective provides insight to new avenues for treatment and research concerning mental illness. To illustrate this point, I focus on three mental illnesses that highlight how they span the life course – ADHD which is normally diagnosed among children; depression which is more likely to develop in young adulthood, and Alzheimer's which usually develops after the age of 65 (early onset Alzheimer's is genetically and neurologically different from "regular" Alzheimer's and is not included as part of this discussion). Each illness is described in terms of its neurological basis (the hypothesized neural circuits involved in the brain as well as the neurochemicals associated with normal functioning); its etiology (factors affecting the normal functioning of the brain and leading to illness); and the neurosociological perspective.

ADHD

Neurological Basis

Generally, ADHD has been described by psychiatrists and therapists as "problems with working memory and frontal-lobe arousal. Most often, ADHD individuals are deficient in the motivational sensations of pleasure or pain" (Ratey 2011: 126) such that they become intensely involved in the moment seeking immediate rewards. Their thirst for the feeling of pleasure overwhelms the rational and reasoned focus offered by the frontal lobe. Ratey (2001: 127) further suggests that there is "neurochemical evidence to show that ADHD is… a reward deficiency syndrome, which results from a deficiency of pleasure neurotransmitters, principally dopamine but also serotonin and endorphins, in the rewards systems of the brain." In talking about the social brain and its development, Cozolino (2006: 41) highlights the role of the frontal cortex in regulating "motor behavior, language, executive functioning, abstract reasoning, and directed attention" as well as the role of the vagal system in coordinating the motor and sensory systems. He states that "the 'tone' of the vagus refers to the vagal system's ability to regulate the heart and other target organs… [and that] inadequate development of vagal tone can impact all levels of psychosocial and cognitive development." More to the point, Cozolino (2006: 91–92) observes that "children with poor vagal tone have difficulty in suppressing emotions in situations that demand their attention, making it difficult for them to engage with their parents, sustain a shared focus with playmates, and maintain attention on important material in the classroom."

There is a general consensus that ADHD symptoms are "caused by fronto-limbic dysfunction….. [W]eak frontal cortical inhibitory control over limbic functions might lead to ADHD" as indicated by the success of stimulant medications (Faraone and Biedermanc 2004: 988). These medications "alleviate the symptoms of ADHD… [suggesting] the hypothesis that this disorder is caused by *underactivity* of dopaminergic transmission." (Carlson 2007: 600) This is supported by mouse studies where the mouse has been bred without a dopamine transporter and "since the dopamine transporter is no longer available to remove dopamine from synapses, dopaminergic tone is increased, and the mouse becomes hyperactive." (Andreasan 2001: 124) Brain imaging studies have found that "people with ADHD have… a 3–5 percent reduction in total brain volume" (Carlson 2007: 600) and, more specifically, "smaller volumes in frontal cortex, cerebellum, and subcortical structures." (Faraone and Biedermanc 2004: 990) While clinicians and neuroscientists agree that medications that affect dopamine supply diminish the symptoms of ADHD, there is no consistent data that clearly indicates which specific region of the brain directly results in ADHD symptoms. Additionally, clinicians and researchers agree that it is "abnormalities in a network of brain regions" (Carlson 2007: 601) that results in ADHD rather than one specific area or process that can be addressed.

The majority of children diagnosed with ADHD will be treated pharmacologically either with stimulants or tricyclic antidepressants. The stimulants "modulate catecholamine pathways… [and]

reduce the overactivity, impulsivity, and inattentiveness of ADHD patients." (Faraone and Biedermanc 2004: 979) The second category of drugs used to treat ADHD is tricyclic antidepressants which "include secondary and tertiary amines with a wide range of receptor actions, efficacy, and side effects." (Faraone and Biedermanc 2004: 980) The drug regiments provide moderate to strong mediation of symptoms, particularly the behavioral symptoms. The majority of the literature concerning ADHD focuses on the impact of the pharmacological protocol in calming restless children by eliminating their hyperactive behaviors, inattentiveness and impulsiveness. However, there are serious side-effects to using these drugs –

> Children on Ritalin show 'a marked drug-related increase in solitary play and a corresponding reduction in their initiation of social interactions,' announced Russell Barkley, a psychologist at the Medical College of Wisconsin, in 1978. This drug, observed Bowling Green State University psychologist Nancy Fiedler, reduced a child's 'curiosity about the environment.' At times, the medicated child 'loses his sparkle,' wrote Canadian pediatrician Till Davy in 1989. Children treated with a stimulant, concluded a team of ULCA psychologists in 1993, often become 'passive, submissive' and 'socially withdrawn.' Some children on the drug even 'seem zombie-like,' noted psychologist James Swanson, director of an ADHD center at the University of California, Irvine. (Whitaker 2010: 223)

Whitaker (2010: 220) further notes that "this prescribing practice is mostly a U.S. phenomenon – children here consume three times the quantity of stimulants consumed by the rest of the world's children combined."

Etiology

ADHD is described as a "heterogeneous disorder" in that it is not a "unitary condition" and "the etiological mechanisms for ADHD has not been worked out in sufficient detail." (Faraone and Biedermanc 2004: 992) Four main causal agents have been identified for ADHD – genetic predisposition, brain damage due to pregnancy and delivery complications, psychosocial factors, and environmental exposure to toxins. Family studies have shown that "parents of ADHD children have a two- to eight-fold increase in the risk for ADHD" (Faraone and Biedermanc 2004: 980) and studies of identical twins find that if one child has ADHD his or her monozygotic twin will also have ADHD (as reported by a parent or teacher). Additionally, research indicates that facets of pregnancy and delivery may lead to ADHD developing in the child. These factors include "toxemia or eclampsia, poor maternal health, maternal age, fetal postmaturity, duration of labor, fetal distress, low birth weight, and antepartum hemorrhage." (Faraone and Biedermanc 2004: 987) It is hypothesized that these factors directly affect frontal lobe development and functioning. Third, psychosocial factors affecting the development of ADHD in children include marital distress, low social class, family size, and other mental disorders in the family. However, despite relatively consistent findings in ten studies such factors are considered "nonspecific triggers of an underlying predisposition or as modifiers of the course of the illness" rather than as causal factors for the development of ADHD in the child (Faraone and Biedermanc 2004: 988). Finally, environmental exposure to toxins as a causal agent focuses on the role that maternal smoking has on the pathophysiology of ADHD. Animal research has shown that extensive exposure to nicotine increases nicotinic receptors in the brain, which assist in managing dopaminergic activity (Faraone and Biedermanc 2004).

Reflecting the potential impact of the psychosocial factors, the American Academy of Pediatrics recommends that clinicians use a combination of therapy and medication in treating ADHD and the National Center for Biotechnology Information includes discussions about the behavioral component of treatment (PubMed 2011). Fabiano et al. (2009: 129) found, using meta-analysis of 114 different studies, that behavioral therapy is highly effective. However, it is important to note that the standard treatment protocol in the United States for ADHD is pharmacological only.

Neurosociological Perspective

There are two important elements of our current understanding of ADHD that are particularly relevant to the neurosociologist – the role of marital distress and family size as part of its etiology and the clinical criteria for diagnosis. The impact of the psychosocial factors of marital distress and family size, which are discounted by neuroscientists and clinicians alike despite consistent findings in the research, imply that social interactions in a child's home may impact the likelihood of whether he or she develops ADHD. If we accept that bonding between mother and child (a specific form of interaction) is affected when the mother is depressed AND if we accept that the lack of bonding affects dopamine levels in the child's brain (Cozolino 2006; Fabiano et al. 2009), then it is possible that these interactions are relevant to understanding how ADHD might develop in children. This suggests two directions for directly addressing whether ADHD develops as well as treatment and research options. First, a preventive measure based on the psychosocial etiology of ADHD highlights the need for educational/training and support programs for mothers (and fathers) to engage in meaningful interactions that results in bonding with their children. The presumption that parenting, as a basic biological function, occurs naturally is not borne out by research. Thousands of cases of child abuse and child abandonment illustrate the need for some type of education, training and support for parents. The thesis is that if children effectively and meaningfully bond with their parents they are less likely to develop ADHD.

The second direction for a neurosociological treatment and research program concerns the clinical criteria for diagnosing children with ADHD. The first issue associated with the clinical criteria is the high level of subjectivity of using the DSM-IV criteria to arrive at a diagnosis. The DSM-IV criteria for diagnosing ADHD focuses on whether the child's behavior has been disruptive for over 6 months with regard to inattention, hyperactivity or inattentiveness. There does not seem to be any consensus concerning what level of hyperactivity, inattentiveness or inattention truly requires a diagnosis of ADHD. Thus, there may be thousands of situations where a child is simply a normal and highly active individual. However, in overcrowded classrooms and among families where there are a high number of stressors the child may be more likely to be perceived as being "disruptive" and then diagnosed as having ADHD. Additionally, there is anecdotal evidence that in the United States childhood has become much more regimented and structured in regard to play and learning activities. If a child does not easily or willingly conform to more structured play and learning activities they are more likely to be diagnosed with ADHD and medicated. The subjectivity of applying the DSM-IV criteria combined with the changing cultural expectations for childhood requires a neurosociological perspective. The sociologist is best prepared to mediate between changing definitions of childhood and how the medical establishment treats children who do not conform. Further, the neurosociologist is more likely to understand how structural conditions (such as overcrowded classrooms and parental work situations) can affect how we interpret others actions.

The second issue related to the DSM-IV clinical criteria for diagnosing ADHD is that a number of the signs of ADHD reflect clinicians' perception of the how well a child interacts with others. For example, some of the signs include "often does not seem to listen when spoken to directly," "often talks excessively," "often blurts out answers before questions have been finished," "often has trouble waiting one's turn," and "often interrupts or intrudes on others (example: butts into conversations or games)." (www.psychnet-uk.com/dsm_iv/attention_deficit_disorder.htm) In other words, children who do not interact appropriately with others may be perceived as unable to control their behavior rather than acting as children might. Yet, we could also interpret these behaviors as children excited about learning or creative children. As childhood in America has become more regimented and regulated, so too have our definitions of appropriate interactions among and with children. The neurosociological perspective can contribute to our understanding of changing interactions in an increasingly networked and regulated society.

Depression

Neurological Basis

Identified as one of two major affective/mood disorders, unipolar depression (hereafter noted as depression) has been linked to low levels of norepinephrine or serotonin as caused by problems with neurotransmitters responsible for their levels. Research concerning the neurological basis for these depletions started with observations of the impact of antidepressants on depressive symptoms, leading to the monoamine hypothesis stating that "depression is caused by insufficient activity of monoaminergic neurons" (Carlson 2007: 572). However, further pharmacological research clearly indicated that the monoamine hypothesis was too simplistic an explanation for describing the mechanisms underlying depression due to three different factors. First, patients responded to treatments for norepinephrine and serotonin in similar ways – when taking either drug the depressive symptoms are ameliorated; when patients stop taking either drug the symptoms return and they relapse back into depression. In other words, both drugs diminish depressive symptoms. As Duman (2004: 422) states, "when patients are successfully treated with either [serotonin] or [norepinephrine] selective reuptake inhibitors, they become vulnerable to depletion of the corresponding monoamine, suggesting that [serotonin] and [norepinephrine] are involved in the maintenance of the antidepressant response." Thus, the mechanism underlying depression is not yet understood. This is partly due to the fact that, like treatments for ADHD, the drugs were originally intended for a different purpose when it was noticed that they also modulated mood disorders. In other words, the drug regimen was implemented before the biological and neurological mechanisms were studied and understood (Whitaker 2010).

The second factor indicating that the relationship between the monoamines and depression is more complex than earlier research showed centers on the time lag between administration of the drug and the point at which the symptoms start to diminish. More to the point, "although antidepressant drugs rapidly increase the levels of monoamines (i.e., within days), the therapeutic action of these treatments is dependent on chronic administration (i.e., several weeks or even months)." (Duman 2004: 422; see also Kandel 2006) A final factor illustrating the complex relationship of the monoamines and depression is that low levels of norepinephrine or serotonin do not necessarily lead to depressive symptoms in everyone. While research has elucidated some of the neurochemical origins of depression the question is in which neurotransmitting system are such levels affected.

Research shows that "happiness and sadness are separate functions and they represent opposite patterns of activity in the hemispheres of the brain." (Ratey 2004: 229; see also Cozolino 2006) Studies have found that people with a tendency to express more negative moods will have greater activity in the right hemisphere. Neuroimaging studies of mood disorders provide evidence for the role of brain abnormalities in producing depressive symptoms, potentially explaining why low levels of norepinephrine, or serotonin, leads to depression for some people but not others. Research has focused on either structural (the organ itself is damaged) or biochemical (the ability of brain to prevent cell death) abnormalities. While Ratey (2004) identifies overactivity in the amygdala and the frontal lobe as being associated with extended sadness, researchers examining structural abnormalities have focused on "the prefrontal cortex, basal ganglia, hippocampus, thalamus, cerebellum, and temporal lobe in the brains of depressed people." (Carlson 2007: 574) Specifically, reductions in width and volume in the frontal lobe have been found in depressives as well as changes in the entire temporal lobe and the basal ganglia. It is worth noting that at least a third of the research focused on elderly depressives, which confounds our understanding of what structural changes are related to depression and which changes are related to age.

Cozolino (2006: 217) argues that research shows that "children of depressed mothers have been reported to show a disruption in the development, connection, and integration of the frontal lobes…." Additionally, it is difficult to separate whether such organic changes are due to depression occurring as opposed to whether such organic changes are causing depression. Studying biochemical abnormalities in brains of depressed people have found several key items. Magnetic resonance research has identified hyperintensities indicating "lesions… in the deep white matter, the periventricular white matter, the basal ganglia, and the pons." (Drevets et al. 2004: 462) Postmortem examination of brains showing such hyperintensities found decreased volume in the hippocampus which has been attributed to decreased "levels of brain-derived neurotrophic factor (BDNF), a chemical that promotes growth and suppresses apoptosis." (Drevets et al. 2004: 574) A loss of BDNF can lead to the structural losses found in brains of depressives. However,

> [d]espite 40 years of concerted research, the primary neurochemical pathology of major depression has not been identified. Dysfunction of many different neurotransmitter systems has been documented in depression, yet no one system or one perturbation has clearly emerged as the fundamental pathology in major depression. (Garlow and Nemeroff 2004: 454)

Ratey (2002: 223) argues that emotions, such as those experienced by depressives, "are the result of multiple body and brain systems." Thus, as this discussion illustrates, it is likely that depression can arise from a number of different brain locations as well as the interaction between the different systems.

The majority of patients treated for depression are subjected to physiologically-based treatments with the primary form being pharmacological. An individual clinically diagnosed with depression will most likely be placed on either a mood stabilizer or on one of several anti-depressants with the goal to find which one of four different types work best – monoamine oxidase inhibitors, tricyclic antidepressants, specific serotonin reuptake inhibitors, or serotonin and norepinephrine reuptake inhibitors (Carlson 2004). As noted above, depression can be caused by one of several different neurotransmitter systems and since each medication can take up to 2 months before it effectively diminishes symptoms, it can take up to 6 months before the correct drug is found to alleviate depression (Garlow and Nemeroff 2004; Duman 2004). Electroconvulsive therapy and transcranial magnetic stimulation are used on those patients for whom the pharmacological treatments are not effective. These treatments literally affect neural networks and affect patients' neurochemistry.

In discussing the impending epidemic of the use of psychotropic drugs to address depression, Whitaker (2010) references a number of studies that highlight the overall lack of efficacy of antidepressants and, even more concerning, how these drugs may actually exacerbate the problems patients' experience. There is a growing number of researchers as well as a growing number of research studies that question whether clinical depression is caused by a biological deficit. Whitaker (2010: 74) references a convincing number of studies, over decades, that illustrate there is "`no clear and convincing evidence that monoamine deficiency accounts for depression; that is, there is no 'real' monoamine deficit." Equally concerning, some researchers argue that anti-depressant drugs actually permanently change the brain thereby requiring patients' to become addicted to the drugs. Patients being treated with pharmaceuticals become trapped in a cycle where they are initially prescribed a drug to deal with an episode of depression. While taking the drug their brains undergo organic changes such that greater dosages are needed to ameliorate the symptoms. If they then try to stop the drug regimen, patients literally go through withdrawal and the depressive symptoms return more severe than previously. This then requires that they begin the drug regimen again (Whitaker 2010: 148–171; 307–312). In summary, increasing numbers of people are being diagnosed and treated with pharmaceuticals for depression despite very strong questions about the efficacy of such treatments.

Etiology

Research indicates that the likelihood of developing depression is significantly higher when close relatives suffer from an affective disorder, although Ratey (2004: 241) states that it is "less genetically based than any other mental illness, and is the one most dependent on environmental factors."

Andreasan (2001: 234) posits that mood disorders are "probably produced by a mixture of factors, spanning the range from personal experience to brain chemistry" and that the contribution of any of these combinations of factors changes from one person to another. Andreasan (2001) discusses the role that stress and sadness play in stimulating the cerebral cortex and limbic system to the endocrine system. When the endocrine system is stimulated it then impacts the hypothalamus and pituitary gland, resulting in a dysfunctional cortisol regulation system. Cozolino (2006) highlights the impact of whether children become securely attached to other people on their stress level on the orbital medial prefrontal cortex, the amygdale, and the autonomic nervous system.

Andreasan (2001) also discusses a genetic basis for depression referencing family studies (whether a close relative has a mood disorder increases the likelihood of another family member experiencing a mood disorder) as well as twin studies (if one identical twin has depression it is highly likely the other twin will as well). Cozolino (2006) and Goleman (2006) both identify the role that mothers' depression has on their children. The depressed mother is less able to attend to, and bond with, her children. Cozolino (2006) argues that the impact of insufficient bonding and attachment between mother and child shapes frontal lobe development, leading to a greater likelihood of depression among children of depressed mothers. This is illustrated by the fact that "babies of depressed mothers have higher levels of stress hormones and lower levels of dopamine and serotonin…." (Goleman 2006: 168)

The Surgeon General's report on mental health in America highlights the genetic, biological, and psychosocial causes of depression. In discussing the biologic factors the authors emphasize that biological changes in the brain could as easily be a result of depression rather than causing it. Also, the report states that although depression does seem to run in families there is no research that clearly proves a genetic connection separate from environmental and psychosocial factors. It is interesting to note that the majority of the etiology discussion focuses on psychosocial factors which include chronic stress, anxiety, how people view and interpret events in their lives (cognitive factors), temperament and personality. The more easy-going a person the less likely they are to succumb to depression. The discussion about the psychosocial factors focuses on how each may possibly affect the production of particular neurochemicals which then may lead to depression. Additionally, the report states that there is a "compelling impact of past parental neglect, physical and sexual abuse, and other forms of maltreatment on both adult emotional well-being and brain function is now firmly established for depression. Early disruption of attachment bonds can lead to enduring problems in developing and maintaining interpersonal relationships and problems with depression and anxiety." (U.S. Department of Health and Human Services 1999) This understanding of depression is reflected in the use of psychotherapy to help "people identify negative cognitive responses and to retrain these patterns so that they are replaced by more positive and affirmative schemas. Psychodynamic psychotherapies may also assist people in exploring the ways that prior experience lead to the tendency to respond with depressive emotions." (Andreasan 2001: 251) The use of "talk therapy" (Ratey 2001) allows the patient to literally "rewire" the way a person reacts and responds to stressful situations. However, it is clear that the majority of people are placed on a drug regimen and it is less clear how many people will also receive some type of "talk therapy" to modify their behavioral responses.

Neurosociological Perspective

The research indicates that the biological basis for depression is complicated and the etiology is equally complex as depression is often comorbid with other mental and physical disorders. The Surgeon General's report focuses extensively on the psychosocial factors of cognition, personality and gender with regard to depression's etiology. Embedded in its discussion, yet not highlighted, is the role that connections to other and social interactions may play in depression's etiology with regard to how social interactions shape our personality and cognitive abilities as well as how connections to others provides the necessary support to cope with life. As discussed in the section on ADHD, Cozolino (2006) argues that depressed mothers are less able to bond with their children and hypothesizes that this affects the child's brain development as well as social development. Our interactions with significant others as children and adolescents shapes how we then respond to life events as young adults. Children who are less comfortable in interacting with others may be more likely to experience a cognitive state of low efficacy and helplessness. One model of cognition argues that such negative cognitions about oneself are associated with the likelihood of developing depression. Further, negative cognition in children may result either in a shy personality while for other children it may result in a more overly aggressive personality. Thus, a neurosociological approach to understanding and treating depression would focus on studying the ways that children learn how to interact and encouraging the development of emotionally meaningful ties to others.

The cognitions, temperament, and personality developed during childhood then affects individuals' connection to others and social interactions as an adult. Children who are comfortable interacting with family members and peer groups are more likely going to be successful interacting with coworkers and potential romantic partners. The greater the number of meaningful social ties to others we develop through social interactions, the larger is our social support network. These social networks are critical to helping individuals cope with challenging life events or situations, such as illness, job loss, and the death of loved ones. Research has shown that individuals without such social support are more likely to develop depression. This suggests that neurosociological research should focus on understanding how to reinforce individual's social support networks. These ideas correspond closely with Durkheim's ideas concerning the role of integration in maintaining social order. He argued the vital role played by the groups to which we belong in forming our social identity linking us to the rest of society. For example, the rash of young adult suicides particularly among gay males in America indicates the likelihood of increasing numbers of cases of clinical depression developing among young adults. The development of preventive treatment programs that focus on creating social support networks for young adults is one avenue for potential research and treatment.

Alzheimer's

Neurological Basis

Alzheimer's is a form of dementia, which is when organic changes in the brain affect all aspects of an individual's ability to live starting from deterioration of cognitive abilities and ending in loss of autonomic functions and, eventually, death. Research concerning the neurological basis of Alzheimer's focuses on "severe degeneration of the hippocampus, entorhinal cortex, neocortex (especially the association cortex of the frontal and temporal lobes), nucleus basalis, locus coeruleus, and raphe nuclei." (Carlson 2007: 539) These changes are linked to excess amounts of the protein ß-amyloid forming the core of amyloid plaques, which are extracellular deposits. These plaques are then "surrounded by degenerating axons and dendrites, along with activated microglia and reactive astrocytes, cells that are involved in destruction of damaged cells. Eventually, the phagocytic glial cells destroy the degenerating axons and dendrites, leaving on a core of ß-amyloid…." (Carlson 2007: 539) Research on the molecular basis of protein production has shown that production of amyloid plaques are most likely to occur when there is a defective form of the ß-amyloid. These plaques literally starve to death different parts of the brain. Neuroimaging and post-mortem studies of Alzheimer's brains focus on the physiological changes to the brain as well as possible origins for these changes. Bobinski et al. (2004: 825) examined the hippocampus, given that memory impairment is one of the first indicators of Alzheimer's, and found that "postmortem histopathological studies show that the hippocampal formation, especially the entorhinal and transentorhinal cortices, is one of the earliest and most severely affected structures ...." Based on a large number of studies – pathological, stereological, post-mortem, and neuroimaging – they argue that "hippocampal changes .... may predict the development of symptoms consistent with the course of [Alzheimer's]" (830–831) and that "memory tests that involve medial temporal lobe structures for execution are the best indices of early cognitive impairment." (831) In other words, Bobinski et al. (2004) posit that specific memory tests and non-invasive imaging of the hippocampus may eventually lead to early diagnosis of Alzheimer's disease.

However, since there is no "specific clinical and/or laboratory diagnostic test for Alzheimer's disease, it is correctly stated that one cannot make a definitive clinical diagnosis of this disorder and confirmation at autopsy is ultimately required." (Perl 2004: 840) Further, as Snowdon's (2001) research demonstrates, structural changes to the brain, such as amyloid plaques or neurofibrillary tangles do not necessarily indicate whether a person actually exhibits symptoms of Alzheimer's disease. In the "Nun Study," Snowdon and his colleagues discovered a confounding factor that determined whether or not an individual developed symptoms of Alzheimer's disease. It is informative to explain this in his own words:

> our autopsies showed that the small (lacunar) brain infarcts had a stunning link to dementia in the sisters *but only if their brains also had enough plaques and tangles to meet our criteria for Alzheimer's disease.* Among the sisters with an "Alzheimer's brain," 93 percent had dementia if, like Sister Agnes, they also had at least one lacunar infarct in the deep white matter, the thalamus, or the neighboring basal ganglia (a collection of structures that control movement). In contrast, only 57 percent of the sisters who had an "Alzheimer's brain" but no strokes had dementia. Our data also suggested that sisters with evidence of a stroke required *fewer* tangles in the neocortex to show signs of dementia than if they had been stroke-free. We concluded that many sisters – in spite of having brain damage from Alzheimer's disease – avoided dementia because they had not suffered small strokes. (2002: 155)

Thus, while it is clear that significant structural changes in the brain are associated with potential development of Alzheimer's they are not necessarily predictive of whether the disease is actually occurring.

The neurological basis of our understanding of Alzheimer's is highlighted in the standard treatment protocol. Once a diagnosis is confirmed, patients' will be encouraged to try one of several pharmaceutical programs whose goal is to slow down the progression of the disease and identifying key chemical agents associated with Alzheimer's is the focus of current research. Specifically, the research and treatments seek to enhance acetylcholine function which helps to improve attention and ease cognitive dysfunction.

Etiology

Kaufer and DeKosky (2004: 772) differentiate between degenerative and nondegenerative forms of dementia whereby the degenerative category "primarily reflects pathophysiological processes that are intrinsic to the central nervous system (CNS) ... [while the] nondegenerative dementias ... are a heterogeneous group of disorders reflecting diverse etiologies: vascular, endocrine, traumatic, demyelinating, neoplastic, infectious, inflammatory, hydrocephalic, systemic, nutritional deficiency, and toxic conditions." Some research has shown that "some forms of Alzheimer's disease appear to run in families and thus appear to be hereditary" (Carlson 2007: 541). Discussion about the genetic basis for Alzheimer's centers on the mutation of genes responsible for transporting cholesterol in the blood as well as genes responsible for the production of ß-amyloid. Also, brain trauma has been shown to correlate with the development of amyloid plaques and neurofibrillary tangles. Andreasan (2001: 82) highlights how "the acetylcholine system plays a major role in the encoding of memory" and how Alzheimer's patients show losses of acetylcholine in the cortex and hippocampus.

While the majority of research centers on the biological causes of Alzheimer's, there is also some evidence that "low occupational level and having a first-degree relative with Alzheimer's" (Andreasan 2001: 265) are correlated with onset of Alzheimer's. Kaufer and DeKosky (2004: 772) identify individual factors associated with the onset of Alzheimer' such as "education, gender, age-related changes, preexisting brain disease, environmental exposures, and medical and psychiatric comorbidity …." while Andreasan identifies experiencing head injuries or having general anesthesia can also contribute to its onset. The Nun Study (Snowdon 2001), as well as other research associated with The Religious Orders Study, found "a positive relation between increased number of years of formal education and cognitive performance, even in people whose brains contained significant numbers of amyloid plaques" (Carlson 2007: 542). More generally, "any intellectually challenging activity stimulates dendritic growth, which adds to the neural connections in the brain …. and that more academic challenge leads to a more flexible brain in old age…" (Ratey 2001: 43). Additionally, research has shown that depressed adults are more likely to develop Alzheimer's disease than are their peers who are not depressed.

Neurosociological Perspective

At first glance, there seems to be little of interest to a neurosociologist given that the majority of research illustrates how Alzheimer's is strongly correlated to significant changes in the brain. It is clear that organic change in certain parts of the brain is associated with the onset of Alzheimer's. It is also evident that some drug regimens are successful in either slowing the progress of the disease or in ameliorating cognitive loss for a period of time. Finally, it is accepted that there is no cure for Alzheimer's and most people consider it a death sentence (as it is for many patients). However, there are several issues that are of particular relevance for neurosociology concerning its etiology.

The Nun Study found that although there were nuns with evidence of organic changes to the brain that would normally indicate the development of Alzheimer's disease, they did not develop Alzheimer's while nuns with relatively few such organic changes were more likely to have developed the disease. The explanation offered (Snowdon 2002) was that the nuns with little organic change to the brain who had developed Alzheimer's had also suffered a stroke previously. Yet, they also found that level of education and type of job (in terms of the number of interactions the job required of the nun) was negatively related to disease development. In other words, the higher the level of education held by the nun and the higher the number of interactions required as part of the job seemed to "protect" the nun from developing Alzheimer's. Additionally, a number of studies have found that depression is as likely a cause of, as well as an outcome, of Alzheimer's disease. Those patients who are have clinically diagnosed depression are significantly more likely to develop Alzheimer's than those who are not depressed. Also, gerontological studies have found that as people age, their social networks decrease in size as they retire from jobs and become less able to travel and meet with friends. These findings imply that decreased social interactions are likely to lead to depression among older adults.

The neurosociologist is uniquely qualified to contribute to our understanding of Alzheimer's. We already understand the impact of structural factors such as education and family on well being as well as how aging changes the quantity and quality of our social networks. The key is to study how such structural factors and social changes literally affect our physiological status. If we are more likely to become depressed as our social networks diminish we are then more likely to develop dementia. The implications for treating such potentially devastating disorders then become important social issues (Mills 1959).

Alternative Research and Treatment Options

The standard of care, in the United States, for ADHD, depression and Alzheimer's has led to what Whitaker (2010) calls a rising epidemic. Treatment protocol for all three illnesses relies heavily on the use of pharmaceuticals whose side effects have created a growing population of permanently disabled people. The United States is ten times more likely to medicate a child diagnosed with ADHD than any other developed country, five times more likely to prescribe pharmaceuticals for individuals diagnosed with depression, and three times more likely to include a drug regimen for a person diagnosed with Alzheimer's. Substantial research, over the past 40 years, clearly identifies significant problems with pharmaceutical regimens in treating mental health disorders. The traditional approach to studying and treating mental illness in the United States is neurological, psychological, or biological. These approaches are more likely to result in diagnosing behaviors that deviate, even slightly, from what is perceived to be normal as a form of mental illness. As we have seen, once diagnosed with a mental illness, the United States is far more likely to treat such illnesses with a panoply of pharmaceuticals that have been found to permanently change the neurochemical, and eventually structural, basis of the brain. Once placed on medication protocol few patients are able to successfully withdraw from the medications. This multi-billion dollar industry is also creating an increasingly disabled population supported by SSI and SSDI.

The neurosociological interpretation of the formally recognized symptoms for each illness focuses on the role of social interactions and meaningful social connections to relevant groups. This approach to understanding mental illness is reflected in how other countries treat mental illness where there are alternative treatment options that do not rely on pharmaceuticals and whose success clearly indicates the role of the social for the addressing the neurological.

In talking about ADHD, Goleman (2006: 180) states that

> Panksepp's research raises an intriguing question: what do you call a child who exhibits hyperactivity, impulsivity, and unfocused, rapid shifting from one activity to another? Some might see these shifts as indicators of attention-deficit/hyperactivity disorder (ADHD), which has reached epidemic proportions among schoolchildren, at least in the United States. But Panksepp, extrapolating to humans from his work with rodents, sees the shifts instead as signs of an active neural system for play. He notes that the psychostimulant medications given to children for ADHD all reduce the activity of the brain's play modules when given to animals, just as they seem to snuff out playfulness in children. He makes a radical, though untested, proposal: let younger children 'vent' their urge to play in an early-morning free-play, rough-and-tumble recess, then bring them into a classroom after their urge to play has been sated, when they can more easily pay attention.

For example, while this chapter does not discuss schizophrenia, a disorder that in the United States is always immediately treated with pharmaceuticals that dull the patient's senses and discourages them from maintaining the drug regiment, psychiatrists in Finland have been using two different types of behavioral therapies. Long term studies in Lapland found that very few schizophrenic patients benefitted from drug treatments and in Tornio they use "open-dialogue therapy" (Whitaker 2010: 340) in which patients, their families, and a team of doctors, nurses, and psychologists work to help the patient manage their perception of reality. More to the point

> Their conception of psychosis is quite distinct in kind, as it doesn't really fit into either the biological or psychological category. Instead, they believe that psychosis arises from severely frayed social relationships. 'Psychosis does not live in the head. It lives in the in-between of family members, and the in-between of people .... It is in the relationship, and the one who is psychotic makes the bad condition visible.' (Whitaker 2010: 341)

European countries utilize treatment protocol that emphasize behavioral therapy ranging from behavior modification programs for young children as well as providing additional attention to those diagnosed with ADHD (England) to intervention programs that bring together the entire family to ensure that the child gets sufficient attention as well as exercise programs that encourage children with ADHD to "burn off" their excess energy prior to requiring them to sit still in a classroom for hours at a time (Norway). Rather than viewing the child diagnosed with ADHD as unhealthy and prescribing medications to make their brains normal or healthier (medicalizing childhood behavior), European countries view such behavior as being within the normal range of behavior for a child requiring modification. The two Seneca Centers in California are one of the last residential facilities that treat behaviorally troubled children without psychiatric drugs. In explaining their approach, Tony Stanton's "own attachment theory" convinced him of the importance of emotional relationships to a child's well-being. Then, in the late 1970s, while he was in charge of a psychiatric ward for children at a county hospital, he assigned a "mentor" to every child rather than medicating them and he saw a number of them become attached to their mentors and "blossom." (Whitaker 2010: 349) More to the point, he found that emotional connections to others are a vital part of a child's mental health in a way that drugs were unable to produce. Similar programs in New Mexico and Alaska also focus on creating and developing children's connections to others in lieu of treating them with drugs. All three programs have had similar success to those in European countries.

Treatment for depression in the United States usually begins with the use of anti-depressants as physicians try to find the right drug to ameliorate the symptoms. Increasingly, treatment protocols will also include behavioral programs such as psychotherapy and counseling. In England, general practitioners are encouraged to write prescriptions for exercise for patients of any age with mild depression. The National Institute for Health and Clinical Excellence found that exercise improves self esteem as well as "enhances cardiovascular function, increases muscle strength, lowers blood pressure, and improves cognitive function. People sleep better, they function better sexually, and they also tend to become more socially engaged" (Whitaker 2010: 345). As stated earlier, the use of "talk therapy" or psychotherapy has been found to be far more effective than the standard drug regimen prescribed. In fact, it is very difficult to successfully "wean" patients' from the drug regimen once they have been on it for more than 6 months. Despite the success of non-pharmaceutical treatment plans, few American insurance plans pay for counseling services such as talk therapy or psychotherapy.

Similarly, several European countries have recently begun exploring how best to deal with what they consider to be an impending epidemic of Alzheimer's as their populations of older citizens continue to survive into their 80s and are more likely to develop the disease. The majority of their discussions, thus far, have ranged from establishing community-based resource centers that would create support groups for patients diagnosed with the disease to ensure that they do not become isolated (England) to bringing a team of professionals (physician, psychologist, social worker) into the patient's home to determine their needs with the goal of keeping them in their home as long as possible with a strong support network including family members and providers (Norway). The general theme of how mental disorders are treated differently in European countries is that the focus is on the social connections, interactions and networks relevant in the patient's life. Rather than regarding the disorder as unhealthy and managing the patient pharmaceutically, they emphasize the need for the individual to remain a functioning member of society by supporting their already existing social connections and reinforcing them. In other words, it is the sociologically relevant aspects of the patients' lives that are used to address mental health disorders.

Evidence from alternative treatment programs, as well as clinical and basic research, indicates that pharmaceutical regimens in treating mental illness actually exacerbate the illness for the majority of patients. Focusing on behavior modification and strengthening social connections between the patient and his or her community illustrates the role that social interactions play in managing mental illness. More to the point, a neurosociological approach has the potential to significantly contribute to more effectively treating mental illnesses such as ADHD, depression, and Alzheimer's with a far less costly approach than found with the current drug regimens.

Conclusion: The Implications for Mental Health in America

Despite the research highlighting how different drugs literally change the brain's chemistry and potentially turn patients into addicts requiring the drugs to ameliorate their symptoms, the American mental health system relies almost exclusively on pharmaceuticals in treating mental illness. This is aptly illustrated by Andreasan's (2001: 336) comment that "in most instances, the psychoactive drugs available in our armamentarium are valued assets. We need more medications and better ones, not fewer. To think otherwise is to revert to the old tradition of stigmatizing mental illnesses and seeing them as moral failures cause by weakness of will or bad parenting." Over the past 40 years significant numbers of people have been added to disability rolls due to chronic mental illness. It is predicted that this trend will only continue to increase when we look at the numbers of children being medicated for ADHD, young adults for depression, and older adults for Alzheimer's. Such treatment programs are the most expensive and result in millions of people unable to work and contribute to the economy as well as the families that are shattered by such illnesses. Caregivers – in the form of parents, spouses, siblings, and professionals – are becoming overwhelmed. The need for sociologists to be engaged in research and discussions concerning mental illness is highlighted the degree to which social interactions as well as the social environment affects the development of these disorders. Neuroscience research is just beginning to examine which types of neural activity and neurons are most likely linked to ADHD, depression and Alzheimer's. These developments in neuroscience, combined with the successful outcomes in addressing mental disorders when the social is emphasized rather than the pharmaceutical, clearly indicates the need for a new area of study and research in sociology. Neurosociology – which I define as the study of how social interactions on the individual and group level impact on mental, emotional, and neurological health – encourages the integration of a neuroscience understanding into sociology *and* a sociological understanding into neuro- and cognitive sciences. By encouraging research on the interaction between the social and neurological, sociologists can begin to contribute to discussions concerning how best to treat and manage such disorders such as is found in other countries.

References and Resources

Alzheimer's Association. (2002). alz.org/downloads/facts_figures_2011.pdf

Andreasan, N. C. (2001). *Brave new brain: Conquering mental illness in the era of the genome*. New York: Oxford University Press.

Aneshensel, C. S., & Phelan, Jo C. (Eds.). (1999). *Handbook of the sociology of mental health*. New York: Springer.

Avison, W. R. (2010). Incorporating children's lives into a life course perspective on stress and mental health. *Journal of Health and Social Behavior, 51*(4), 361–375.

Bobinski, M., DeLeon, M. J., Convit, A., & DeSanti, S. (2004). Neuropathological and neuroimaging studies of the hippocampus in normal aging and in Alzheimer's disease. In *Neurobiology of mental illness* (2nd ed., pp. 821–835). New York: Oxford University Press.

Cacioppo, J. T., & Berntson, G. G. (2002). Social neuroscience. In J. T. Cacioppo et al. (Eds.), *Foundations in social neuroscience* (pp. 3–10). Cambridge, MA: MIT Press.

Carlson, N. R. (2007). *Physiology of behavior* (9th ed.). Boston: Pearson.

Charney, D. S., & Nestler, E. J. (Eds.). (2004 [1999]). *Neurobiology of mental illness* (2nd ed.). New York: Oxford University Press.

Cockerham, W. C. (2010). *Sociology of mental disorder* (8th ed.). Upper Saddle River: Prentice Hall.

Cohen, S., Doyle, W. J., Skoner, D. P., Rabin, B. S., & Gwaltney, J. M., Jr. (2002). Social ties and susceptibility to the common cold. In J. T. Cacioppo et al. (Eds.), *Foundations in social neuroscience* (pp. 1269–1278). Cambridge, MA: MIT Press.

Cozolino, L. (2006). *The neuroscience of human relationships: Attachment and the developing brain*. New York: W. W. Norton & Company.

Drevets, W. C., Gadde, K. M., Ranga, K., & Krishnan, R. (2004). Neuroimaging studies of mood disorders. In D. S. Charney & E. J. Nestler (Eds.), *Neurobiology of mental illness* (2nd ed., pp. 461–490). New York: Oxford University Press.

Duman, R. S. (2004). The neurochemistry of depressive disorders: Preclinical studies. In D. S. Charney & E. J. Nestler (Eds.), *Neurobiology of mental illness* (2nd ed., pp. 421–439). New York: Oxford University Press.

Fabiano, G. A., Pelham, W. E., Jr., Coles, E. K., Gnagy, E. M., Chronis-Tuscano, A., & O'Connor, B. C. (2009). A meta-analysis of behavioral treatments for attention-deficit/hyperactivity disorder. *Clinical Psychology Review, 29*(2), 129–140.

Faraone, S. V., & Biedermanc, J. (2004). Neurobiology of attention deficit hyperactivity disorder. In D. S. Charney & E. J. Nestler (Eds.), *Neurobiology of mental illness* (2nd ed., pp. 979–999). New York: Oxford University Press.

Franks, D. D. (2006). The neuroscience of emotions. In J. E. Stets & J. H. Turner (Eds.), *Handbook of the sociology of emotions*. New York: Springer.

Franks, D. D. (2010). *Neurosociology: The nexus between neuroscience and social psychology*. New York: Springer.

Franks, D. D., & Smith, T. (Eds.). (1999). *Mind, brain and society: Toward a neurosociology of emotion: Social perspectives on emotion* (Vol. 5). Stanford: JAI Press.

Garlow, S. J., & Nemeroff, C. B. (2004). The neurochemistry of depressive disorders: Clinical studies. In D. S. Charney & E. J. Nestler (Eds.), *Neurobiology of mental illness* (2nd ed., pp. 440–460). New York: Oxford University Press.

Goldberg, S. (2004). *The four-minute neurologic exam*. Miami: MedMaster.

Goleman, D. (2006). *Social intelligence: The new science of human relationships*. New York: Bantam Books.

Hochhausen, L., Le, H.-N., & Perry, D. F. (2011). Community-based mental health service utilization among low-income Latina. *Community Mental Health Journal, 47*(1), 14–23.

Hopcroft, R. L. (2002). The evolution of sex discrimination. *Psychology, Evolution & Gender, 4*(1), 43–67.

Hopcroft, R. L. (2009). Gender inequality in interaction – An evolutionary account. *Social Forces, 87*(4), 1845–1871.

Hopcroft, R. L. (2010). *Sociology: A biosocial introduction*. Boulder: Paradigm Publishers.

Hopcroft, RL. (2011, Spring). *Evolution, Biology & Society Newsletter*. www2.asanet.org/sectionevol/

Kaufer, D. I., & DeKosky, S. T. (2004). Diagnostic classifications: Relationship to the neurobiology of dementia. In D. S. Charney & E. J. Nestler (Eds.), *Neurobiology of mental illness* (2nd ed., pp. 771–782). New York: Oxford University Press.

Knox, S. S., & Uvnäs-Moberg, K. (2002). Social isolation and cardiovascular disease: An atherosclerotic pathway? In J. T. Cacioppo et al. (Eds.), *Foundations in social neuroscience* (pp. 1241–1254). Cambridge, MA: MIT Press.

Lindrooth, R. C., Lo Sasso, A. T., & Ithai, Z. (2005). The effect of expanded mental health benefits on treatment initiation and specialist utilization. *Health Services Research, 40*(4), 1092–1107.

Marcel, J.-C., & Mucchielli, L. (2010). Maurice Halbwachs's *mémoire collective*. In A. Erll & A. Nünning (Eds.), *A companion to cultural memory studies* (pp. 141–149). Berlin: Walter De Gruyter.

Mazur, A. (1985). A biosocial model of status in face-to-face primate groups. *Social Forces, 64*(2), 377–402.

Mazur, A. (2005). *Biosociology of dominance and deference*. Lanham: Rowman & Littlefield Publishers.

Mazur, A. (2009). A hormonal interpretation of Collins's micro-sociological theory of violence. *Journal for the Theory of Social Behaviour, 39*(4), 434–447.

Medline Plus. (2011). *Mental health*. http://www.nlm.nih.gov/medlineplus/mentalhealth.html

Mills, C. W. (1959). *The sociological imagination*. New York: Oxford University Press.

National Alliance on Mental Illness (NAMI). (2011). http://www.nami.org/template.cfm?section=about_mental_illness

Perl, D. P. (2004). Abnormalities in brain structure on postmortem analysis of dementia. In *Neurobiology of mental illness* (2nd ed., pp. 836–848). New York: Oxford University Press.

PubMed. (2011). *Attention deficit hyperactivity disorder*. http://www.ncbi.nlm.nih.gov/pubmedhealth/PMH0002518/

Ratey, J. J. (2002). *A user's guide to the brain: Perception, attention, and the four theaters of the brain*. New York: Vintage Books.

Rogers, A., & Pilgrim, D. (2010). *A sociology of mental health and illness*. Buckingham: Open University Press.

Scheid, T. L., & Brown, T. N. (Eds.). (2010). *A handbook for the study of mental health: Social contexts, theories, and systems*. New York: Cambridge University Press.

Smith, T. S., & Stevens, G. T. (1996). Emergence, self-organization, and social interaction: Arousal-dependent structure in social systems. *Sociological Theory, 14*(2), 131–153.

Smith, T. S., & Stevens, G. T. (2002). Hyperstructures and the biology of interpersonal dependence: Rethinking reciprocity and altruism. *Sociological Theory, 20*(1), 106–130.

Snowdon, D. (2002). *Aging with grace: What the nun study teaches us about leading longer, healthier, and more meaningful lives*. New York: Bantam Books.

Tausig, M., Michello, J., & Subedi, S. (2003). *Sociology of mental illness* (2nd ed.). Upper Saddle River: Prentice Hall.

TenHouten, W. D. (1996). Outlines of a socioevolutionary theory of the emotions. *The International Journal of Sociology and Social Policy, 16*(9–10), 190–208.

TenHouten, W. D. (2005). *Time and society*. New York: SUNY Press.

TenHouten, W. D., & Kaplin, C. (1973). *Science and its mirror image: A theory of inquiry*. New York: Harper and Row Publishers.

Turner, J. H. (2000). *On the origins of human emotions: A sociological inquiry into the evolution of human affect*. Palo Alto: Stanford University Press.

U.S. Department of Health and Human Services. (1999). *Mental health: A report of the surgeon general—executive summary*. Rockville: U.S. Department of Health and Human Services, Substance Abuse and Mental Health Services Administration, Center for Mental Health Services, National Institutes of Health, National Institute of Mental Health.

Uneo, K. (2010). Mental health differences between young adults with and without same-sex contact: A simultaneous examination of underlying mechanisms. *Journal of Health and Social Behavior, 51*(3), 391–407.

Whitaker, R. (2010). *Anatomy of an epidemic: Magic bullets, psychiatric drugs, and the astonishing rise of mental illness in America*. New York: Crown Publishers.

Zhang, W., Chen, Q., McCubbin, H., McCubbin, L., & Foley, S. (2011). Predictors of mental and physical health: Individual and neighborhood levels of education, social well-being, and ethnicity. *Health & Place, 17*(1), 238–247.

Index

D.D. Franks and J.H. Turner (eds.), *Handbook of Neurosociology*, Handbooks of Sociology and Social Research, DOI 10.1007/978-94-007-4473-8,

FSC
www.fsc.org
MIX
Papier | Fördert
gute Waldnutzung
FSC® C083411

Kolibri 360 GmbH
Ferdinand-Jühlke-Straße 7
99095 Erfurt, Deutschland
produktsicherheit@kolibri360.de